Giuliano Benenti. Born in Voghera (Pavia), Italy, November 7, 1969. He is a researcher in Theoretical Physics at Università dell' Insubria, Como. He received his Ph.D. in physics at Universita di Milano, Italy and was a postdoctoral fellow at CEA, Saclay, France. His main research interests are in the fields of classical and quantum chaos, open quantum systems, mesoscopic physics, disordered systems, phase transitions, many-body systems and quantum information theory.

Giulio Casati. Born in Brenna (Como), Italy, December 9, 1942. He is a professor of Theoretical Physics at Università dell' Insubria, Como, former professor at Milano University, and distinguished visiting professor at NUS, Singapore. A member of the Academia Europea, and director of the Center for Nonlinear and Complex Systems, he was awarded the F. Somaini Italian prize for physics in 1991. As editor of several volumes on classical and quantum chaos, he has done pioneering research in nonlinear dynamics, classical and quantum chaos with applications to atomic, solid state, nuclear physics and, more recently, to quantum computers.

Giuliano Strini. Born in Roma, Italy, September 9, 1937. He is an associate professor in Experimental Physics and has been teaching a course on Quantum Computation at Universita di Milano, for several years. From 1963, he has been involved in the construction and development of the Milan Cyclotron. His publications concern nuclear reactions and spectroscopy, detection of gravitational waves, quantum optics and, more recently, quantum computers. He is a member of the Italian Physical Society, and also the Optical Society of America.

Principles of Quantum Computation and Information

Volume II: Basic Tools and Special Topics

Giuliano Benenti and Giulio Casati

Universitá degli Studi dell Insubria, Italy
Istituto Nazionale per la Fisica della Materia, Italy

Giuliano Strini

Universitá di Milano, Italy

W® World Scientific

NEW JERSEY · LONDON · SINGAPORE · BEIJING · SHANGHAI · HONG KONG · TAIPEI · CHENNAI

Published by

World Scientific Publishing Co. Pte. Ltd.

5 Toh Tuck Link, Singapore 596224

USA office: 27 Warren Street, Suite 401-402, Hackensack, NJ 07601

UK office: 57 Shelton Street, Covent Garden, London WC2H 9HE

British Library Cataloguing-in-Publication Data
A catalogue record for this book is available from the British Library.

ISBN-13 978-981-256-345-3
ISBN-10 981-256-345-8
ISBN-13 978-981-256-528-0 (pbk)
ISBN-10 981-256-528-0 (pbk)

Printed in Singapore.

To Silvia and Arianna
g.b.

To my wife for her love and encouragement
g.c.

To my family and friends
g.s.

About the Cover

This acrostic is the famous *sator* formula. It can be translated as:

'*Arepo the sower holds the wheels at work*'

The text may be read in four different ways:

    (i)   horizontally, from left to right (downward) and from right to left (upward);

    (ii)   vertically, downward (left to right) and upward (right to left).

The resulting phrase is always the same.

It has been suggested that it might be a form of secret message.

    This acrostic was unearthed during archeological excavation work at Pompeii, which was buried, as well known, by the eruption of Vesuvius in 79 A.D. The formula can be found throughout the Roman Empire, probably also spread by legionnaires. Moreover, it has been found in Mesopotamia, Egypt, Cappadocia, Britain and Hungary.

    The *sator* acrostic may have a mystical significance and might have been used as a means for persecuted Christians to recognize each other (it can be rearranged into the form of a cross, with the opening words of the Lord's prayer, *A Paternoster O*, both vertically and horizontally, intersecting at the letter N, the Latin letters A and O corresponding to the Greek letters alpha and omega, beginning and end of all things).

Preface

Purpose of the book

This book is addressed to undergraduate and graduate students in physics, mathematics and computer science. It is written at a level comprehensible to readers with the background of a student near the end of an undergraduate course in one of the above three disciplines. Note that no prior knowledge of either quantum mechanics or classical computation is required to follow this book. Indeed, the first two chapters are a simple introduction to classical computation and quantum mechanics. Our aim is that these chapters should provide the necessary background for an understanding of the subsequent chapters.

The book is divided into two volumes. In volume I, after providing the necessary background material in classical computation and quantum mechanics, we develop the basic principles and discuss the main results of quantum computation and information. Volume I would thus be suitable for a one-semester introductory course in quantum information and computation, for both undergraduate and graduate students. It is also our intention that volume I be useful as a general education for other readers who wish to learn the basic principles of quantum computation and information and who have the basic background in physics and mathematics acquired in undergraduate courses in physics, mathematics or computer science.

Volume II deals with various important aspects, both theoretical and experimental, of quantum computation and information. The areas include quantum data compression, accessible information, entanglement concentration, limits to quantum computation due to decoherence, quantum error correction, and the first experimental implementations of quantum information protocols. This volume also includes a selection of special topics:

chaos and the quantum-to-classical transition, quantum trajectories, quantum computation and quantum chaos, and the Zeno effect. For an understanding of this volume, a knowledge of the material discussed in the first volume is necessary.

General approach

Quantum computation and information is a new and rapidly developing field. It is therefore not easy to grasp the fundamental concepts and central results without having to face many technical details. Our purpose in this book is to provide the reader interested in the field with a useful and not overly heavy guide. Mathematical rigour is therefore not our primary concern. Instead, we have tried to present a simple and systematic treatment, such that the reader might understand the material presented without the need for consulting other texts. Moreover, we have not tried to cover all aspects of the field, preferring to concentrate on the fundamental concepts. Nevertheless, the two volumes should prove useful as a reference guide to researchers just starting out in the field.

To gain complete familiarity with the subject, it is important to practice problem solving. The book contains a large number of exercises (with solutions), which are an essential complement to the main text. In order to develop a solid understanding of the arguments dealt with here, it is indispensable that the student try to solve a large part of them.

Note to the reader

Some of the material presented is not necessary for understanding the rest of the book and may be omitted on a first reading. We have adopted two methods of highlighting such parts:

1) The sections or subsections with an asterisk before the title contain more advanced or complementary material. Such parts may be omitted without risk of encountering problems in reading the rest of the book.

2) Comments, notes or examples are printed in a small typeface.

Acknowledgments

We are indebted to several colleagues for criticism and suggestions. In particular, we wish to thank Alberto Bertoni, Gabriel Carlo, David Cory, Jürgen Eschner, Paolo Facchi, Rosario Fazio, Giuseppe Florio, Bertrand Georgeot, Luigi Lugiato, Paolo Mataloni, Sandro Morasca, Simone Montangero, Massimo Palma, Saverio Pascazio, Christian Roos, Davide Rossini, Nicoletta Sabadini, Marcos Saraceno, Fabio Sciarrino, Stefano Serra Capizzano, Lorenza Viola and Robert Walters, who read preliminary versions of the book. We are also grateful to Federico Canobbio and Sisi Chen. Special thanks is due to Philip Ratcliffe, for useful remarks and suggestions, which substantially improved our book. Obviously no responsibility should be attributed to any of the above regarding possible flaws that might remain, for which the authors alone are to blame.

Contents – Volume I

Contents – Volume II

Chapter 5

Quantum Information Theory

Classical information theory deals with the transmission of messages (say, binary strings) over communication channels. Its fundamental questions are: How much can a message be compressed and still be transmitted reliably? Can we protect this message against errors that will appear in noisy communication channels? In this chapter, we discuss the above questions in the light of quantum mechanics, which opens up new possibilities for information theory. Before doing so, we need to introduce a few useful tools. The density-matrix formalism is the natural framework in which to treat open and composite quantum systems. We also introduce the concept of generalized measurement and discuss a simple example in which it proves to be useful.

Following this, we review the main results of classical information theory. It turns out that it is possible to compress a message into a shorter string of letters, the compression factor being the Shannon entropy. This is the content of Shannon's celebrated noiseless coding theorem. We discuss the natural extension of this result to quantum mechanics. To this end one may consider a message whose letters are quantum states, transmitted through a quantum communication channel. Such quantum states may be treated as though they were (quantum) information and one might thus ask to what extent this quantum message can be compressed. Schumacher's quantum noiseless coding theorem states that the optimal compression factor is given by the von Neumann entropy. Therefore, the von Neumann entropy is the appropriate measure of quantum information, just as the Shannon entropy is for classical information. If Alice codes a classical message by means of quantum states, it is natural to ask how much information Bob can gain on the message by performing (generalized) measurements on the quantum states received. This is not an easy question since the

transmitted quantum states are not necessarily orthogonal and they cannot therefore be perfectly distinguished. The Holevo bound establishes an
upper limit on the information accessible to Bob.

We also discuss how to quantify the entanglement content of a generic
pure state and, briefly, how to concentrate entanglement. This last is an
important issue since maximally entangled states are required for faithful
teleportation of quantum states. Entanglement measures and the Peres
separability criterion for mixed states are also briefly addressed. We post-
pone the discussion of the transmission of quantum information over noisy
quantum channels to Chapt. 7, where we shall consider this subject in the
context of quantum-error correcting codes. A special-topic section on the
different definitions of entropy used in physics closes the chapter.

The present chapter requires more formal development than those pre-
ceding. This is quite natural since we are concerned with the most general
results on the properties of quantum information. Nonetheless, in order
to illustrate these general concepts, we shall describe significant concrete
examples in detail.

5.1   The density matrix

In practice, the state of a physical system is often not perfectly determined.
For example, if we consider a beam of atoms emitted by a thermal source,
we do not know the kinetic energy of each atom, but only the distribution
of their kinetic energies. In this case, we say that our information on the
system is *incomplete*. We only know that the system is in a state taken
from the ensemble

$$\big\{|\psi_1\rangle, |\psi_2\rangle, \ldots, |\psi_l\rangle\big\}, \tag{5.1}$$

with probabilities $\{p_1, p_2, \ldots, p_l\}$, satisfying the condition of unit total
probability, $\sum_i p_i = 1$. We say that we have a *statistical mixture* (also
known as a *mixed state*) of the states $|\psi_k\rangle$, with weights p_k. By contrast,
the single states $|\psi_k\rangle$ are known as *pure states*. We note that the states
$|\psi_k\rangle$ are not necessarily orthogonal.

As remarked in Sec. 2.4, the statistical mixture of the states $|\psi_k\rangle$, with
weights p_k, should not be confused with the linear superposition

$$|\psi\rangle = \sum_k c_k |\psi_k\rangle, \qquad |c_k|^2 = p_k. \tag{5.2}$$

It is actually impossible to describe a statistical mixture by means of an "average state vector". As we shall see, it is instead possible to describe it using an "average operator": the density operator.

The probability $p(i)$ that a measurement of the observable A yields outcome a_i is given by

$$p(i) = \sum_{k=1}^{l} p_k \langle \psi_k | P_i | \psi_k \rangle, \qquad (5.3)$$

where P_i is the projector onto the subspace associated with the eigenvalue a_i of A. In this expression, the probabilities $\langle \psi_k | P_i | \psi_k \rangle$ that $A = a_i$ on the pure states $|\psi_k\rangle$ are computed according to the measurement postulate discussed in Sec. 2.4. As a result, the mean value of any observable A is

$$\langle A \rangle = \sum_{i=1}^{n} a_i p(i) = \sum_{k=1}^{l} p_k \sum_{i=1}^{n} a_i \langle \psi_k | P_i | \psi_k \rangle = \sum_{k=1}^{l} p_k \langle \psi_k | A | \psi_k \rangle. \quad (5.4)$$

Probabilities therefore appear twice:

 (i) in the initial (lack of) information on the system, characterized by the weights p_k;
(ii) in the measurement process, characterized by the probabilities $\langle \psi_k | P_i | \psi_k \rangle$ to obtain outcomes a_i from the measurement of the observable A when the system is described by the state $|\psi_k\rangle$. These latter probabilities are intrinsically quantum mechanical.

The question now is how to take into account the partial information we have on the system and to simultaneously include in our description the laws of both quantum mechanics and probability theory.

It is very useful to introduce the density operator ρ, defined as

$$\rho \equiv \sum_{k} p_k |\psi_k\rangle\langle\psi_k|. \qquad (5.5)$$

Given a generic orthonormal basis $\{|i\rangle\}$, with $i = 1, 2, \ldots, n$ (n is the dimension of the Hilbert space $\mathcal{H}$ associated with the system), we naturally associate the operator ρ with a matrix representation. The corresponding matrix, known as the *density matrix*, has elements

$$\rho_{ij} \equiv \langle i|\rho|j\rangle. \qquad (5.6)$$

Note that it is also customary to call the density operator ρ in Eq. (5.5) a density matrix.

The mean value of any observable A can be computed by means of the density operator as follows:

$$\text{Tr}(\rho A) = \sum_{i=1}^{n} \langle i|\rho A|i\rangle = \sum_{k=1}^{l}\sum_{i=1}^{n} p_k \langle i|\psi_k\rangle\langle\psi_k|A|i\rangle, \tag{5.7}$$

which is equal to $\langle A\rangle$, as given in Eq. (5.4). The equality between (5.7) and (5.4) follows trivially, if we take into account the completeness relation $\sum_i |i\rangle\langle i| = I$. It is also easy to check that the probability $p(i)$ that a measurement of the observable A gives outcome a_i, given by Eq. (5.3), is equal to

$$p(i) = \text{Tr}(\rho P_i). \tag{5.8}$$

Therefore, the density operator ρ completely characterizes the system; from it we can predict the probabilities of the possible outcomes of any experiment performed on the system.

As discussed in Sec. 2.4, if a system is described by the state vector $|\psi_k\rangle$ and we measure the observable A, obtaining outcome a_i, then the state of the system immediately after the measurement is

$$|\psi_k'\rangle = \frac{P_i|\psi_k\rangle}{\sqrt{\langle\psi_k|P_i|\psi_k\rangle}}, \tag{5.9}$$

with P_i the projector onto the subspace associated with the eigenvalue a_i. Therefore, if the system is in a mixed state described by the density matrix $\rho = \sum_k p_k|\psi_k\rangle\langle\psi_k|$ and we obtain the outcome a_i from the measurement of A, then the new density matrix after the measurement is given by

$$\rho' = \sum_{k=1}^{l} p(k|i)\,|\psi_k'\rangle\langle\psi_k'| = \sum_{k=1}^{l} p(k|i)\frac{P_i|\psi_k\rangle\langle\psi_k|P_i}{\langle\psi_k|P_i|\psi_k\rangle}, \tag{5.10}$$

where $p(k|i)$ is the probability that the system is described by the state vector $|\psi_k'\rangle$, given that the measurement of the observable A resulted in a_i. Elementary probability theory tells us that $p(k,i) = p(i)\,p(k|i)$, where $p(k,i)$ is the joint probability to have the state $|\psi_k'\rangle$ and the outcome a_i. Likewise, we have $p(k,i) = p_k\,p(i|k)$ and therefore $p(k|i) = p(k,i)/p(i) = p(i|k)\,p_k/p(i)$. Observe now that the probability of obtaining result a_i, given that the system was in the state $|\psi_k\rangle$, is $p(i|k) = \langle\psi_k|P_i|\psi_k\rangle$. Finally, we can read $p(i)$ from Eq. (5.8) and insert it in Eq. (5.10), together with the expression found for $p(i|k)$. We then obtain

$$\rho' = \frac{P_i \rho P_i}{\text{Tr}(\rho P_i)}. \tag{5.11}$$

It is also possible to describe the dynamical evolution of a mixed system in the density-operator picture. We have

$$\frac{d}{dt}\rho(t) = \frac{d}{dt}\sum_{k=1}^{l} p_k |\psi_k(t)\rangle\langle\psi_k(t)|$$

$$= \sum_{k=1}^{l} p_k \left[\left(\frac{d}{dt}|\psi_k(t)\rangle\right)\langle\psi_k(t)| + |\psi_k(t)\rangle\left(\frac{d}{dt}\langle\psi_k(t)|\right) \right]. \tag{5.12}$$

The temporal evolution of the state vectors $|\psi_k\rangle$ is governed by the Schrödinger equation, which reads

$$i\hbar\frac{d}{dt}|\psi_k(t)\rangle = H|\psi_k(t)\rangle. \tag{5.13}$$

and equivalently, for the dual vector $\langle\psi_k|$,

$$-i\hbar\frac{d}{dt}\langle\psi_k(t)| = \langle\psi_k(t)|H. \tag{5.14}$$

If we insert Eqs. (5.13) and (5.14) into Eq. (5.12), we obtain

$$\frac{d}{dt}\rho(t) = \frac{1}{i\hbar}\big(H\rho(t) - \rho(t)H\big) = \frac{1}{i\hbar}\big[H, \rho(t)\big]. \tag{5.15}$$

This equation, known as the von Neumann equation, governs the temporal evolution of the density operator ρ.

As we saw in Sec. 2.4, the state vector $|\psi_k(t)\rangle$ at time t is related to the state vector $|\psi_k(t_0)\rangle$ at time t_0 by a unitary operator $U(t, t_0)$: we have $|\psi_k(t)\rangle = U(t, t_0)|\psi_k(t_0)\rangle$. Therefore, the density matrix $\rho(t)$ at time t is related to the density matrix $\rho(t_0)$ at time t_0 as follows:

$$\rho(t) = \sum_{k=1}^{l} p_k |\psi_k(t)\rangle\langle\psi_k(t)| = \sum_{k=1}^{l} p_k U(t, t_0)|\psi_k(t_0)\rangle\langle\psi_k(t_0)|U^\dagger(t, t_0)$$

$$= U(t, t_0)\,\rho(t_0)\,U^\dagger(t, t_0). \tag{5.16}$$

Since, as is clear from the above discussion, the postulates of quantum mechanics can be reformulated in the density-operator picture, it is completely equivalent to describe a pure system by means of either the wave function $|\psi(t)\rangle$ or the density matrix $\rho(t) = |\psi(t)\rangle\langle\psi(t)|$. A nice property of the density-matrix picture is that the arbitrary global phase factor (associated with the wave function) disappears in this formulation: the state vectors $|\psi(t)\rangle$ and $e^{i\delta}|\psi(t)\rangle$, with δ a real number, give exactly the same density matrix. More importantly, as we shall see later in this volume, the density matrix is extremely useful in the description of mixed states and of composite quantum systems.

The density operator ρ satisfies the following properties:

1. ρ is *Hermitian*. Indeed, if we expand any pure state $|\psi_k\rangle$ over an orthonormal basis $\{|i\rangle\}$; that is,

$$|\psi_k\rangle = \sum_{i=1}^{n} c_i^{(k)}|i\rangle, \tag{5.17}$$

then we have

$$\rho_{ij} = \sum_{k=1}^{l} p_k \langle i|\psi_k\rangle\langle\psi_k|j\rangle = \sum_{k=1}^{l} p_k \sum_{l,m=1}^{n} c_l^{(k)} c_m^{(k)\star} \langle i|l\rangle\langle m|j\rangle$$

$$= \sum_{k=1}^{l} p_k\, c_i^{(k)} c_j^{(k)\star}. \tag{5.18}$$

The last equality follows from the orthonormality condition, which implies $\langle i|l\rangle = \delta_{il}$ and $\langle m|j\rangle = \delta_{mj}$. From the above expression, it is easy to check that ρ is Hermitian, since

$$\rho_{ji}^{\star} = \sum_{k=1}^{l} p_k\, c_j^{(k)\star} c_i^{(k)} = \rho_{ij}. \tag{5.19}$$

2. ρ has *unit trace*. Using expansion (5.18), we obtain

$$\mathrm{Tr}\,\rho = \sum_{i=1}^{n} \rho_{ii} = \sum_{k=1}^{l} p_k \sum_{i=1}^{n} |c_i^{(k)}|^2 = \sum_{k=1}^{l} p_k = 1. \tag{5.20}$$

3. ρ is a *non-negative* operator; that is, for any vector $|\varphi\rangle$ in the Hilbert space $\mathcal{H}$, we have $\langle\varphi|\rho|\varphi\rangle \geq 0$. Indeed, we have

$$\langle\varphi|\rho|\varphi\rangle = \langle\varphi|\left(\sum_{k=1}^{l} p_k|\psi_k\rangle\langle\psi_k|\right)|\varphi\rangle = \sum_{k=1}^{l} p_k|\langle\varphi|\psi_k\rangle|^2 \geq 0. \tag{5.21}$$

It is important to note that $\mathrm{Tr}\,\rho^2 < 1$ for a mixed state, while $\mathrm{Tr}\,\rho^2 = 1$ for a pure state. This is a simple criterion for determining whether a state is pure or mixed. To prove it, let us consider the spectral decomposition of the Hermitian operator ρ:

$$\rho = \sum_{j=1}^{n} \lambda_j |j\rangle\langle j|, \tag{5.22}$$

where the normalized vectors $|j\rangle$ are orthogonal and $\lambda_j \geq 0$ since ρ is non-negative. In the $\{|j\rangle\}$ basis the matrix representation of ρ is diagonal and given by

$$\rho = \begin{bmatrix} \lambda_1 & 0 & \cdots & 0 \\ 0 & \lambda_2 & \cdots & 0 \\ \vdots & \vdots & \vdots\vdots\vdots & \vdots \\ 0 & 0 & \cdots & \lambda_n \end{bmatrix}. \tag{5.23}$$

Since, as we have shown above, $\mathrm{Tr}\,\rho = 1$, then we have $\sum_j \lambda_j = 1$. Using the spectral decomposition (5.22), it is easy to compute ρ^2 and we obtain

$$\rho^2 = \sum_{j=1}^{n} \lambda_j^2 |j\rangle\langle j|, \tag{5.24}$$

with matrix representation

$$\rho^2 = \begin{bmatrix} \lambda_1^2 & 0 & \cdots & 0 \\ 0 & \lambda_2^2 & \cdots & 0 \\ \vdots & \vdots & \vdots\vdots\vdots & \vdots \\ 0 & 0 & \cdots & \lambda_n^2 \end{bmatrix}. \tag{5.25}$$

Thus, we have $\mathrm{Tr}\,\rho^2 = \sum_j \lambda_j^2$. Since $\sum_j \lambda_j = 1$ and $\lambda_j \geq 0$, then $0 \leq \lambda_j \leq 1$. Therefore, $\mathrm{Tr}\,\rho^2 = 1$ if and only if $\lambda_j = 1$ for just one $j = \bar{j}$ and $\lambda_j = 0$ otherwise. This corresponds to a pure state, described by the density matrix $\rho = |\bar{j}\rangle\langle\bar{j}|$. It is also easy to check that for a pure state $\rho^2 = \rho$. By definition, we have a mixed state if the diagonal representation (5.22) of ρ involves more than one pure state and in this case $\lambda_j < 1$ for all j. Therefore, $\sum_j \lambda_j^2 < \sum_j \lambda_j = 1$. This proves that $\mathrm{Tr}\,\rho^2 < 1$ for a mixed state.

Let us discuss the physical meaning of the matrix elements of the density operator ρ. From Eq. (5.18) we can see that the diagonal term

$$\rho_{ii} = \sum_k p_k |c_i^{(k)}|^2 = \mathrm{Tr}(\rho P_i), \quad \text{where } P_i = |i\rangle\langle i|, \tag{5.26}$$

represents the probability that the system is left in the state $|i\rangle$ after measuring the observable whose eigenstates are $\{|i\rangle\}$. For this reason we say that ρ_{ii} represents the *population* of the state $|i\rangle$.

The off-diagonal terms ρ_{ij} represent interference between the states $|i\rangle$ and $|j\rangle$. Such interference is present for any state $|\psi_k\rangle$ of the statistical mixture containing a linear superposition of $|i\rangle$ and $|j\rangle$. We can see from Eq. (5.18) that ρ_{ij} is a weighted sum of the interference terms $c_i^{(k)} c_j^{(k)\star}$. We stress that the individual terms $c_i^{(k)} c_j^{(k)\star}$ appearing in the sum (5.18) are complex quantities and, therefore, ρ_{ij} can be equal to zero even though the individual terms are not. If $\rho_{ij} \neq 0$, then, even after averaging over the statistical mixture, a quantum-coherence effect between the states $|i\rangle$ and $|j\rangle$ will remain. For this reason the off-diagonal elements of the density matrix are known as *coherences*.

We point out that the distinction between diagonal and off-diagonal terms; that is, between populations and coherences, depends on the choice of the basis $\{|i\rangle\}$. Actually, since ρ is Hermitian, non-negative and has unit trace, it is always possible to find an orthonormal basis $\{|m\rangle\}$ such that

$$\rho = \sum_m \alpha_m |m\rangle\langle m|, \quad 0 \le \alpha_m \le 1, \quad \sum_m \alpha_m = 1. \tag{5.27}$$

This implies that the density matrix ρ can always be seen as a statistical mixture of the states $\{|m\rangle\}$, without coherences between them, even though these states are not, in general, eigenstates of a physical observable.

5.1.1 *The density matrix for a qubit*

We now apply the density-operator formalism to the qubit. As we have seen in Sec. 3.1, the pure state of a qubit is represented by a point on a sphere of unit radius, known as the Bloch sphere. This point is singled out by the spherical coordinates θ and ϕ and corresponds to the state

$$|\psi(\theta, \phi)\rangle = \cos \tfrac{\theta}{2} |0\rangle + e^{i\phi} \sin \tfrac{\theta}{2} |1\rangle, \tag{5.28}$$

with $|0\rangle$ and $|1\rangle$ eigenstates of the Pauli matrix σ_z. The corresponding density operator is given by

$$\rho(\theta, \phi) = |\psi(\theta, \phi)\rangle \langle \psi(\theta, \phi)| \tag{5.29}$$

and its matrix representation in the $\{|0\rangle, |1\rangle\}$ basis is

$$\rho(\theta, \phi) = \begin{bmatrix} \cos^2 \frac{\theta}{2} & \sin \frac{\theta}{2} \cos \frac{\theta}{2} e^{-i\phi} \\ \sin \frac{\theta}{2} \cos \frac{\theta}{2} e^{i\phi} & \sin^2 \frac{\theta}{2} \end{bmatrix}. \tag{5.30}$$

It is easy to check that $\rho^2(\theta, \phi) = \rho(\theta, \phi)$, as it must be for a pure state.

Exercise 5.1 Show that any 2×2 Hermitian matrix can be expanded over the basis $\{I, \sigma_x, \sigma_y, \sigma_z\}$, the coefficients of this expansion being real.

We now consider the density matrix ρ for the mixed state of a qubit. Since it has to be a 2×2 Hermitian matrix, we can expand it over the basis $\{I, \sigma_x, \sigma_y, \sigma_z\}$ (see exercise 5.1); exercise 5.1); that is,

$$\rho = aI + b\sigma_x + c\sigma_y + d\sigma_z, \tag{5.31}$$

the coefficients a, b, c, d being real. Since the condition $\text{Tr}(\rho) = 1$ must be satisfied for a density matrix, $\text{Tr}(I) = 2$ and $\text{Tr}(\sigma_x) = \text{Tr}(\sigma_y) = \text{Tr}(\sigma_z) = 0$, we have $a = \frac{1}{2}$. We can therefore express ρ as follows:

$$\rho = \frac{1}{2}(I + x\sigma_x + y\sigma_y + z\sigma_z) = \frac{1}{2} \begin{bmatrix} 1+z & x-iy \\ x+iy & 1-z \end{bmatrix}, \tag{5.32}$$

with $x = 2b$, $y = 2c$, $z = 2d$. We have seen that a density matrix is non-negative and therefore ρ must have eigenvalues λ_1, $\lambda_2 \geq 0$. Thus, we have $\det \rho = \lambda_1 \lambda_2 \geq 0$. We can compute explicitly from Eq. (5.32) $\det \rho = \frac{1}{4}(1 - |\boldsymbol{r}|^2)$, with $\boldsymbol{r} \equiv (x, y, z)$. We have $\det \rho \geq 0$ if and only if $0 \leq |\boldsymbol{r}| \leq 1$. There is a one-to-one correspondence between the density matrices for a qubit and the points on the unit ball $0 \leq |\boldsymbol{r}| \leq 1$, which is known as the Bloch ball. The vector $\boldsymbol{r}$ is known as the Bloch vector. For a pure state, the density matrix ρ has eigenvalues $\lambda_1 = 1$ and $\lambda_2 = 0$. Thus, $\det \rho = 0$, which in turn implies $|\boldsymbol{r}| = 1$. We conclude that pure states are located on the boundary of the Bloch ball.

As an example of mixed state, we consider the case of qubit, which may point in any direction of the space with equal probability. We integrate over all the possible directions and obtain

$$\rho = \frac{1}{4\pi} \int_0^{2\pi} d\phi \int_0^\pi d\theta \sin\theta \begin{bmatrix} \cos^2 \frac{\theta}{2} & \sin \frac{\theta}{2} \cos \frac{\theta}{2} e^{-i\phi} \\ \sin \frac{\theta}{2} \cos \frac{\theta}{2} e^{i\phi} & \sin^2 \frac{\theta}{2} \end{bmatrix}$$

$$= \frac{1}{2} \begin{bmatrix} 1 & 0 \\ 0 & 1 \end{bmatrix} = \frac{1}{2} I, \tag{5.33}$$

where we need the normalization factor $\frac{1}{4\pi}$ because $\int_0^{2\pi} d\phi \int_0^{\pi} d\theta \sin\theta = 4\pi$. We note that $\rho^2 = \frac{1}{4}I$ and therefore $\text{Tr}(\rho^2) = \frac{1}{2} < 1$, as it must for a mixed state. Taking into account formula (5.32), we can easily see that the density matrix $\rho = \frac{1}{2}I$ corresponds to the centre of the Bloch ball. We say that a qubit described by this mixed state is unpolarized, because $\langle \sigma_i \rangle = 0$, for $i = x, y, z$. Indeed, we have

$$\langle \sigma_i \rangle = \text{Tr}(\rho \sigma_i) = \text{Tr}(\tfrac{1}{2}I\sigma_i) = \tfrac{1}{2}\text{Tr}\,\sigma_i = 0. \qquad (5.34)$$

Exercise 5.2 Show that

$$\text{Tr}(\sigma_i \sigma_j) = 2\delta_{ij}, \qquad (5.35)$$

for $i, j = x, y, z$.

For a generic density matrix, the qubit polarization along the direction singled out by the unit vector $\hat{n}$ is given by

$$\langle \sigma_{\hat{n}} \rangle = \text{Tr}(\rho \sigma_{\hat{n}}) = \text{Tr}\left[\tfrac{1}{2}(I + \boldsymbol{r} \cdot \boldsymbol{\sigma})\,\hat{n} \cdot \boldsymbol{\sigma}\right]. \qquad (5.36)$$

Here, $\sigma_{\hat{n}} \equiv \hat{n} \cdot \boldsymbol{\sigma}$, $\boldsymbol{\sigma} = (\sigma_x, \sigma_y, \sigma_z)$ and the density matrix ρ has been expressed as in Eq. (5.32). Taking into account the result of exercise 5.2, we have

$$\langle \sigma_{\hat{n}} \rangle = \hat{n} \cdot \boldsymbol{r}. \qquad (5.37)$$

Thus, the vector $\boldsymbol{r}$ parametrizes the polarization of the qubit. As we shall see in Sec. 5.5, if many identically prepared systems are available; that is, the same density matrix ρ describes each system, then we can determine $\boldsymbol{r}$ (and therefore the density matrix ρ) by measuring $\hat{n} \cdot \boldsymbol{\sigma}$ along three independent axes.

Finally, we note that the decomposition of the density matrix into ensembles of pure states (Eq. (5.5)) is not unique. For example, let us consider the density matrix

$$\rho = \tfrac{2}{3}|0\rangle\langle 0| + \tfrac{1}{3}|1\rangle\langle 1|. \qquad (5.38)$$

This density matrix is obtained if the system is in the state $|0\rangle$ with probability $\frac{2}{3}$ or in the state $|1\rangle$ with probability $\frac{1}{3}$. However, this is not the only ensemble of pure states giving the density matrix (5.38). There are infinitely many other possibilities, for instance we can consider the situation in which the states

$$|a\rangle = \sqrt{\tfrac{2}{3}}|0\rangle + \sqrt{\tfrac{1}{3}}|1\rangle, \quad |b\rangle = \sqrt{\tfrac{2}{3}}|0\rangle - \sqrt{\tfrac{1}{3}}|1\rangle \qquad (5.39)$$

are prepared with equal probabilities $p_a = p_b = \frac{1}{2}$. We have

$$\rho = \tfrac{1}{2}|a\rangle\langle a| + \tfrac{1}{2}|b\rangle\langle b| = \tfrac{2}{3}|0\rangle\langle 0| + \tfrac{1}{3}|1\rangle\langle 1|. \tag{5.40}$$

A further example is provided by the density matrix $\rho = \frac{1}{2}I$. It can correspond to the decomposition (5.33), but also to the statistical mixture of equal portions of the states $|0\rangle_{\hat{n}}$ and $|1\rangle_{\hat{n}}$, where these states are the eigenvectors of the matrix $\sigma_{\hat{n}}$, with $\hat{n}$ an arbitrary unit vector. For instance, considering $\hat{n} = (1, 0, 0)$, we obtain

$$\rho = \tfrac{1}{2}|0\rangle_{x\,x}\langle 0| + \tfrac{1}{2}|1\rangle_{x\,x}\langle 1| = \tfrac{1}{2}|0\rangle\langle 0| + \tfrac{1}{2}|1\rangle\langle 1| = \tfrac{1}{2}I. \tag{5.41}$$

Since the probabilities of the different outcomes of any conceivable experiment are governed by the density matrix ρ (see Eq. (5.8)), it is impossible to distinguish between different mixtures leading to the same ρ. Therefore, we say that, in contrast to the case of a pure state, our information on the system is *incomplete*.

Exercise 5.3 Show that two density matrices for a qubit commute if their Bloch vectors are parallel.

5.1.2 *Composite systems*

Let us consider a pure state of a bipartite system. As we saw in Sec. 2.5, this state resides in the Hilbert space $\mathcal{H} = \mathcal{H}_1 \otimes \mathcal{H}_2$, which is the tensor product of the Hilbert spaces associated with the subsystems 1 and 2. We can therefore express a generic state $|\psi\rangle \in \mathcal{H}$ as

$$|\psi\rangle = \sum_{i,\alpha} c_{i\alpha}|i\rangle_1|\alpha\rangle_2, \quad \text{with} \quad \sum_{i,\alpha}|c_{i\alpha}|^2 = 1, \tag{5.42}$$

where $\{|i\rangle_1\}$ and $\{|\alpha\rangle_2\}$ are basis sets for $\mathcal{H}_1$ and $\mathcal{H}_2$, respectively. The corresponding density operator is

$$\begin{aligned}
\rho = |\psi\rangle\langle\psi| &= \sum_{i,\alpha}\sum_{j,\beta} c_{i\alpha}c_{j\beta}^{\star}|i\rangle_1|\alpha\rangle_2\,{}_1\langle j|_2\langle\beta| \\
&= \sum_{i,\alpha}\sum_{j,\beta} \rho_{i\alpha;j\beta}|i\rangle_1|\alpha\rangle_2\,{}_1\langle j|_2\langle\beta|,
\end{aligned} \tag{5.43}$$

with the matrix elements of ρ defined by

$$\rho_{i\alpha;j\beta} \equiv {}_1\langle i|\,_2\langle\alpha|\rho|j\rangle_1\,|\beta\rangle_2. \tag{5.44}$$

Let us assume that the total system is described by the density matrix ρ and we wish to compute the mean value of an operator A_1 acting only on subsystem 1. First of all, we trivially extend the operator A_1 to the entire Hilbert space $\mathcal{H}$ by defining the operator

$$\tilde{A} \equiv A_1 \otimes I_2, \qquad (5.45)$$

with I_2 the identity operator in $\mathcal{H}_2$. Thus, we have

$$\langle A_1 \rangle = \text{Tr}(\rho \tilde{A}) = \sum_{k,\gamma} {}_1\langle k| \, {}_2\langle \gamma| \rho \tilde{A} |k\rangle_1 \, |\gamma\rangle_2 \qquad (5.46)$$

$$= \sum_{k,\gamma} {}_1\langle k| \, {}_2\langle \gamma| \Big(\sum_{i,\alpha} \sum_{j,\beta} \rho_{i\alpha;j\beta} |i\rangle_1 |\alpha\rangle_2 \, {}_1\langle j| {}_2\langle \beta| \Big) \Big(A_1 \otimes I_2 \Big) |k\rangle_1 |\gamma\rangle_2.$$

Taking into account the orthonormality relations ${}_1\langle k|i\rangle_1 = \delta_{ik}$, ${}_2\langle \gamma|\alpha\rangle_2 = \delta_{\alpha\gamma}$ and ${}_2\langle \beta|\gamma\rangle_2 = \delta_{\beta\gamma}$, we can remove three out of the six sums appearing in the above equation. This gives

$$\langle A_1 \rangle = \sum_{i,j,\alpha} \rho_{i\alpha;j\alpha} \, {}_1\langle j|A_1|i\rangle_1. \qquad (5.47)$$

It is now useful to introduce the *reduced* density matrix

$$\rho_1 \equiv \text{Tr}_2 \, \rho, \qquad (5.48)$$

where Tr_2 denotes the partial trace over subsystem 2:

$$\text{Tr}_2 \, \rho \equiv \sum_{\alpha} {}_2\langle \alpha|\rho|\alpha\rangle_2. \qquad (5.49)$$

We note that it is also possible to define similarly a reduced density matrix for subsystem 2:

$$\rho_2 \equiv \text{Tr}_1 \, \rho \equiv \sum_{i} {}_1\langle i|\rho|i\rangle_1. \qquad (5.50)$$

The matrix elements of ρ_1 in the $\{|i\rangle_1\}$ basis are given by

$$(\rho_1)_{ij} = {}_1\langle i|\rho_1|j\rangle_1 = \sum_{\alpha} \rho_{i\alpha;j\alpha}. \qquad (5.51)$$

After insertion of this equality into Eq. (5.47), we obtain

$$\langle A_1 \rangle = \sum_{i,j} {}_1\langle i|\rho_1|j\rangle_1 \, {}_1\langle j|A_1|i\rangle_1 = \sum_{i} {}_1\langle i|\rho_1 A_1|i\rangle_1 = \text{Tr}(\rho_1 A_1). \quad (5.52)$$

Therefore, it is possible to compute the expectation value of an operator acting only on subsystem 1 as if the system were isolated and described by the reduced density matrix ρ_1. We can conclude that ρ_1, obtained after partial tracing over subsystem 2, describes the state of subsystem 1.[1]

Exercise 5.4 Show that the reduced density matrix is Hermitian, non-negative and has unit trace.

It is important to point out that, even though ρ corresponds to a pure state of the composite system, it is not assured that the reduced density matrices ρ_1 and ρ_2 describe a pure state. A very significant example is provided by the states of the Bell basis, introduced in Sec. 3.4.1. For instance, let us consider the state $|\psi^-\rangle = \frac{1}{\sqrt{2}}(|01\rangle - |10\rangle)$. This state has density operator

$$\rho = \tfrac{1}{2}\big(|01\rangle\langle 01| + |10\rangle\langle 10| - |01\rangle\langle 10| - |10\rangle\langle 01|\big). \tag{5.53}$$

Its matrix representation in the basis $\{|00\rangle, |01\rangle, |10\rangle, |11\rangle\}$ is given by

$$\rho = \tfrac{1}{2}\begin{bmatrix} 0 & 0 & 0 & 0 \\ 0 & 1 & -1 & 0 \\ 0 & -1 & 1 & 0 \\ 0 & 0 & 0 & 0 \end{bmatrix}. \tag{5.54}$$

We can readily check that

$$\rho_1 = \mathrm{Tr}_2\,\rho = \tfrac{1}{2}\big(|0\rangle\langle 0| + |1\rangle\langle 1|\big) = \tfrac{1}{2}I_1. \tag{5.55}$$

Indeed, we have

$$\begin{aligned}
(\rho_1)_{00} &= \rho_{00;00} + \rho_{01;01} = 0 + \tfrac{1}{2} = \tfrac{1}{2}, \\
(\rho_1)_{01} &= \rho_{00;10} + \rho_{01;11} = 0 + 0 = 0, \\
(\rho_1)_{10} &= \rho_{10;00} + \rho_{11;01} = 0 + 0 = 0, \\
(\rho_1)_{11} &= \rho_{10;10} + \rho_{11;11} = \tfrac{1}{2} + 0 = \tfrac{1}{2}.
\end{aligned} \tag{5.56}$$

---

[1]The temporal evolution of the density matrix $\rho(t)$ is governed by the von Neumann equation (5.15). At any given time, we have seen how to compute $\rho_1(t)$ from $\rho(t)$. However, the problem of finding an equation describing the evolution of $\rho_1(t)$ is much more complex. We shall discuss this issue in Sec. 6.2.

Likewise, we obtain $\rho_2 = \frac{1}{2}I_2$. Its matrix representation is given by

$$
\begin{aligned}
(\rho_2)_{00} &= \rho_{00;00} + \rho_{10;10} = 0 + \tfrac{1}{2} = \tfrac{1}{2}, \\
(\rho_2)_{01} &= \rho_{00;01} + \rho_{10;11} = 0 + 0 = 0, \\
(\rho_2)_{10} &= \rho_{01;00} + \rho_{11;10} = 0 + 0 = 0, \\
(\rho_2)_{11} &= \rho_{01;01} + \rho_{11;11} = \tfrac{1}{2} + 0 = \tfrac{1}{2}.
\end{aligned}
\tag{5.57}
$$

We note that the same expressions for the reduced density matrices are also obtained for the other states of the Bell basis. Thus, ρ_1 and ρ_2 clearly correspond to mixed states: we have $\rho_1^2 = \rho_2^2 = \frac{1}{4}I$ and therefore $\text{Tr}(\rho_1^2) = \text{Tr}(\rho_2^2) = \frac{1}{2} < 1$. As is easy to check, this example also shows that the density matrix ρ for the entire system is *not* equal to the tensor product $\rho_1 \otimes \rho_2$ of the reduced density matrices. This means that the quantum correlations between systems 1 and 2 are not included in $\rho_1 \otimes \rho_2$.

It is instructive to discuss a simple argument illustrating that the non-locality in the EPR phenomenon cannot be used to transmit information faster than light. Assume that Alice and Bob share the Bell state $|\psi^-\rangle = \frac{1}{\sqrt{2}}(|0\rangle|1\rangle - |1\rangle|0\rangle)$ and Alice wishes to employ it to instantaneously communicate a message to Bob, who may be located arbitrarily far away. We know that it is also possible to write $|\psi^-\rangle = \frac{1}{\sqrt{2}}(|0\rangle_x|1\rangle_x - |1\rangle_x|0\rangle_x)$. Thus, it would be tempting for Alice to measure σ_z or σ_x on her half of a Bell state to communicate a bit of classical information. That is to say, she would measure σ_z to transmit 0 and σ_x to transmit 1. In both cases, Alice obtains outcomes 0 or 1 with equal probabilities $\frac{1}{2}$. Thus, her measurement generates the global density matrix

$$
\rho^{(z)} = \tfrac{1}{2}\big(|0\rangle\langle 0| \otimes |1\rangle\langle 1| + |1\rangle\langle 1| \otimes |0\rangle\langle 0|\big),
\tag{5.58}
$$

if she measures σ_z, or

$$
\rho^{(x)} = \tfrac{1}{2}\big(|0\rangle_{x\ x}\langle 0| \otimes |1\rangle_{x\ x}\langle 1| + |1\rangle_{x\ x}\langle 1| \otimes |0\rangle_{x\ x}\langle 0|\big),
\tag{5.59}
$$

if she measures σ_x. In the first case, the reduced density matrix for Bob is $\rho_B = \text{Tr}_A\, \rho^{(z)}$, in the latter $\rho_B = \text{Tr}_A\, \rho^{(x)}$ (here Tr_A denotes the trace over Alice's degrees of freedom). In any instance, it is easy to check that $\rho_B = \frac{1}{2}I$. Since no measurement performed by Bob can distinguish between the two different preparations of the same density matrix ρ_B, the message sent by Alice is unreadable.

Exercise 5.5 Consider the teleportation protocol described in Sec. 4.5 and show that Bob cannot receive any information on the qubit to be teleported before Alice sends him the classical bits.

Exercise 5.6 Show that for a pure bipartite separable state the reduced density matrices ρ_1 and ρ_2 correspond to pure states and the total density matrix of the system is given by $\rho = \rho_1 \otimes \rho_2$.

5.1.3 * *The quantum copying machine*

As an example application of the density-matrix formalism in quantum information, we shall now describe the copying machine of Bužek and Hillery. We saw in Sec. 4.2 that it is impossible to create a perfect duplicate of an arbitrary qubit. This is the content of the no-cloning theorem. However, such a theorem does not forbid the existence of a quantum copying machine that *approximately* copies quantum mechanical states. Bužek and Hillery devised a machine that produces two identical copies of the original qubit, the quality of the copies being independent of the input state. It can be shown that the quantum copying machine of Bužek and Hillery is optimal, in the sense that it maximizes the average fidelity between the input and output states (see Gisin and Massar, 1997 and Bruß *et al.*, 1998). The fidelity is a measure of the quality of the copy and is defined by

$$F = \langle\psi|\rho|\psi\rangle, \tag{5.60}$$

where $|\psi\rangle$ is the state to be copied and ρ is the density matrix describing the copies (see Sec. 6.5.8 for a general discussion of the fidelity of quantum motion or quantum Loschmidt echo). We have $0 \leq F \leq 1$ and the maximum value $F = 1$ is taken when $\rho = |\psi\rangle\langle\psi|$. We note that Eq. (5.60) generalizes the definition of fidelity of two pure states $|\psi\rangle, |\phi\rangle$, given by $F = |\langle\psi|\phi\rangle|^2$ (see exercise 3.1). If $\rho = \sum_k p_k |\phi_k\rangle\langle\phi_k|$, then Eq. (5.60) gives $F = \sum_k p_k F_k$, where $F_k = |\langle\psi|\phi_k\rangle|^2$. Therefore, F is the weighted sum of the pure-state fidelities F_k.

Let us describe the workings of the Bužek–Hillery copying machine. Given a qubit in a generic unknown pure state

$$|\psi\rangle = \alpha|0\rangle + \beta|1\rangle, \tag{5.61}$$

we consider the copying network shown in Fig. 5.1. This circuit can be decomposed into two parts: (i) the preparation of a specific state of the quantum copier and (ii) the copying process. It can be seen from Fig. 5.1 that only part (ii) depends on the state $|\psi\rangle$ to be copied. Let us first look at the preparation stage. The gates labelled by θ_i denote the application of the rotation matrices

$$R_y(-2\theta_i) = \begin{bmatrix} \cos\theta_i & \sin\theta_i \\ -\sin\theta_i & \cos\theta_i \end{bmatrix}, \tag{5.62}$$

where, as discussed in Sec. 3.3.1, $R_y(-2\theta_i)$ corresponds to a counterclockwise rotation through an angle $-2\theta_i$ about the y-axis of the Bloch sphere and the angles θ_i are chosen as

$$\cos 2\theta_1 = \frac{1}{\sqrt{5}}, \qquad \cos 2\theta_2 = \frac{\sqrt{5}}{3}, \qquad \cos 2\theta_3 = \frac{2}{\sqrt{5}}. \tag{5.63}$$

At the end of the preparation stage, the two qubits initially in the state $|00\rangle$ are transformed into the state vector

$$|\Phi\rangle = \frac{1}{\sqrt{6}}\left(2|00\rangle + |01\rangle + |11\rangle\right). \tag{5.64}$$

It can be checked that the copying part of the circuit in Fig. 5.1 transforms the state $|\psi\rangle|\Phi\rangle$ into

$$|A_0\rangle|0\rangle + |A_1\rangle|1\rangle, \tag{5.65}$$

with

$$\begin{aligned} |A_0\rangle &= \alpha\sqrt{\tfrac{2}{3}}\,|00\rangle + \beta\sqrt{\tfrac{1}{6}}\left(|10\rangle + |01\rangle\right), \\ |A_1\rangle &= \beta\sqrt{\tfrac{2}{3}}\,|11\rangle + \alpha\sqrt{\tfrac{1}{6}}\left(|10\rangle + |01\rangle\right). \end{aligned} \tag{5.66}$$

Since the three qubits are now entangled, we must trace over two of them to obtain the (mixed) state describing the third. Let us first trace over the bottom qubit of Fig. 5.1, obtaining the density matrix

$$|A_0\rangle\langle A_0| + |A_1\rangle\langle A_1|. \tag{5.67}$$

Then, by further tracing over one of the first two qubits, it is possible to check that each of the two qubits at the output of the quantum copier (the two top qubits in Fig. 5.1) is described by the same reduced density operator

$$\rho = \tfrac{2}{3}|\psi\rangle\langle\psi| + \tfrac{1}{6}I. \tag{5.68}$$

It is easy to check that, independently of the initial state $|\psi\rangle$, the (optimal) fidelity of the copy ρ is given by

$$F = \langle\psi|\rho|\psi\rangle = \langle\psi|\left(\tfrac{2}{3}|\psi\rangle\langle\psi| + \tfrac{1}{6}I\right)|\psi\rangle = \tfrac{5}{6}. \tag{5.69}$$

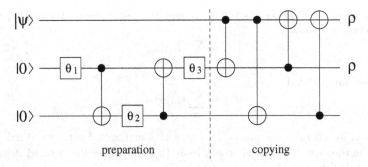

Fig. 5.1 A quantum copying network: two imperfect copies of the state $|\psi\rangle$, described by the density matrix ρ, are recovered at the output. The θ_i-symbols stand for the rotation matrices $R_y(-2\theta_i)$ defined by Eq. (5.62). Note that, to simplify notation, on the left-hand side of the circuit we show the state vectors instead of the corresponding density matrices $|\psi\rangle\langle\psi|$ and $|0\rangle\langle0|$.

Exercise 5.7 Check that the quantum circuit in Fig. 5.1 produces two copies described by the density matrix (5.68).

5.2 The Schmidt decomposition

In this section, we shall demonstrate the existence of a very useful decomposition, known as the Schmidt decomposition, for any pure state of a bipartite quantum system.

Theorem 5.1 The Schmidt decomposition theorem: *Given a pure state* $|\psi\rangle \in \mathcal{H} = \mathcal{H}_1 \otimes \mathcal{H}_2$ *of a bipartite quantum system, there exist orthonormal states* $\{|i\rangle_1\}$ *for* $\mathcal{H}_1$ *and* $\{|i'\rangle_2\}$ *for* $\mathcal{H}_2$ *such that*

$$|\psi\rangle = \sum_{i=1}^{k} \sqrt{p_i}\, |i\rangle_1 |i'\rangle_2 = \sqrt{p_1}\, |1\rangle_1 |1'\rangle_2 + \cdots + \sqrt{p_k}\, |k\rangle_1 |k'\rangle_2, \quad (5.70)$$

with p_i *positive real numbers satisfying the condition* $\sum_{i=1}^{k} p_i = 1$.

It is important to stress that the states $\{|i\rangle_1\}$ and $\{|i'\rangle_2\}$ depend on the particular state $|\psi\rangle$ that we wish to expand.

Proof. Given an arbitrary state vector $|\psi\rangle$ in $\mathcal{H}$, we can always write

$$|\psi\rangle = \sum_{i,\alpha} c_{i\alpha} |i\rangle_1 |\alpha\rangle_2, \quad (5.71)$$

with $\{|i\rangle_1\}$ and $\{|\alpha\rangle_2\}$ basis sets for $\mathcal{H}_1$ and $\mathcal{H}_2$, respectively. We can express this decomposition as

$$|\psi\rangle = \sum_i |i\rangle_1 |\tilde{i}\rangle_2, \tag{5.72}$$

where we have defined

$$|\tilde{i}\rangle_2 = \sum_\alpha c_{i\alpha} |\alpha\rangle_2. \tag{5.73}$$

Of course, in general, these states $|\tilde{i}\rangle_2$ are neither orthogonal nor normalized. To prove the theorem, we must choose a basis $\{|i\rangle_1\}$ in which the reduced density matrix ρ_1 is diagonal:

$$\rho_1 = \sum_i p_i \, |i\rangle_1 {}_1\langle i|. \tag{5.74}$$

Since ρ_1 is a density matrix, it is Hermitian, non-negative and has unit trace. Thus, we have $p_i \geq 0$ and $\sum_i p_i = 1$. The expansion (5.74) involves only terms with $p_i > 0$, corresponding to the non-zero eigenvalues of ρ_1. We can also compute ρ_1 from the partial trace

$$
\begin{aligned}
\rho_1 &= \mathrm{Tr}_2\Big(|\psi\rangle\langle\psi|\Big) = \mathrm{Tr}_2\Big(\sum_{i,j} |i\rangle_1 |\tilde{i}\rangle_2 \, {}_2\langle j|_2\langle \tilde{j}|\Big)\\
&= \sum_{i,j} |i\rangle_1 {}_1\langle j| \sum_k {}_2\langle k|\tilde{i}\rangle_2 \, {}_2\langle \tilde{j}|k\rangle_2 = \sum_{i,j} |i\rangle_1 {}_1\langle j| \sum_k {}_2\langle \tilde{j}|k\rangle_2 \, {}_2\langle k|\tilde{i}\rangle_2\\
&= \sum_{i,j} {}_2\langle \tilde{j}|\tilde{i}\rangle_2 \, |i\rangle_1 {}_1\langle j|, \tag{5.75}
\end{aligned}
$$

where $\{|k\rangle\}_2$ is an orthonormal basis for $\mathcal{H}_2$ and the last equality follows by taking into account the completeness relation $\sum_k |k\rangle_2 {}_2\langle k| = I_2$. The equality between (5.74) and (5.75) requires ${}_2\langle \tilde{j}|\tilde{i}\rangle_2 = p_i \delta_{ij}$. Thus, the vectors $|\tilde{i}\rangle_2$ are orthogonal and can be normalized as follows:

$$|i'\rangle_2 = \frac{1}{\sqrt{p_i}} |\tilde{i}\rangle_2. \tag{5.76}$$

After inserting (5.76) into (5.72), we obtain the Schmidt decomposition (5.70). Thus, we have explicitly constructed the orthonormal states $\{|i\rangle_1\}$ and $\{|i'\rangle_2\}$ which allow us to write down the Schmidt decomposition.

We can also take the partial trace over the first subsystem, obtaining

$$
\begin{aligned}
\rho_2 &= \mathrm{Tr}_1(|\psi\rangle\langle\psi|) = \mathrm{Tr}_1\Big(\sum_{i,j} \sqrt{p_i}\sqrt{p_j}\, |i\rangle_1 |i'\rangle_2 \, {}_1\langle j|_2\langle j'|\Big)\\
&= \sum_i p_i \, |i'\rangle_2 {}_2\langle i'|. \tag{5.77}
\end{aligned}
$$

$\square$

Therefore, the reduced density matrices ρ_1 and ρ_2 have the same non-zero eigenvalues. Their number is also the number k of terms in the Schmidt decomposition (5.70) and is known as the *Schmidt number* (or the *Schmidt rank*) of the state $|\psi\rangle$. It is clear that a separable state, which by definition can be written as

$$|\psi\rangle = |\phi\rangle_1 |\xi\rangle_2, \tag{5.78}$$

has Schmidt number equal to one. Thus, we have the following entanglement criterion: a bipartite pure state is entangled if and only if its Schmidt number is greater than one.

We stress that the Schmidt number is a criterion for entanglement, not a measure of entanglement. In order to clarify this point, let us consider the following two states:

$$\begin{aligned} |\phi^+\rangle &= \tfrac{1}{\sqrt{2}}\Big(|0\rangle_1|0\rangle_2 + |1\rangle_1|1\rangle_2\Big), \\ |\psi\rangle &= \sqrt{1-2\epsilon^2}\,|0\rangle_1|0\rangle_1 + \epsilon\,|1\rangle_1|1\rangle_2 + \epsilon\,|2\rangle_1|2\rangle_2, \end{aligned} \tag{5.79}$$

with $\epsilon \ll 1$. It is clear that the Schmidt number of the Bell state $|\phi^+\rangle$ is 2; that is, it is smaller than the Schmidt number of $|\psi\rangle$, which is 3. On the other hand, one sees intuitively that the entanglement content of a Bell state is much larger than that of the state $|\psi\rangle$ since we have assumed $\epsilon \ll 1$. This intuition can be formalized by introducing, as a measure of entanglement, a quantity borrowed from condensed matter physics, the *participation ratio*:

$$\xi = \frac{1}{\sum_{i=1}^{k} p_i^2}. \tag{5.80}$$

This quantity is bounded between 1 and k: it is close to 1 (that is, to separability) if a single term dominates the Schmidt decomposition, whereas $\xi = k$ if all terms in the decomposition have the same weight ($p_1 = \cdots = p_k = \tfrac{1}{k}$). We have $\xi = 2$ for $|\phi^+\rangle$, which is larger than the value $\xi \approx 1 + 4\epsilon^2$ obtained for the state $|\psi\rangle$.

Exercise 5.8 Show that there are states

$$|\psi\rangle = \sum_{\alpha,\beta,\gamma} c_{\alpha\beta\gamma} |\alpha\rangle_1 |\beta\rangle_2 |\gamma\rangle_3 \tag{5.81}$$

that cannot be written as

$$|\psi\rangle = \sum_i \sqrt{p_i}\,|i\rangle_1 |i'\rangle_2 |i''\rangle_3. \tag{5.82}$$

This means that the Schmidt decomposition cannot be extended to systems composed of more than two parts.

5.3   Purification

Given a quantum system described by the density matrix ρ_1, it is possible to introduce another system, which we call system 2, such that the state $|\psi\rangle$ of the composite system is a pure state and $\rho_1 = \text{Tr}_2(|\psi\rangle\langle\psi|)$. This procedure, known as *purification*, allows us to associate a pure state $|\psi\rangle$ with a density matrix ρ_1. We note that the added system 2 does not necessarily have a physical significance. We just have a useful mathematical tool at our disposal to work with pure states instead of density matrices.

A generic pure state of the global system $1 + 2$ is given by

$$|\psi\rangle = \sum_{i,\,\alpha} c_{i\alpha} |i\rangle_1 |\alpha\rangle_2, \tag{5.83}$$

with $\{|i\rangle_1\}$ and $\{|\alpha\rangle_2\}$ basis sets for the Hilbert spaces associated with the subsystems 1 and 2. The corresponding density matrix is

$$\rho = \sum_{i,\alpha} \sum_{j,\beta} c_{i\alpha} c_{j\beta}^\star |i\rangle_1 |\alpha\rangle_2\, {}_1\langle j|\, {}_2\langle\beta|. \tag{5.84}$$

Given a generic density matrix for system 1,

$$\rho_1 = \sum_{k,l} (\rho_1)_{k,l} |k\rangle_1\, {}_1\langle l|, \tag{5.85}$$

we say that the state $|\psi\rangle$ defined by Eq. (5.84) is a purification of ρ_1 if

$$\rho_1 = \text{Tr}_2\Big(|\psi\rangle\langle\psi|\Big) = \sum_{\gamma} {}_2\langle\gamma| \Big(\sum_{i,\alpha} \sum_{j,\beta} c_{i\alpha}\, c_{j\beta}^\star |i\rangle_1 |\alpha\rangle_2\, {}_1\langle j|\, {}_2\langle\beta| \Big) |\gamma\rangle_2$$

$$= \sum_{\gamma} \sum_{i,j} c_{i\gamma}\, c_{j\gamma}^\star |i\rangle_1\, {}_1\langle j|, \tag{5.86}$$

where we have used the orthonormality relations ${}_2\langle\beta|\gamma\rangle_2 = \delta_{\beta\gamma}$ and ${}_2\langle\gamma|\alpha\rangle_2 = \delta_{\gamma\alpha}$. The equality between (5.85) and (5.86) implies

$$(\rho_1)_{ij} = \sum_{\gamma} c_{i\gamma}\, c_{j\gamma}^\star. \tag{5.87}$$

Here the matrix elements $(\rho_1)_{ij}$ are given and we wish to determine the coefficients $c_{i\gamma}$. It is clear that system (5.87) always admits a solution, provided the Hilbert space of system 2 is large enough (we shall show below that it is sufficient to consider a system 2 whose Hilbert space dimension is the same as that of system 1).

As an example, we consider a qubit whose density matrix ρ_1 is known. It turns out that the addition of a second qubit (known as an *ancillary* qubit) is sufficient for the purification of the density matrix ρ_1. Indeed, in this case the condition (5.87) gives the following set of equations:

$$(\rho_1)_{00} = c_{00}c_{00}^\star + c_{01}c_{01}^\star,$$
$$(\rho_1)_{01} = c_{00}c_{10}^\star + c_{01}c_{11}^\star = (\rho_1)_{10}^\star, \qquad (5.88)$$
$$(\rho_1)_{11} = c_{10}c_{10}^\star + c_{11}c_{11}^\star.$$

We can select a solution to this system if we put $c_{01} = 0$. It is then easy to find that

$$c_{00} = \sqrt{(\rho_1)_{00}}, \quad c_{01} = 0,$$
$$c_{10} = \frac{(\rho_1)_{01}^\star}{\sqrt{(\rho_1)_{00}}}, \quad c_{11} = \sqrt{\frac{(\rho_1)_{00}(\rho_1)_{11} - |(\rho_1)_{01}|^2}{(\rho_1)_{00}}}. \qquad (5.89)$$

Thus, a possible purification is given by

$$|\psi\rangle = \sqrt{(\rho_1)_{00}} \, |0\rangle_1 |0\rangle_2 + \frac{(\rho_1)_{01}^\star}{\sqrt{(\rho_1)_{00}}} \, |1\rangle_1 |0\rangle_2$$
$$+ \sqrt{\frac{(\rho_1)_{00}(\rho_1)_{11} - |(\rho_1)_{01}|^2}{(\rho_1)_{00}}} \, |1\rangle_1 |1\rangle_2. \qquad (5.90)$$

We point out that, given a two-qubit system, it is possible to generate any density matrix ρ_1 for one of the two qubits by means of unitary operations on the two-qubit system. For this purpose, it is sufficient to prepare the state (5.90). Qubit 1 is then described by the desired density matrix ρ_1, obtained after tracing over qubit 2.

We note that if we express the reduced density matrix using its diagonal representation,

$$\rho_1 = \sum_i p_i |i\rangle_1 \, _1\langle i|, \qquad (5.91)$$

it is sufficient to consider a system 2 having the same state space as system 1. Indeed, a purification for the density matrix (5.91) is given by

$$|\psi\rangle = \sum_i \sqrt{p_i} \, |i\rangle_1 \, |i'\rangle_2. \qquad (5.92)$$

The close connection between purification and Schmidt decomposition is self-evident.

Exercise 5.9   Find a purification for the density matrix describing the state of two qubits (*Hint*: use two ancillary qubits and assume $c_{12} = c_{13} = c_{14} = c_{23} = c_{24} = c_{34} = 0$).

5.4   The Kraus representation

Let us consider a bipartite system $1 + 2$. The system undergoes a unitary evolution and we wish to describe the evolution of subsystem 1 alone. We assume that initially the two subsystems are not entangled (we shall see later in this section that there is no lack of generality in this assumption) and described by the density matrix

$$\rho_{12} = \rho_1 \otimes |0\rangle_{2\,2}\langle 0|. \tag{5.93}$$

Namely, subsystem 2 is in a pure state, which we call $|0\rangle_2$. There is no loss of generality in the assumption that subsystem 2 is initially in a pure state. As we saw in the previous section, if this is not the case, we can enlarge the Hilbert space of subsystem 2 in order to purify it. The temporal evolution of the total system is governed by the unitary time-evolution operator U, which leads to the new density matrix

$$\rho'_{12} = U\rho_{12}U^\dagger = U\left(\rho_1 \otimes |0\rangle_{2\,2}\langle 0|\right)U^\dagger. \tag{5.94}$$

The quantum circuit implementing this transformation is shown in Fig. 5.2. As explained in Sec. 5.1.2, since we are interested in the new density matrix ρ'_1 describing subsystem 1, we must trace over the second subsystem:

$$\begin{aligned}
\rho'_1 = \mathrm{Tr}_2(\rho'_{12}) &= \mathrm{Tr}_2\left[U\left(\rho_1 \otimes |0\rangle_{2\,2}\langle 0|\right)U^\dagger\right] \\
&= \sum_k {}_2\langle k|U|0\rangle_2\, \rho_1\, {}_2\langle 0|U^\dagger|k\rangle_2,
\end{aligned} \tag{5.95}$$

where $\{|k\rangle_2\}$ is a basis set for the Hilbert space $\mathcal{H}_2$ associated with subsystem 2 and ${}_2\langle k|U|0\rangle_2$ is an operator acting on the Hilbert space $\mathcal{H}_1$ associated with subsystem 1. If we define the *Kraus operators*

$$E_k \equiv {}_2\langle k|U|0\rangle_2, \tag{5.96}$$

then we can rewrite Eq. (5.95) as

$$\rho_1' = \sum_k E_k \rho_1 E_k^\dagger. \tag{5.97}$$

Since U is unitary, the operators E_k satisfy the property

$$\sum_k E_k^\dagger E_k = \sum_k {}_2\langle 0|U^\dagger|k\rangle_2 \, {}_2\langle k|U|0\rangle_2 = {}_2\langle 0|U^\dagger U|0\rangle_2 = I_1, \tag{5.98}$$

where I_1 denotes the identity operator in the Hilbert space $\mathcal{H}_1$. Note that we have used the completeness relation $\sum_k |k\rangle_2 \, {}_2\langle k| = I_2$. Equation (5.97) defines a linear map from linear operators to linear operators:

$$S : \rho_1 \to \rho_1' = \sum_k E_k \rho_1 E_k^\dagger. \tag{5.99}$$

If the completeness relation (5.98) is satisfied, map S is known as a *quantum operation* or a *superoperator* and Eq. (5.97) is known as the *Kraus representation* (or the *operator-sum representation*) of the superoperator S. Note that, if $U(t)$ denotes the time-evolution operator from time 0 to time t, then E_k depends on time and Eq. (5.99) can be written as $S(t) : \rho_1(0) \to \rho_1(t) = \sum_k E_k(t) \rho_1(0) E_k^\dagger(t)$, where $\rho_1(t)$ is the density matrix describing subsystem 1 at time t.

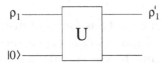

Fig. 5.2 A quantum circuit implementing the transformation (5.94). The state ρ_1' is obtained after partial tracing over the other subsystem (lower line in the figure) the overall density operator $U(\rho_1 \otimes |0\rangle\langle 0|)U^\dagger$.

A superoperator maps density operators to density operators, since:

1. ρ_1' is Hermitian if ρ_1 is Hermitian:

$$\left(\rho_1'\right)^\dagger = \left(\sum_k E_k \rho_1 E_k^\dagger\right)^\dagger = \sum_k \left(E_k^\dagger\right)^\dagger \rho_1^\dagger E_k^\dagger = \sum_k E_k \rho_1 E_k^\dagger = \rho_1'; \tag{5.100}$$

2. ρ_1' has unit trace if ρ_1 has unit trace:

$$\mathrm{Tr}(\rho_1') = \mathrm{Tr}\left(\sum_k E_k \rho_1 E_k^\dagger\right) = \sum_k \mathrm{Tr}(\rho_1 E_k^\dagger E_k)$$
$$= \mathrm{Tr}(\rho_1 \sum_k E_k^\dagger E_k) = \mathrm{Tr}(\rho_1) = 1; \qquad (5.101)$$

3. ρ_1' is non-negative if ρ_1 is non-negative:

$$_1\langle\psi|\rho_1'|\psi\rangle_1 = \sum_k {}_1\langle\psi|E_k\rho_1 E_k^\dagger|\psi\rangle_1 = \sum_k {}_1\langle\varphi_k|\rho_1|\varphi_k\rangle_1 \geq 0, \quad (5.102)$$

where $|\psi\rangle_1$ is any vector in $\mathcal{H}_1$ and $|\varphi_k\rangle_1 \equiv E_k^\dagger|\psi\rangle_1$.

Unitary representation. So far, we have shown that the unitary evolution of a composite system naturally gives rise to an operator-sum representation describing the evolution of a subsystem. We now tackle the converse problem: given a Kraus representation for the evolution of system 1, we shall show that it is possible to introduce an auxiliary system 2 so that the evolution of the total system $1 + 2$ is unitary. In this manner, we construct the unitary representation corresponding to a given superoperator. We define an operator U, acting as follows on states of the form $|\psi\rangle_1|0\rangle_2$:

$$U|\psi\rangle_1|0\rangle_2 \equiv \sum_k E_k|\psi\rangle_1|k\rangle_2, \qquad (5.103)$$

where $\{|k\rangle_2\}$ is an orthonormal basis for subsystem 2, whose dimension is given by the number of Kraus operators appearing in the operator-sum representation. The operator U preserves the inner product. Indeed, for arbitrary states $|\psi\rangle_1$ and $|\phi\rangle_1$ we have

$$_1\langle\psi|_2\langle 0|U^\dagger U|\phi\rangle_1|0\rangle_2 = \left(\sum_k {}_1\langle\psi|E_k^\dagger {}_2\langle k|\right)\left(\sum_l E_l|\phi\rangle_1|l\rangle_2\right)$$
$$= \sum_k {}_1\langle\psi|E_k^\dagger E_k|\phi\rangle_1 = {}_1\langle\psi|\phi\rangle_1, \qquad (5.104)$$

where we have used the orthonormality relation $_2\langle k|l\rangle_2 = \delta_{kl}$ and the completeness relation (5.98). As the operator U preserves the inner product when acting on the subspace whose states are of the form $|\psi\rangle_1|0\rangle_2$, then it can be extended to a unitary operator acting on the entire Hilbert space

$\mathcal{H}_1 \otimes \mathcal{H}_2$ of the joint system.[2]

Composing superoperators. Two superoperators $\mathcal{S}_A$ and $\mathcal{S}_B$ can be composed to give a new superoperator $\mathcal{S} = \mathcal{S}_B \mathcal{S}_A$, defined by $\mathcal{S}(\rho_1) = \mathcal{S}_B(\mathcal{S}_A(\rho_1))$. If $\mathcal{S}_A$ describes the evolution of the density matrix for system 1 from time t_0 to time t_1 and $\mathcal{S}_B$ from t_1 to t_2, then $\mathcal{S} = \mathcal{S}_B \mathcal{S}_A$ describes the evolution from t_0 to t_2. It can be shown that a superoperator is invertible if and only if it is unitary. Thus, superoperators are a *semigroup* instead of a group. Physically, this means that an arrow of time has been introduced for subsystem 1. We can describe the evolution from t_0 to $t_1 > t_0$ by means of a superoperator but not from t_1 to t_0. There is a loss of information from system 1 to system 2 (known as the *environment*) and we cannot run the evolution of system 1 backward if we know its state but ignore the state of the environment. This phenomenon is known as *decoherence* and in the next chapter we shall show that superoperators provide a very general theoretical framework for its description.

There is a freedom in the operator-sum representation; that is, different representations can give rise to the same superoperator. The following theorem holds: two superoperators $\mathcal{S}(\rho_1) = \sum_k E_k \rho_1 E_k^\dagger$ and $\mathcal{S}'(\rho_1) = \sum_k F_k \rho_1 F_k^\dagger$ coincide if and only if there exists a unitary matrix W such that $F_i = \sum_j W_{ij} E_j$. The proof of this theorem can be found in Schumacher (1996).

We note that, in order to build a unitary representation of a superoperator, the dimension of the Hilbert space $\mathcal{H}_2$ must be at least as large as the number of operators E_k appearing in the Kraus representation. If the Hilbert space $\mathcal{H}_1$ has dimension N, it is possible to prove that all superoperators $\mathcal{S}(\rho_1)$ can be generated by an operator-sum representation containing at most N^2 operators E_k (see, e.g., Preskill, 1998a). Therefore, it will be sufficient to consider a Hilbert space $\mathcal{H}_2$ of dimension N^2.

We may also ask how many real parameters are required to parametrize a generic superoperator $\mathcal{S} : \rho_1 \to \rho_1'$ on a Hilbert space of dimension N. A superoperator maps a density operator into another density operator; that is, an $N \times N$ Hermitian matrix to another $N \times N$ Hermitian matrix. A basis for the space $\mathcal{M}_\mathcal{H}$ of the $N \times N$ Hermitian matrices has N^2 elements. Therefore, the

---

[2]It is easy to verify that the unitary transformation (5.103) really induces an operator-sum representation on subsystem 1. Indeed, the evolution of a pure state $\rho_1 = |\psi\rangle_1{}_1\langle\psi|$ is as follows:

$$\rho_1 \to \rho_1' = \text{Tr}_2\left(U|\psi\rangle_1|0\rangle_2{}_1\langle\psi|{}_2\langle 0|U^\dagger\right) = \sum_k E_k|\psi\rangle_1{}_1\langle\psi|E_k^\dagger = \sum_k E_k\rho_1 E_k^\dagger.$$

And since a generic density matrix can be expressed as an ensemble of pure states, $\rho_1 = \sum_i p_i|\psi_i\rangle_1{}_1\langle\psi_i|$, we recover the Kraus representation (5.97) for arbitrary ρ_1.

most general linear transformation acting on the space $\mathcal{M}_\mathcal{H}$ has $(N^2)^2 = N^4$ free real parameters. We should then take into account the completeness relation $\sum_k E_k^\dagger E_k = I_1$, which gives N^2 constraints for these parameters (note that we have N^2 constraints and not $2N^2$ since the terms $E_k E_k^\dagger$ are Hermitian; in other words, the Hermitian conjugate of the completeness relation is again the same completeness relation). Hence, a generic superoperator is parametrized by $N^4 - N^2$ real parameters. For instance, in the case of a single qubit ($N = 2$) we need 12 real parameters, while in the two-qubit case ($N = 4$) we need 240 real parameters.

We may now state the following fundamental theorem (for a proof see Schumacher, 1996):

Theorem 5.2   The Kraus representation theorem: *A map* $\mathcal{S} : \rho_1 \rightarrow \rho_1'$ *satisfying the following requirements: it*

1. *is linear; that is,*

$$\mathcal{S}(p_1\rho_1 + p_2\rho_2) = p_1\mathcal{S}(\rho_1) + p_2\mathcal{S}(\rho_2), \qquad (5.105)$$

2. *preserves hermiticity,*
3. *preserves trace,*
4. *is completely positive,*

has an operator-sum representation (5.97) and a unitary representation (5.103) on a larger Hilbert space.

We say that $\mathcal{S}$ is positive if, for a non-negative ρ_1, $\rho_1' = \mathcal{S}(\rho_1)$ is also non-negative. The complete positivity of $\mathcal{S}$ is a stronger requirement. It means that, for any extension of the Hilbert space $\mathcal{H}_1$ to $\mathcal{H}_1 \otimes \mathcal{H}_E$, the superoperator $\mathcal{S} \otimes \mathcal{I}_E$ is positive. That is, if we add any system E that has a trivial dynamics (the identity $\mathcal{I}_E$ means that no state of E is changed), independently of the dynamics of system 1, the resulting superoperator $\mathcal{S} \otimes \mathcal{I}_E$ must be positive. This requirement is physically motivated since, in general, it cannot be excluded that the two systems are initially entangled. If this is the case and we call ρ_{1E} the density matrix corresponding to the initially entangled state, then $\rho_{1E}' \equiv (\mathcal{S} \otimes \mathcal{I}_E)(\rho_{1E})$ must be a valid density matrix. This implies the positivity of $\mathcal{S} \otimes \mathcal{I}_E$ for any E, namely the complete positivity of $\mathcal{S}$.

Exercise 5.10   Consider the state

$$|\psi\rangle_{1E} = \tfrac{1}{\sqrt{2}}\left(|0\rangle_1|1\rangle_E + |1\rangle_1|0\rangle_E\right). \qquad (5.106)$$

Let ρ_1 denote the density operator for the first qubit and show that the transposition operator

$$\mathcal{T}(\rho_1) \equiv \rho_1^T \qquad (5.107)$$

is positive but not completely positive. For this purpose, it will be sufficient to show that $\mathcal{T} \otimes \mathcal{I}_E$ is not positive.

The Kraus representation theorem tells us that, if the evolution $\rho_1' = \mathcal{S}(\rho_1)$ of the density matrix ρ_1 preserves hermiticity and trace, is linear and completely positive, then this evolution can be realized by the unitary transformation (5.103), acting on a larger Hilbert space $\mathcal{H}_1 \otimes \mathcal{H}_2$. Note that in (5.103) subsystems 1 and 2 are not initially entangled. Thus, if we are only interested in the evolution of the density matrix ρ_1, there is no lack of generality in assuming that subsystem 1 is not initially entangled with subsystem 2.

Examples. As a simple example illustrating the Kraus representation, we now consider two single-qubit subsystems 1 and 2, with initial density operator given by $\rho_{12} = |0\rangle_2\,_2\langle 0| \otimes \rho_1$, whose matrix representation is

$$\rho_{12} = \begin{bmatrix} \rho_1 & \mathbf{0} \\ \mathbf{0} & \mathbf{0} \end{bmatrix}, \qquad (5.108)$$

where ρ_1 and $\mathbf{0}$ are 2×2 submatrices and all elements of $\mathbf{0}$ are zero. Note that we have taken the qubit in the state $|0\rangle_2$ as the most significant qubit to obtain a simple block representation for the matrix (5.108). The evolution of the global system $1 + 2$ is governed by a unitary 4×4 matrix U and we obtain the new reduced density matrix ρ_1' after tracing over the degrees of freedom of subsystem 2 (see the quantum circuit implementing this transformation in Fig. 5.2). We have

$$\rho_1' = \text{Tr}_2\left(U\,\rho_{12}\,U^\dagger\right) = \text{Tr}_2\left(\begin{bmatrix} A & B \\ C & D \end{bmatrix}\begin{bmatrix} \rho_1 & \mathbf{0} \\ \mathbf{0} & \mathbf{0} \end{bmatrix}\begin{bmatrix} A^\dagger & C^\dagger \\ B^\dagger & D^\dagger \end{bmatrix}\right)$$

$$= A\,\rho_1\,A^\dagger + C\,\rho_1\,C^\dagger, \qquad (5.109)$$

where the unitary matrix U has been expressed in terms of the 2×2 submatrices A, B, C and D. If we require that the transformation maps density matrices to density matrices, then $\text{Tr}(\rho_1') = \text{Tr}\left(\rho_1(A^\dagger A + C^\dagger C)\right) = \text{Tr}(\rho_1)$ for any ρ_1. It follows that we must have

$$A^\dagger A + C^\dagger C = I_1. \qquad (5.110)$$

Thus, we have explicitly constructed a Kraus representation for a single qubit, with the Kraus operators A and C satisfying property (5.98).

We point out that, in the most general case for a single qubit, the number of Kraus operators appearing in the operator-sum representation is $N^2 = 4$ ($N = 2$ being the dimension of the Hilbert space for a single qubit), corresponding to two qubits for subsystem 2. For instance, if we consider the unitary evolution

$$U|\psi\rangle_1|0\rangle_2 = \sqrt{1-p}\,I_1|\psi\rangle_1|0\rangle_2$$
$$+ \sqrt{\tfrac{1}{3}p}\Big[(\sigma_x)_1|\psi\rangle_1|1\rangle_2 + (\sigma_y)_1|\psi\rangle_1|2\rangle_2 + (\sigma_z)_1|\psi\rangle_1|3\rangle_2\Big], \quad (5.111)$$

with $0 \le p \le 1$, then the four Kraus operators are given by $E_k = {}_2\langle k|U|0\rangle_2$. We readily obtain

$$E_0 = \sqrt{1-p}\,I_1, \quad E_1 = \sqrt{\tfrac{1}{3}p}\,(\sigma_x)_1, \quad E_2 = \sqrt{\tfrac{1}{3}p}\,(\sigma_y)_1, \quad E_3 = \sqrt{\tfrac{1}{3}p}\,(\sigma_z)_1,$$
$$(5.112)$$

and it is easy to check that the operators E_k satisfy the normalization condition $\sum_k E_k^\dagger E_k = I_1$. The evolution of the reduced density matrix ρ_1, corresponding to the unitary evolution (5.111), is given by

$$\rho_1 \to \rho_1' = \sum_{k=0}^{3} E_k \rho_1 E_k^\dagger$$
$$= (1-p)\,I_1 + \tfrac{1}{3}p\Big[(\sigma_x)_1\rho_1(\sigma_x)_1 + (\sigma_y)_1\rho_1(\sigma_y)_1 + (\sigma_z)_1\rho_1(\sigma_z)_1\Big].$$
$$(5.113)$$

This example corresponds to the so-called depolarizing channel and will be discussed, together with several other examples of Kraus representations for a single qubit, in Chap. 6.

5.5 Measurement of the density matrix for a qubit

We saw in Sec. 3.1 that the coordinates (x, y, z) singling out a pure state on the Bloch sphere can be measured, provided a large number of states prepared in the same manner are available. We now show that the same conclusions hold for mixed states. Following Sec. 5.1, we write the density matrix for a qubit as

$$\rho = \tfrac{1}{2} \begin{bmatrix} 1+z & x-iy \\ x+iy & 1-z \end{bmatrix}. \tag{5.114}$$

The measurement procedure is shown in Fig. 5.3: a unitary transformation U maps ρ into a new density matrix $\rho' = U\rho U^\dagger$ and the detector D measures σ_z. The possible outcomes of this measurement are $i = 0, 1$ (we associate $i = 0$ with $\sigma_z = +1$ and $i = 1$ with $\sigma_z = -1$), obtained with probabilities

$$p_i = \mathrm{Tr}(\rho' P_i), \tag{5.115}$$

where the projector operators P_i read in the $\{|0\rangle, |1\rangle\}$ basis as follows:

$$P_0 = \begin{bmatrix} 1 & 0 \\ 0 & 0 \end{bmatrix}, \qquad P_1 = \begin{bmatrix} 0 & 0 \\ 0 & 1 \end{bmatrix}. \tag{5.116}$$

We can also write

$$p_i = \mathrm{Tr}(U\rho U^\dagger P_i) = \mathrm{Tr}(\rho U^\dagger P_i U) = \mathrm{Tr}(\rho Q_i), \tag{5.117}$$

where we have defined the new operators

$$Q_i \equiv U^\dagger P_i U. \tag{5.118}$$

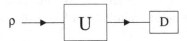

Fig. 5.3 A schematic drawing of the measurement of the density matrix. The unitary transformation U comes before a standard measurement performed by the detector D.

In order to measure the coordinate z, we take $U = I$, so that $Q_0 = P_0$ and $Q_1 = P_1$. It is easy to compute p_0, p_1 and to check that

$$p_0 - p_1 = z. \tag{5.119}$$

To compute x, we take $U = R_y(-\tfrac{\pi}{2})$; that is, the Bloch sphere is rotated clockwise through an angle $\tfrac{\pi}{2}$ about the y-axis (we follow the definition given in Sec. 3.3.1 for the rotation matrices). In this manner, the x-axis is transformed into the z-axis and the coordinate x can be computed by measuring σ_z. Hence, we consider

$$U = R_y\left(-\frac{\pi}{2}\right) = \frac{1}{\sqrt{2}}\begin{bmatrix} 1 & 1 \\ -1 & 1 \end{bmatrix}, \tag{5.120}$$

and therefore

$$Q_0 = U^\dagger P_0 U = \frac{1}{2}\begin{bmatrix} 1 & 1 \\ 1 & 1 \end{bmatrix}, \quad Q_1 = U^\dagger P_1 U = \frac{1}{2}\begin{bmatrix} 1 & -1 \\ -1 & 1 \end{bmatrix}. \tag{5.121}$$

We can readily check that

$$p_0 - p_1 = x. \tag{5.122}$$

Likewise, we can compute y. We take

$$U = R_x\left(-\frac{\pi}{2}\right) = \frac{1}{\sqrt{2}}\begin{bmatrix} 1 & i \\ i & 1 \end{bmatrix} \tag{5.123}$$

and therefore

$$Q_0 = \frac{1}{2}\begin{bmatrix} 1 & i \\ -i & 1 \end{bmatrix}, \quad Q_1 = \frac{1}{2}\begin{bmatrix} 1 & -i \\ i & 1 \end{bmatrix}, \tag{5.124}$$

which implies

$$p_1 - p_0 = y. \tag{5.125}$$

Of course, we must repeat the entire procedure (preparation of the initial state, unitary transformation and measurement) a large number of times to obtain good estimates of x, y and z. We note that the method can be generalized to measure density matrices of larger dimensions.

5.6   Generalized measurements

A generalized measurement is described by a set $\{M_i\}$ of measurement operators, not necessarily self-adjoint, that satisfy the completeness relation

$$\sum_i M_i^\dagger M_i = I. \tag{5.126}$$

If the state vector of the system before the measurement is $|\psi\rangle$, then with probability

$$p_i = \langle\psi|M_i^\dagger M_i|\psi\rangle \tag{5.127}$$

the measurement gives outcome i and the post-measurement state of the system is

$$|\psi_i'\rangle = \frac{M_i|\psi\rangle}{\sqrt{\langle\psi|M_i^\dagger M_i|\psi\rangle}}. \tag{5.128}$$

We note that the completeness equation (5.126) assures the fact that the probabilities sum to unity; that is, $\sum_i p_i = \sum_i \langle\psi|M_i^\dagger M_i|\psi\rangle = 1$. We note that the projective measurements described in Sec. 2.4 are a special case of generalized measurements, in which the operators M_i are orthogonal projectors; that is, $M_i^\dagger = M_i$ and $M_i M_j = \delta_{ij} M_i$. Therefore, in this case, the completeness relation becomes $\sum_i M_i = I$. It turns out that projective measurements together with unitary operations are equivalent to generalized measurements, provided ancillary qubits are added. This simply means that generalized measurements are equivalent to projective measurements on a larger Hilbert space. This statement is known as Neumark's theorem and is discussed, e.g., in Peres (1993).

In the following, we show that, if we restrict our attention to a subsystem of a given system, a projective measurement performed on the system cannot in general be described as a projective measurement on the subsystem. Let us consider the unitary evolution (5.103) of a composite system $1+2$, initially in the state $|\psi\rangle_1|0\rangle_2$:

$$U|\psi\rangle_1|0\rangle_2 = \sum_k E_k|\psi\rangle_1|k\rangle_2, \tag{5.129}$$

where $\{|k\rangle_2\}$ is an orthonormal basis for subsystem 2. A projective measurement, described by the projectors $P_i = I_1 \otimes |i\rangle_{2\ 2}\langle i|$, with $\sum_i P_i = I_1 \otimes I_2$, gives outcome i with probability

$$p_i = \text{Tr}\left(\rho_{12} P_i\right) = \text{Tr}\left(\sum_{k,k'} E_k|\psi\rangle_1|k\rangle_{2\ 1}\langle\psi|_2\langle k'|E_{k'}^\dagger|i\rangle_{2\ 2}\langle i|\right)$$

$$= {}_1\langle\psi|E_i^\dagger E_i|\psi\rangle_1, \tag{5.130}$$

where $\rho_{12} = U|\psi\rangle_1|0\rangle_{2\ 1}\langle\psi|_2\langle 0|U^\dagger$ and the Kraus operators E_i satisfy the condition $\sum_i E_i^\dagger E_i = I_1$ and, in general, are not projectors. Therefore, a standard projective measurement performed on the system can be described as a generalized measurement on subsystem 1.

5.6.1 * *Weak measurements*

An interesting kind of generalized measurement is the weak measurement; that is, a measurement that disturbs the state of the system very little. In this section, following Brun (2002), we provide a concrete example of weak measurement, obtained from a projective measurement performed on an environment weakly coupled to the system. We assume that both the system and the environment can be described as qubits. The system qubit is initially in a generic state $\alpha|0\rangle_S + \beta|1\rangle_S$, while the environment qubit is in the state $|0\rangle_E$. We also assume that the evolution of the two qubits is described by the unitary transformation $\{[R_z(\theta)]_S \otimes I_E\}(\cos\theta I - i\sin\theta\,\text{CNOT})$, with $\theta \ll 1$ (note that in the CNOT gate the system acts as the control and the environment as the target qubit). After this, the two qubits are (up to an overall phase factor $\exp(i\frac{\theta}{2})$) in the state

$$|\Psi\rangle = \big(\alpha|0\rangle_S + \beta\cos\theta\,|1\rangle_S\big)|0\rangle_E - i\beta\sin\theta\,|1\rangle_S|1\rangle_E. \qquad (5.131)$$

If we measure the environment in the z-basis, we obtain outcomes 0 or 1 with probabilities $p_0 = |\alpha|^2 + |\beta|^2\cos^2\theta \approx 1 - |\beta|^2\theta^2$ and $p_1 = |\beta|^2\sin^2\theta \approx |\beta|^2\theta^2 \ll 1$. In both cases, the system and the environment are no longer entangled after the measurement. If the outcome 0 occurs, then the system is left in the state

$$|\psi_0\rangle_S = \frac{\alpha|0\rangle_S + \beta\cos\theta\,|1\rangle_S}{\sqrt{|\alpha|^2 + |\beta|^2\cos^2\theta}} \approx \alpha\Big(1 + \tfrac{1}{2}|\beta|^2\theta^2\Big)|0\rangle_S + \beta\Big(1 - \tfrac{1}{2}|\alpha|^2\theta^2\Big)|1\rangle_S.$$
$$(5.132)$$

In this case, the system is weakly perturbed and the (weak) information obtained from the measurement of the environment is that it is now more probable that the system is found in the state $|0\rangle_S$. If instead the outcome 1 is obtained from this measurement, then the system is left in the state

$$|\psi_1\rangle_S = |1\rangle_S. \qquad (5.133)$$

Therefore, in this example the measurement is weak in the sense that most of the time (with probability $p_0 \approx 1 - |\beta|^2\theta^2$) the system is weakly perturbed. However, in rare occasions (with probability $p_1 \approx |\beta|^2\theta^2$) the system changes abruptly (in the language of quantum trajectories, to be discussed in Sec. 6.6.1, we say that a *jump* occurs). We stress that the system–environment interaction plus the projective measurement acting on the environment can be conveniently described as a generalized measurement acting on the system, with the measurement operators

$$M_0 = |0\rangle_S{}_S\langle0| + \cos\theta\,|1\rangle_S{}_S\langle1|, \qquad M_1 = \sin\theta\,|1\rangle_S{}_S\langle1|, \qquad (5.134)$$

satisfying the completeness relation $M_0^\dagger M_0 + M_1^\dagger M_1 = I_S$.

Exercise 5.11   Show that, if the weak measurement (5.134) is repeated a very large number of times, then the effect is the same as a strong measurement: given a state $|\psi\rangle_S = \alpha|0\rangle_S + \beta|1\rangle_S$, the system is at the end left in the state $|0\rangle_S$ with probability $p_0 \approx |\alpha|^2$ or in the state $|1\rangle_S$ with probability $p_1 \approx |\beta|^2$.

It is important to point out that the state of the system after the measurement of the environment qubit depends on the selected measurement basis. Let us consider, for instance, what happens if we measure the environment in the x-basis. For this purpose, it is useful to rewrite the state (5.131) as follows:

$$|\Psi\rangle = \tfrac{1}{\sqrt{2}}\left(\alpha|0\rangle_S + \beta e^{-i\theta}|1\rangle_S\right)|+\rangle_E + \tfrac{1}{\sqrt{2}}\left(\alpha|0\rangle_S + \beta e^{i\theta}|1\rangle_S\right)|-\rangle_E, \quad (5.135)$$

where $|\pm\rangle_E = \tfrac{1}{\sqrt{2}}(|0\rangle_E \pm |1\rangle_E)$ are the eigenstates of $(\sigma_x)_E$ corresponding to the eigenvalues ±1. The two measurement outcomes $(\sigma_x)_E = \pm1$ leave the system in the new states

$$|\psi_+\rangle_S = \alpha|0\rangle_S + \beta e^{-i\theta}|1\rangle_S, \qquad |\psi_-\rangle_S = \alpha|0\rangle_S + \beta e^{i\theta}|1\rangle_S. \quad (5.136)$$

It can be clearly seen that in both cases the state of the system is weakly perturbed: a small relative phase $\pm\theta$ is added. The sign of this phase is chosen randomly due to the inherent randomness of the quantum measurement process. If we repeat the entire procedure (system–environment interaction plus environment measurement) several times we do not have jumps but a slow *diffusion* in the relative phase θ between the coefficients in front of the states $|0\rangle_S$ and $|1\rangle_S$. Note that, also in this case in which the measurement is performed in the x-basis, we can give a convenient description in terms of generalized measurement, with the measurement operators

$$M_0 = \tfrac{1}{\sqrt{2}}\left(|0\rangle\langle0| + e^{-i\theta}|1\rangle\langle1|\right), \qquad M_1 = \tfrac{1}{\sqrt{2}}\left(|0\rangle\langle0| + e^{i\theta}|1\rangle\langle1|\right). \quad (5.137)$$

Of course, this measurement weakly disturbs the state but also gives a small amount of information on it: we only know that, as a result of the weak measurement, a relative phase θ has been added or subtracted.

5.6.2   *POVM measurements*

POVM's ("Positive Operator-Valued Measurements") are well suited to describing experiments where the system is measured only once and therefore we are not interested in the state of the system after the measurement. This is, for instance, the case of a photon detected by a photomultiplier: the photon is destroyed in the measurement process and therefore the measurement cannot be repeated. A POVM is described by a set of positive (more precisely, non-negative) operators F_i (POVM elements), such that

$$\sum_i F_i = I. \tag{5.138}$$

If the measurement is performed on a system described by the state vector $|\psi\rangle$, the probability of obtaining outcome i is

$$p_i = \langle\psi|F_i|\psi\rangle. \tag{5.139}$$

POVM's can be seen as generalized measurements, provided we define $F_i = M_i^\dagger M_i$. Indeed, it is evident that this definition assures that F_i is a non-negative operator. It is also clear that projective measurements are POVM's since in this case $F_i = M_i^\dagger M_i = M_i$, with M_i projectors and $\sum_i F_i = \sum_i M_i = I$. However, we stress that in the POVM formalism we do not make any assumption on the post-measurement state of the system.[3]

An example of POVM is shown in Fig. 5.4. The system qubit is initially in the state ρ, the environment qubit in the state $|0\rangle$. In Fig. 5.4, R denotes the rotation matrix

$$R = \begin{bmatrix} r & t \\ -t & r \end{bmatrix} \tag{5.140}$$

(we assume $0 < r < 1$ and $t = \sqrt{1 - r^2}$) and K a modified Hadamard matrix:

$$K = \frac{1}{\sqrt{2}} \begin{bmatrix} 1 & 1 \\ -1 & 1 \end{bmatrix}. \tag{5.141}$$

It is easy to check by direct matrix multiplication that the circuit in Fig. 5.4

---

[3]The POVM formalism can also be used when the system is prepared in a mixed state ρ. In this case, the probability of obtaining outcome i is $p_i = \text{Tr}(F_i\rho)$; see, *e.g.*, Peres (1993).

implements the unitary transformation

$$U = \frac{1}{\sqrt{2}} \begin{bmatrix} 1 & r & -t & 0 \\ -1 & r & -t & 0 \\ 0 & t & r & 1 \\ 0 & -t & -r & 1 \end{bmatrix}. \tag{5.142}$$

The two detectors D_1 and D_0 drawn in Fig. 5.4 perform a standard projective measurement with possible outcomes 0 and 1. In general, we have four possible outcomes: 00, 01, 10 and 11 (in integer notation, 0, 1, 2 and 3), associated with the projectors

$$P_0 = \begin{bmatrix} 1 & 0 & 0 & 0 \\ 0 & 0 & 0 & 0 \\ 0 & 0 & 0 & 0 \\ 0 & 0 & 0 & 0 \end{bmatrix}, \quad P_1 = \begin{bmatrix} 0 & 0 & 0 & 0 \\ 0 & 1 & 0 & 0 \\ 0 & 0 & 0 & 0 \\ 0 & 0 & 0 & 0 \end{bmatrix},$$

$$P_2 = \begin{bmatrix} 0 & 0 & 0 & 0 \\ 0 & 0 & 0 & 0 \\ 0 & 0 & 1 & 0 \\ 0 & 0 & 0 & 0 \end{bmatrix}, \quad P_3 = \begin{bmatrix} 0 & 0 & 0 & 0 \\ 0 & 0 & 0 & 0 \\ 0 & 0 & 0 & 0 \\ 0 & 0 & 0 & 1 \end{bmatrix}. \tag{5.143}$$

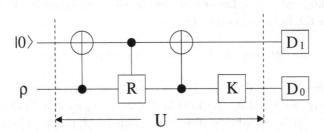

Fig. 5.4 A quantum circuit implementing a POVM measurement. The letters D_0 and D_1 denote two detectors performing standard projective single qubit measurements.

The probability of obtaining outcome i is given by

$$p_i = \mathrm{Tr}\left(U \rho_{\text{in}}^{(\text{tot})} U^\dagger P_i\right) = \mathrm{Tr}\left(\rho_{\text{in}}^{(\text{tot})} Q_i\right), \tag{5.144}$$

where $\rho_{\text{in}}^{(\text{tot})}$ is the initial two-qubit state and we have introduced the operators $Q_i = U^\dagger P_i U$. We assume that $\rho_{\text{in}}^{(\text{tot})} = |0\rangle\langle 0| \otimes \rho$, with matrix representation

$$\rho_{\text{in}}^{(\text{tot})} = \begin{bmatrix} \rho & \mathbf{0} \\ \mathbf{0} & \mathbf{0} \end{bmatrix}, \tag{5.145}$$

where ρ and $\mathbf{0}$ are 2×2 submatrices and $\mathbf{0}$ has all matrix elements equal to 0. Given this initial state, we have

$$p_i = \text{Tr}\left(\rho_{\text{in}}^{(\text{tot})} Q_i\right) = \text{Tr}\left(\rho F_i\right), \tag{5.146}$$

where F_i is the 2×2 submatrix of Q_i corresponding to the value 0 of the most significant qubit. In particular, if $\rho = |\psi\rangle\langle\psi|$ is a pure state, then $p_i = \langle\psi|F_i|\psi\rangle$. We obtain

$$F_0 = \frac{1}{2}\begin{bmatrix} 1 & r \\ r & r^2 \end{bmatrix}, \quad F_1 = \frac{1}{2}\begin{bmatrix} 1 & -r \\ -r & r^2 \end{bmatrix}, \quad F_2 = \begin{bmatrix} 0 & 0 \\ 0 & 1-r^2 \end{bmatrix}, \tag{5.147}$$

where we have added in F_2 the contributions coming from Q_2 and Q_3 since they are identical. The F_i constitute a POVM. Indeed, they are non-negative operators and fulfill the condition $\sum_i F_i = I$.

POVM measurements are useful, for instance, to avoid misidentification of non-orthogonal states. Let us consider the following example: Alice sends Bob one of the following two states:

$$|\psi_1\rangle = \sin\theta\,|0\rangle + \cos\theta\,|1\rangle, \qquad |\psi_2\rangle = \sin\theta\,|0\rangle - \cos\theta\,|1\rangle, \tag{5.148}$$

where we assume $0 < \theta < \frac{\pi}{4}$. Then Bob performs on the received state a measurement described by the POVM elements F_0, F_1 and F_2 defined by Eq. (5.147). Bob's probability of obtaining outcome i, provided he received the state $|\psi_k\rangle$ $(k = 1, 2)$, is

$$p(i|k) = \langle\psi_k|F_i|\psi_k\rangle. \tag{5.149}$$

We choose $r = \tan\theta$. We have $p(1|1) = 0$ and $p(0|2) = 0$. Therefore, the outcome $i = 1$ excludes that the state $|\psi_1\rangle$ was sent, whereas $i = 0$ excludes $|\psi_2\rangle$. Finally, if we obtain outcome $i = 2$, we cannot conclude anything. Bob cannot always distinguish which one of the two non-orthogonal states $|\psi_1\rangle$ and $|\psi_2\rangle$ was sent. However, taking advantage of POVM measurements, he can avoid misidentification.

5.7   The Shannon entropy

The first basic task of classical information theory is to quantify the information contained in a message. This problem was solved by Shannon in 1948. A message is a string of letters chosen from an alphabet $\mathcal{A} = \{a_1, a_2, \ldots, a_k\}$. We assume that the letters in the message are statistically independent and that the letter a_i occurs with a *priori* probability p_i, where $\sum_{i=1}^{k} p_i = 1$. The assumption that the letters are statistically independent has been made to simplify the discussion. In practice, this is not the case in many important examples. For instance, there are strong correlations between consecutive letters in an English text. However, the ideas developed in this section can be extended to include more complicated situations with correlations. Thus, in what follows statistically independence of the letters will always be assumed and it should not be forgotten that the case of a real language (such as English) is somewhat different.

The Shannon entropy associated with the probability distribution $\{p_1, p_2, \ldots, p_k\}$ is defined by

$$H(p_1, p_2, \ldots, p_k) \equiv -\sum_{i=1}^{k} p_i \log p_i. \tag{5.150}$$

Note that, here as in the rest of this book, all the logarithms are base 2 unless otherwise indicated. We shall show that the Shannon entropy quantifies how much information we gain, on average, when we learn the value of a letter of the message. Let us consider the special case $k = 2$ and define $p_1 = p$ (where $0 \leq p \leq 1$). Since $p_2 = 1 - p$, the Shannon *binary* entropy is a function of p alone and we can write

$$H_{\text{bin}}(p) \equiv H(p_1, p_2) = -p \log p - (1 - p) \log(1 - p). \tag{5.151}$$

In the following we shall simply write $H(p)$ instead of $H_{\text{bin}}(p)$. The Shannon binary entropy $H(p)$ is plotted in Fig. 5.5: it is equal to zero when $p = 0$ or $p = 1$ and attains its maximum value $H = 1$ when $p = \frac{1}{2}$. This is consistent with our interpretation of $H(p)$ as the average information content of each letter in the message. Indeed, information is a measure of our a *priori* ignorance. If we already know that we shall receive the letter a_1 with certainty ($p = 1$), then no information is gained from the reception of this letter. The same conclusion holds when $p = 0$ and we always receive a_2. If, on the other hand, both letters are equiprobable, our a *priori* ignorance is maximum and therefore when we receive a letter, we gain the maximum possible information $H(\frac{1}{2}) = 1$. In this case, we say that we have received one unit of information, known as a *bit*. Typically, we write the letters as binary digits; that is, $a_1 = 0$ and $a_2 = 1$.

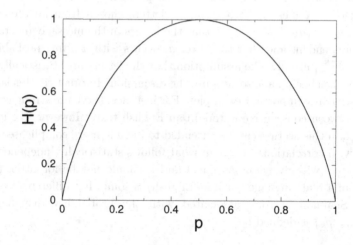

Fig. 5.5   The Shannon binary entropy $H(p) = -p \log p - (1-p) \log(1-p)$.

Exercise 5.12   Show that the Shannon entropy $H(p_1, \ldots, p_k)$ is maximum when $p_1 = \cdots = p_k = 1/k$.

5.8   Classical data compression

5.8.1   *Shannon's noiseless coding theorem*

We now show that the Shannon entropy is a good *measure* of information. Let us consider the following fundamental problem: how much can a message be *compressed* while still obtaining essentially the same information? In other words, what are the minimal physical resources required in order to store a message without loosing its information content?

As an example, we consider a message written using an alphabet with four letters, $\mathcal{A} = \{a_1, a_2, a_3, a_4\}$. We assume that these letters occur with probabilities $p_1 = \frac{1}{2}$, $p_2 = \frac{1}{4}$, $p_3 = p_4 = \frac{1}{8}$. To specify a letter out of four we need 2 bits of information. It is instead more convenient to encode the letters as follows:

$$a_1 \rightarrow c_1 \equiv 0, \quad a_2 \rightarrow c_2 \equiv 10, \quad a_3 \rightarrow c_3 \equiv 110, \quad a_4 \rightarrow c_4 \equiv 111.$$
$$(5.152)$$

To send one coded letter we need, on average, $\sum_{i=1}^{4} p_i l_i$ bits, where l_i is the length, in bits, of the coded letter c_i (we have $l_1 = 1$, $l_2 = 2$, $l_3 = l_4 = 3$). Since $\sum_i p_i l_i = \frac{7}{4} < 2$, we have compressed the information. Note that the good strategy, here as in any other useful compression code, is to encode the most probable strings in the shortest sequences and the less probable strings in the longest sequences.

Shannon proved that the optimal compression rate is given by the Shannon entropy. If Alice sends Bob a string of n letters taken from the alphabet $\mathcal{A} = \{a_1, \ldots, a_k\}$ and each letter a_i occurs with the *a priori* probability p_i, then, for large n, Alice can reliably communicate her message by sending only $nH(p_1, \ldots, p_k)$ bits of information. This is the content of the Shannon's noiseless coding theorem.

Theorem 5.3 Shannon's noiseless coding theorem: *Given a message in which the letters have been chosen independently from the ensemble $\mathcal{A} = \{a_1, \ldots, a_k\}$ with* a priori *probabilities $\{p_1, \ldots, p_k\}$, there exists, asymptotically in the length of the message, an optimal and reliable code compressing the message to $H(p_1, \ldots, p_k)$ bits per letter.*

Note that in the example considered above the optimal compression rate is $H = -\sum_{i=1}^{4} p_i \log p_i = \frac{7}{4}$. Since $\sum_i p_i l_i = \frac{7}{4} = H$, the optimal compression established by the Shannon's theorem has been attained.

A proof of Shannon's theorem can be found in Cover and Thomas (1991). Here, we shall limit ourselves to explaining the basic argument. First of all, it is useful to introduce the concept of *typical sequence*. A particular n-letter message, $x_1, x_2, \ldots, x_n$, where $x_i \in \mathcal{A}$, occurs with *a priori* probability

$$p(x_1, x_2, \ldots, x_n) = p(x_1) \, p(x_2) \cdots p(x_n), \qquad (5.153)$$

where we have assumed that the different letters of the message are independent and identically distributed according to the probability distribution $\{p_1, p_2, \ldots, p_k\}$. A typical sequence contains approximately np_1 times the letter a_1, np_2 times the letter a_2, ... and np_k times the letter a_k. The number of such strings is given by $n! / \prod_{i=1}^{k} (np_i)!$, which represents the number of distinct strings having np_1 times a_1, np_2 times a_2 and so on.

It is easy to show (see exercise 5.13) that

$$\frac{n!}{\prod_{i=1}^{k}(np_i)!} \approx 2^{nH(p_1,\ldots,p_k)}. \tag{5.154}$$

Exercise 5.13 Using Stirling's formula, $\log n! = n \log n - n/\ln 2 + O(\log n)$, where ln denotes the natural logarithm (having base e), prove Eq. (5.154).

The probability of obtaining any given typical sequence $x_1, x_2, \ldots, x_n$ is

$$p(x_1, x_2, \ldots, x_n) \approx 2^{-nH(p_1,\ldots,p_k)}. \tag{5.155}$$

Indeed, from Eq. (5.153) we obtain

$$-\frac{1}{n} \log p(x_1, \ldots, x_n) = -\frac{1}{n} \sum_{i=1}^{n} \log p(x_i) \approx H(p_1, \ldots, p_k), \tag{5.156}$$

where the last (approximate) equality is guaranteed by the law of large numbers and is obtained as follows: for all $j = 1, 2, \ldots, k$, the frequency n_j/n of the letter j in the message is substituted by the *a priori* probability p_j (n_j is the number of times that j appears in the message). The law of large numbers also tells us that, if we fix $\epsilon > 0$ and we say that a sequence is ϵ-typical when

$$\left| -\frac{1}{n} \log p(x_1, \ldots, x_n) - H(p_1, \ldots, p_k) \right| < \epsilon, \tag{5.157}$$

then, for any $\delta > 0$, the probability that a given sequence is ϵ-typical is larger than $1 - \delta$, for sufficiently large n. Therefore, most of the sequences are ϵ-typical in the limit of large n.

Since there are 2^{nH} typical sequences (asymptotically in n), each occurring with probability 2^{-nH}, we can identify which one of these sequences actually occurred using nH bits. Moreover, it can be shown that this asymptotic compression to H bits per letter is optimal. Note that it is sufficient to code only the typical sequences since the probability that a message is atypical becomes negligible for large n.

5.8.2 *Examples of data compression*

It is clear that an "asymptotic" data compression strategy; that is, a strategy based on the compression of long typical sequences is not practical:

to compress a long n-letter message, we must accumulate all n letters before identifying the typical sequence and compressing it. Fortunately, there exist quite efficient methods to encode smaller strings of letters.

A first example was shown in Sec. 5.8.1. Here we consider further examples. First of all, we apply the encoding (5.152) to a four-letter alphabet, with $p_1 = 0.9$, $p_2 = 0.05$, $p_3 = p_4 = 0.025$. The optimal compression is determined by $H(p_1, p_2, p_3, p_4) \approx 0.62$, while the code gives $\sum p_i l_i = 1.15$ and therefore data compression in this case, even though useful, is not optimal.

Let us apply the same code to the case in which the four letters are equiprobable, $p_i = \frac{1}{4}$ for $i = 1, \ldots, 4$. In this case, no compression is possible, because $H = 2$ and we send exactly two bits to specify a letter. Furthermore, if we try to apply the previous code, we obtain $\sum p_i l_i = 2.25 > 2$ and therefore the code is in this case detrimental to the efficiency of data transmission.

Finally, let us consider the Huffman code, shown in Table 5.1. We consider a binary alphabet $\{0, 1\}$ and the encoding procedure is applied to strings four bits long. There are $2^4 = 16$ such strings ($0 \equiv 0000$, $1 \equiv 0001$, $\ldots, 15 \equiv 1111$). Let P_i denote the probability that the string i occurs, with $i = 0, \ldots, 15$. We have $P_0 = p_0^4$, $P_1 = p_0^3 p_1$, $\ldots$, $P_{15} = p_1^4$. If we consider, for instance, the case with $p_0 = \frac{3}{4}$ and $p_1 = \frac{1}{4}$, we find that the best possible compression for a four-letter message is given by $4H(p_0, p_1) \approx 3.25$, while the Huffman code gives on average $\sum_{i=0}^{15} P_i l_i \approx 3.27$ bits, which is very close to the optimal value. This shows the power of data compression codes.

The enormous practical importance of data compression in fields such as telecommunication is self-evident. Data compression allows us to increase the transmission rate or the storage capacity of a computer. To achieve such results, we simply exploit the *redundancies* that any message contains: for instance, the letters of an (English) text are not equiprobable but appear with different frequencies. Shannon's theorem tells us that, as far as the letters of a message are not equiprobable, data compression is possible.[4]

5.9   The von Neumann entropy

The quantum analogue of the Shannon entropy is the von Neumann entropy. If a quantum system is described by the density matrix ρ, its von Neumann

---

[4]The notion of differing probabilities should not be confused with correlations, which are present in a real language but are not being considered here.

Table 5.1   Data encoding by means of the
Huffman code, with $p_0 = \frac{3}{4}$, $p_1 = \frac{1}{4}$.

Message	Huffman's encoding
0000	10
0001	000
0010	001
0011	11000
0100	010
0101	11001
0110	11010
0111	1111000
1000	011
1001	11011
1010	11100
1011	111111
1100	11101
1101	111110
1110	111101
1111	1111001

entropy $S(\rho)$ is defined as

$$S(\rho) \equiv -\operatorname{Tr}(\rho \log \rho). \tag{5.158}$$

To see the analogy with the Shannon entropy, let us consider the following situation: Alice has at her disposal an alphabet $\mathcal{A} = \{\rho_1, \rho_2, \ldots, \rho_k\}$, where the letters ρ_i are density matrices describing quantum states (pure or mixed). The letters are chosen at random with probabilities p_i, where $\sum_{i=1}^{k} p_i = 1$. Let us assume that Alice sends a letter (a quantum state) to Bob and that Bob only knows that the letter has been taken from the ensemble $\{\rho_i, p_i\}$. Thus, he describes this quantum system by means of the density matrix

$$\rho = \sum_{i=1}^{k} p_i \rho_i. \tag{5.159}$$

Therefore,

$$S(\rho) = -\operatorname{Tr}(\rho \log \rho) = -\sum_{i=1}^{k} \lambda_i \log \lambda_i = H(\lambda_1, \ldots, \lambda_k), \tag{5.160}$$

where the λ_i are the eigenvalues of the density matrix ρ and $H(\lambda_1, \ldots, \lambda_k)$ is the Shannon entropy associated with the ensemble $\{\lambda_i\}$.

The von Neumann entropy satisfies the following properties:

1. For a pure state, $S(\rho) = 0$. Indeed, in this case only one eigenvalue of ρ is different from zero, say $\lambda_1 = 1$, so that $-\sum_i \lambda_i \log \lambda_i = -\lambda_1 \log \lambda_1 = 0$.

2. The entropy is not modified by a unitary change of basis; that is, $S(U\rho U^\dagger) = S(\rho)$. Actually $S(\rho)$ depends only on the eigenvalues of ρ, which are basis-independent. This property means that the von Neumann entropy is invariant under unitary temporal evolution.

3. If the density operator ρ acts on a N-dimensional Hilbert space, then $0 \leq S(\rho) \leq \log N$. It is easy to see that $S(\rho) \geq 0$ since $0 \leq \lambda_i \leq 1$ and therefore $-\lambda_i \log \lambda_i \geq 0$. To show that $S(\rho) \leq \log N$, we use $S(\rho) = H(\lambda_1, \ldots, \lambda_N)$ and remember that the Shannon entropy $H(\lambda_1, \ldots, \lambda_N)$ takes its maximum value $\log N$ when $\lambda_1 = \cdots = \lambda_N = 1/N$ (see exercise 5.12). Hence, $S_{\max} = -\frac{1}{N} \sum_{i=1}^{N} \log \frac{1}{N} = \log N$.

The following examples give a flavour of the similarities and the differences between the von Neumann entropy and the Shannon entropy.

5.9.1 *Example 1: source of orthogonal pure states*

In the simplest case, Alice has at her disposal a source of two orthogonal pure states for a qubit. These states constitute a basis for the single qubit Hilbert space and we call them $|0\rangle$ and $|1\rangle$. The corresponding density matrices are $\rho_0 = |0\rangle\langle 0|$ and $\rho_1 = |1\rangle\langle 1|$. We assume that the source generates the states $|0\rangle$ or $|1\rangle$ with the *a priori* probabilities $p_0 = p$ and $p_1 = 1 - p$, respectively. Therefore, we can write

$$\rho = p_0|0\rangle\langle 0| + p_1|1\rangle\langle 1| = \begin{bmatrix} p_0 & 0 \\ 0 & p_1 \end{bmatrix}, \tag{5.161}$$

and the von Neumann entropy is given by

$$S(\rho) = -\operatorname{Tr}(\rho \log \rho) = -\operatorname{Tr}\left(\begin{bmatrix} p_0 & 0 \\ 0 & p_1 \end{bmatrix} \begin{bmatrix} \log p_0 & 0 \\ 0 & \log p_1 \end{bmatrix} \right)$$

$$= -p_0 \log p_0 - p_1 \log p_1 = H(p_0, p_1). \tag{5.162}$$

Therefore, in this case, in which the letters of the alphabet correspond to orthogonal pure states, the von Neumann entropy coincides with the Shannon entropy. Thus, the situation is in practice classical, from the point of view of information theory. This is quite natural since orthogonal states are perfectly distinguishable.

5.9.2   *Example 2: source of non-orthogonal pure states*

Let us consider the case in which the pure states $|\tilde{0}\rangle$ and $|\tilde{1}\rangle$ generated by
a source are not orthogonal. It is always possible to choose an appropriate
basis set $\{|0\rangle, |1\rangle\}$ (see Fig. 5.6) so that

$$|\tilde{0}\rangle = \cos\theta\,|0\rangle + \sin\theta\,|1\rangle = \begin{bmatrix} C \\ S \end{bmatrix}, \tag{5.163a}$$

$$|\tilde{1}\rangle = \sin\theta\,|0\rangle + \cos\theta\,|1\rangle = \begin{bmatrix} S \\ C \end{bmatrix}, \tag{5.163b}$$

where we have defined $C \equiv \cos\theta$ and $S \equiv \sin\theta$. We consider, without any
loss of generality, $0 \le \theta \le \pi/4$. Note that the inner product of these two
states is in general non-zero and given by

$$\langle\tilde{0}|\tilde{1}\rangle = \sin 2\theta. \tag{5.164}$$

The density matrices corresponding to the states $|\tilde{0}\rangle$ and $|\tilde{1}\rangle$ read

$$\rho_0 = |\tilde{0}\rangle\langle\tilde{0}| = \begin{bmatrix} C^2 & CS \\ CS & S^2 \end{bmatrix}, \qquad \rho_1 = |\tilde{1}\rangle\langle\tilde{1}| = \begin{bmatrix} S^2 & CS \\ CS & C^2 \end{bmatrix}. \tag{5.165}$$

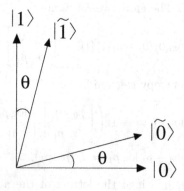

Fig. 5.6   A representation of two non-orthogonal quantum states $|\tilde{0}\rangle$ and $|\tilde{1}\rangle$ in an appropriately chosen basis $\{|0\rangle, |1\rangle\}$ for a qubit.

If the source generates the state $|\tilde{0}\rangle$ with probability p and the state $|\tilde{1}\rangle$

with probability $(1 - p)$, the corresponding density matrix is

$$\rho = p\rho_0 + (1 - p)\rho_1 = \begin{bmatrix} S^2 + p\cos 2\theta & CS \\ CS & C^2 - p\cos 2\theta \end{bmatrix}. \tag{5.166}$$

The eigenvalues of the density matrix are

$$\lambda_\pm = \tfrac{1}{2}\left(1 \pm \sqrt{1 + 4p(p-1)\cos^2 2\theta}\right). \tag{5.167}$$

They are represented in Fig. 5.7 as a function of the probability p and for different values of θ. We note that for $\theta = 0$ the states are orthogonal and the eigenvalues of the density matrix are p and $1 - p$; namely, we recover the classical case. For the other values of θ the eigenvalues "repel" each other, as can be seen from Fig. 5.7. As we shall show in the next section, this has important consequences for quantum data compression.

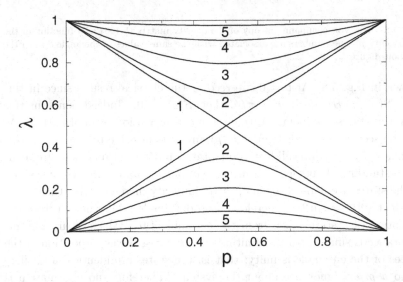

Fig. 5.7 The eigenvalues of the density matrix (5.166) as a function of the probability p. The values of the angle θ are: 1: $\theta = 0$, 2: $\theta = 0.2\frac{\pi}{4}$, 3: $\theta = 0.4\frac{\pi}{4}$, 4: $\theta = 0.6\frac{\pi}{4}$ and 5: $\theta = 0.8\frac{\pi}{4}$. The value $\theta = 0$ corresponds to orthogonal states.

Starting from the eigenvalues of the density matrix (5.166), it is easy to compute the von Neumann entropy

$$S(\rho) = -\lambda_+ \log \lambda_+ - \lambda_- \log \lambda_-, \tag{5.168}$$

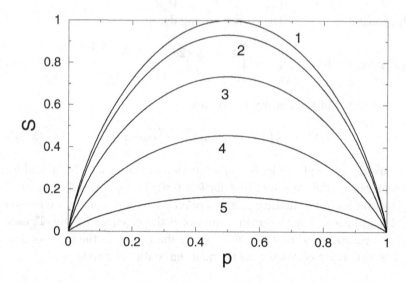

Fig. 5.8 The Von Neumann entropy of the density matrix (5.166) as a function of the probability p. The numbers are associated with the same values of the angle θ as in the previous figure.

shown in Fig. 5.8. At $\theta = 0$, we recover the classical results since in this case $S(\rho) = H(p)$. If $\theta = \pi/4$, then $S(\rho) = 0$. Indeed, since in this case the states are identical, there is no transmission of information. As can be seen in Fig. 5.8, $S(\rho) \le H(p)$ and it is possible to prove that this inequality has general validity. A qualitative interpretation follows from our understanding of entropy as a measure of our ignorance about the system. If the states are non-orthogonal, their similarity increases with their inner product $\langle \tilde{0} | \tilde{1} \rangle = \sin 2\theta$. Therefore, Bob obtains less information from the reception of a state taken from the ensemble $\{ | \tilde{0} \rangle, | \tilde{1} \rangle \}$ since his *a priori* ignorance is smaller. In the limiting case $\theta = \pi/4$ the superposition of the states of the ensemble is unity; that is, the states are identical and there is no *a priori* ignorance about the system. Therefore, no information is transmitted in this case.

5.10 Quantum data compression

5.10.1 *Schumacher's quantum noiseless coding theorem*

The Schumacher's quantum noiseless coding theorem is an extension to the quantum case of the Shannon's noiseless coding theorem discussed in Sec. 5.8. Alice sends Bob a message of n letters, each letter being chosen at random from the alphabet (ensemble of pure states) $\mathcal{A} = \{|\psi_1\rangle, |\psi_2\rangle, \ldots, |\psi_k\rangle\}$. The state $|\psi_i\rangle$ is extracted with *a priori* probability p_i and $\sum_i p_i = 1$. Therefore, each letter in the message is described by the density matrix

$$\rho = \sum_{i=1}^{k} p_i |\psi_i\rangle\langle\psi_i|, \qquad (5.169)$$

and the density matrix for the entire message is

$$\rho^n = \rho^{\otimes n}, \qquad (5.170)$$

where $\rho^{\otimes n}$ denotes the tensor product $\rho \otimes \rho \otimes \cdots \otimes \rho$. It is clear that we have assumed that all the letters in the message are statistically independent and described by the same density matrix ρ. Schumacher's theorem tells us that it is possible to compress the message, namely to encode it in a shorter message, the optimal compression rate being the von Neumann entropy.

Theorem 5.4 Schumacher's quantum noiseless coding theorem: *Given a message whose letters are pure quantum states drawn independently from the ensemble $\mathcal{A} = \{|\psi_1\rangle, \ldots, |\psi_k\rangle\}$ with a priori probabilities $\{p_1, \ldots, p_k\}$, there exists, asymptotically in the length of the message, an optimal and reliable code compressing the message to $S(\rho)$ qubits per letter, where $\rho = \sum_{i=1}^{k} p_i |\psi_i\rangle\langle\psi_i|$.*

The proof of this theorem can be found in Schumacher (1995) and closely follows the techniques used in the proof of the Shannon's noiseless coding theorem. Here, we simply illustrate the basic ideas of the proof. Let us first write the spectral decomposition of the density operator ρ:

$$\rho = \sum_{i=1}^{k} \lambda_i |a_i\rangle\langle a_i|. \qquad (5.171)$$

Clearly, we have $H(\lambda_1, \ldots, \lambda_k) = S(\rho)$. The ensemble $\mathcal{A}' = \{|a_1\rangle, \ldots, |a_k\rangle\}$ constitutes an alphabet of orthogonal pure quantum states. Following the definition of ϵ-typical sequence given Sec. 5.8, we say that a state $|x_1\rangle \otimes \cdots \otimes |x_n\rangle$, with $|x_i\rangle \in \mathcal{A}'$, is ϵ-typical when

$$\left| -\frac{1}{n} \log \left[\lambda(x_1) \cdots \lambda(x_n) \right] - S(\rho) \right| < \epsilon, \qquad (5.172)$$

where $\lambda(x_i) = \lambda_j$ if $|x_i\rangle$ is the letter $|a_j\rangle$. We define the ϵ-*typical subspace* as the subspace spanned by the ϵ-typical states. It can be shown that the dimension of this subspace is $\approx 2^{nS(\rho)}$. If P_{typ} denotes the projector on this subspace, then, for any $\delta > 0$, we have $\text{Tr}(P_{\text{typ}} \rho^n) > 1 - \delta$, provided n is large enough. Therefore, for $n \to \infty$ the density matrix ρ^n has its support on a typical subspace of dimension $2^{nS(\rho)}$. A typical n-state message can then be encoded using $nS(\rho)$ qubits.

5.10.2   *Compression of an n-qubit message*

In this section, we follow the presentation of Schumacher (1998). Let us consider the binary alphabet $\mathcal{A} = \{|\psi_0\rangle, |\psi_1\rangle\}$, where $|\psi_0\rangle \equiv |\tilde{0}\rangle$ and $|\psi_1\rangle \equiv |\tilde{1}\rangle$ are the qubit states defined by Eqs. (5.163a–5.163b). Assume that Alice wishes to send the following n-qubit message to Bob:

$$|\Psi_K\rangle = |\psi_{k_1}\rangle \otimes |\psi_{k_2}\rangle \otimes \cdots \otimes |\psi_{k_n}\rangle, \qquad (5.173)$$

where $K = \{k_1, k_2, \ldots, k_n\}$ singles out the message ($k_i = 0, 1$). The states $|\tilde{0}\rangle$ and $|\tilde{1}\rangle$ are drawn from the alphabet $\mathcal{A}$ with probabilities p and $1 - p$, respectively. Any n-letter message $|\Psi_K\rangle$ belongs to the Hilbert space

$$\mathcal{H}^n = \mathcal{H}^{\otimes n}, \qquad (5.174)$$

where $\mathcal{H}$ is the Hilbert space for a single qubit. Thus, $\mathcal{H}^n$ has dimension 2^n. It is possible to diagonalize the density matrix

$$\rho = p |\tilde{0}\rangle\langle\tilde{0}| + (1 - p) |\tilde{1}\rangle\langle\tilde{1}| \qquad (5.175)$$

and then construct the typical subspace as explained in the previous subsection. A generic message $|\psi_K\rangle$ can then be decomposed into a component belonging to the typical subspace (we call it $\mathcal{H}_{\text{typ}}$) and another belonging to its orthogonal complement, known as the atypical subspace ($\mathcal{H}_{\text{atyp}}$). We can write

$$|\Psi_K\rangle = \alpha_K |\tau_K\rangle + \beta_K |\tau_K^\perp\rangle, \qquad (5.176)$$

where $|\tau_K\rangle \in \mathcal{H}_{\text{typ}}$ and $|\tau_K^\perp\rangle \in \mathcal{H}_{\text{atyp}}$.

Alice performs a measurement to determine if $|\Psi_K\rangle$ belongs to the typical subspace or not. If this is the case, the message is encoded and sent to Bob. Since the typical subspace has dimension $\approx 2^{nS(\rho)}$, we need only $nS(\rho)$ qubit for the encoding (we shall see an example of encoding for $n = 3$ qubits in Sec. 5.10.4). If instead $|\Psi_K\rangle$ belongs to the atypical subspace, we substitute it with some reference state $|R\rangle$ living in the typical subspace. Finally, Bob decodes the $nS(\rho)$ qubits received from Alice and obtains a state described by the density matrix

$$\tilde{\rho}_K = |\alpha_K|^2\, |\tau_K\rangle\langle\tau_K| + |\beta_K|^2\, |R\rangle\langle R|. \tag{5.177}$$

How reliable is the transmission of quantum information by means of this procedure? A method to answer this question is to compute the *fidelity F*, defined as

$$F = \langle\Psi_K|\tilde{\rho}_K|\Psi_K\rangle. \tag{5.178}$$

We have $0 \leq F \leq 1$, where the maximum value $F = 1$ is obtained when the initial and final states coincide ($\tilde{\rho}_K = |\Psi_K\rangle\langle\Psi_K|$), while $F = 0$ when the initial and final states are orthogonal. The average fidelity $\bar{F}$ is obtained after averaging over all the possible messages $|\Psi_K\rangle$, each weighted with the probability p_K of its occurrence:

$$
\begin{aligned}
\bar{F} &= \sum_K p_K \langle\Psi_K|\tilde{\rho}_K|\Psi_K\rangle \\
&= \sum_K p_K \langle\Psi_K|\Big(|\alpha_K|^2|\tau_K\rangle\langle\tau_K| + |\beta_K|^2|R\rangle\langle R|\Big)|\Psi_K\rangle \\
&= \sum_K p_K|\alpha_K|^4 + \sum_K |\beta_K|^2|\langle\Psi_K|R\rangle|^2.
\end{aligned}
\tag{5.179}
$$

It is possible to show that the average fidelity tends to 1 as $n \to \infty$. This means that, in this limit, messages have unit overlap with the typical subspace. Hence, we can code only the typical subspace and still achieve good fidelity.

Comments

(i) Alice could send Bob classical information and Bob could use this information to reconstruct Alice's n-qubit message (5.173). Indeed, she could send the sequence $K = \{k_1, k_2, \ldots, k_n\}$, as this sequence uniquely determine the message $|\Psi_K\rangle$. According to Shannon's noiseless coding theorem, this sequence can be compressed by a factor given by the Shannon entropy H. However, this compression is not optimal if the alphabet is made of non-orthogonal quantum states. For instance, if the states $|\tilde{0}\rangle$ and $|\tilde{1}\rangle$ are taken with equal probability $p_0 = p = \frac{1}{2}$ and $p_1 = 1 - p = \frac{1}{2}$, then $H(\frac{1}{2}) = 1$, whereas

$$
\begin{aligned}
S(\rho) = &-\tfrac{1}{2}(1 + \sin 2\theta) \log\!\big(\tfrac{1}{2}(1 + \sin 2\theta)\big) \\
&-\tfrac{1}{2}(1 - \sin 2\theta) \log\!\big(\tfrac{1}{2}(1 - \sin 2\theta)\big),
\end{aligned} \tag{5.180}
$$

see Eqs. (5.167–5.168), which is smaller than $H(\frac{1}{2})$ as far as $\theta \neq 0$.

(ii) The price to pay to compress the quantum information by a factor $S < H$ is that Bob can reliably reconstruct the quantum state that Alice sent to him, but cannot know exactly what state he received. Indeed, each letter received is taken from a source of non-orthogonal quantum states and, as we know, non-orthogonal states cannot be distinguished with perfect reliability. Nevertheless, the compression of quantum information may be useful for several foreseen applications. For instance, one could compress the quantum memory of a quantum computer or transfer compressed quantum information between different quantum processors.

5.10.3   *Example 1: two-qubit messages*

This simple example illustrates the difference between the compression of classical and quantum messages in the case in which the letters are represented by non-orthogonal quantum states. Let us consider the alphabet $\mathcal{A} = \{|\tilde{0}\rangle, |\tilde{1}\rangle\}$ defined by Eqs. (5.163a–5.163b). We assume that the state $|\tilde{0}\rangle$ is drawn from the alphabet $\mathcal{A}$ with probability p and the state $|\tilde{1}\rangle$ with probability $1 - p$. Alice generates a two-qubit message but she can only afford to send Bob a single qubit. Bob receives this qubit and guesses that the second letter of the message is some reference state, say $|\tilde{0}\rangle$. What is the fidelity of his guess? Let us first compute the fidelities $F_K = |\langle\psi_2|\tilde{0}\rangle|^2$ of the four possible messages, $|\psi_2\rangle$ being the actual state of the second qubit. We have $F_K = 1$ if $|\psi_2\rangle = |\tilde{0}\rangle$ and $F_K = \sin^2 2\theta$ if $|\psi_2\rangle = |\tilde{1}\rangle$.

K	Message	p_K	Bob's guess	F_K
0	$|\tilde{0}\tilde{0}\rangle$	p^2	$|\tilde{0}\tilde{0}\rangle$	1
1	$|\tilde{0}\tilde{1}\rangle$	$p(1-p)$	$|\tilde{0}\tilde{0}\rangle$	$\sin^2 2\theta$
2	$|\tilde{1}\tilde{0}\rangle$	$p(1-p)$	$|\tilde{1}\tilde{0}\rangle$	1
3	$|\tilde{1}\tilde{1}\rangle$	$(1-p)^2$	$|\tilde{1}\tilde{0}\rangle$	$\sin^2 2\theta$

We can readily compute the average fidelity

$$\bar{F} = \sum_K p_K F_K = p\cos^2 2\theta + \sin^2 2\theta, \qquad (5.181)$$

which is shown in Fig. 5.9 for various values of θ. We note that $\theta = 0$ (transmission of orthogonal states) corresponds to the classical case. Indeed, we can define a classical fidelity $f_{c,K}$, which is equal to 1 if a message is correctly transmitted (in our example, for $K = 0$ and $K = 2$) and equal to 0 otherwise (for $K = 1$ and $K = 3$). It turns out that the average classical fidelity $\bar{f}_c = \sum_K p_K f_{c,K} = p$ is equal to the quantum fidelity for $\theta = 0$. For $\theta \neq 0$, the states $|\tilde{0}\rangle$ and $|\tilde{1}\rangle$ are no longer orthogonal and therefore the fidelity is higher (we have $F_1 = F_3 = \sin^2 2\theta > 0$, while $f_{c,1} = f_{c,3} = 0$). In the limiting case $\theta = \pi/4$, the states $|\tilde{0}\rangle$ and $|\tilde{1}\rangle$ coincide and therefore $F = 1$ for any value of p. Note that in this case no information is transmitted since the states $|\tilde{0}\rangle$ and $|\tilde{1}\rangle$ cannot be distinguished by any measurement.

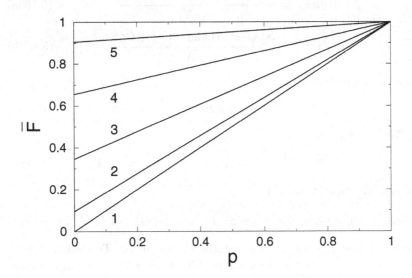

Fig. 5.9 The average fidelity $\bar{F}$ for a two-qubit message when only the first qubit is sent (see text). The values of the angle θ are: 1: $\theta = 0$, 2: $\theta = 0.2\frac{\pi}{4}$, 3: $\theta = 0.4\frac{\pi}{4}$, 4: $\theta = 0.6\frac{\pi}{4}$ and 5: $\theta = 0.8\frac{\pi}{4}$.

5.10.4 *Example 2: three-qubit messages*

In order to clarify the principles of quantum data compression, it is useful to consider the example of a message consisting of three qubits chosen from the ensemble $\{|\tilde{0}\rangle, |\tilde{1}\rangle\}$ with *a priori* probabilities $\{p, 1 - p\}$, where $|\tilde{0}\rangle$ is the letter generated with probability $p \geq \frac{1}{2}$. Let us assume that Alice can only afford to send Bob two qubits and that Alice and Bob wish to devise a strategy to maximize the average fidelity for the transmission of a three-qubit messages.

Each letter of the message is described by the density matrix $\rho = p|\tilde{0}\rangle\langle\tilde{0}| + (1 - p)|\tilde{1}\rangle\langle\tilde{1}|$, whose eigenvalues $\lambda_{\pm}$ were written down in Eq. (5.167). The corresponding eigenstates are given by

$$|\pm\rangle = \frac{1}{\sqrt{(\lambda_{\pm} + p\cos 2\theta - C^2)^2 + C^2 S^2}} \begin{bmatrix} \lambda_{\pm} + p\cos 2\theta - C^2 \\ CS \end{bmatrix}, \quad (5.182)$$

where we have again used the shorthand notation $C = \cos\theta$ and $S = \sin\theta$. It is also useful to write down the inner products

$$\langle\tilde{0}|\pm\rangle = \frac{C[\lambda_{\pm} + p\cos 2\theta - C^2] + CS^2}{\sqrt{N_{\pm}}}, \quad (5.183a)$$

$$\langle\tilde{1}|\pm\rangle = \frac{S[\lambda_{\pm} + p\cos 2\theta - C^2] + C^2 S}{\sqrt{N_{\pm}}}, \quad (5.183b)$$

where we have defined

$$N_{\pm} = (\lambda_{\pm} + p\cos 2\theta - C^2)^2 + C^2 S^2. \quad (5.184)$$

We call $|\Psi_K\rangle$ the 8 possible messages,

$$|\Psi_0\rangle = |\tilde{0}\tilde{0}\tilde{0}\rangle, \quad |\Psi_1\rangle = |\tilde{0}\tilde{0}\tilde{1}\rangle, \quad \ldots, \quad |\Psi_7\rangle = |\tilde{1}\tilde{1}\tilde{1}\rangle, \quad (5.185)$$

and $|\chi_J\rangle$ the eigenstates of $\rho^{\otimes 3}$:

$$|\chi_0\rangle = |+++\rangle, \quad |\chi_1\rangle = |++-\rangle, \quad \ldots, \quad |\chi_7\rangle = |---\rangle, \quad (5.186)$$

where $|+\rangle$ and $|-\rangle$ are the eigenstates (5.182) of ρ. The states $\{|\chi_J\rangle\}$ constitute a basis for the three-qubit Hilbert space and we can therefore decompose the possible messages as follows:

$$|\Psi_K\rangle = \sum_J c_{KJ} |\chi_J\rangle, \quad (5.187)$$

where we have defined $c_{KJ} = \langle\chi_J|\Psi_K\rangle$.

Since $\lambda_+ > \lambda_-$ for $p > \frac{1}{2}$, then in the spectral decomposition of the density matrix ρ the weight λ_+ of the eigenstate $|+\rangle$ is higher than the weight λ_- of the eigenstate $|-\rangle$. The most likely subspace is spanned by the most likely states, namely

$$\{|\chi_0\rangle=|+++\rangle, \ |\chi_1\rangle=|++-\rangle, \ |\chi_2\rangle=|+-+\rangle, \ |\chi_4\rangle=|-++\rangle\}, \quad (5.188)$$

while the unlikely subspace is spanned by

$$\{|\chi_3\rangle=|+--\rangle, \ |\chi_5\rangle=|-+-\rangle, \ |\chi_6\rangle=|--+\rangle, \ |\chi_7\rangle=|---\rangle\}. \quad (5.189)$$

The states $|\Psi_K\rangle$ of the message can be decomposed into a component $|\tau_K\rangle$ along the likely subspace and a component $|\tau_K^\perp\rangle$ along the unlikely subspace; that is, $|\Psi_K\rangle = \alpha_K|\tau_K\rangle + \beta_K|\tau_K^\perp\rangle$. The coefficients α_K and β_K are given by

$$\begin{aligned}\alpha_K &= \sqrt{|c_{K0}|^2 + |c_{K1}|^2 + |c_{K2}|^2 + |c_{K4}|^2}, \\ \beta_K &= \sqrt{|c_{K3}|^2 + |c_{K5}|^2 + |c_{K6}|^2 + |c_{K7}|^2},\end{aligned} \quad (5.190)$$

where the coefficients c_{Ki} can be easily computed by exploiting expressions (5.183a) and (5.183b) for the inner products $\langle\tilde{0}|\pm\rangle$ and $\langle\tilde{1}|\pm\rangle$.

In order to code the message, Alice employs the following strategy. She applies a unitary transformation U that rotates the basis states spanning the likely subspace ($|\chi_0\rangle$ $|\chi_1\rangle$, $|\chi_2\rangle$ and $|\chi_4\rangle$) into the states $|i_1\rangle|i_2\rangle|0\rangle$ (with $i_1, i_2 = 0, 1$), whereas the unlikely states $|\chi_3\rangle$ $|\chi_5\rangle$, $|\chi_6\rangle$ and $|\chi_7\rangle$ are rotated into $|i_1\rangle|i_2\rangle|1\rangle$. She then performs a measurement of the third qubit: if she obtains 0, her state $|\Psi_K\rangle$ has been projected onto the likely subspace. In this case, she sends the first two qubits to Bob. If instead she obtains outcome 1, her state has been projected onto the unlikely subspace and she sends Bob the first two qubits of $U|R\rangle$, where $|R\rangle$ is some reference state belonging to the likely subspace. For instance, she takes $|R\rangle$ equal to the most likely state $|\chi_0\rangle$. Bob appends to the two qubits received an ancillary qubit, prepared in the state $|0\rangle$. He then applies the operator U^{-1} to these three qubits and ends up with a state described by the density matrix

$$\tilde{\rho}_K = |\alpha_K|^2|\tau_K\rangle\langle\tau_K| + |\beta_K|^2|R\rangle\langle R|. \quad (5.191)$$

The average fidelity is then given by

$$\bar{F} = \sum_{K=0}^{7} p_K \langle\Psi_K|\tilde{\rho}_K|\Psi_K\rangle = \sum_{K=0}^{7} p_K\left(|\alpha_K|^4 + |\beta_K|^2 \, |\langle\Psi_K|R\rangle|^2\right), \quad (5.192)$$

where p_K is the probability that the message $|\Psi_K\rangle$ is generated. The graph of $\bar{F}$ as a function of p is shown in Fig. 5.10, for various values of θ. This figure exhibits several interesting features. First of all, for $\theta = 0$ we recover the classical case, in which the average fidelity $\bar{f}_c$ is obtained after summing the probabilities of all messages correctly transmitted. The calculation is similar to that performed in the previous subsection for two-qubit messages and gives

$$\bar{f}_c = p^3 + 3p^2(1-p) = 3p^2 - 2p^3. \tag{5.193}$$

We note that for $p = \frac{1}{2}$ we have $\bar{f}_c = \frac{1}{2}$. Indeed, in this case we have 8 messages occurring with the same probability and only 4 are correctly transmitted. The average quantum fidelity $\bar{F}$ is instead larger than $\frac{1}{2}$ when $\theta > 0$. This is because our *a priori* ignorance for non-orthogonal states is smaller than for orthogonal states. In the limiting case $\theta = \pi/4$ the states $|\tilde{0}\rangle$ and $|\tilde{1}\rangle$ superimpose. Thus, $\bar{F}(\pi/4) = 1$ but there is no transmission of information.

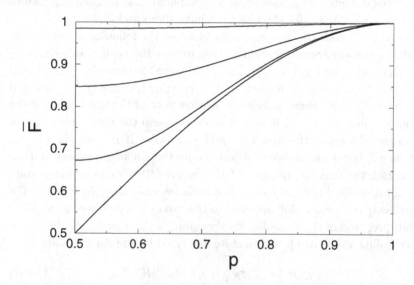

Fig. 5.10   The average fidelity $\bar{F}$ for the transmission of a three-qubit message by means of a two-qubit code (see details in the text). From bottom to top: $\theta = 0$, $\pi/16$, $\pi/10$ and $\pi/6$.

5.11 Accessible information

We assume that Alice sends Bob a message whose letters are chosen independently from the alphabet $\mathcal{A} = \{a_1, \ldots, a_k\}$ with *a priori* probabilities $\{p_1, \ldots, p_k\}$. The letters of the alphabet are coded by quantum states that are not necessarily orthogonal. In this section we consider the following problem: how much information can Bob gain on the message by performing measurements on the quantum states received? This problem is non-trivial since non-orthogonal quantum states cannot be perfectly distinguished. It is important to emphasize that, as we saw in Chapter 4, this property lies at the heart of quantum cryptography.

First of all, a few definitions are needed. If X is a random variable that takes the value x with probability $p(x)$ ($x \in \{a_1, \ldots, a_k\}$ and $p(x) \in \{p_1, \ldots, p_k\}$), then the Shannon entropy $H(p_1, \ldots, p_k)$ is also called $H(X)$ and we write

$$H(X) \equiv -\sum_x p(x) \log p(x) = -\sum_{i=1}^{k} p_i \log p_i. \tag{5.194}$$

Note that $H(X)$ indicates a function not of X but of the information content of the random variable X.

Joint entropy: the joint entropy of a pair of random variables X and Y having values x and y with probabilities $p(x)$ and $p(y)$, respectively, is defined by

$$H(X,Y) \equiv -\sum_{x,y} p(x,y) \log p(x,y), \tag{5.195}$$

where $p(x,y)$ is the probability that $X = x$ and $Y = y$.

Conditional entropy: The conditional entropy $H(Y|X)$ is defined by

$$H(Y|X) \equiv H(X,Y) - H(X). \tag{5.196}$$

It is a measure of our residual ignorance about Y, provided we already know the value of X. Similarly, we can define $H(X|Y) \equiv H(X,Y) - H(Y)$. It is easy to show that

$$H(Y|X) = -\sum_{x,y} p(x,y) \log p(y|x), \tag{5.197}$$

where $p(y|x) = p(x,y)/p(x)$ is the probability that $Y = y$, provided $X = x$. Indeed,

$$
\begin{aligned}
H(X,Y) - H(X) &= -\sum_{x,y} p(x,y) \log p(x,y) + \sum_{x} p(x) \log p(x) \\
&= -\sum_{x,y} p(x,y) \log\big(p(x)p(y|x)\big) + \sum_{x,y} p(x,y) \log p(x) \\
&= -\sum_{x,y} p(x,y) \log p(y|x), \qquad\qquad (5.198)
\end{aligned}
$$

where we have used $\sum_y p(x,y) = p(x)$. Similarly, we obtain

$$
H(X|Y) = -\sum_{x,y} p(x,y) \log p(x|y). \qquad\qquad (5.199)
$$

Mutual information: The mutual information $I(X{:}Y)$ is defined by

$$
I(X{:}Y) \equiv H(X) + H(Y) - H(X,Y). \qquad\qquad (5.200)
$$

This quantity is a measure of how much information X and Y have in common. It can be easily shown that

$$
I(X{:}Y) = -\sum_{x,y} p(x,y) \log \frac{p(x)p(y)}{p(x,y)}. \qquad\qquad (5.201)
$$

From this expression it is clear that, if X and Y are independent, namely $p(x,y) = p(x)p(y)$, then $I(X{:}Y) = 0$. The mutual information is related to the conditional entropy as follows:

$$
I(X{:}Y) = H(X) - H(X|Y) = H(Y) - H(Y|X). \qquad\qquad (5.202)
$$

We note that, as is clear from its definition (5.200), the mutual information is symmetric:

$$
I(Y{:}X) = I(X{:}Y). \qquad\qquad (5.203)
$$

Let us now return to the problem introduced at the beginning of this section. If X and Y denote the random variables associated with the letters generated by Alice and with Bob's measurement outcomes, respectively, then the *accessible information* is defined as the maximum of $I(X{:}Y)$ over all possible measurement schemes.

5.11.1 *The Holevo bound*

The Holevo bound (proved by Holevo in 1973) establishes an upper bound on the accessible information.

Theorem 5.5 The Holevo Bound: *If Alice prepares a (mixed) state ρ_X chosen from the ensemble $\mathcal{A} = \{\rho_0, \ldots, \rho_k\}$ with a priori probabilities $\{p_1, \ldots, p_k\}$ and Bob performs a POVM measurement on that state, with POVM elements $\{F_1, \ldots, F_l\}$ and measurement outcome described by the random variable Y, then the mutual information $I(X{:}Y)$ is bounded as follows:*

$$I(X{:}Y) \leq S(\rho) - \sum_{i=1}^{k} p_i S(\rho_i) \equiv \chi(\mathcal{E}), \qquad (5.204)$$

where $\rho = \sum_{i=1}^{k} p_i \rho_i$ and $\chi(\mathcal{E})$ is known as the Holevo information of the ensemble $\mathcal{E} \equiv \{\rho_1, \ldots, \rho_k; p_1, \ldots, p_k\}$.

A proof of this theorem can be found in Nielsen and Chuang (2000). Here, we shall limit ourselves to discuss the Holevo bound in a few concrete examples.

5.11.2 *Example 1: two non-orthogonal pure states*

If Alice sends Bob pure orthogonal quantum states drawn from the ensemble $\{|\psi_1\rangle, \ldots, |\psi_k\rangle\}$, then Bob can unambiguously distinguish these states by means of projective measurements described by the POVM elements (in this case, simple projectors) $\{F_1 = |\psi_1\rangle\langle\psi_1|, \ldots, F_k = |\psi_k\rangle\langle\psi_k|\}$. It is easy to check that $I(X{:}Y) = H(X)$ (we have $H(X|Y) = 0$) and therefore this case is no different from the transmission of classical information over a noiseless channel: if we send the letter a_x, we recover the same letter; that is, $a_y = a_x$.

The simplest example that cannot be reduced to classical information theory is that in which Alice sends Bob states generated by a source of non-orthogonal pure quantum states. We assume that the states $|\tilde{0}\rangle$ and $|\tilde{1}\rangle$, defined by Eqs. (5.163a–5.163b), are generated with probabilities $p_0 = p$ and $p_1 = 1 - p$, respectively.

Since the single letters are represented in this case by pure states, their von Neumann entropy is equal to zero: $S(\rho_0) = S(|\tilde{0}\rangle\langle\tilde{0}|) = 0$ and $S(\rho_1) = S(|\tilde{1}\rangle\langle\tilde{1}|) = 0$. Therefore, the Holevo information $\chi(\mathcal{E})$ reduces to

$$\chi(\mathcal{E}) = S(\rho), \qquad (5.205)$$

where $\rho = p\rho_0 + (1 - p)\rho_1$. Hence, the Holevo bound gives

$$I(X{:}Y) \leq S(\rho). \tag{5.206}$$

A plot of $S(\rho)$ was already shown in Fig. 5.8. It reveals that, for non-orthogonal states ($\theta \neq 0$), $S(\rho) < H(X)$ and therefore $I(X{:}Y) < H(X)$. It is possible to show that this strict inequality also has general validity for mixed states $\{\rho_i\}$, provided they do not have orthogonal support (for orthogonal support, $I(X{:}Y) = H(X)$).

It is instructive to consider the following special case: we assume that Bob performs a projective measurement on the received qubits along the direction $\hat{n}$ (that is, he measures $\hat{n} \cdot \boldsymbol{\sigma}$) and we show that in this case the Holevo bound is satisfied. For this purpose, we compute the mutual information. Bob's measurement along the direction $\hat{n}$ is described by the POVM elements (projectors)

$$F_0 = \tfrac{1}{2}(I + \hat{n} \cdot \boldsymbol{\sigma}), \qquad F_1 = \tfrac{1}{2}(I - \hat{n} \cdot \boldsymbol{\sigma}). \tag{5.207}$$

For instance, if $\hat{n} = (0, 0, 1)$, then $F_0 = |0\rangle\langle0|$ and $F_1 = |1\rangle\langle1|$. We compute the conditional probability

$$p(y|x) = \mathrm{Tr}(\rho_x F_y), \quad (x, y = 0, 1), \tag{5.208}$$

which is the probability that Bob's measurement gives outcome y, provided the state ρ_x was sent by Alice. For this purpose, we write down the Bloch-sphere representation of the density matrices associated with the states $|\tilde{0}\rangle$ and $|\tilde{1}\rangle$ (see Sec. 5.1.1):

$$\rho_0 = |\tilde{0}\rangle\langle\tilde{0}| = \tfrac{1}{2}(I + \boldsymbol{r}_0 \cdot \boldsymbol{\sigma}), \qquad \rho_1 = |\tilde{1}\rangle\langle\tilde{1}| = \tfrac{1}{2}(I + \boldsymbol{r}_1 \cdot \boldsymbol{\sigma}), \tag{5.209}$$

where the Cartesian components of the Bloch vectors $\boldsymbol{r}_0$ and $\boldsymbol{r}_1$ are given by

$$\boldsymbol{r}_0 = (\sin 2\theta, 0, \cos 2\theta), \qquad \boldsymbol{r}_1 = (\sin 2\theta, 0, -\cos 2\theta). \tag{5.210}$$

Taking into account that $\mathrm{Tr}(\sigma_i) = 0$ and $\mathrm{Tr}(\sigma_i\sigma_j) = 2\delta_{ij}$ for $i, j = x, y, z$ (see exercise 5.2), it is now straightforward to compute the conditional probabilities:

$$\begin{aligned}
p(0|0) &= \mathrm{Tr}(\rho_0 F_0) = \tfrac{1}{2}(1 + \boldsymbol{r}_0 \cdot \hat{n}), \\
p(1|0) &= \mathrm{Tr}(\rho_0 F_1) = \tfrac{1}{2}(1 - \boldsymbol{r}_0 \cdot \hat{n}), \\
p(0|1) &= \mathrm{Tr}(\rho_1 F_0) = \tfrac{1}{2}(1 + \boldsymbol{r}_1 \cdot \hat{n}), \\
p(1|1) &= \mathrm{Tr}(\rho_1 F_1) = \tfrac{1}{2}(1 - \boldsymbol{r}_1 \cdot \hat{n}).
\end{aligned} \tag{5.211}$$

If, for the sake of simplicity, we assume that the measurement direction lies in the (x, z) plane of the Bloch sphere; that is, $\hat{n} = (\sin\bar{\theta}, 0, \cos\bar{\theta})$ (see Fig. 5.11), we have

$$p(0|0) = \tfrac{1}{2}\left[1 + \cos(\bar{\theta} - 2\theta)\right], \qquad p(1|0) = \tfrac{1}{2}\left[1 - \cos(\bar{\theta} - 2\theta)\right],$$
$$p(0|1) = \tfrac{1}{2}\left[1 - \cos(\bar{\theta} + 2\theta)\right], \qquad p(1|1) = \tfrac{1}{2}\left[1 + \cos(\bar{\theta} + 2\theta)\right]. \tag{5.212}$$

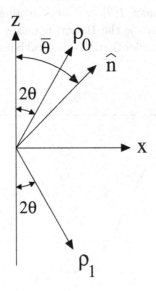

Fig. 5.11   A geometric visualization of the Bloch sphere vectors ρ_0 and ρ_1 and of the measurement axis $\hat{n}$.

We now compute $p(x, y) = p(x)p(y|x)$, where, as stated at the beginning of this subsection, we assume that the states $|\tilde{0}\rangle$ and $|\tilde{1}\rangle$ are generated with probabilities $p(X = 0) = p$ and $p(X = 1) = 1 - p$, respectively. We thus have

$$p(0,0) = \tfrac{1}{2}p\left[1 + \cos(\bar{\theta} - 2\theta)\right],$$
$$p(0,1) = \tfrac{1}{2}p\left[1 - \cos(\bar{\theta} - 2\theta)\right],$$
$$p(1,0) = \tfrac{1}{2}(1 - p)\left[1 - \cos(\bar{\theta} + 2\theta)\right],$$
$$p(1,1) = \tfrac{1}{2}(1 - p)\left[1 + \cos(\bar{\theta} + 2\theta)\right]. \tag{5.213}$$

Then we compute $p(y) = \sum_x p(x,y)$ and obtain

$$p(Y{=}0) = \tfrac{1}{2}\left[1 + p\cos(\bar{\theta} - 2\theta) - (1-p)\cos(\bar{\theta} + 2\theta)\right],$$
$$p(Y{=}1) = \tfrac{1}{2}\left[1 - p\cos(\bar{\theta} - 2\theta) + (1-p)\cos(\bar{\theta} + 2\theta)\right]. \tag{5.214}$$

Finally, we insert the expressions derived for $p(x)$, $p(y)$ and $p(x,y)$ into Eq. (5.201), obtaining the mutual information $I(X{:}Y)$.

As an example, in Fig. 5.12 we show the mutual information $I(X{:}Y)$ for $\theta = \pi/10$ and $p = 0.8$. Within the chosen measurement scheme, the only free parameter that may be varied in order to maximize I is $\bar{\theta}$. The maximum value $I_{\max} \equiv \max_{\bar{\theta}} I(\bar{\theta}) \approx 0.40$ is attained for $\bar{\theta} \approx 0.14\pi$. We stress that this value is below the Holevo bound $\chi = S(\rho) \approx 0.526$. Of course, this value is also smaller than the classical bound $I(X{:}Y) \leq H(X) \approx 0.722$.

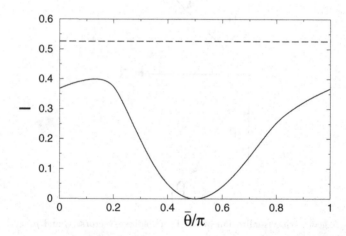

Fig. 5.12   The mutual information $I(X{:}Y)$ for a message coded by means of the non-orthogonal states (5.163a–5.163b), with $\theta = \pi/10$ and $p = 0.8$. The angle $\bar{\theta}$ determines the measurement direction $\hat{\boldsymbol{n}} = (\sin\bar{\theta}, 0, \cos\bar{\theta})$. The dashed line shows the Holevo bound $\chi \approx 0.526$.

Exercise 5.14   Alice sends Bob the state $|0\rangle$ with probability p or the state $|1\rangle$ with probability $1 - p$. For this purpose, they employ a quantum channel whose action on a state with Bloch vector (x, y, z) is given by

$$x \to x' = ax, \quad y \to y' = ay, \quad z \to z' = z, \tag{5.215}$$

where $0 < a < 1$ (note that this quantum channel corresponds to the phase-flip channel, which will be discussed in Sec. 6.1.5). Finally, Bob performs a standard projective measurement along the direction $\hat{n}$. Compute Alice and Bob's mutual information.

Exercise 5.15 Repeat the previous exercise for the case in which the action of the quantum channel on a state with Bloch vector (x, y, z) is given by

$$x \rightarrow x' = x\cos\theta, \quad y \rightarrow y' = y\cos\theta, \quad z \rightarrow z' = \sin^2\theta + z\cos^2\theta,$$
$$(5.216)$$

where $0 \le \theta \le \frac{\pi}{2}$ (this quantum channel corresponds to the amplitude-damping channel, which will be discussed in Sec. 6.1.8).

5.11.3 * *Example 2: three non-orthogonal pure states*

Let us now consider the case in which Alice's alphabet is $\mathcal{A} = \{|\phi_0\rangle, |\phi_1\rangle, |\phi_2\rangle\}$, where

$$|\phi_0\rangle = |0\rangle, \quad |\phi_1\rangle = \cos\theta\,|0\rangle + \sin\theta\,|1\rangle, \quad |\phi_2\rangle = \cos\theta\,|0\rangle - \sin\theta\,|1\rangle.$$
$$(5.217)$$

A graphical representation of these three non-orthogonal quantum states is shown in Fig. 5.13. We call $\rho_0 = |\phi_0\rangle\langle\phi_0|$, $\rho_1 = |\phi_1\rangle\langle\phi_1|$ and $\rho_2 = |\phi_2\rangle\langle\phi_2|$ the density operators associated with these quantum states. We assume that each letter of Alice's message is one of the three states of this alphabet, chosen with *a priori* probabilities $\{p_0, p_1, p_2\}$. In the following we assume that $p_0 = p_1 = p_2 = p = 1/3$ and $\theta = 2\pi/3$. Under these conditions, the matrix representations of the density operators ρ_0, ρ_1 and ρ_2 in the $\{|0\rangle, |1\rangle\}$ basis read

$$\rho_0 = \begin{bmatrix} 1 & 0 \\ 0 & 0 \end{bmatrix}, \quad \rho_1 = \frac{1}{4}\begin{bmatrix} 1 & -\sqrt{3} \\ -\sqrt{3} & 3 \end{bmatrix}, \quad \rho_2 = \frac{1}{4}\begin{bmatrix} 1 & \sqrt{3} \\ \sqrt{3} & 3 \end{bmatrix}. \quad (5.218)$$

The density matrix that describes the above ensemble of pure quantum states is

$$\rho = p_0\rho_0 + p_1\rho_1 + p_2\rho_2 = \tfrac{1}{2}I, \quad (5.219)$$

and therefore $S(\rho) = 1$. Since the letters of Alice's message are pure states, the Holevo bound on mutual information gives $I(X{:}Y) \le S(\rho) = 1$.

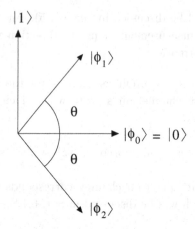

Fig. 5.13   A graphical representation of the three quantum states of the alphabet $\mathcal{A} = \{|\phi_0\rangle, |\phi_1\rangle, |\phi_2\rangle\}$.

Bob measures the received qubits by means of the POVM scheme described in Sec. 5.6.2, with POVM elements

$$F_0 = \frac{1}{2}\begin{bmatrix} 1 & r \\ r & r^2 \end{bmatrix}, \quad F_1 = \frac{1}{2}\begin{bmatrix} 1 & -r \\ -r & r^2 \end{bmatrix}, \quad F_2 = \begin{bmatrix} 0 & 0 \\ 0 & 1-r^2 \end{bmatrix}. \quad (5.220)$$

As in the previous example, we compute the conditional probabilities

$$p(y|x) = \mathrm{Tr}(\rho_x F_y), \quad (x, y = 0, 1, 2), \quad (5.221)$$

namely the probability that Bob's POVM measurement gives outcome y, provided the state x was sent by Alice. We obtain

$$\begin{aligned} p(0|0) &= \tfrac{1}{2}, & p(0|1) &= \tfrac{1}{8}(1 - \sqrt{3}\,r)^2, & p(0|2) &= \tfrac{1}{8}(1 + \sqrt{3}\,r)^2, \\ p(1|0) &= \tfrac{1}{2}, & p(1|1) &= \tfrac{1}{8}(1 + \sqrt{3}\,r)^2, & p(1|2) &= \tfrac{1}{8}(1 - \sqrt{3}\,r)^2, & (5.222) \\ p(2|0) &= 0, & p(2|1) &= \tfrac{3}{4}(1 - r^2), & p(2|2) &= \tfrac{3}{4}(1 - r^2). \end{aligned}$$

We now compute $p(x, y) = p(x)p(y|x)$. In this case, $p(x, y) = \tfrac{1}{3}p(y|x)$ since $p(X{=}0) = p(X{=}1) = p(X{=}2) = \tfrac{1}{3}$. We then compute $p(y) = \sum_x p(x, y)$ and obtain

$$p(Y{=}0) = p(Y{=}1) = \tfrac{1}{4}(1 + r^2), \qquad p(Y{=}2) = \tfrac{1}{2}(1 - r^2). \quad (5.223)$$

Finally, we insert the above expressions for $p(x)$, $p(y)$ and $p(x, y)$ into Eq. (5.201), thus obtaining the mutual information $I(X{:}Y)$. The graph of I as a function of the parameter r is shown in Fig. 5.14. Its maximum value $I_{\max} \approx 0.585$ is attained for $r \approx 0.577$. Note that $I_{\max}$ is well below the Holevo bound $\chi = S(\rho) = 1$.

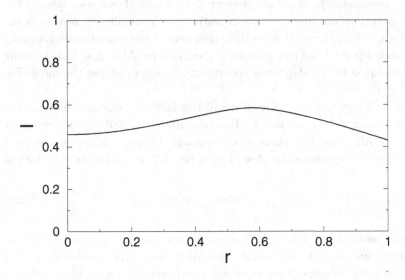

Fig. 5.14 The mutual information $I(X{:}Y)$ for a message coded by means of the three non-orthogonal states given by Eq. (5.217), with $\theta = \frac{2}{3}\pi$ and $p = \frac{1}{3}$. Bob's measurement is described by the POVM operators (5.220). The Holevo bound gives $I \leq 1$.

5.12 Entanglement concentration and von Neumann entropy

As we saw in Chaps. 3 and 4, entanglement is not only one of the most intriguing features predicted by the quantum theory but also a fundamental resource for quantum information and communication. In particular, in Chap. 4 we saw that entanglement enables apparently impossible tasks, such as dense coding and quantum teleportation. We wish to stress that teleportation is also interesting from the viewpoint of quantum computation since it is a powerful tool for transferring quantum states between different systems, as would be necessary in a quantum computer with several independent units. Since faithful teleportation requires that Alice and Bob share a maximally entangled EPR pair, it is important to devise methods to *distill* maximally entangled states starting from partially entangled pairs (later in this section we shall discuss how to quantify entanglement). These entanglement concentration techniques act on qubits that can be located very far away and therefore only rely on the so-called LOCC; that is, on *local operations*, possibly supplemented by *classical communication*. Local operations are unitary transformations or (generalized) measurements performed by Alice or Bob on their members of the shared non-maximally entangled pair. Classical communication enables Alice and Bob to share the results of the local quantum operations, in order to select the successful, maximally entangled cases.

It is instructive to study in detail the following example of entanglement concentration, devised by Bandyopadhyay (2000). We assume that initially Alice and Bob share a pure entangled state. Taking into account the Schmidt decomposition described in Sec. 5.2, we can write this state as

$$|\psi\rangle_{AB} = \alpha|00\rangle_{AB} + \beta|11\rangle_{AB}, \qquad (5.224)$$

where, without any loss of generality, we may assume α, β to be real and positive, and $\alpha \geq \beta$. We assume that Alice knows the coefficients α, β of the Schmidt decomposition in advance and prepares an ancillary qubit in the state

$$|\chi\rangle_A = \alpha|0\rangle_A + \beta|1\rangle_A. \qquad (5.225)$$

Hence, the combined state of the three qubits is given by

$$|\Psi\rangle = |\chi\rangle_A \otimes |\psi\rangle_{AB} = (\alpha|0\rangle_A + \beta|1\rangle_A) \otimes (\alpha|00\rangle_{AB} + \beta|11\rangle_{AB})$$
$$= \alpha^2|000\rangle_{A_1A_2B} + \alpha\beta|011\rangle_{A_1A_2B} + \alpha\beta|100\rangle_{A_1A_2B} + \beta^2|111\rangle_{A_1A_2B}. \qquad (5.226)$$

The first two qubits, denoted by A_1 and A_2, belong to Alice and the third (B) to Bob. Alice performs a CNOT gate on the two qubits in her possession, A_1 being the control and A_2 the target qubits. The resulting state is

$$|\Psi\rangle = \alpha^2|000\rangle_{A_1A_2B} + \alpha\beta|011\rangle_{A_1A_2B} + \alpha\beta|110\rangle_{A_1A_2B} + \beta^2|101\rangle_{A_1A_2B}. \qquad (5.227)$$

After interchanging the position of the first two qubits and writing the wave

-function normalization in an appropriate manner, we have

$$|\Psi\rangle = \sqrt{\alpha^4 + \beta^4}\, |0\rangle_{A_2} \otimes \left(\frac{\alpha^2}{\sqrt{\alpha^4 + \beta^4}}\, |00\rangle_{A_1 B} + \frac{\beta^2}{\sqrt{\alpha^4 + \beta^4}}\, |11\rangle_{A_1 B} \right)$$

$$+ \sqrt{2\alpha^2\beta^2}\, |1\rangle_{A_2} \otimes \frac{1}{\sqrt{2}} \left(|01\rangle_{A_1 B} + |10\rangle_{A_1 B} \right). \quad (5.228)$$

Alice then performs a standard projective measurement of qubit A_2 on the basis $\{|0\rangle, |1\rangle\}$. It is straightforward to see from Eq. (5.228) that she obtains outcome 0 with probability $(\alpha^4 + \beta^4)$ or outcome 1 with probability $(2\alpha^2\beta^2)$. In the latter case, Alice and Bob realize an EPR pair (the qubits A_1 and B). In the first case, they obtain less entangled states. Thus, given N non-maximally entangled states (5.224), the above technique produces $2\alpha^2\beta^2 N$ maximally entangled states. As shown by Bandyopadhyay (2000), it is possible to iterate the procedure for the remaining $N(1 - 2\alpha^2\beta^2)$ states to improve its efficiency (the efficiency being defined as the fraction of EPR pairs extracted). We note that classical communication is also required for Alice to transmit the results of her measurements to Bob, in order to select the successful cases.

It might appear paradoxical that local operations plus classical communication allow a concentration of entanglement, which is a purely quantum non-local property. However, there is no real surprise if we remember that quantum mechanics is a probabilistic theory and that a non-vanishing maximally entangled component is present in the state (5.224). This component is quantified by the fidelity

$$F = \left| {}_{AB}\langle \phi^+ | \psi \rangle_{AB} \right|^2 = \left| \tfrac{1}{\sqrt{2}} \big(\langle 00| + \langle 11| \big)\big(\alpha|00\rangle + \beta|11\rangle \big) \right|^2 = \tfrac{1}{2}(\alpha + \beta)^2,$$
$$(5.229)$$

where we have considered the EPR state $|\phi^+\rangle_{AB} = \frac{1}{\sqrt{2}}\big(|00\rangle + |11\rangle \big)$. Therefore, the above entanglement concentration protocol selects this maximally entangled component.

The previous example naturally raises the following questions: What is the optimal entanglement concentration? Can we measure entanglement? Nowadays, we can answer these questions unambiguously but only for bipartite pure states. First of all, a few definitions are needed.

Entanglement cost: Let us assume that Alice and Bob share many EPR pairs, say $|\phi^+\rangle_{AB}$, and that they wish to prepare a large number n of copies of a given bipartite pure state $|\psi\rangle_{AB}$, using only local operations and classical communication. If we call $k_{\min}$ the minimum number of EPR pairs necessary to accomplish this task, we define the entanglement cost as the limiting ratio $k_{\min}/n$, for $n \to \infty$.

Distillable entanglement: Let us consider the reverse process; that is, Alice and Bob share a large number n of copies of a pure state $|\psi\rangle_{AB}$ and they wish to concentrate entanglement, again using only local operations supplemented by classical communication. If $k'_{\max}$ denotes the maximum number of EPR pairs that can be obtained in this manner, we define the distillable entanglement as the ratio $k'_{\max}/n$ in the limit $n \to \infty$.

It is clear that $k'_{\max} \leq k_{\min}$. Otherwise, we could employ local operations and classical communication to create entanglement, which is a non-local, purely quantum resource (it would be sufficient to prepare n states $|\psi\rangle_{AB}$ from $k_{\min}$ EPR pairs and then distill $k'_{\max} > k_{\min}$ EPR states). Furthermore, it is possible to show that, asymptotically in n, the entanglement cost and the distillable entanglement coincide and that the ratios $k_{\min}/n$ and $k'_{\max}/n$ are given by the reduced single-qubit von Neumann entropies. Indeed, we have

$$\lim_{n \to \infty} \frac{k_{\min}}{n} = \lim_{n \to \infty} \frac{k'_{\max}}{n} = S(\rho_A) = S(\rho_B), \qquad (5.230)$$

where $S(\rho_A)$ and $S(\rho_B)$ are the von Neumann entropies of the reduced density matrices $\rho_A = \mathrm{Tr}_B \left(|\psi\rangle_{AB}\,_{AB}\langle\psi| \right)$ and $\rho_B = \mathrm{Tr}_A \left(|\psi\rangle_{AB}\,_{AB}\langle\psi| \right)$, respectively. Therefore, the process that changes n copies of $|\psi\rangle_{AB}$ into k copies of $|\phi^+\rangle_{AB}$ is asymptotically reversible. Moreover, it is possible to show that it is faithful; namely,the change takes place with unit fidelity when $n \to \infty$. The proof of this result is based on Schumacher's quantum data compression and can be found in Bennett *et al.* (1996a). We can therefore quantify the entanglement of a bipartite pure state $|\psi\rangle_{AB}$ as

$$E(|\psi\rangle_{AB}) = S(\rho_A) = S(\rho_B). \qquad (5.231)$$

It ranges from 0 for a separable state to 1 for maximally entangled two-qubit states (the EPR states). Hence, it is common practice to say that the entanglement of an EPR pair is 1 *ebit*. More generally, a maximally entangled state of two subsystems has d equally weighted terms in its Schmidt decomposition (d is the dimension of the Hilbert space of the smaller subsystem) and therefore its entanglement content is $\log d$ ebits.

A natural extension of the discussion of this section is to consider bipartite mixed states, with $\rho_{AB} = \sum_i p_i |\psi\rangle_{AB}\,_{AB}\langle\psi|$, instead of pure states.

However, mixed-state entanglement is not as well understood as pure-state bipartite entanglement and is the focus of ongoing research (for a review, see, e.g., Bruß, 2002, Alber *et al.*, 2001a and Plenio and Virmani, 2007).

5.13   The Peres separability criterion

Given a quantum state, pure or mixed, is it separable or entangled? As we know (see Sec. 2.5), this question has a clear answer if we refer to pure states and to bipartite entanglement: a pure state $|\psi\rangle_{AB}$ of a bipartite system $A + B$ is separable if and only if it can be written as $|\psi\rangle_{AB} = |\alpha\rangle_A \otimes |\beta\rangle_B$, with states $|\alpha\rangle_A$ and $|\beta\rangle_B$ describing the components of the two systems.

A mixed state is said to be separable if it can be prepared by two parties (Alice and Bob) in a "classical" manner; that is, by means of local operations and classical communication. This means that Alice and Bob agree over the phone on the local preparation of the two subsystems A and B. Therefore, a mixed state is separable if and only if it can be written as

$$\rho_{AB} = \sum_k p_k \, \rho_{Ak} \otimes \rho_{Bk}, \quad \text{with } p_k \geq 0 \text{ and } \sum_k p_k = 1, \qquad (5.232)$$

where ρ_{Ak} and ρ_{Bk} are density matrices for the two subsystems. A separable system always satisfies Bell's inequalities; that is, it only contains classical correlations.

Given a density matrix ρ_{AB}, it is in general a non-trivial task to prove whether a decomposition as in (5.232) exists or not. We therefore need separability criteria that are easier to test. Several such criteria have been proposed but we shall limit ourselves to considering the Peres separability criterion.

The Peres criterion provides a necessary condition for the existence of decomposition (5.232), in other words, a violation of this criterion is a sufficient condition for entanglement. This criterion is based on the *partial transpose* operation. Introducing an orthonormal basis $\{|i\rangle_A |\alpha\rangle_B\}$ in the Hilbert space $\mathcal{H}_{AB}$ associated with the bipartite system $A + B$, the density matrix ρ_{AB} has matrix elements $(\rho_{AB})_{i\alpha;j\beta} = {}_A\langle i|_B\langle \alpha|\rho_{AB}|j\rangle_A|\beta\rangle_B$. The partial transpose density matrix is constructed by only taking the transpose in either the Latin or Greek indices (recall that Latin indices refer to Alice's subsystem and Greek indices to Bob's). For instance, the partial transpose with respect to Alice is given by

$$\left(\rho_{AB}^{T_A}\right)_{i\alpha;j\beta} = \left(\rho_{AB}\right)_{j\alpha;i\beta}. \qquad (5.233)$$

Since a separable state ρ_{AB} can always be written in the form (5.232) and the density matrices ρ_{Ak} and ρ_{Bk} have non-negative eigenvalues, then the overall density matrix ρ_{AB} also has non-negative eigenvalues. The partial transpose of a separable state reads

$$\rho_{AB}^{T_A} = \sum_k p_k \, \rho_{Ak}^T \otimes \rho_{Bk}. \tag{5.234}$$

Since the transpose matrices $\rho_{Ak}^T = \rho_{Ak}^\star$ are Hermitian non-negative matrices with unit trace, they are also legitimate density matrices for Alice. It follows that none of the eigenvalues of $\rho_{AB}^{T_A}$ is non-negative. This is a necessary condition for decomposition (5.232) to hold. It is then sufficient to have at least one negative eigenvalue of $\rho_{AB}^{T_A}$ to conclude that the state ρ_{AB} is entangled.

As an example, we consider the so-called Werner state

$$(\rho_W)_{AB} = \tfrac{1}{4}(1-p)\,I + p\,|\psi^-\rangle\langle\psi^-|, \tag{5.235}$$

where $0 \le p \le 1$, I is the identity in the Hilbert space $\mathcal{H}_{AB}$ and $|\psi^-\rangle = \frac{1}{\sqrt{2}}\big(|01\rangle - |10\rangle\big)$ is a state of the Bell basis (see Sec. 3.4.1). In the basis $\{|00\rangle, |01\rangle, |10\rangle, |11\rangle\}$ the density matrix $(\rho_W)_{AB}$ reads

$$(\rho_W)_{AB} = \begin{bmatrix} \frac{1-p}{4} & 0 & 0 & 0 \\ 0 & \frac{1+p}{4} & -\frac{p}{2} & 0 \\ 0 & -\frac{p}{2} & \frac{1+p}{4} & 0 \\ 0 & 0 & 0 & \frac{1-p}{4} \end{bmatrix}. \tag{5.236}$$

Taking the partial transpose yields

$$(\rho_W)_{AB}^{T_A} = \begin{bmatrix} \frac{1-p}{4} & 0 & 0 & -\frac{p}{2} \\ 0 & \frac{1+p}{4} & 0 & 0 \\ 0 & 0 & \frac{1+p}{4} & 0 \\ -\frac{p}{2} & 0 & 0 & \frac{1-p}{4} \end{bmatrix}. \tag{5.237}$$

This latter matrix has eigenvalues $\lambda_1 = \lambda_2 = \lambda_3 = \frac{1+p}{4}$ and $\lambda_4 = \frac{1-3p}{4}$. As $\lambda_4 < 0$ for $\frac{1}{3} < p \le 1$, we may conclude that the Werner state is entangled for these values of the parameter p.

It can be shown (M. Horodecki et al., 1996) that for composite states of dimension 2×2 and 2×3, the Peres criterion provides a necessary and sufficient condition for separability; that is, the state ρ_{AB} is separable if and only if $\rho_{AB}^{T_A}$ is non-negative. This result teaches us, for instance, that the

Werner state is separable for $0 \le p \le \frac{1}{3}$. However, for higher dimensional systems, states exist for which all eigenvalues of the partial transpose non-negative, but that are non-separable (P. Horodecki, 1997). These states are known as *bound entangled states* since they cannot be distilled by means of local operations and classical communication to form a maximally entangled state (M. Horodecki *et al.*, 1998).

We stress that the Peres criterion is more sensitive than Bell's inequality for detecting quantum entanglement; that is, there are states detected as entangled by the Peres criterion that do not violate Bell's inequalities (Peres, 1996).

Exercise 5.16 Show that for a separable two-qubit state ρ_{AB} the following inequality is satisfied:

$$\langle (\Delta \Sigma_x)^2 \rangle + \langle (\Delta \Sigma_y)^2 \rangle + \langle (\Delta \Sigma_z)^2 \rangle \ge 4, \qquad (5.238)$$

where $\Sigma_i = \sigma_i^{(A)} \otimes I^{(B)} + I^{(A)} \otimes \sigma_i^{(B)}$, $\langle (\Delta \Sigma_i)^2 \rangle = \langle (\Sigma_i)^2 \rangle - \langle \Sigma_i \rangle^2$ ($i = x, y, z$) and the angle brackets denote the expectation value over ρ_{AB}. Show that this criterion allows us to conclude that the Werner state (5.235) is entangled when $\frac{1}{3} < p \le 1$.

5.14 * Entropies in physics

The concept of entropy is very closely connected to those of energy, information and chaos. It is a fundamental concept in both information science and physics. There exist many entropy-like quantities. Here, we shall briefly describe those we consider to be the most significant in physics (at least in relation to this book) while endeavouring to elucidate the possible links between the different definitions of entropy. We shall also discuss the relation between these entropies and the Shannon entropy.

5.14.1 * Thermodynamic entropy

In order to define the thermodynamic entropy, consider first the integral

$$\int_A^B \frac{\delta Q}{T}, \qquad (5.239)$$

extended over a *reversible transformation* from A to B, where A and B are two equilibrium states of a given system and δQ is the amount of heat absorbed *reversibly* by the system at temperature T. It can be proved that the above integral depends only on the initial and final states of the transformation, and not on the transformation itself; that is, it is the same for all reversible paths (transformations) joining A to B.

This property enables us to define a state function $S(A)$, known as the thermodynamic entropy.[5] The entropy $S(A)$ of any equilibrium state A of the system is defined by

$$S(A) = \int_O^A \frac{\delta Q}{T}, \qquad (5.240)$$

where the integration path is any reversible transformation from O to A and O is some chosen reference equilibrium state. Note that the entropy $S(A)$ is only defined up to an arbitrary additive constant. Indeed, if we choose a different reference state O' instead of O and define $S'(A) = \int_{O'}^A \frac{\delta Q}{T}$, then $S'(A) = S(A) + S'(O)$. Therefore, the additive constant $S'(O)$ is independent of the state A. The difference in the entropy of two states is, on the other hand, completely defined. We have

$$S(B) - S(A) = \int_A^B \frac{\delta Q}{T}. \qquad (5.241)$$

It follows that, for any infinitesimal reversible transformation, the change in entropy is

$$dS = \frac{\delta Q}{T}. \qquad (5.242)$$

Note that, in contrast to δQ, dS is an exact differential.

Nerst's theorem, also referred to as the third law of thermodynamics, allows us to determine the additive constant appearing in the definition of entropy. This theorem states that *the entropy of every system at absolute zero can always be taken equal to zero* (note that here we assume that the ground state of the system is non-degenerate). It is therefore convenient to choose the state of the system at $T = 0$ as the reference state in (5.240), so that its entropy is set equal to zero. The entropy of any equilibrium state A is now defined as follows:

$$S(A) = \int_{T=0}^A \frac{\delta Q}{T}. \qquad (5.243)$$

Note that formula (5.243) is restricted to equilibrium states. However, for systems composed of several parts, it is possible to define the entropy

---

[5]The thermodynamic entropy was introduced by Clausius in 1865.

even for non-equilibrium states, in the case in which each part is itself in an equilibrium state A_i with corresponding entropy S_i. The global entropy of the system is then given by the sum of the entropies of all the parts: $S = \sum_i S_i$.

An important property of entropy arises from Eq. (5.241). For a thermally isolated system (that is, $\delta Q = 0$) reversible transformations do not change the entropy of the system: $S(B) = S(A)$. On the other hand, it is possible to show that for *irreversible transformations* we have

$$S(B) - S(A) \geq \int_A^B \frac{\delta Q}{T}. \tag{5.244}$$

Therefore, for $\delta Q = 0$ we find

$$S(B) \geq S(A), \tag{5.245}$$

that is, for any transformation occurring in a thermally isolated system, the entropy of the final state can never be less than that of the initial state. Thus, the state of maximum entropy is the most stable state for an isolated system.

Let us consider a transformation from an initial state A to a final state B of a system in contact with an environment that is maintained at a constant temperature T. Applying Eq. (5.244), we obtain

$$Q = \int_A^B \delta Q \leq T\left[S(B) - S(A)\right]. \tag{5.246}$$

The first law of thermodynamics, see Eq. (1.36), tells us that the work W performed by the system is given by

$$W = -\Delta E + Q, \tag{5.247}$$

where $\Delta E = E(B) - E(A)$ is the variation of the internal energy of the system. From Eqs. (5.246) and (5.247) we obtain

$$W \leq E(A) - E(B) + T\left[(S(B) - S(A)\right]. \tag{5.248}$$

This inequality sets an upper limit on the amount of work that can be extracted from the transformation $A \to B$. If such a transformation is reversible, then the equality sign holds and the work performed saturates the upper limit. It is useful to define the function

$$F = E - TS. \tag{5.249}$$

Then Eq. (5.248) becomes

$$W \leq F(A) - F(B) = -\Delta F. \qquad (5.250)$$

For a reversible transformation we have $W = -\Delta F$. Therefore, the quantity F, known as the *free energy* of the system, plays a role analogous to that of the internal energy E in a purely mechanical system (indeed, in such a case, $Q = 0$, so that $W = -\Delta E$).

5.14.2 * Statistical entropy

One of the principal purposes of equilibrium statistical mechanics is to explain the laws of thermodynamics starting from the laws of molecular dynamics. The question is: given the laws of motion and the interactions between the molecules, what are the macroscopic properties of matter composed of these molecules?

In the second half of the nineteenth century, Boltzmann and Clausius tried to derive the second law of thermodynamics from mechanics. This followed the line of Maxwell, who had already put forward the idea that "the second law of thermodynamics has only a statistical certainty".

As we discussed in the previous subsection, in a thermally isolated system entropy can never decrease. Thus, the system evolution is such that it never becomes more ordered. A familiar demonstration of this principle is the flow of heat from hot to cold bodies until a uniform temperature is reached.

Boltzmann related the notion of entropy to the logarithm of the number of possible different microscopic states compatible with a given macroscopic state. For instance, let us consider $N/2$ white molecules and $N/2$ black molecules ($N \gg 1$) inside a single vessel and distinguish the microscopic state of each molecule by the fact that it is located in the left or right half of the vessel. It is clear that there is a single microscopic state corresponding to the macroscopic state "all white molecules in the left half and all black molecules in the right half of the vessel" while there are many more microscopic states corresponding to the macroscopic state "the white and black molecules are equally distributed between the left and the right halves of the vessel". Therefore, the entropy, or "disorder", is much larger in the latter macroscopic state (see too the discussion on Maxwell's demon in Sec. 1.5.1). More precisely, Boltzmann defined the thermodynamic entropy as

$$S(E) = k_B \ln \omega(E), \qquad (5.251)$$

where k_B is the Boltzmann constant and $\omega(E)$ is the measure of the energy surface $H(q,p) = E$, where H is the system Hamiltonian and (q,p) denotes the phase-space coordinates and momenta of the N molecules.

We point out that Boltzmann's definition of entropy (5.251) assumes that when a system is in thermodynamic equilibrium, all microscopic states satisfying the macroscopic conditions of the system are equiprobable. This implies that in equilibrium the density $\rho(q,p)$ of points in phase space is described by the *microcanonical ensemble*:

$$\rho(q,p) = \frac{1}{\omega(E)} \delta\big(H(q,p) - E\big). \qquad (5.252)$$

Although Eq. (5.251) is a bridge between the microscopic and the macroscopic descriptions of matter, it only refers to states in thermodynamic equilibrium. In order to obtain a definition of entropy that is also applicable out of equilibrium, it is convenient to consider the *canonical ensemble*, which is appropriate for the description of systems in contact with a heat reservoir at temperature T. In this case, taking into account the first and the second principles of thermodynamics, one obtains (see, for instance, Toda *et al.*, 1983)

$$S(T) = k_B(\ln Z + \beta \bar{E}), \qquad (5.253)$$

where $\bar{E}$ is the average energy of the system, $\beta = \frac{1}{k_B T}$ and

$$Z = \int dq dp \, e^{-\beta H(q,p)} \qquad (5.254)$$

is the partition function. Taking into account that

$$\bar{E} = \int dq dp \, \rho(q,p) H(q,p), \qquad (5.255)$$

with

$$\rho(q,p) = \frac{1}{Z} e^{-\beta H(q,p)}, \qquad (5.256)$$

we obtain the following from (5.253):

$$S = k_B \left(\ln Z + \beta \frac{1}{Z} \int dq dp \, e^{-\beta H(q,p)} H(q,p) \right)$$

$$= k_B \left(\ln Z - \frac{1}{Z} \int dq dp \, e^{-\beta H(q,p)} \ln(\rho Z) \right)$$

$$= -k_B \int dq dp \, \rho(q,p) \ln \rho(q,p). \qquad (5.257)$$

Note that the statistical entropy $S = -k_B \int dq dp \, \rho(q,p) \ln \rho(q,p)$ can be directly defined as the average value of $-\ln \rho(q,p)$. In particular, in the case of the microcanonical ensemble, $\rho(q,p)$ is given by (5.252) and therefore the statistical entropy (5.257) reduces to the thermodynamic entropy (5.251).

It can be shown that if on the energy surface we consider a distribution $\rho'(q,p)$ different from the microcanonical distribution (5.252), then the entropy $S = -k_B \int dq dp \, \rho'(q,p) \ln \rho'(q,p)$ is smaller than the thermodynamic entropy (5.251). We therefore conclude that the microcanonical distribution maximizes the statistical entropy.

It is interesting that the expression $-\int dq dp \, \rho \ln \rho$ can be viewed as a measure of the "degree of uncertainty" associated with the measure $d\mu = \rho \, dq dp$. Such uncertainty is small when μ is peaked and large when ρ is spread over the energy surface. We therefore obtain a simple statistical interpretation of the entropy of, say, a gas: it has the meaning of the degree of uncertainty in the microscopic state of the gas corresponding to a given macroscopic state. Hence, we can exploit the fact that the microcanonical ensemble maximizes the expression $-\int dq dp \, \rho \ln \rho$ to justify its use in statistical physics.

Finally, we wish to point out that the expression $-\int dq dp \, \rho \ln \rho$ is the analogue, for continuous variables, of the Shannon entropy $-\sum_i p_i \ln p_i$.

5.14.3   * *Dynamical Kolmogorov–Sinai entropy*

The Kolmogorov–Sinai (KS) entropy refers to the dynamical behaviour of a system: it characterizes its dynamical stability and provides a measure of the rate at which memory of the initial conditions is lost. We shall not be concerned here with rigorous mathematical details and, instead, we give below a simple operative definition for computation of the KS entropy.

Let us consider a partition Q of the energy surface into N cells and attach an index j to each cell ($j = 1, \ldots, N$). We then follow the evolution of an orbit at discrete times $t_0 = 0$, $t_1 = T$, $t_2 = 2T$ and so on. We associate the sequence of symbols (or letters) $i_0, i_1, i_2, \ldots$ with the orbit if the orbit resides in cell i_0 at time t_0, in i_1 at t_1, in i_2 at t_2 and so on. Given a sequence (word) of m symbols $s_1, \ldots, s_m$, we call $p^{(Q)}_{s_1, \ldots, s_m}$ the probability that such a sequence appears in the orbit. The probability $p^{(Q)}_{s_1, \ldots, s_m}$ can be computed in practice by following the orbit up to very long times and counting the number of recurrences of the sequence $s_1, \ldots, s_m$ in the sequence $i_0, i_1, i_2, \ldots$ associated with the orbit. After repeating the same calculation for all m-letter words, we obtain the quantity

$$K^{(Q)}(m) = -\sum_{s_1, \ldots, s_m = 1}^{N} p^{(Q)}_{s_1, \ldots, s_m} \ln p^{(Q)}_{s_1, \ldots, s_m}. \tag{5.258}$$

The KS entropy h is finally defined as

$$h = \sup_{Q} \lim_{m \to \infty} \frac{K^{(Q)}(m)}{m}. \tag{5.259}$$

In practice, the entropy of a dynamical system is computed numerically starting from a regular partition of the energy surface into D-dimensional hypercubes of volume ϵ^D. In this manner, the entropy of the ϵ-partition can be computed as

$$h(\epsilon) = \lim_{m \to \infty} \frac{K^{(\epsilon)}(m)}{m}. \tag{5.260}$$

The dynamical entropy is then obtained as

$$h = \lim_{\epsilon \to 0} h(\epsilon). \tag{5.261}$$

The practical advantage of this procedure is evident: it is not necessary to consider all possible partitions, but simply take hypercubes of small enough volume. At any rate, the numerical calculation of the quantity $h(\epsilon)$ is very difficult. Indeed, the Shannon–McMillan theorem states that the number of "typical" m-words increases as $\exp(h(\epsilon)m)$. Thus, for chaotic systems ($h(\epsilon) > 0$) the number of typical m-words grows exponentially with m, so that it is very hard to compute $K^{(\epsilon)}(m)$. Intuitively, such an exponential proliferation is related to the exponential instability of orbits. This intuition is made rigorous by Pesin's theorem, which states that

$$h = \sum_{\lambda_i > 0} \lambda_i, \tag{5.262}$$

where the sum extends over all positive Lyapunov exponents.[6] This result is also useful for computing the KS entropy in a simple manner.

It is interesting to investigate what relation (if any) exists between the Boltzmann–Gibbs statistical entropy

$$S(t) = -k_B \int dq dp\, \rho(q,p;t) \ln \rho(q,p;t) \tag{5.263}$$

and the Kolmogorov–Sinai entropy. First of all we observe that, according to Liouville's theorem, the phase-space volume occupied by the distribution $\rho(q,p;t)$ is conserved. This implies that the entropy $S(t)$ does not vary with time at all. However, the shape of the volume becomes increasingly complicated, due to the chaotic dynamics. Thus, after smoothing of the probability distribution, the volume occupied increases. The simplest manner to perform such *coarse graining* is to divide the energy surface into N cells, each of the same extension. Let $p_i(t)$ denote then the probability that the state of the system falls inside the cell i at time t. The coarse-grained statistical entropy is now defined as

$$S_c(t) = -\sum_i p_i(t) \ln p_i(t). \tag{5.264}$$

For a chaotic system the coarse-grained distribution converges to the microcanonical distribution; that is, $\lim_{t\to\infty} p_i(t) = \frac{1}{N}$. Therefore, the equilibrium value of the coarse-grained entropy (5.264) is $S_c(\infty) = \lim_{t\to\infty} S_c(t) = \ln N$. Let us assume that the initial, far-from-equilibrium distribution $\rho(q,p;0)$ is strongly peaked in phase space, for instance it is localized in a single cell. In this case, the coarse-grained entropy $S_c(t)$ evolves from the initial value $S_c(0) = 0$ to the equilibrium value $S_c(\infty) = \ln N$. There are no rigorous mathematical results connecting the KS entropy to the coarse-grained statistical entropy. Nevertheless, qualitative analytical arguments as well as numerical results (see Latora and Baranger, 1999) show that, for systems characterized by uniform (in phase space) exponential instability, after an initial transient stage $S_c(t)$ increases linearly with a slope given by the Kolmogorov–Sinai entropy. A simple example illustrating the connection between the growth rate of the coarse-grained statistical entropy S_c and the KS entropy h is shown in Fig. 5.15.

---

[6]For a dynamical system evolving in an n-dimensional phase space, there are n Lyapunov exponents; for their definition see, *e.g.*, Ott (2002). The largest Lyapunov exponent is defined in Sec. 1.4.2.

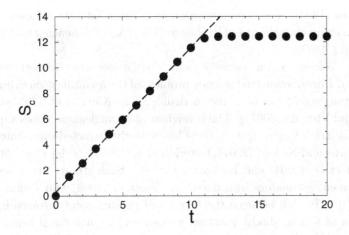

Fig. 5.15 Time evolution of the coarse-grained entropy S_c (circles) for the classical sawtooth map (see Sec. 3.15.3) on the torus $0 \leq \theta < 2\pi$, $-\pi \leq J < \pi$, with $K = \sqrt{2}$. The coarse graining is obtained by dividing the torus into $N = 2.5 \times 10^5$ square cells. The initial density distribution uniformly covers a single cell centred at $(\theta, J) = \left(\frac{\pi}{5}, \frac{3\pi}{5}\right)$. The dashed line has slope given by the Kolmogorov–Sinai entropy $h \approx 1.13$ (note that the sawtooth map is a conservative chaotic system and there is a single positive Lyapunov exponent λ, so that, according to Pesin's theorem, $h = \lambda$).

5.15 A guide to the bibliography

A very useful introduction to the density operator formalism can be found in Cohen-Tannoudji *et al.* (1977).

The Bužek–Hillery copying machine was introduced by Bužek and Hillery (1996) (see also Gisin and Massar, 1997 and Bruß *et al.*, 1998). Quantum cloning in spin networks is discussed in De Chiara *et al.* (2004). A review of quantum cloning machines, including the experimental demonstrations of optimal quantum cloning, is given in Scarani *et al.* (2005).

The Kraus representation is discussed in Kraus (1983).

Interesting discussions of quantum measurements can be found in Braginsky and Khalili (1992), Gardiner and Zoller (2000), Namiki *et al.* (1997) and Peres (1993).

Modern information theory started with the work of Shannon (1948), while general references are Cover and Thomas (1991) and Gray (1990).

The quantum noiseless coding theorem is due to Schumacher (1995), see also Barnum *et al.* (1996).

A simplified derivation of the Holevo bound can be found in Fuchs and Caves (1994). The communication of classical information over noisy quantum channels is discussed in Holevo (1998) and Schumacher and Westmoreland (1997).

Basic references on entanglement distillation are Bennett *et al.* (1996a,c). Introductions to the open problem of the quantification of mixed-state entanglement can be found in Bruß (2002), Alber *et al.* (2001a) and Plenio and Virmani (2007)). The behaviour of entanglement across a quantum phase transition in spin systems has recently attracted much interest, see Osborne and Nielsen (2002), Osterloh *et al.* (2002), Vidal *et al.* (2003), Roscilde *et al.* (2004) and references therein. Such studies are based on entanglement estimators introduced in Wootters (1998) and Coffman *et al.* (2000). The link between the amount of entanglement involved in the evolution of a many-body quantum system and its numerical simulation by means of the density-matrix renormalization group is discussed in Vidal (2004) and Verstraete *et al.* (2004). The role of entanglement in the speedup of quantum computation is investigated in Jozsa and Linden (2003), Vidal (2003) and Orús and Latorre (2004). The Peres criterion was found by Peres (1996); see also M. Horodecki *et al.* (1996) and Alber *et al.* (2001a).

Thermodynamic and statistical entropies are discussed in statistical mechanics textbooks, such as Huang (1987) and Toda *et al.* (1983). An introduction to the Kolmogorov–Sinai entropy is Kornfeld *et al.* (1982).

Chapter 6

Decoherence

In practice, any quantum system is *open*; namely, it is never perfectly isolated from the *environment*. The word decoherence, used in its broader meaning, denotes any quantum-noise process due to the unavoidable coupling of the system to the environment. Decoherence theory has a fundamental interest beyond quantum information science since it provides explanations of the emergence of classicality in a world governed by the laws of quantum mechanics. The core of the problem is the superposition principle, according to which any superposition of quantum states is an acceptable quantum state. This entails consequences that are absurd according to classical intuition, such as the superposition of "live cat" and "dead cat" considered in Schrödinger's well-known cat paradox. The interaction with the environment can destroy the coherence between the states appearing in a superposition (for instance, the "live-cat" and "dead-cat" states).

In quantum information processing, decoherence is a threat to the actual implementation of any quantum computation or communication protocol. Indeed, decoherence invalidates the quantum superposition principle, which lies at the heart of the potential power of any quantum algorithm. On the other hand, decoherence is also an essential ingredient for quantum information processing, which must end up with a measurement by converting quantum states into classical outcomes. We shall see that decoherence plays a key role in the quantum measurement process.

In this chapter, we shall describe decoherence using various tools, from the quantum-operation formalism introduced in the previous chapter to the master-equation and the quantum-trajectory approaches. We shall start with simple single-qubit noise models and end with a detailed description of the effects of various noise sources (coupling to the environment, noisy gates, imperfections in the quantum computer hardware) on the stability of quantum computation. In parallel, we shall discuss the fundamental issue of the quantum to classical transition, focusing on the role played by decoherence and chaotic dynamics.

6.1 Decoherence models for a single qubit

In this section, we shall study quantum-noise (decoherence) processes that can act on a single qubit. A general formulation of the problem, in terms of Kraus operators, will be given in Sec. 6.1.1. Before doing so, it is instructive to consider a very simple decoherence model, drawn in Fig. 6.1. Here the environment consists of a single qubit and the system–environment interaction is represented by a CNOT gate. Let us assume that initially the system is in a pure state, $|\psi\rangle = \alpha|0\rangle + \beta|1\rangle$, corresponding to the density matrix $\rho = |\psi\rangle\langle\psi|$, whose matrix representation in the $\{|0\rangle, |1\rangle\}$ basis is given by

$$\rho = \begin{bmatrix} |\alpha|^2 & \alpha\beta^\star \\ \alpha^\star\beta & |\beta|^2 \end{bmatrix}. \tag{6.1}$$

The diagonal terms of ρ are known as *populations* (see Sec. 5.1), and give the probabilities to obtain, from a polarization measurement along the z-axis, outcomes 0 or 1, respectively. The off-diagonal terms, known as *coherences*, appear when the state $|\psi\rangle$ is a superposition of the states $|0\rangle$ and $|1\rangle$. They are completely destroyed by the decoherence process drawn in Fig. 6.1. Indeed, this quantum circuit changes the initial global system–environment state,

$$|\Psi\rangle = |\psi\rangle \otimes |0\rangle = (\alpha|0\rangle + \beta|1\rangle)|0\rangle, \tag{6.2}$$

into the final state

$$|\Psi'\rangle = \alpha|00\rangle + \beta|11\rangle. \tag{6.3}$$

Note that the CNOT interaction has entangled the qubit with the environment, as the state $|\Psi'\rangle$ is non-separable. The final density matrix ρ' of the system is obtained after tracing over the environment:

$$\rho' = \mathrm{Tr}_{\mathrm{env}} |\Psi'\rangle\langle\Psi'| = \begin{bmatrix} |\alpha|^2 & 0 \\ 0 & |\beta|^2 \end{bmatrix}. \tag{6.4}$$

This decoherence process has a particularly appealing interpretation: it is evident from Eq. (6.3) that the environment has learnt, through the CNOT interaction, what the state of the system is. Indeed, if the state of the system is $|0\rangle$, the state of the environment remains $|0\rangle$; on the other hand, if the state of the system is $|1\rangle$, the state of the environment is flipped and becomes $|1\rangle$. Therefore, the CNOT gate is basically a measurement performed by the environment on the system. The information on the relative phases of the coefficients α and β appearing in the initial state $|\psi\rangle$ is now hidden in the system–environment quantum correlations. Since we do not keep records of the state of the environment, this information is lost for us. In short, information leaks from the system into the external world.

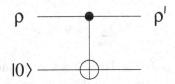

Fig. 6.1　Quantum circuit modelling complete decoherence.

6.1.1　*The quantum black box*

Let us consider a two-level system (qubit) interacting with a generic physical system. This system is known as a quantum *black box*, and its action on the qubit is described in terms of a quantum operation $\mathcal{S}$:

$$\rho \to \rho' = \mathcal{S}(\rho) = \sum_k E_k\, \rho\, E_k^\dagger, \quad \text{with} \quad \sum_k E_k^\dagger E_k = I, \qquad (6.5)$$

where the Kraus operators are denoted by E_k. It is convenient to write the states ρ and ρ' in the Bloch-sphere representation (5.32):

$$\rho = \tfrac{1}{2}(I + \boldsymbol{r} \cdot \boldsymbol{\sigma}) \quad \text{and} \quad \rho' = \tfrac{1}{2}(I + \boldsymbol{r}' \cdot \boldsymbol{\sigma}), \qquad (6.6)$$

where the Bloch vectors $\boldsymbol{r} = (x, y, z)$ and $\boldsymbol{r}' = (x', y', z')$ are such that $|\boldsymbol{r}|, |\boldsymbol{r}'| \in [0, 1]$. The transformation

$$\boldsymbol{r} \to \boldsymbol{r}' = M\boldsymbol{r} + \boldsymbol{c} \qquad (6.7)$$

is known as an *affine map*. To find the matrix M and the vector $\boldsymbol{c}$ as functions of the Kraus operators E_k, it is convenient to expand the Kraus operators over the basis $\{I, \sigma_1 \equiv \sigma_x, \sigma_2 \equiv \sigma_y, \sigma_3 \equiv \sigma_z\}$:

$$E_k = \gamma_k I + \sum_{l=1}^{3} a_{kl}\sigma_l. \qquad (6.8)$$

After a lengthy but straightforward calculation (see exercise 6.1), we obtain

$$M_{jk} = \sum_{l=1}^{3}\left[a_{lj}a_{lk}^{\star} + a_{lj}^{\star}a_{lk} + \left(|\gamma_l|^2 - \sum_{p=1}^{3}|a_{lp}|^2\right)\delta_{jk} \right.$$
$$\left. + i\sum_{p=1}^{3}\epsilon_{jkp}\left(\gamma_l a_{lp}^{\star} - \gamma_l^{\star}a_{lp}\right)\right], \quad (6.9)$$

$$c_j = 2i\sum_{k,l,m=1}^{3}\epsilon_{jlm}a_{kl}a_{km}^{\star}. \qquad (6.10)$$

In these expressions, ϵ_{jkl} is the Levi-Civita antisymmetric tensor, with $\epsilon_{jkl} = 0$ if the three indices are not all different, $\epsilon_{123} = \epsilon_{231} = \epsilon_{312} = 1$ and $\epsilon_{213} = \epsilon_{321} = \epsilon_{132} = -1$.

Exercise 6.1 Check Eqs. (6.9) and (6.10).

In order to clarify the meaning of the affine map (6.7), we may take advantage of the *polar decomposition*

$$M = OS, \qquad (6.11)$$

where S is a symmetric, non-negative matrix and O an orthogonal matrix. Hence, in the affine map S deforms the Bloch sphere into an ellipsoid, while O rotates it and c displaces its centre.

Exercise 6.2 Show that the polar decomposition (6.11) is possible for any real matrix M.

We need to determine 12 parameters to describe the action of a generic quantum black box on a two-level system: 6 parameters to determine the symmetric 3×3 matrix S, 3 for the orthogonal matrix O and 3 for the displacement c. Note that the values taken by these parameters must be such that r' is still a Bloch vector; that is, ρ' is still a density matrix.

The number of independent parameters in the single-qubit case is in agreement with the general result of Sec. 5.4: we need $N^4 - N^2$ independent real parameters to characterize a quantum operation acting on an N-level quantum system. In particular, we have 12 parameters in the single-qubit case ($N = 2$).

6.1.2 *Measuring a quantum operation acting on a qubit*

We wish to measure the 12 parameters that determine the quantum operation $\mathcal{S}$ mapping the single-qubit density matrix ρ to $\rho' = \mathcal{S}(\rho)$. For this purpose, we consider the experiment drawn schematically in Fig. 6.2. The source S emits a large number of qubits, whose states are described by the density matrix ρ. The qubits enter the quantum black box and come out in states described by a different density matrix, ρ'. A detector D measures the density matrix ρ', following the procedure described in Sec. 5.5.

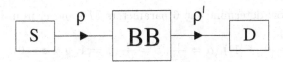

Fig. 6.2 A schematic diagram of the measurement procedure used to determine the effect of a quantum black box (BB) on a qubit.

The affine map (6.7) reads

$$
\begin{bmatrix} x' \\ y' \\ z' \end{bmatrix} = \begin{bmatrix} M_{11} & M_{12} & M_{13} \\ M_{21} & M_{22} & M_{23} \\ M_{31} & M_{32} & M_{33} \end{bmatrix} \begin{bmatrix} x \\ y \\ z \end{bmatrix} + \begin{bmatrix} c_x \\ c_y \\ c_z \end{bmatrix}, \tag{6.12}
$$

where we assume the parameters M_{ij} and c_i to be time-independent; namely, the quantum black box always acts in the same manner on every two-level system. We wish to determine these parameters. To this end it is sufficient to consider pure initial states $|\psi\rangle = \alpha|0\rangle + \beta|1\rangle$. The corresponding density matrix $\rho = |\psi\rangle\langle\psi|$ is given by Eq. (6.1). Note that in the Bloch-sphere representation

$$
\rho = \frac{1}{2} \begin{bmatrix} 1+z & x-iy \\ x+iy & 1-z \end{bmatrix}, \tag{6.13}
$$

and so the coordinates (x, y, z) are related to α and β as follows:

$$
\alpha\beta^\star = \tfrac{1}{2}(x - iy), \qquad |\alpha|^2 = \tfrac{1}{2}(1 + z), \qquad |\beta|^2 = \tfrac{1}{2}(1 - z). \tag{6.14}
$$

To determine the 12 parameters M_{ij} and c_i, we need to prepare different, appropriate, initial states, for instance:

(i) $|\psi_1\rangle = |0\rangle$ ($\alpha = 1$, $\beta = 0$, $x = y = 0$, $z = 1$). As described in Sec. 5.5, if we have at our disposal a large number of identically prepared qubits in the state $|\psi_1\rangle$ and entering the quantum black box, we can measure the final density matrix ρ_1' and determine its Bloch coordinates x_1', y_1' and z_1', up to statistical errors. From Eq. (6.12), we obtain

$$x_1' = M_{13} + c_x, \qquad y_1' = M_{23} + c_y, \qquad z_1' = M_{33} + c_z. \qquad (6.15)$$

(ii) $|\psi_2\rangle = |1\rangle$ ($\alpha = 0$, $\beta = 1$, $x = y = 0$, $z = -1$). In this case,

$$x_2' = -M_{13} + c_x, \qquad y_2' = -M_{23} + c_y, \qquad z_2' = -M_{33} + c_z. \qquad (6.16)$$

We can now determine the 6 parameters M_{i3} and c_i from Eqs. (6.15) and (6.16).

(iii) $|\psi_3\rangle = \frac{1}{\sqrt{2}}(|0\rangle + |1\rangle)$ ($\alpha = \frac{1}{\sqrt{2}}$, $\beta = \frac{1}{\sqrt{2}}$, $x = 1$, $y = z = 0$).

$$x_3' = M_{11} + c_x, \qquad y_3' = M_{21} + c_y, \qquad z_3' = M_{31} + c_z. \qquad (6.17)$$

(iv) $|\psi_4\rangle = \frac{1}{\sqrt{2}}(|0\rangle + i|1\rangle)$ ($\alpha = \frac{1}{\sqrt{2}}$, $\beta = \frac{i}{\sqrt{2}}$, $y = 1$, $x = z = 0$).

$$x_4' = M_{12} + c_x, \qquad y_4' = M_{22} + c_y, \qquad z_4' = M_{32} + c_z. \qquad (6.18)$$

The remaining 6 unknown parameters M_{i1} and M_{i2} are computed using Eqs. (6.17) and (6.18).

In principle, the method described in this section can be extended to quantum black boxes acting on many-qubit systems; already though with two qubits (a Hilbert space of dimension $N = 4$), there are $N^4 - N^2 = 240$ real parameters to be determined.

6.1.3   *Quantum circuits simulating noise channels*

A useful representation of quantum operations is obtained using quantum circuits, in which the environment is represented by ancillary qubits.

Let us consider the circuit drawn in Fig. 6.3. We have a single-qubit system plus an environment with two ancillary qubits. We assume that initially the system is described by the density matrix ρ, while the ancillary qubits are in the pure state

$$|\psi\rangle = \alpha|00\rangle + \beta|01\rangle + \gamma|10\rangle + \delta|11\rangle, \qquad (6.19)$$

with the normalization condition $|\alpha|^2 + |\beta|^2 + |\gamma|^2 + |\delta|^2 = 1$. The initial total density matrix (system plus environment) is given by

$$\rho_{\text{in}}^{(\text{tot})} = |\psi\rangle\langle\psi| \otimes \rho = \begin{bmatrix} |\alpha|^2\rho & .. & .. & .. \\ .. & |\beta|^2\rho & .. & .. \\ .. & .. & |\gamma|^2\rho & .. \\ .. & .. & .. & |\delta|^2\rho \end{bmatrix}, \tag{6.20}$$

where, to simplify the expression, we have denoted by .. the matrix blocks whose expressions are not needed in subsequent calculations.

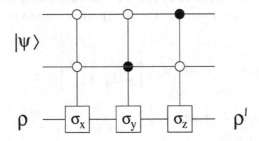

Fig. 6.3 A quantum circuit implementing the deformation of the Bloch sphere into an ellipsoid centred at the origin of the Bloch sphere with axes directed along x, y and z.

The quantum circuit in Fig. 6.3 implements the unitary transformation

$$U = \begin{bmatrix} \sigma_x & 0 & 0 & 0 \\ 0 & \sigma_y & 0 & 0 \\ 0 & 0 & \sigma_z & 0 \\ 0 & 0 & 0 & I \end{bmatrix}, \tag{6.21}$$

where I is the 2×2 identity matrix. This means that, as shown in Fig. 6.3, σ_x, σ_y, σ_z, or I are applied to the bottom qubit if the two upper qubits are in the states $|00\rangle$, $|01\rangle$, $|10\rangle$, or $|11\rangle$. The final three-qubit state (system plus environment) is described by the density matrix

$$\rho_{\text{fin}}^{(\text{tot})} = U\rho_{\text{in}}^{(\text{tot})}U^\dagger. \tag{6.22}$$

Note that the system is now, in general, entangled with the environment. After tracing over environmental qubits, we obtain the final state of the system:

$$\rho' = \mathrm{Tr}_{\mathrm{env}}\, \rho_{\mathrm{fin}}^{(\mathrm{tot})} = \mathcal{S}(\rho)$$
$$= |\alpha|^2 \sigma_x \rho \sigma_x^\dagger + |\beta|^2 \sigma_y \rho \sigma_y^\dagger + |\gamma|^2 \sigma_z \rho \sigma_z^\dagger + |\delta|^2 \rho. \qquad (6.23)$$

If we introduce the Kraus operators

$$
\begin{aligned}
E_1 &= |\alpha|\sigma_x, & E_2 &= |\beta|\sigma_y, \\
E_3 &= |\gamma|\sigma_z, & E_0 &= |\delta|I,
\end{aligned}
\qquad (6.24)
$$

we have $\rho' = \sum_{i=0}^{3} E_i \rho E_i^\dagger$. The transformation induced by the Kraus operators (6.24) can be clearly visualized in the Bloch-sphere representation. Let us consider the Bloch vectors r and r' associated with the density matrices ρ and ρ', respectively (see Eq. (6.6)). It is easy to check by direct computation that

$$\sigma_x \rho \sigma_x^\dagger = \sigma_x \rho \sigma_x = \frac{1}{2} \begin{bmatrix} 1-z & x+iy \\ x-iy & 1+z \end{bmatrix}, \qquad (6.25\mathrm{a})$$

$$\sigma_y \rho \sigma_y^\dagger = \sigma_y \rho \sigma_y = \frac{1}{2} \begin{bmatrix} 1-z & -(x+iy) \\ -(x-iy) & 1+z \end{bmatrix}, \qquad (6.25\mathrm{b})$$

$$\sigma_z \rho \sigma_z^\dagger = \sigma_z \rho \sigma_z = \frac{1}{2} \begin{bmatrix} 1+z & -(x-iy) \\ -(x+iy) & 1-z \end{bmatrix}. \qquad (6.25\mathrm{c})$$

Taking into account that $|\delta|^2 = 1 - |\alpha|^2 - |\beta|^2 - |\gamma|^2$, we obtain

$$
\begin{aligned}
x' &= \left[1 - 2\left(|\beta|^2 + |\gamma|^2\right)\right] x, \\
y' &= \left[1 - 2\left(|\gamma|^2 + |\alpha|^2\right)\right] y, \\
z' &= \left[1 - 2\left(|\alpha|^2 + |\beta|^2\right)\right] z.
\end{aligned}
\qquad (6.26)
$$

These expressions tell us that the Bloch sphere is deformed into an ellipsoid, centred at the origin of the Bloch sphere and whose axes are directed along x, y and z. As we shall see in the following, depending on the choice of the parameters $|\alpha|$, $|\beta|$ and $|\gamma|$, many interesting noise channels can be obtained.

We note that, as can be clearly seen from Eq. (6.26), the state ρ' only depends on the amplitudes of the coefficients α, β, γ and δ in the state (6.19) and not on their phases. This implies that, for any density matrix describing the initial state of the two ancillary qubits and having diagonal terms equal to $|\alpha|^2$, $|\beta|^2$, $|\gamma|^2$ and $|\delta|^2$, the circuit of Fig. 6.3 would implement the quantum operation (6.23).

We shall show in the following subsections that commonly investigated noise channels such as the bit-flip or the phase-flip channel can be obtained as special cases of the circuit in Fig. 6.3.

6.1.4 *The bit-flip channel*

The bit-flip channel is obtained by taking $\beta = \gamma = 0$ in the state (6.19). In this case Eq. (6.23) reduces to

$$\rho' = \mathcal{S}(\rho) = |\alpha|^2 \sigma_x \rho \sigma_x^\dagger + \left(1 - |\alpha|^2\right) \rho, \qquad (6.27)$$

and the transformation $\rho \to \rho' = \sum_k E_k \rho E_k^\dagger$ can be implemented by means of the Kraus operators

$$E_0 = \sqrt{1 - |\alpha|^2} I, \qquad E_1 = |\alpha| \sigma_x. \qquad (6.28)$$

We point out that this noise channel flips the state of a qubit (from $|0\rangle$ to $|1\rangle$ and *vice versa*) with probability $|\alpha|^2$. Indeed, the state $\rho = |0\rangle\langle 0|$ is mapped into $\rho' = |\alpha|^2 |1\rangle\langle 1| + (1 - |\alpha|^2)|0\rangle\langle 0|$, while $\rho = |1\rangle\langle 1|$ becomes $\rho' = |\alpha|^2 |0\rangle\langle 0| + (1 - |\alpha|^2)|1\rangle\langle 1|$.

The deformation of the Bloch-sphere coordinates, given by Eq. (6.26), simplifies as follows:

$$x' = x, \qquad y' = \left(1 - 2|\alpha|^2\right)y, \qquad z' = \left(1 - 2|\alpha|^2\right)z. \qquad (6.29)$$

Hence, the Bloch sphere is mapped into an ellipsoid with x as the symmetry axis. Note that the x component is not modified by the bit-flip channel since the eigenstates $|\pm\rangle = \frac{1}{\sqrt{2}}(|0\rangle \pm |1\rangle)$ of the Kraus operator E_1 are directed along the x-axis of the Bloch sphere and $\mathcal{S}(|\pm\rangle\langle\pm|) = |\pm\rangle\langle\pm|$.

A quantum circuit implementing the bit-flip channel is shown in Fig. 6.4. Note that a single auxiliary qubit is sufficient to describe such a quantum operation (in this quantum circuit we take $|\psi\rangle = \alpha|0\rangle + \delta|1\rangle$ with $|\delta| = \sqrt{1 - |\alpha|^2}$).

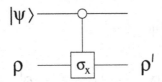

Fig. 6.4   A quantum circuit implementing the bit-flip channel.

6.1.5   *The phase-flip channel*

The phase-flip channel is obtained by taking $\alpha = \beta = 0$ in the state (6.19). In this case Eq. (6.23) reduces to

$$\rho' = \mathcal{S}(\rho) = |\gamma|^2 \sigma_z \rho \sigma_z^\dagger + \left(1 - |\gamma|^2\right) \rho, \qquad (6.30)$$

and the superoperator $\rho \rightarrow \rho' = \mathcal{S}(\rho)$ can be realized by means of the Kraus operators

$$E_0 = \sqrt{1 - |\gamma|^2}\, I, \qquad E_1 = |\gamma| \sigma_z. \qquad (6.31)$$

This noise channel introduces a phase error with probability $|\gamma|^2$. Indeed, if we apply the superoperator $\mathcal{S}$ to a pure state $|\varphi_+\rangle = \mu|0\rangle + \nu|1\rangle$, described by the density matrix $\rho = |\varphi_+\rangle\langle\varphi_+|$, we obtain

$$\rho' = \mathcal{S}(\rho) = |\gamma|^2 |\varphi_-\rangle\langle\varphi_-| + \left(1 - |\gamma|^2|\right) |\varphi_+\rangle\langle\varphi_+|, \qquad (6.32)$$

where $|\varphi_-\rangle = \mu|0\rangle - \nu|1\rangle$. Note that $|\varphi_-\rangle$ differ from $|\varphi_+\rangle$ only in the relative sign of the coefficients in front of the basis states $|0\rangle$ and $|1\rangle$. Therefore, this noise channel has no classical analogue.

The deformation of the Bloch-sphere coordinates, given by Eq. (6.26), simplifies as follows:

$$x' = \left(1 - 2|\gamma|^2\right) x, \qquad y' = \left(1 - 2|\gamma|^2\right) y, \qquad z' = z. \qquad (6.33)$$

Hence, the Bloch sphere is mapped into an ellipsoid with z as symmetry axis. Note that the component z is not modified by the phase-flip channel since the eigenstates of the Kraus operator E_1 ($|0\rangle$ and $|1\rangle$) are directed along the z-axis of the Bloch sphere, and we have $\mathcal{S}(|0\rangle\langle0|) = |0\rangle\langle0|$, $\mathcal{S}(|1\rangle\langle1|) = |1\rangle\langle1|$.

A quantum circuit implementing the phase-flip channel is shown in Fig. 6.5 (in this quantum circuit, the initial state of the auxiliary qubit is $|\psi\rangle = \gamma|0\rangle + \delta|1\rangle$, with $|\delta| = \sqrt{1 - |\gamma|^2}$).

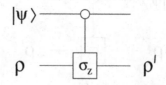

Fig. 6.5   A quantum circuit implementing the phase-flip channel.

As we saw in Sec. 5.4, there is a freedom in the Kraus representation; namely, we can choose different sets of Kraus operators giving rise to the same quantum operation. An example is shown in Fig. 6.6: this circuit leads (see exercise 6.3) to the Kraus operators

$$F_0 = \begin{bmatrix} 1 & 0 \\ 0 & \cos\theta \end{bmatrix}, \quad F_1 = \begin{bmatrix} 0 & 0 \\ 0 & \sin\theta \end{bmatrix} \tag{6.34}$$

and we have

$$\rho' = S(\rho) = \sum_{i=0}^{1} E_i \rho E_i^\dagger = \sum_{i=0}^{1} F_i \rho F_i^\dagger, \tag{6.35}$$

provided $\cos\theta = 1 - 2|\gamma|^2$. It is easy to check that $\{E_0, E_1\}$ and $\{F_0, F_1\}$ are connected by a unitary transformation, as it must be the case for different sets of Kraus operators representing the same superoperator. Indeed, we have $F_i = \sum_{j=0}^{1} W_{ij} E_j$, where the unitary matrix W reads

$$W = \begin{bmatrix} \cos\frac{\theta}{2} & \sin\frac{\theta}{2} \\ \sin\frac{\theta}{2} & -\cos\frac{\theta}{2} \end{bmatrix}. \tag{6.36}$$

Exercise 6.3 Show that the quantum circuit of Fig. 6.6 induces a quantum operation $\rho \to \rho' = F_0 \rho F_0^\dagger + F_1 \rho F_1^\dagger$, where F_0 and F_1 are the Kraus operators written in Eq. (6.34).

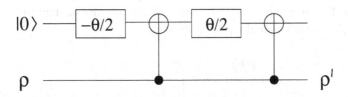

Fig. 6.6 A second circuit implementing the phase-flip channel. The gates labelled $\pm\theta/2$ stand for the rotation matrices $R_y(\mp\theta)$ (see Eq. (5.62)).

6.1.6 *The bit-phase-flip channel*

The bit-phase-flip channel is defined by setting $\alpha = \gamma = 0$ in the state (6.19), so that Eq. (6.23) reduces to

$$\rho' = S(\rho) = |\beta|^2 \sigma_y \rho \sigma_y^\dagger + (1 - |\beta|^2) \rho, \tag{6.37}$$

and the superoperator $\rho \rightarrow \rho' = \mathcal{S}(\rho)$ can be expressed in terms of the Kraus operators

$$E_0 = \sqrt{1 - |\beta|^2}\, I, \qquad E_1 = |\beta|\sigma_y. \qquad (6.38)$$

This channel induces both bit flip and phase flip. Indeed, it maps the state $\mu|0\rangle + \nu|1\rangle$ into $\mu|1\rangle - \nu|0\rangle$ with probability $|\beta|^2$.

The transformation of the Bloch-sphere coordinates is given by:

$$x' = (1 - 2|\beta|^2)x, \qquad y' = y, \qquad z' = (1 - 2|\beta|^2)z. \qquad (6.39)$$

Hence, the Bloch sphere is mapped into an ellipsoid symmetric about the y-axis.

A quantum circuit implementing the bit-phase-flip channel is shown in Fig. 6.7 (in this quantum circuit, the initial state of the auxiliary qubit is $|\psi\rangle = \beta|0\rangle + \delta|1\rangle$ with $|\delta| = \sqrt{1 - |\beta|^2}$).

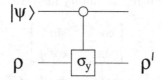

Fig. 6.7   A quantum circuit implementing the bit-phase-flip channel.

Exercise 6.4   Study the quantum-noise operation implemented by the circuit of Fig. 6.8, where U is a generic 2×2 unitary matrix and $|\psi\rangle = \alpha|0\rangle + \beta|1\rangle$ a generic single-qubit pure state.

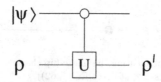

Fig. 6.8   A quantum circuit implementing a single-qubit quantum-noise operation.

6.1.7   *The depolarizing channel*

The depolarizing channel is defined by setting $|\alpha|^2 = |\beta|^2 = |\gamma|^2 = p/3$ in the state (6.19). We can apply the results of Sec. 6.1.3 to this special case and obtain

$$\rho' = \tfrac{1}{3}p \left[\sigma_x \rho \sigma_x^\dagger + \sigma_y \rho \sigma_y^\dagger + \sigma_z \rho \sigma_z^\dagger\right] + (1-p)\rho. \qquad (6.40)$$

In the Bloch-sphere representation,

$$\boldsymbol{r}' = \left(1 - \tfrac{4}{3}p\right)\boldsymbol{r}. \qquad (6.41)$$

Therefore, the Bloch vector is contracted by a factor $(1-\tfrac{4}{3}p)$, independently of its direction. The centre of the Bloch sphere, $\boldsymbol{r} = (0,0,0)$, is the fixed point of this noise channel. If $p = \tfrac{3}{4}$, then $\boldsymbol{r}' = (0,0,0)$ for any $\boldsymbol{r}$. This corresponds to complete depolarization since, as we saw in Sec. 5.1.1, for this state the qubit polarization along any direction is equal to zero.

The quantum operation (6.40) can be implemented by means of the Kraus operators

$$E_0 = \sqrt{1-p}\,I, \quad E_1 = \sqrt{\tfrac{1}{3}p}\,\sigma_x, \quad E_2 = \sqrt{\tfrac{1}{3}p}\,\sigma_y, \quad E_3 = \sqrt{\tfrac{1}{3}p}\,\sigma_z. \tag{6.42}$$

6.1.8 *Amplitude damping*

The amplitude-damping channel is defined by

$$\rho' = \mathcal{S}(\rho) = E_0 \rho E_0^\dagger + E_1 \rho E_1^\dagger, \qquad (6.43)$$

where the two Kraus operators E_0 and E_1 read

$$E_0 = \begin{bmatrix} 1 & 0 \\ 0 & \sqrt{1-p} \end{bmatrix}, \qquad E_1 = \begin{bmatrix} 0 & \sqrt{p} \\ 0 & 0 \end{bmatrix}. \qquad (6.44)$$

Using this definition, it is straightforward to obtain

$$\rho' = \begin{bmatrix} \rho_{00} + p\rho_{11} & \sqrt{1-p}\,\rho_{01} \\ \sqrt{1-p}\,\rho_{10} & (1-p)\rho_{11} \end{bmatrix}, \qquad (6.45)$$

where ρ_{ij} are the matrix elements of the density operator ρ in the basis $\{|0\rangle, |1\rangle\}$. Equation (6.45) implies that the Bloch-sphere coordinates change as follows:

$$x' = \sqrt{1-p}\,x, \quad y' = \sqrt{1-p}\,y, \quad z' = p + (1-p)z. \qquad (6.46)$$

Therefore, the Bloch sphere is deformed into an ellipsoid, with axes directed along x, y and z and centre at $(0,0,p)$. It is clearly seen from Eq. (6.46) that p represents the probability that the state $|1\rangle$ decays to the state

$|0\rangle$ (*damping probability*). Indeed, if we start from $\rho = |1\rangle\langle 1|$; that is, $r = (0, 0, -1)$, we obtain $r' = (0, 0, p - (1 - p))$, corresponding to $\rho' = p|0\rangle\langle 0| + (1 - p)|1\rangle\langle 1|$.

It is instructive to consider the case in which the amplitude-damping channel is applied repeatedly. In this case, we obtain

$$\rho_{11}^{(n)} = (1 - p)^n \rho_{11} = e^{n \ln(1-p)} \rho_{11}, \qquad (6.47)$$

where n denotes the number of applications of the channel. Therefore, the probability $p_1^{(n)}$ to find the qubit in the state $|1\rangle$ drops exponentially with the number n of channel iterations:

$$p_1^{(n)} = (1 - p)^n p_1^{(0)} = e^{n \ln(1-p)} p_1^{(0)}. \qquad (6.48)$$

This means that, for $n \to \infty$, the system is driven to $\rho^{(\infty)} = |0\rangle\langle 0|$. We should stress that, even though quantum noise generally transforms pure states into mixed states, in this case, whatever the initial state is (pure or mixed), we always end up with the pure state $|0\rangle$.

Of course, it is possible to give a continuous time version of this result. If $p = \Gamma(\Delta t)$, $t = n(\Delta t)$ is time and we let $\Delta t \to 0$, then

$$p_1(t) = \lim_{\Delta t \to 0} (1 - \Gamma \Delta t)^{t/\Delta t} p_1(0) = e^{-\Gamma t} p_1(0). \qquad (6.49)$$

Therefore, Γ represents the transition rate for $|1\rangle \to |0\rangle$.

Exercise 6.5   Show that the amplitude-damping channel can be modelled by means of the circuit in Fig. 6.9, where the gates labelled $\pm\theta/2$ stand for the rotation matrices $R_y(\mp\theta)$ (see Eq. 5.62), with $\cos\theta = \sqrt{p}$, $\sin\theta = \sqrt{1-p}$.

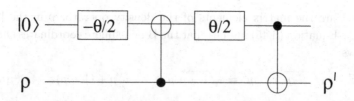

Fig. 6.9   A quantum circuit implementing the amplitude-damping channel.

6.1.9 *Phase damping*

The most general single-qubit density matrix can be written as

$$\rho = \begin{bmatrix} p & \alpha \\ \alpha^\star & 1-p \end{bmatrix}, \qquad (6.50)$$

where the diagonal, real elements p and $1 - p$ $(0 \le p \le 1)$ represent the probabilities of finding the qubit in the state $|0\rangle$ or $|1\rangle$, respectively. The off-diagonal elements (*quantum coherences*) have no classical analogue. Note that we have $|\alpha| \le \sqrt{p(1-p)}$. As we shall see in the following, the effect of the phase-damping channel is to induce a decay of the off-diagonal terms, a process known ad *decoherence*.[1] Therefore, as we shall discuss in detail later in this chapter, the phase-damping channel plays a central role in the transition from the quantum to the classical world.

Two phenomenological models leading to decoherence are the simple example discussed at the beginning of Sec. 6.1 and the quantum circuits implementing the phase-flip channel, introduced in Sec. 6.1.5. Of course, these models are phenomenological and do not represent the physical mechanisms inducing decoherence any better than a resistance in an electric circuit represents the scattering processes undergone by conduction electrons. It is therefore useful to justify decoherence by means of a simple model, leaving a more complete and formal development for the subsequent sections. Our qubit is described by the density matrix (6.50), and we assume that quantum coherences are initially non-zero $(\alpha \ne 0)$. We model the effect of the interaction with the environment as a rotation (*phase kick*) through an angle θ about the z-axis of the Bloch sphere. This rotation is described, as we saw in Sec. 3.3.1, by the matrix

$$R_z(\theta) = \begin{bmatrix} e^{-i\frac{\theta}{2}} & 0 \\ 0 & e^{i\frac{\theta}{2}} \end{bmatrix}. \qquad (6.51)$$

We assume that the rotation angle is drawn from the random distribution

$$p(\theta) = \frac{1}{\sqrt{4\pi\lambda}} e^{-\frac{\theta^2}{4\lambda}}. \qquad (6.52)$$

Therefore, the new density matrix ρ', obtained after averaging over θ, is

---

[1] Here it is useful to remind the reader that, more generally, the word decoherence is used to refer to any quantum-noise process due to coupling of the system with the environment.

given by

$$\rho' = \int_{-\infty}^{+\infty} d\theta \, p(\theta) R_z(\theta) \rho R_z^\dagger(\theta) = \begin{bmatrix} p & \alpha e^{-\lambda} \\ \alpha^\star e^{-\lambda} & 1-p \end{bmatrix}. \quad (6.53)$$

This means that the Bloch-sphere coordinates are mapped by the phase-damping channel as follows:

$$x' = e^{-\lambda} x, \qquad y' = e^{-\lambda} y, \qquad z' = z. \quad (6.54)$$

Since these transformations coincide with those of Eq. (6.33) (provided we set $1 - 2|\gamma|^2 = e^{-\lambda}$ in that equation), the phase-damping channel is the same as the phase-flip channel.

Exercise 6.6 Check Eqs. (6.53) and (6.54).

Notice that, in the case in which the phase-damping channel is applied repeatedly, coherences drop to zero exponentially: $\alpha_n = e^{-\lambda n}\alpha$, where n denotes the number of applications of the channel. Similarly to what was discussed for the amplitude-damping channel, it is possible to give a continuous time version of the coherences decay. If $\lambda = \Gamma(\Delta t)$, $t = n(\Delta t)$ is time variable and we let $\Delta t \to 0$, then $\alpha(t) = e^{-\Gamma t}\alpha(0)$. Therefore, Γ represents the *decoherence rate* associated with this noise channel.

Exercise 6.7 Study the transformation of the Bloch sphere induced by the circuit of Fig. 6.10, where

$$D = \begin{bmatrix} C_0 & 0 & -S_0 & 0 \\ 0 & C_1 & 0 & -S_1 \\ S_0 & 0 & C_0 & 0 \\ 0 & S_1 & 0 & C_1 \end{bmatrix}, \quad (6.55)$$

with $C_i \equiv \cos\theta_i$, $S_i \equiv \sin\theta_i$, $(i = 0, 1)$, and

$$U = \exp\left[-i\frac{\xi}{2}(\boldsymbol{n} \cdot \boldsymbol{\sigma})\right], \quad (6.56)$$

where $\boldsymbol{n}$ is a unit vector and ξ a real number.

Exercise 6.8 Determine how many quantum-noise operations, characterized by 4×4 unitary matrices $U_1, U_2, \ldots$ (see Fig. 6.11) do we need to generate a generic affine map $\rho \to \rho' = \mathcal{S}(\rho)$?

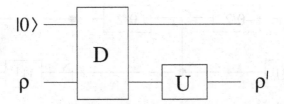

Fig. 6.10   A quantum circuit modelling the noise channel described in exercise 6.7.

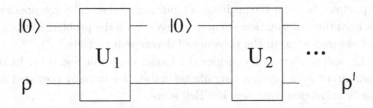

Fig. 6.11   A quantum circuit modelling the noise channel described in exercise 6.8.

Exercise 6.9   Show that the final density matrix ρ' in the quantum circuit of Fig. 6.12 is independent of the unitary matrix V (in this circuit, U is a generic 4×4 unitary matrix).

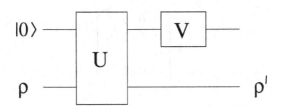

Fig. 6.12   A quantum circuit modelling the noise channel described in exercise 6.9.

Exercise 6.10   Study the quantum-noise operation implemented by the quantum circuit of Fig. 6.13, in the cases in which the unitary matrix U is given by (i) $U = \sigma_x$, (ii) $U = \frac{1}{\sqrt{2}}(I \pm i\sigma_j)$ $(j = x, y, z)$.

6.1.10   De-entanglement

Entanglement is arguably the most peculiar feature of quantum systems, with no analogue in classical mechanics. Furthermore, it is an important

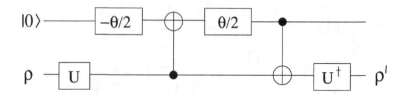

Fig. 6.13   A quantum circuit modelling the noise channel described in exercise 6.10.

physical resource for quantum communication and computation. It is therefore important, both for the problem of quantum to classical correspondence and for quantum information science, to investigate the problem of the stability of entanglement in the presence of decoherence effects.

In this section, we shall consider the model drawn in Fig. 6.14. In this quantum circuit, the two most significant qubits are initially prepared in a maximally entangled state, say the Bell state

$$|\psi^+\rangle = \tfrac{1}{\sqrt{2}}\big(|01\rangle + |10\rangle\big). \qquad (6.57)$$

This corresponds to the density matrix

$$\rho = \tfrac{1}{2}\big(|01\rangle\langle01| + |10\rangle\langle10| + |01\rangle\langle10| + |10\rangle\langle01|\big), \qquad (6.58)$$

whose matrix representation in the basis $\{|00\rangle, |01\rangle, |10\rangle, |11\rangle\}$ is

$$\rho = \tfrac{1}{2}\begin{bmatrix} 0 & 0 & 0 & 0 \\ 0 & 1 & 1 & 0 \\ 0 & 1 & 1 & 0 \\ 0 & 0 & 0 & 0 \end{bmatrix}. \qquad (6.59)$$

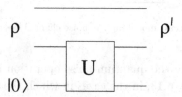

Fig. 6.14   A quantum circuit modelling de-entanglement (loss of the entanglement between the two upper qubits due to the coupling with the third qubit, which represents the environment).

To be concrete, we consider the case in which the interaction with the environment (*i.e.*, the third qubit) is modelled by the phase-flip channel described in Sec. 6.1.5. We have

$$\rho' = \sum_{i=0}^{1} F_i^{(2)} \rho \, (F_i^{(2)})^\dagger, \qquad (6.60)$$

where the Kraus operators $F_0^{(2)}$ and $F_1^{(2)}$ are defined as

$$F_0^{(2)} = I \otimes F_0 = \begin{bmatrix} 1 & 0 \\ 0 & 1 \end{bmatrix} \otimes \begin{bmatrix} 1 & 0 \\ 0 & \cos\theta \end{bmatrix} = \begin{bmatrix} 1 & 0 & 0 & 0 \\ 0 & \cos\theta & 0 & 0 \\ 0 & 0 & 1 & 0 \\ 0 & 0 & 0 & \cos\theta \end{bmatrix},$$

$$(6.61)$$

$$F_1^{(2)} = I \otimes F_1 = \begin{bmatrix} 1 & 0 \\ 0 & 1 \end{bmatrix} \otimes \begin{bmatrix} 0 & 0 \\ 0 & \sin\theta \end{bmatrix} = \begin{bmatrix} 0 & 0 & 0 & 0 \\ 0 & \sin\theta & 0 & 0 \\ 0 & 0 & 0 & 0 \\ 0 & 0 & 0 & \sin\theta \end{bmatrix},$$

F_0 and F_1 being the Kraus operators (6.34) introduced in Sec. 6.1.5 for the phase-flip channel. It can be seen by direct calculation that

$$\rho' = \frac{1}{2} \begin{bmatrix} 0 & 0 & 0 & 0 \\ 0 & 1 & \cos\theta & 0 \\ 0 & \cos\theta & 1 & 0 \\ 0 & 0 & 0 & 0 \end{bmatrix}. \qquad (6.62)$$

For $\cos\theta = 1$, we have $\rho' = \rho$; that is, the final state is identical to the initial, maximally entangled, Bell state. For $\cos\theta = 0$, the final state is separable. Indeed, we have

$$\rho' = \tfrac{1}{2}\big(|01\rangle\langle01| + |10\rangle\langle10|\big), \qquad (6.63)$$

and this density matrix corresponds to the statistical mixture of the separable states $|01\rangle$ and $|10\rangle$, taken with equal probabilities. In the intermediate case $0 < \cos\theta < 1$, we have partial loss of entanglement, corresponding to the partial loss of quantum coherence discussed in Sec. 6.1.5.

Exercise 6.11 Study the errors introduced in the teleportation and dense-coding protocols when the partially entangled state (6.62) is used instead of a Bell state.

It is interesting that, if we only consider a member of the Bell pair, its state is preserved with unit fidelity by the phase-flip channel. Indeed, the reduced density matrix describing this state is $\frac{1}{2}I$, which is not modified by this noise channel. On the other hand, as seen above, the Bell pair (6.59) is corrupted, even though the phase-flip noise acts only on a member of the pair (the second line in Fig. 6.14). We can intuitively understand this result by saying that it is more difficult to preserve both the state of a system and the entanglement of the system with the outside world (here, the other member of the Bell pair) than just the state of the system. This intuition can be formalized, see Schumacher (1995).

6.2   The master equation

The master equation describes the continuous temporal evolution of open quantum systems. In this section, we shall discuss two derivations of this equations. The first (Sec. 6.2.1) is based on a microscopic model and gives a clear physical picture of the approximations made to arrive at the master equation. The second (Sec. 6.2.2) clarifies the link between the master-equation approach and the quantum-operation formalism.

Before going into technical details, let us state the main approximations involved in deriving the master equation:

(i) *Born approximation* – The environment is large and practically unaffected by interaction with system.

(ii) *Markov approximation* – The system density matrix $\rho(t)$ evolves under a first-order differential equation in time. Therefore, the knowledge of the density matrix $\rho(t_0)$ at a given time t_0 is sufficient to determine $\rho(t)$ at any time $t > t_0$. We stress that this is a non-trivial requirement since the system interacts with the environment, and so, in general, the environmental state at time t_0 depends on the system density matrix $\rho(t')$ at earlier times $t' < t_0$. In other words, the environment acquires information on the system, but this information can flow back, at least in part, to the system. Therefore, the knowledge of the system density matrix $\rho(t_0)$ at time t_0 is in general not sufficient to determine $\rho(t)$ at later times. Indeed, we have

$$\begin{aligned}
\rho(t_0 + dt) &= \mathrm{Tr}_{\mathrm{env}}[\rho_{\mathrm{tot}}(t_0 + dt)] \\
&= \mathrm{Tr}_{\mathrm{env}}[U(t_0 + dt, t_0)\, \rho_{\mathrm{tot}}(t_0)\, U^\dagger(t_0 + dt, t_0)], \quad (6.64)
\end{aligned}$$

where ρ_{tot} is the density matrix of the system plus environment, whose evolution from time t_0 to time $t_0 + dt$ is driven by the unitary operator $U(t_0 + dt, t_0)$. As stated above, $\rho_{\text{tot}}(t_0)$ depends on $\rho(t)$, for all times $t \leq t_0$. This means that we cannot fully determine $\rho(t_0 + dt)$ from $\rho(t_0)$ alone. In the Markovian approximation, we assume that the environment is memoryless; that is, its state at time t_0 is essentially unaffected by the history of the system. This means that the information flow is essentially one-way, namely from the system to the environment. The Markovian approximation provides a good description of quantum noise if the memory of any effect that the system has on the environment is limited to a time scale much shorter than the time scales of interest for the dynamics of the system.

6.2.1 * *Derivation of the master equation*

Let us consider a system in interaction with the environment (also known as a *bath* or *reservoir*). The most general Hamiltonian describing this situation reads as follows:

$$H = H_S \otimes I_R + I_S \otimes H_R + H_{SR} \equiv H_0 + H_{SR}, \tag{6.65}$$

where H_S, H_R and H_{SR} describe the system, the reservoir and the interaction, respectively.

We call χ the density matrix describing the system plus reservoir, and

$$\rho = \text{Tr}_R \chi \tag{6.66}$$

the reduced density matrix describing the system. As we saw in Sec. 5.1, the evolution of χ is governed by the von Neumann equation:

$$i\hbar\dot{\chi} = [H, \chi]. \tag{6.67}$$

We assume that interaction is weak so that we shall be able to separate the fast motion, due to $H_0 = H_S + H_R$, from the slow motion, due to the interaction H_{SR}. For this purpose, we exploit the so-called *interaction picture*, defining

$$i\hbar\dot{U}_S = H_S U_S, \qquad U_S(0) = I_S,$$

$$i\hbar\dot{U}_R = H_R U_R, \qquad U_R(0) = I_R, \tag{6.68}$$

$$U = U_S \otimes U_R, \tag{6.69}$$

$$\tilde{\chi} = U^\dagger \chi U, \qquad \tilde{H}_{SR} = U^\dagger H_{SR} U. \tag{6.70}$$

After substitution of (6.68–6.70) into (6.67), we obtain

$$i\hbar\dot{\tilde{\chi}} = \left[\tilde{H}_{SR}, \tilde{\chi}\right], \tag{6.71}$$

which is equivalent to the integro-differential equation

$$\tilde{\chi}(t) = \tilde{\chi}(0) + \frac{1}{i\hbar}\int_0^t d\tau \left[\tilde{H}_{SR}(\tau), \tilde{\chi}(\tau)\right], \tag{6.72}$$

where $\tilde{\chi}(0) = \chi(0)$. We now insert this expression into the right-hand side of (6.71) and obtain

$$\dot{\tilde{\chi}}(t) = \frac{1}{i\hbar}\left[\tilde{H}_{SR}(t), \chi(0)\right] - \frac{1}{\hbar^2}\int_0^t d\tau \left[\tilde{H}_{SR}(t), \left[\tilde{H}_{SR}(\tau), \tilde{\chi}(\tau)\right]\right]. \tag{6.73}$$

Let us compute $\text{Tr}_R(\tilde{\chi})$. We have

$$\text{Tr}_R(\tilde{\chi}) = \text{Tr}_R(U^\dagger \chi U) = U_S^\dagger \text{Tr}_R(U_R^\dagger \chi U_R) U_S = U_S^\dagger \rho U_S \equiv \tilde{\rho}, \tag{6.74}$$

where we have used the definitions (6.66) and (6.69) and the cyclic property of the trace, leading to $\text{Tr}_R(U_R^\dagger \chi U_R) = \text{Tr}_R(\chi U_R U_R^\dagger) = \text{Tr}_R(\chi)$. We can now trace Eq. (6.73) over the environmental degrees of freedom and obtain

$$\dot{\tilde{\rho}}(t) = \frac{1}{i\hbar}\text{Tr}_R\left\{\left[\tilde{H}_{SR}(t), \chi(0)\right]\right\}$$
$$- \frac{1}{\hbar^2}\int_0^t d\tau\,\text{Tr}_R\left\{\left[\tilde{H}_{SR}(t), \left[\tilde{H}_{SR}(\tau), \tilde{\chi}(\tau)\right]\right]\right\}. \tag{6.75}$$

This equation is exact. We now need a few assumptions:

(i) First of all, we assume that at time $t = 0$ the system and the environment are not entangled:

$$\chi(0) = \rho(0) \otimes \rho_R, \tag{6.76}$$

where ρ and ρ_R denote the system and environment density matrices, respectively.

(ii) We also assume that

$$\text{Tr}_R\left\{\left[\tilde{H}_{SR}, \chi(0)\right]\right\} = 0. \tag{6.77}$$

If this is not the case, $\text{Tr}_R\{[\tilde{H}_{SR}, \chi(0)]\}$ is an operator acting on the system alone. Therefore, it is possible to show that we can always redefine H_S and H_{SR} (while keeping the global Hamiltonian constant) in order to fulfill Eq. (6.77).

(iii) *Born approximation* – We assume that the coupling is so weak and the reservoir so large that its state is essentially unaffected by the interaction. Therefore,

$$\tilde{\chi}(\tau) \approx \tilde{\rho}(\tau) \otimes \tilde{\rho}_R, \tag{6.78}$$

so that Eq. (6.75) becomes

$$\dot{\tilde{\rho}}(t) = -\frac{1}{\hbar^2} \int_0^t d\tau \, \mathrm{Tr}_R \left\{ \left[\tilde{H}_{SR}(t), \left[\tilde{H}_{SR}(\tau), \tilde{\rho}(\tau) \tilde{\rho}_R \right] \right] \right\}. \tag{6.79}$$

(iv) *Markov approximation* – In Eq. (6.79), we replace $\tilde{\rho}(\tau)$ with $\tilde{\rho}(t)$ and obtain

$$\dot{\tilde{\rho}}(t) = -\frac{1}{\hbar^2} \int_0^t d\tau \, \mathrm{Tr}_R \left\{ \left[\tilde{H}_{SR}(t), \left[\tilde{H}_{SR}(\tau), \tilde{\rho}(t) \tilde{\rho}_R \right] \right] \right\}. \tag{6.80}$$

It is important to point out that this equation is no longer integro-differential, but simply differential, since $\tilde{\rho}(t)$ appears on the right-hand side instead of $\tilde{\rho}(\tau)$.

We now expand H_{SR} over a basis of Hermitian operators $\{\sigma_i\}$ acting on the system:

$$H_{SR} = \sum_{i=0}^{M-1} \sigma_i B_i, \tag{6.81}$$

where the operators B_i act on the environment. Notice that, if the Hilbert space of the system has dimension N, we have $M = N^2$ since the $N \times N$ matrices constitute a linear vector space of dimension N^2. For instance, if $N = 2$, we can take $\sigma_0 = I$, $\sigma_1 = \sigma_x$, $\sigma_2 = \sigma_y$ and $\sigma_3 = \sigma_z$. We have

$$\tilde{H}_{SR}(t) = U^\dagger(t) \, H_{SR} \, U(t) = \sum_i \tilde{\sigma}_i(t) \, \tilde{B}_i(t), \tag{6.82}$$

where

$$\tilde{\sigma}_i(t) \equiv U_S^\dagger(t) \, \sigma_i \, U_S(t), \qquad \tilde{B}_i(t) \equiv U_R^\dagger(t) \, B_i \, U_R(t). \tag{6.83}$$

We now insert (6.82) and (6.83) into (6.80) and obtain

$$\dot{\tilde{\rho}}(t) = -\frac{1}{\hbar^2} \sum_{i,j} \int_0^t d\tau \, \mathrm{Tr}_R \left\{ \left[\tilde{\sigma}_i(t) \tilde{B}_i(t), \left[\tilde{\sigma}_j(\tau) \tilde{B}_j(\tau), \tilde{\rho}(t) \tilde{\rho}_R \right] \right] \right\}. \tag{6.84}$$

We expand the commutators on the right-hand side of this equation and obtain

$$\dot{\tilde{\rho}}(t) = -\frac{1}{\hbar^2} \sum_{i,j} \int_0^t d\tau \, \text{Tr}_R \left\{ \tilde{\sigma}_i(t)\tilde{B}_i(t)\tilde{\sigma}_j(\tau)\tilde{B}_j(\tau)\tilde{\rho}(t)\tilde{\rho}_R \right.$$

$$- \tilde{\sigma}_i(t)\tilde{B}_i(t)\tilde{\rho}(t)\tilde{\rho}_R\tilde{\sigma}_j(\tau)\tilde{B}_j(\tau) - \tilde{\sigma}_j(\tau)\tilde{B}_j(\tau)\tilde{\rho}(t)\tilde{\rho}_R\tilde{\sigma}_i(t)\tilde{B}_i(t)$$

$$\left. + \tilde{\rho}(t)\tilde{\rho}_R\tilde{\sigma}_j(\tau)\tilde{B}_j(\tau)\tilde{\sigma}_i(t)\tilde{B}_i(t) \right\}. \quad (6.85)$$

Taking advantage of the cyclic property of the trace, we have

$$\dot{\tilde{\rho}}(t) = -\frac{1}{\hbar^2} \sum_{i,j} \int_0^t d\tau \, \text{Tr}_R \left\{ \tilde{\sigma}_i(t)\tilde{\sigma}_j(\tau)\tilde{\rho}(t)\tilde{B}_i(t)\tilde{B}_j(\tau)\tilde{\rho}_R \right.$$

$$- \tilde{\sigma}_i(t)\tilde{\rho}(t)\tilde{\sigma}_j(\tau)\tilde{B}_j(\tau)\tilde{B}_i(t)\tilde{\rho}_R - \tilde{\sigma}_j(\tau)\tilde{\rho}(t)\tilde{\sigma}_i(t)\tilde{B}_i(t)\tilde{B}_j(\tau)\tilde{\rho}_R$$

$$\left. + \tilde{\rho}(t)\tilde{\sigma}_j(\tau)\tilde{\sigma}_i(t)\tilde{B}_j(\tau)\tilde{B}_i(t)\tilde{\rho}_R \right\}. \quad (6.86)$$

After defining

$$\tilde{\Gamma}_{ij}(t,\tau) \equiv \text{Tr}_R \left\{ \tilde{B}_i(t)\tilde{B}_j(\tau)\tilde{\rho}_R \right\}, \quad (6.87)$$

we obtain

$$\dot{\tilde{\rho}}(t) = -\frac{1}{\hbar^2} \sum_{i,j} \int_0^t d\tau \left\{ \tilde{\sigma}_i(t)\tilde{\sigma}_j(\tau)\tilde{\rho}(t)\tilde{\Gamma}_{ij}(t,\tau) - \tilde{\sigma}_j(t)\tilde{\rho}(t)\tilde{\sigma}_i(\tau)\tilde{\Gamma}_{ij}(t,\tau) \right.$$

$$\left. - \tilde{\sigma}_j(\tau)\tilde{\rho}(t)\tilde{\sigma}_i(t)\tilde{\Gamma}_{ij}(t,\tau) + \tilde{\rho}(t)\tilde{\sigma}_i(\tau)\tilde{\sigma}_j(t)\tilde{\Gamma}_{ij}(t,\tau) \right\}. \quad (6.88)$$

We now assume that the bath correlation functions $\tilde{\Gamma}_{ij}$ are memoryless; namely,

$$\tilde{\Gamma}_{ij}(t,\tau) = \tilde{\gamma}_{ij}\delta(t-\tau). \quad (6.89)$$

This implies that

$$\dot{\tilde{\rho}}(t) = -\frac{1}{\hbar^2} \sum_{i,j} \frac{\tilde{\gamma}_{ij}}{2} \left\{ \tilde{\sigma}_i(t)\tilde{\sigma}_j(t)\tilde{\rho}(t) - \tilde{\sigma}_j(t)\tilde{\rho}(t)\tilde{\sigma}_i(t) \right.$$

$$\left. - \tilde{\sigma}_j(t)\tilde{\rho}(t)\tilde{\sigma}_i(t) + \tilde{\rho}(t)\tilde{\sigma}_i(t)\tilde{\sigma}_j(t) \right\}. \quad (6.90)$$

Finally, we use Eq. (6.68) and define $\gamma_{ij} \equiv U\tilde{\gamma}_{ij}U^\dagger$ to derive the master equation

$$\dot{\rho} = -\frac{i}{\hbar}[H_S, \rho] + \frac{1}{\hbar^2}\sum_{i,j}\frac{\gamma_{ji}}{2}\Big\{[\sigma_i, \rho\sigma_j] + [\sigma_i\rho, \sigma_j]\Big\}. \qquad (6.91)$$

6.2.2 * The master equation and quantum operations

It is instructive to derive the master equation within the framework of the quantum-operation formalism. In the Kraus representation, the density matrices $\rho(t)$ and $\rho(t+dt)$, which describe the system at times t and $t+dt$, are related as follows:

$$\rho(t+dt) = \mathcal{S}(t; t+dt)\rho(t) = \sum_{k=0}^{M-1} E_k\,\rho(t)\,E_k^\dagger, \qquad (6.92)$$

where $\mathcal{S}(t; t+dt)$ is the superoperator mapping $\rho(t)$ into $\rho(t+dt)$ and the operators E_k are the Kraus operators, whose number M is $\leq N^2$ (N is the dimension of the Hilbert space). Note that the operators E_k in (6.92), in contrast with Sec. 5.4, refer to infinitesimal transformations. In order to assure that $\mathcal{S}(t; t)$ is equal to the identity, we may write the Kraus operators as follows:

$$\begin{aligned} E_0 &= I + \frac{1}{\hbar}(-iH + K)\,dt, \\ E_k &= L_k\sqrt{dt}, \quad (k = 1, \dots, M-1), \end{aligned} \qquad (6.93)$$

where H and K are Hermitian operators, and the operators L_k are known as the *Lindblad operators*. The normalization condition $\sum_k E_k^\dagger E_k = I$ gives

$$\left[I + \frac{1}{\hbar}(iH + K)dt\right]\left[I + \frac{1}{\hbar}(-iH + K)dt\right] + \sum_{k=1}^{M-1} L_k^\dagger L_k\,dt = I, \qquad (6.94)$$

that is,

$$\frac{2}{\hbar}K\,dt + \sum_{k=1}^{M-1} L_k^\dagger L_k\,dt + O((dt)^2) = 0. \qquad (6.95)$$

Therefore,

$$K = -\frac{\hbar}{2}\sum_{k=1}^{M-1} L_k^\dagger L_k. \qquad (6.96)$$

We now insert (6.93) and (6.96) into (6.92) and obtain

$$\rho(t + dt) = \rho(t) - \frac{i}{\hbar} \left[H, \rho(t) \right] dt$$
$$+ \sum_{k=1}^{M-1} \left(L_k \rho(t) L_k^\dagger - \tfrac{1}{2} L_k^\dagger L_k \rho(t) - \tfrac{1}{2} \rho(t) L_k^\dagger L_k \right) dt + O\big((dt)^2\big). \quad (6.97)$$

If we assume that

$$\rho(t + dt) = \rho(t) + \dot{\rho}(t) dt + O\big((dt)^2\big), \quad (6.98)$$

we obtain the GKSL (Gorini, Kossakowski, Sudarshan and Lindblad) master equation (see Gorini *et al.*, 1976 and Lindblad, 1976):

$$\dot{\rho} = -\frac{i}{\hbar} [H, \rho] + \sum_{k=1}^{M-1} \left(L_k \rho L_k^\dagger - \tfrac{1}{2} L_k^\dagger L_k \rho - \tfrac{1}{2} \rho L_k^\dagger L_k \right). \quad (6.99)$$

We should stress that the expansion (6.98) is possible under the Markovian approximation previously discussed. We should also point out that the quantum-operation formalism is more general than the master-equation approach. Indeed, a quantum process described in terms of an operator-sum representation is, in general, non-Markovian and therefore cannot be described by means of a Markovian master equation.

Exercise 6.12 As an example application of the GKSL master equation, we consider a two-level atom in a thermal radiation field. In such a situation, the master equation reads (see, *e.g.*, Gardiner and Zoller, 2000)

$$\dot{\rho} = -\frac{i}{\hbar} [H, \rho] + \gamma \left(\bar{n} + 1 \right) \left(\sigma_+ \rho \sigma_- - \tfrac{1}{2} \sigma_- \sigma_+ \rho - \tfrac{1}{2} \rho \sigma_- \sigma_+ \right)$$
$$+ \gamma \bar{n} \left(\sigma_- \rho \sigma_+ - \tfrac{1}{2} \sigma_+ \sigma_- \rho - \tfrac{1}{2} \rho \sigma_+ \sigma_- \right), \quad (6.100)$$

where

$$H = \tfrac{1}{2} \hbar \omega_0 \sigma_z, \quad (6.101)$$

so that ω_0 is the frequency of the radiation that the atom will emit or absorb, $\sigma_+ = \frac{1}{2}(\sigma_x + i\sigma_y)$ and $\sigma_- = \frac{1}{2}(\sigma_x - i\sigma_y) = \sigma_+^\dagger$ are the so-called raising and lowering operators and $\bar{n}$ represents the mean occupation number at temperature T (we have $\bar{n} = 1/[\exp(\hbar\omega_0/k_B T) - 1]$, where k_B is the Boltzmann constant). Note that in (6.100) the Lindblad operators are $L_1 = \sqrt{\gamma(\bar{n} + 1)}\, \sigma_+$ and $L_2 = \sqrt{\gamma \bar{n}}\, \sigma_-$. While L_1 drives the transition

$|1\rangle \rightarrow |0\rangle$, L_2 induces the jump $|0\rangle \rightarrow |1\rangle$. Solve the master equation (6.100). In particular, discuss the approach to equilibrium.

It is possible to show that the master equations (6.91) and (6.99) are equivalent. First of all, for the expansion (6.81) we choose a basis $\{\sigma_0, \ldots, \sigma_{N^2-1}\}$ such that $\sigma_0 = I/\sqrt{N}$, $\mathrm{Tr}(\sigma_i) = 0$, $\mathrm{Tr}(\sigma_i^\dagger \sigma_j) = \delta_{ij}$ $(i, j = 1, \ldots, N^2 - 1)$. In this basis, the master equation can be expressed in the form given by Gorini $et\ al.$ (1976):

$$\dot{\rho} = -\frac{i}{\hbar}[H, \rho] + \frac{1}{2}\sum_{i,j=1}^{N^2-1} A_{ij}\left\{[\sigma_i, \rho\sigma_j^\dagger] + [\sigma_i\rho, \sigma_j^\dagger]\right\}, \qquad (6.102)$$

where $H = H^\dagger$ and A is a positive complex matrix. The term $-(i/\hbar)[H, \rho]$ describes the Hamiltonian part in the evolution of the density matrix, while the other terms in the right-hand side of (6.102) describe dissipation and decoherence. Note that H is not necessarily the same as the system Hamiltonian since it may include a non-dissipative contribution coming from the interaction with the environment.

Let us introduce the matrix-valued vectors

$$\boldsymbol{\sigma} \equiv \begin{bmatrix} \sigma_1 \\ \vdots \\ \sigma_{N^2-1} \end{bmatrix}, \qquad \boldsymbol{w} \equiv \begin{bmatrix} w_1 \\ \vdots \\ w_{N^2-1} \end{bmatrix}, \qquad (6.103)$$

where

$$\boldsymbol{w}^\dagger \equiv S\boldsymbol{\sigma}^\dagger \qquad (6.104)$$

and the matrix S is such that

$$\tilde{A} \equiv SAS^\dagger \qquad (6.105)$$

is diagonal. Let $\{\lambda_i\}$ denote the eigenvalues of A. Since A is a positive matrix, $\lambda_i \geq 0$ for all i. We order these eigenvalues in such a manner that $\lambda_i > 0$ for $i = 1, \ldots, M$ $(M \leq N^2 - 1)$ and $\lambda_i = 0$ for $i = M+1, \ldots, N^2 - 1$. After defining the vector

$$\boldsymbol{L} \equiv \begin{bmatrix} L_1 \\ \vdots \\ L_{N^2-1} \end{bmatrix} \equiv \begin{bmatrix} \sqrt{\lambda_1}w_1^\dagger \\ \vdots \\ \sqrt{\lambda_{N^2-1}}w_{N^2-1}^\dagger \end{bmatrix}, \qquad (6.106)$$

Eq. (6.102) reduces to the GKSL master equation (6.99), $L_1, \ldots, L_M$ being the Lindblad operators.

Exercise 6.13   Using Eqs. (6.103–6.106), show the equivalence of the master equations (6.102) and (6.99).

6.2.3   *The master equation for a single qubit*

In this section, we study, in the Markovian approximation, the most general evolution of the density matrix for a single qubit. The master equation (6.102) is given by

$$\dot{\rho}(t) = -\frac{i}{\hbar}\,[H, \rho(t)] + \frac{1}{2}\sum_{i,j=1}^{3} A_{ij}\{[\sigma_i, \rho(t)\sigma_j] + [\sigma_i\rho(t), \sigma_j]\}, \qquad (6.107)$$

where the $\{\sigma_i\}$ are the Pauli matrices ($\sigma_1 = \sigma_x$, $\sigma_2 = \sigma_y$, $\sigma_3 = \sigma_z$; $\sigma_i^\dagger = \sigma_i$) and A is a Hermitian matrix. We assume that both the Hamiltonian H and the matrix A are time-independent. In other words, we assume that the environment is stationary and not modified by the interaction with the system. As shown in Sec. 6.2.1, the parameters A_{ij} can be identified with the bath correlation functions.

The Hamiltonian H describes the reversible part of the qubit dynamics. It can be written as follows:

$$H = \frac{\hbar}{2}\left(\omega_0\sigma_3 + \Delta\sigma_1 + \Delta'\sigma_2\right). \qquad (6.108)$$

Therefore, the Hamiltonian part of the evolution depends on the three real parameters ω_0, Δ and Δ'. Since the dissipative part of the evolution is governed by the 3×3 Hermitian matrix A, it depends on 9 real parameters. Therefore, the evolution (6.107) of the single-qubit density matrix is determined by 12 independent real parameters. These parameters correspond to the 12 parameters appearing in the Kraus representation of the most general quantum-noise process acting on a single qubit.

It is useful to gain an intuitive understanding of the effect of these parameters on the evolution of the single-qubit density matrix. For this purpose, we employ the Bloch-sphere representation, in which

$$\rho(t) = \frac{1}{2}\begin{bmatrix} 1+z(t) & x(t)-iy(t) \\ x(t)+iy(t) & 1-z(t) \end{bmatrix} = \frac{1}{2}\left[I + \boldsymbol{r}(t)\cdot\boldsymbol{\sigma}\right], \qquad (6.109)$$

where $\boldsymbol{r}(t) = \big(x(t), y(t), z(t)\big)$. We can derive a first-order differential equation for the Bloch vector:

$$\dot{\boldsymbol{r}}(t) = M\boldsymbol{r}(t) + \boldsymbol{c}. \qquad (6.110)$$

Indeed, if we insert (6.109) into (6.107), for the Hamiltonian part of the evolution we obtain

$$
\begin{bmatrix} \dot{x} \\ \dot{y} \\ \dot{z} \end{bmatrix}_H = \begin{bmatrix} 0 & -\omega_0 & \Delta' \\ \omega_0 & 0 & -\Delta \\ -\Delta' & \Delta & 0 \end{bmatrix} \begin{bmatrix} x \\ y \\ z \end{bmatrix}, \tag{6.111}
$$

and, for the dissipative part,

$$
\begin{bmatrix} \dot{x} \\ \dot{y} \\ \dot{z} \end{bmatrix}_D = \begin{bmatrix} -2(A_{22}+A_{33}) & (A_{12}+A_{21}) & (A_{13}+A_{31}) \\ (A_{12}+A_{21}) & -2(A_{33}+A_{11}) & (A_{23}+A_{32}) \\ (A_{13}+A_{31}) & (A_{23}+A_{32}) & -2(A_{11}+A_{22}) \end{bmatrix} \begin{bmatrix} x \\ y \\ z \end{bmatrix}
$$
$$
+ \begin{bmatrix} 2i(A_{23}-A_{32}) \\ 2i(A_{31}-A_{13}) \\ 2i(A_{12}-A_{21}) \end{bmatrix}. \tag{6.112}
$$

After introducing the new parameters

$$
\gamma_1 = 2(A_{22}+A_{33}), \quad \gamma_2 = 2(A_{33}+A_{11}), \quad \gamma_3 = 2(A_{11}+A_{22}),
$$
$$
\alpha = (A_{12}+A_{21}), \quad \beta = (A_{13}+A_{31}), \quad \gamma = (A_{23}+A_{32}), \tag{6.113}
$$
$$
c_1 = 2i(A_{23}-A_{32}), \quad c_2 = 2i(A_{31}-A_{13}), \quad c_3 = 2i(A_{12}-A_{21}),
$$

we obtain

$$
\begin{bmatrix} \dot{x} \\ \dot{y} \\ \dot{z} \end{bmatrix} = \begin{bmatrix} -\gamma_1 & \alpha-\omega_0 & \beta+\Delta' \\ \alpha+\omega_0 & -\gamma_2 & \gamma-\Delta \\ \beta-\Delta' & \gamma+\Delta & -\gamma_3 \end{bmatrix} \begin{bmatrix} x \\ y \\ z \end{bmatrix} + \begin{bmatrix} c_1 \\ c_2 \\ c_3 \end{bmatrix}, \tag{6.114}
$$

which is of the form (6.110), with $\boldsymbol{c} = (c_1, c_2, c_3)$ and

$$
M = M_H + M_D,
$$

$$
M_H = \begin{bmatrix} 0 & -\omega_0 & \Delta' \\ \omega_0 & 0 & -\Delta \\ -\Delta' & \Delta & 0 \end{bmatrix}, \quad M_D = \begin{bmatrix} -\gamma_1 & \alpha & \beta \\ \alpha & -\gamma_2 & \gamma \\ \beta & \gamma & -\gamma_3 \end{bmatrix}. \tag{6.115}
$$

The matrix M_H, corresponding to the Hamiltonian part of the master equation (6.107), generates unitary evolution. The matrix M_D, corresponding to the dissipative part of this equation, is symmetric and can therefore be diagonalized. Its eigenvalues give the contraction rates of the

Bloch sphere along the directions identified by the corresponding eigenvectors. Hence, M_D deforms the Bloch sphere into an ellipsoid. The term $\boldsymbol{c}$ in (6.110) induces a rigid shift of the Bloch sphere. It is clear from (6.110) that the evolution of the Bloch vector over an infinitesimal time dt is given by

$$\boldsymbol{r}(t+dt) = (I + M\,dt)\boldsymbol{r}(t) + \boldsymbol{c}\,dt + O\big((dt)^2\big). \qquad (6.116)$$

This is an affine map. The evolution of the Bloch vector in a time t can be obtained by applying the infinitesimal evolution (6.116) $\frac{t}{dt}$ times, in the limit $dt \to 0$. Since the composition of two affine maps is again an affine map, also the generic evolution of the Bloch vector in a finite time interval is an affine map. This conclusion allows us to obtain a precise correspondence between the 12 parameters appearing in the single-qubit master equation and the 12 parameters needed to characterize a generic quantum operation acting on a two-level system.

We now discuss a few special cases:

(i) $M_D = 0$, $\boldsymbol{c} = \boldsymbol{0}$ (Hamiltonian case).

Equation (6.114) reduces to

$$\dot{x} = -\omega_0 y + \Delta' z, \quad \dot{y} = \omega_0 x - \Delta z, \quad \dot{z} = -\Delta' x + \Delta y. \qquad (6.117)$$

The solution of these equations corresponds to a rotation of the Bloch sphere about the axis

$$\boldsymbol{n} = \left(\frac{\Delta}{\sqrt{\omega_0^2 + \Delta^2 + \Delta'^2}}, \frac{\Delta'}{\sqrt{\omega_0^2 + \Delta^2 + \Delta'^2}}, \frac{\omega_0}{\sqrt{\omega_0^2 + \Delta^2 + \Delta'^2}} \right),$$
$$\qquad (6.118)$$

with frequency $\Omega = \sqrt{\Delta^2 + \Delta'^2 + \omega_0^2}$.

Exercise 6.14 Solve Eq. (6.117).

(ii) M_D diagonal, $M_H = 0$, $\boldsymbol{c} = \boldsymbol{0}$.

Equation (6.114) becomes

$$\dot{x} = -\gamma_1 x, \qquad \dot{y} = -\gamma_2 y, \qquad \dot{z} = -\gamma_3 z. \qquad (6.119)$$

These equations are readily solved, and we obtain

$$x(t) = x(0)e^{-\gamma_1 t}, \quad y(t) = y(0)e^{-\gamma_2 t}, \quad z(t) = z(0)e^{-\gamma_3 t}. \qquad (6.120)$$

Therefore, the Bloch sphere collapses exponentially fast onto its centre.

(iii) M_D diagonal, $M_H = 0$, $\mathbf{c} \neq \mathbf{0}$.

The differential equations that govern the evolution of the Bloch vector are given by

$$\dot{x} = -\gamma_1 x + c_1, \qquad \dot{y} = -\gamma_2 y + c_2, \qquad \dot{z} = -\gamma_3 z + c_3. \qquad (6.121)$$

The solution is

$$x(t) = x(0)e^{-\gamma_1 t} + \frac{c_1}{\gamma_1}\left(1 - e^{-\gamma_1 t}\right),$$

$$y(t) = y(0)e^{-\gamma_2 t} + \frac{c_2}{\gamma_2}\left(1 - e^{-\gamma_2 t}\right), \qquad (6.122)$$

$$z(t) = z(0)e^{-\gamma_3 t} + \frac{c_3}{\gamma_3}\left(1 - e^{-\gamma_3 t}\right).$$

As in the previous case, the Bloch sphere shrinks exponentially fast onto a single point. However, this point is no longer the centre of the Bloch sphere but has coordinates $\left(\frac{c_1}{\gamma_1}, \frac{c_2}{\gamma_2}, \frac{c_3}{\gamma_3}\right)$.

6.3 Quantum to classical transition

6.3.1 *Schrödinger's cat*

The problem of the emergence of classical behaviour in a world governed by the laws of quantum mechanics has fascinated scientists since the dawn of quantum theory. The heart of the problem is the superposition principle, which entails consequences that appear unacceptable according to classical intuition. This point is clearly elucidated by Schrödinger's cat paradox. Inside a box we have a radioactive source, a detector, a hammer, a vial of poison and a cat. The source is a two-level atom, initially in its excited state $|1\rangle$. The atom can decay to the ground state $|0\rangle$ by emission of a photon, which triggers the detector. The click of the detector induces the hammer to break the vial of poison and kill the cat. We assume that initially the state of the composite atom–cat system is

$$|\psi_0\rangle = |1\rangle|\text{live}\rangle. \qquad (6.123)$$

Since the poison kills the cat if the atom decays to the state $|0\rangle$, we obtain, after a time corresponding to the half-life of the atom, the state

$$|\psi\rangle = \frac{1}{\sqrt{2}}\left(|1\rangle|\text{live}\rangle + |0\rangle|\text{dead}\rangle\right), \qquad (6.124)$$

which is a superposition of the live- and dead-cat states. We emphasize that the cat and the atom are now entangled. Let us consider the density matrix of the state (6.124):

$$\rho = |\psi\rangle\langle\psi| = \tfrac{1}{2}\Big(|1\rangle|\text{live}\rangle\langle 1|\langle\text{live}| + |0\rangle|\text{dead}\rangle\langle 0|\langle\text{dead}|$$

$$+ |1\rangle|\text{live}\rangle\langle 0|\langle\text{dead}| + |0\rangle|\text{dead}\rangle\langle 1|\langle\text{live}|\Big). \quad (6.125)$$

In the basis $\{|0\rangle|\text{live}\rangle, |0\rangle|\text{dead}\rangle, |1\rangle|\text{live}\rangle, |1\rangle|\text{dead}\rangle\}$ we have

$$\rho = \tfrac{1}{2}\begin{bmatrix} 0 & 0 & 0 & 0 \\ 0 & 1 & 1 & 0 \\ 0 & 1 & 1 & 0 \\ 0 & 0 & 0 & 0 \end{bmatrix}. \quad (6.126)$$

This density matrix contains non-zero matrix elements not only along the diagonal but also off-diagonal. These latter elements, known as *coherences*, have no classical analogue.

Decoherence plays a key role in understanding the transition from the quantum to classical world. The atom–cat system is never perfectly isolated from the environment, so that, instead of the state (6.124), we must consider the state

$$|\Psi\rangle = \tfrac{1}{\sqrt{2}}\Big(|1\rangle|\text{live}\rangle|E_1\rangle + |0\rangle|\text{dead}\rangle|E_0\rangle\Big), \quad (6.127)$$

where $|E_0\rangle$ and $|E_1\rangle$ are states of the environment. If $|E_0\rangle$ and $|E_1\rangle$ are orthogonal, then, after tracing over the environment, we obtain a diagonal density matrix:

$$\rho_{\text{dec}} = \tfrac{1}{2}\Big(|1\rangle|\text{live}\rangle\langle 1|\langle\text{live}| + |0\rangle|\text{dead}\rangle\langle 0|\langle\text{dead}|\Big). \quad (6.128)$$

This diagonal density matrix corresponds to a mixed state and is compatible with a classical description of the system in terms of probabilities. The cat is dead with probability $p = 1/2$ and alive with the same probability, and we discover its state upon observation. Note that this situation is different from that described by (6.125). In that case, the atom–cat system is in a non-classical superposition state and only collapses onto a "classical" state (corresponding to the live or dead cat) after a measurement.

6.3.2 *Decoherence and destruction of cat states*

In this subsection, by means of a simple model, we shall show that a very weak interaction with the environment can lead to very fast coherence decay. These studies are of interest not only to understand the quantum to classical correspondence but also from the viewpoint of quantum computation. Since a quantum computer is never perfectly isolated from the external world, it is important to estimate the degree of isolation required to reliably implement a given quantum algorithm.

Let us first consider a free particle moving along a line. The wave function $\psi(x)$ describing such a system resides in the infinite-dimensional Hilbert space $\mathcal{L}^2(\mathrm{R})$.[2] A particle localized at x_0 is described by a Gaussian wave packet:

$$\psi(x) = \langle x|\psi\rangle = \frac{1}{\sqrt{\sqrt{\pi}\,\delta}} \exp\left[-\frac{(x-x_0)^2}{2\delta^2}\right]. \qquad (6.129)$$

It is easy to check (see exercise 6.15) that, for any state such, the mean values of position and momentum are

$$\langle x \rangle = x_0, \qquad\qquad \langle p \rangle = 0, \qquad (6.130)$$

and the variances are

$$\langle (\Delta x)^2 \rangle = \langle (x - \langle x \rangle)^2 \rangle \qquad \langle (\Delta p)^2 \rangle = \langle (p - \langle p \rangle)^2 \rangle$$
$$= \frac{\delta^2}{2}, \qquad\qquad\qquad = \frac{\hbar^2}{2\delta^2}. \qquad (6.131)$$

In the momentum-space representation the wave function $\psi(p)$ is given by

---

[2] $\mathcal{L}^2(\mathrm{R})$ is the space of the functions $f(x)$ such that $\int_{-\infty}^{+\infty} dx\,|f(x)|^2 < +\infty$. In this space the inner product of two vectors f_1 and f_2 is defined as $\langle f_1|f_2\rangle = \int_{-\infty}^{+\infty} dx\, f_1^\star(x) f_2(x)$. Any wave vector $|\psi\rangle$ has unit norm; namely,

$$\||\psi\rangle\| = \int_{-\infty}^{+\infty} dx\,|\langle x|\psi\rangle|^2 = \int_{-\infty}^{+\infty} dx\,|\psi(x)|^2 = 1.$$

If the system is described by the wave function $\psi(x)$, the average value of any observable O is given by

$$\langle O \rangle = \int_{-\infty}^{+\infty} dx\, \psi^\star(x) O \psi(x).$$

the Fourier transform of $\psi(x)$ and again has a Gaussian shape:

$$\psi(p) = \langle p|\psi \rangle = \frac{1}{\sqrt{2\pi\hbar}} \int_{-\infty}^{+\infty} dx \exp\left(-\frac{ipx}{\hbar}\right) \psi(x)$$

$$= \sqrt{\frac{\delta}{\sqrt{\pi}\hbar}} \exp\left(-\frac{\delta^2 p^2}{2\hbar^2}\right) \exp\left(-\frac{ipx_0}{\hbar}\right). \qquad (6.132)$$

The density matrix corresponding to the Gaussian wave packet (6.129) reads

$$\langle x|\rho|x' \rangle = \langle x|\psi \rangle \langle \psi|x' \rangle = \frac{1}{\sqrt{\pi}\delta} \exp\left(-\frac{(x-x_0)^2 + (x'-x_0)^2}{2\delta^2}\right) \qquad (6.133)$$

and is drawn in Fig. 6.15.

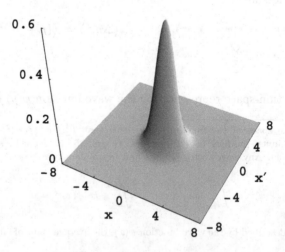

Fig. 6.15 The density matrix corresponding to a Gaussian wave packet centred at $x_0 = 0$. In this and in the other figures of this subsection we set $\delta = 1$.

Exercise 6.15 Check that the Gaussian wave packet

$$\psi(x,t) = \frac{1}{\sqrt{\sqrt{\pi}\left(1+i\frac{\hbar t}{m\delta^2}\right)\delta}} \exp\left\{-\frac{\left(x-x_0-\frac{p_0 t}{m}\right)^2}{2\delta^2\left(1+i\frac{\hbar t}{m\delta^2}\right)}\right.$$
$$\left.+\frac{i}{\hbar}\left[p_0(x-x_0)-\frac{p_0^2}{2m}t\right]\right\} \quad (6.134)$$

is solution of the Schrödinger equation for a free particle moving in one dimension:

$$i\hbar\frac{\partial}{\partial t}\psi(x,t) = H\psi(x,t) = -\frac{\hbar^2}{2m}\frac{\partial^2}{\partial x^2}\psi(x,t). \quad (6.135)$$

Show that

$$\langle x\rangle = x_0 + \frac{p_0}{m}t, \qquad \langle p\rangle = p_0; \quad (6.136)$$

namely, that the wave packet moves with constant velocity p_0/m. Finally, check that

$$\langle(\Delta x)^2\rangle = \frac{\delta^2}{2}\left[1+\frac{\hbar^2 t^2}{m^2\delta^4}\right], \qquad \langle(\Delta p)^2\rangle = \frac{\hbar^2}{2\delta^2}. \quad (6.137)$$

This implies that

$$\Delta x\,\Delta p = \frac{\hbar}{2}\sqrt{1+\frac{\hbar^2 t^2}{m^2\delta^4}}, \quad (6.138)$$

where we have defined $\Delta x = \sqrt{\langle(\Delta x)^2\rangle}$ and $\Delta p = \sqrt{\langle(\Delta p)^2\rangle}$. Therefore, (6.134) is a minimum uncertainty wave packet ($\Delta x\Delta p = \hbar/2$) only at time $t=0$.

We now consider the superposition of two Gaussian wave packets centred at $+x_0$ and $-x_0$, respectively. We assume that the distance $2x_0$ between these two packets is much larger than their width δ. These states are known as *cat states*, for a reason that will soon become clear. If $\psi_+(x)$ and $\psi_-(x)$ denote the two Gaussian packets, we have

$$\psi_{\text{cat}}(x) \equiv \frac{1}{\sqrt{2}}\left[\psi_+(x)+\psi_-(x)\right]$$
$$= \frac{1}{\sqrt{2\sqrt{\pi}\delta}}\left\{\exp\left[-\frac{(x-x_0)^2}{2\delta^2}\right]+\exp\left[-\frac{(x+x_0)^2}{2\delta^2}\right]\right\}. \quad (6.139)$$

An example probability distribution for a cat state is shown in Fig. 6.16. The corresponding density matrix, drawn in Fig. 6.17, has four components:

$$\langle x|\rho_{\text{cat}}|x'\rangle = \langle x|\psi\rangle_{\text{cat cat}}\langle\psi|x'\rangle = \psi_{\text{cat}}(x)\psi_{\text{cat}}^{\star}(x') = \psi_{\text{cat}}(x)\psi_{\text{cat}}(x')$$
$$= \tfrac{1}{2}\left[\psi_{+}(x)\psi_{+}(x') + \psi_{-}(x)\psi_{-}(x') + \psi_{+}(x)\psi_{-}(x') + \psi_{-}(x)\psi_{+}(x')\right].$$
$$(6.140)$$

The peaks along the diagonal ($x = x'$) correspond to the two possible locations of the particle, $x = x_0$ or $x = -x_0$. The off-diagonal peaks ($x = -x'$) are purely quantum and demonstrate that the particle is neither localized in x_0 nor in $-x_0$. We have a coherent superposition of the states $\psi_{+}(x)$ and $\psi_{-}(x)$; as in Schrödinger's cat paradox, we have a superposition of the states $|1\rangle|\text{live}\rangle$ and $|0\rangle|\text{dead}\rangle$.

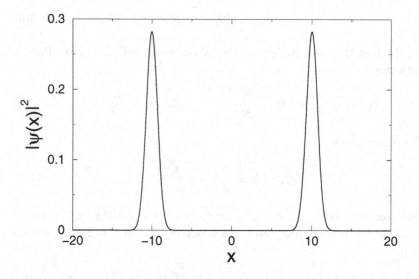

Fig. 6.16   The probability distribution for the cat state (6.139), with $x_0 = 10\delta$.

In order to investigate the physical origin of decoherence, it is useful to compare the Fourier transforms of a Gaussian and of a cat state. For the Gaussian wave packet (6.129) we have

$$\left|\langle p|\psi\rangle\right|^2 = \frac{\delta}{\sqrt{\pi}\hbar}\exp\left(-\frac{\delta^2 p^2}{\hbar^2}\right),\qquad (6.141)$$

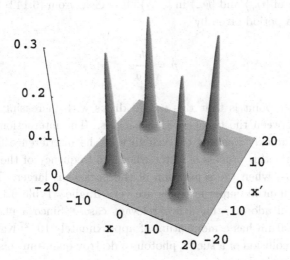

Fig. 6.17 The density matrix corresponding to the cat state (6.139), with $x_0 = 10\delta$.

while for a cat state we obtain

$$\left|\langle p|\psi\rangle_{\text{cat}}\right|^2 = \frac{2\delta}{\sqrt{\pi}\hbar} \exp\left(-\frac{\delta^2 p^2}{\hbar^2}\right) \cos^2\left(\frac{px_0}{\hbar}\right). \qquad (6.142)$$

Both (6.141) and (6.142) are shown in Fig. 6.18.

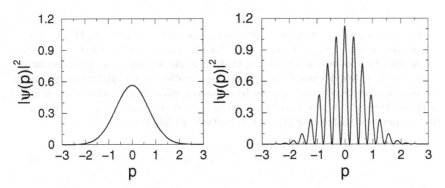

Fig. 6.18 The probability distributions $|\psi(p)|^2$ of a Gaussian state (left) and a cat state (right), with $x_0 = 10\delta$. In this figure we set $\delta = \hbar = 1$.

We emphasize the presence of interference fringes in the case of the cat state. These fringes are of pure quantum origin and are due to the coherent superposition of $|\psi_+\rangle$ and $|\psi_-\rangle$ in $|\psi_{\text{cat}}\rangle$. It is clear from (6.142) that these fringes have a period given by

$$\tilde{p} = \frac{h}{x_0}. \tag{6.143}$$

The important point is that this period drops with increasing the separation $2x_0$ between the two Gaussian packets. The interaction with the environment quickly weakens the visibility of the interference fringes. It is intuitive that this process is faster when the frequency of the fringes is higher; that is, when the separation of the packets is larger. Therefore, "non-local" quantum superpositions are very fragile. Table 6.1 gives the order of magnitude of $\tilde{p}$ for a few relevant cases. Since a photon with wavelength 600 nm has a momentum of approximately 10^{-27} Kg m/s, it is sufficient the collision of a such a photon to destroy quantum coherences in the cases shown in Table 6.1.

Table 6.1   Relevant orders of magnitude for different cat states.

system	mass	x_0	$\tilde{p}(\text{Kg m/s})$
atom	30 a.m.u.	600 nm	10^{-27}
dust particle	10^{-9} g	10^{-2} cm	7×10^{-30}
cat	1 Kg	10 cm	7×10^{-33}

It is instructive to present a simple microscopic model illustrating the destruction of cat states (here we follow Cohen-Tannoudji, unpublished lecture notes). To simplify the discussion, we assume that the particle subjected to the decoherence process is heavy enough to neglect the variation of its kinetic energy over the decoherence time scale. We model a system–environment interaction event as the scattering of a heavy particle (the system) with a light particle. We write the composite initial state of these two particles as follows:

$$\Phi_i(\boldsymbol{X}, \boldsymbol{x}) = \psi(\boldsymbol{X}) \otimes \phi_{\boldsymbol{p}_i}(\boldsymbol{x}), \tag{6.144}$$

where $\boldsymbol{X}$ and $\boldsymbol{x}$ denote the positions of the heavy and the light particles, moving in three-dimensional space, and $\boldsymbol{p}_i$ is the initial momentum of the light particle. Let us describe the scattering of these two particles. For this purpose, it is convenient to write the Fourier expansion of the wave function $\psi(X)$:

$$\psi(X) = \frac{1}{(2\pi\hbar)^{3/2}} \int dP \exp\left(\frac{i\, P \cdot X}{\hbar}\right) \tilde{\psi}(P). \qquad (6.145)$$

We insert (6.145) into (6.144) and obtain

$$\Phi_i(X, x) = \frac{1}{(2\pi\hbar)^{3/2}} \int dP \exp\left(\frac{i\, P \cdot X}{\hbar}\right) \tilde{\psi}(P) \otimes \phi_{p_i}(x). \qquad (6.146)$$

The scattering of the two particles changes momenta: $P_i = P \to P_f$ and $p_i \to p_f$. Momentum is conserved, so that

$$P_i + p_i = P_f + p_f. \qquad (6.147)$$

Therefore, the effect of scattering on the wave function (6.146) is described by the following transformation:

$$\exp\left(\frac{i\, P \cdot X}{\hbar}\right) \otimes \phi_{p_i}(x) \to \exp\left(\frac{i\,(P + p_i - p_f) \cdot X}{\hbar}\right) \otimes \phi_{p_f}(x). \qquad (6.148)$$

In order to obtain the final state of the composite system $\Phi_f(X, x)$, we must integrate over all possible states after scattering. We have

$$\Phi_f(X, x) = \frac{1}{(2\pi\hbar)^{3/2}} \int dP \int dp_f A(p_f, p_i) \exp\left(\frac{i\, P \cdot X}{\hbar}\right) \tilde{\psi}(P)$$

$$\times \exp\left(\frac{i\,(p_i - p_f) \cdot X}{\hbar}\right) \otimes \phi_{p_f}(x)$$

$$= \psi(X) \int dp_f A(p_f, p_i) \exp\left(\frac{i\,(p_i - p_f) \cdot X}{\hbar}\right) \otimes \phi_{p_f}(x), \qquad (6.149)$$

where we have assumed that the transition amplitude $A(p_f, p_i)$ is independent of the state of the heavy particle. It is important to stress that the collision has transformed the separable state (6.144) into an entangled state. In other words, the system is now entangled with the environment. The key point is that, while the global (system plus environment) final state $\Phi_f(X, x)$ is a pure state, this is not the case for the state of the system alone. Therefore, it must be described by means of a density matrix, obtained after tracing over the environmental degrees of freedom:

$$\langle X'|\rho_f|X''\rangle = \int dx \, \Phi_f(X',x)\Phi_f^*(X'',x)$$

$$= \int dx \, \psi(X')\psi^*(X'') \int dp'_f \, A(p'_f,p_i)\exp\left(\frac{i\,(p_i - p'_f)\cdot X'}{\hbar}\right) \otimes \phi_{p'_f}(x)$$

$$\times \int dp''_f \, A^*(p''_f,p_i)\exp\left(-\frac{i\,(p_i - p''_f)\cdot X''}{\hbar}\right) \otimes \phi_{p''_f}^*(x)$$

$$= \langle X'|\rho_i|X''\rangle \int dp_f \, |A(p_f,p_i)|^2 \exp\left(\frac{i\,(p_i - p_f)\cdot (X' - X'')}{\hbar}\right), \qquad (6.150)$$

where we have used the orthogonality relation

$$\int dx \, \phi_{p'_f}(x)\phi_{p''_f}^*(x) = \delta(p'_f - p''_f) \qquad (6.151)$$

and factored out the initial density matrix of the system:

$$\langle X'|\rho_i|X''\rangle = \psi(X')\psi^*(X''). \qquad (6.152)$$

We must evaluate the integral appearing in the last line of (6.150). First of all, we assume that the collision is elastic, so that energy is conserved; that is,

$$\frac{P_i^2}{2M} + \frac{p_i^2}{2m} = \frac{P_f^2}{2M} + \frac{p_f^2}{2m}, \qquad (6.153)$$

where M and m are the masses of the heavy and the light particle, respectively. It is reasonable to assume that the kinetic energy of the heavy particle is essentially unchanged ($\frac{P_i^2}{2M} \approx \frac{P_f^2}{2M}$). Thus, $\frac{p_i^2}{2m} = \frac{p_f^2}{2m}$ and so $|p_i| \approx |p_f|$. This implies that the integral in the last line of (6.150) averages to zero when $|X' - X''| \gg \hbar/|p_i|$, as in this case its argument oscillates rapidly. Therefore, the matrix elements of ρ_f for $|X' - X''| \gg \hbar/|p_i|$ are eliminated after a single scattering event. This means that we may limit ourselves to the case $|X' - X''| \ll \hbar/|p_i|$. In this limit we may expand the exponent appearing in (6.150) as follows:

$$\exp\left(\frac{i\,(p_i - p_f)\cdot (X' - X'')}{\hbar}\right) \approx 1 + \frac{i\,(p_i - p_f)\cdot (X' - X'')}{\hbar}$$

$$- \frac{1}{2}\left[\frac{(p_i - p_f)\cdot (X' - X'')}{\hbar}\right]^2. \qquad (6.154)$$

We now insert this expression into (6.150). The first term gives

$$\langle X'|\rho_i|X''\rangle \int dp_f \, |A(p_f,p_i)|^2 = \langle X'|\rho_i|X''\rangle. \qquad (6.155)$$

We can transfer $\langle X'|\rho_i|X''\rangle$ to the left-hand side of (6.150) to define the variation of the density matrix due to the collision as follows:

$$\delta\langle X'|\rho|X''\rangle \equiv \langle X'|\rho_f|X''\rangle - \langle X'|\rho_i|X''\rangle. \tag{6.156}$$

Assuming invariance of A with respect to reflections, namely $A(p_f, p_i) = A(-p_f, -p_i)$, we see that the linear term in (6.154) does not contribute to the integral in (6.150). Finally, we insert the second-order term of (6.154) into (6.150) and obtain

$$\delta\langle X'|\rho|X''\rangle \approx -\tfrac{1}{2} \int dp_f \, |A(p_i, p_f)|^2 \left[\frac{(p_i - p_f)}{\hbar} \cdot (X' - X'')\right]^2 \langle X'|\rho_i|X''\rangle$$

$$\propto |X' - X''|^2 \langle X'|\rho_i|X''\rangle. \tag{6.157}$$

Therefore, given a cat state, the interaction with the environment drops coherences and leaves the diagonal terms of the density matrix practically unchanged, corresponding to "classical" probability distributions.

6.4 * Decoherence and quantum measurements

The role of measurement is to convert quantum states into classical outcomes. In this section, we shall discuss the role of decoherence in the quantum measurement process. This issue is important in the problem of the transition from quantum physics to the classical world. Moreover, it is of interest in quantum information processing since any quantum protocol must end with a measurement.

We seek a purely unitary model of measurement that does not require the collapse of the wave packet. In this model, the first stage, known as *premeasurement*, is to establish *correlations* between the system and the measurement apparatus. In order to clarify this point, we consider a simple example, sketched in Fig. 6.19. A one-qubit system, prepared in a generic pure state $|\psi\rangle_S = \alpha|0\rangle_S + \beta|1\rangle_S$, interacts with a measurement apparatus, initially in the state $|0\rangle_A$. We assume that this interaction induces a CNOT gate. Therefore, the premeasurement process maps the initial state

$$|\Phi_0\rangle = |\psi\rangle_S|0\rangle_A = (\alpha|0\rangle_S + \beta|1\rangle_S)|0\rangle_A \tag{6.158}$$

into the state

$$|\Phi\rangle = \alpha|00\rangle_{SA} + \beta|11\rangle_{SA}, \tag{6.159}$$

corresponding to the density matrix

$$\rho_{SA} = \begin{bmatrix} |\alpha|^2 & 0 & 0 & \alpha\beta^* \\ 0 & 0 & 0 & 0 \\ 0 & 0 & 0 & 0 \\ \alpha^*\beta & 0 & 0 & |\beta|^2 \end{bmatrix}. \tag{6.160}$$

We stress that this final state is entangled. This means that purely quantum correlations between the system and the measurement apparatus have been established. As a consequence, in the density matrix there are non-zero off-diagonal terms.

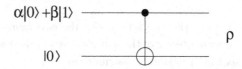

Fig. 6.19 A single-qubit premeasurement. The top line represents the system, the bottom line the measurement apparatus.

The presence of entanglement in the state (6.159) engenders ambiguity in the measurement process. The problem arises if we wish to associate the possible states of the apparatus with those appearing in (6.159), $|0\rangle_A$ and $|1\rangle_A$. At first glance, it would be tempting to say that the measured quantity is σ_z and that the possible outcomes are ± 1, corresponding to the states $|0\rangle_A$ and $|1\rangle_A$. However, this interpretation is not correct. To illustrate this point, let us take $\alpha = \beta = \frac{1}{\sqrt{2}}$ in (6.159). In this case we may write

$$|\Phi\rangle = \tfrac{1}{\sqrt{2}}\big(|00\rangle + |11\rangle\big) = \tfrac{1}{\sqrt{2}}\big(|++\rangle + |--\rangle\big), \tag{6.161}$$

where $|\pm\rangle = \frac{1}{\sqrt{2}}(|0\rangle \pm |1\rangle)$ are the eigenstates of σ_x. Therefore, the above interpretation would lead to an ambiguity in the definition of the measured quantity. Moreover, the direction of the information flow is not uniquely determined either. The action of the CNOT gate is

$$\text{CNOT}\,(|x\rangle_S\,|y\rangle_A) = |x\rangle_S\,|y \oplus x\rangle_A, \tag{6.162}$$

where $x, y = 0, 1$. Since the first qubit is unchanged, the direction of the information transfer is from the first qubit to the second qubit. Thus, it appears reasonable to identify the first (control) qubit with the system and the second (target) qubit with the measurement apparatus. However, this identification is not always correct since we have

$$\text{CNOT}\,(|\pm\rangle_S\,|+\rangle_A) = |\pm\rangle_S\,|+\rangle_A,$$
$$\text{CNOT}\,(|\pm\rangle_S\,|-\rangle_A) = |\mp\rangle_S\,|-\rangle_A. \tag{6.163}$$

Thus, for the states of the $\{|+\rangle, |-\rangle\}$ basis the second qubit is unchanged while the first qubit is flipped when the state of the second is $|-\rangle$ (this is the backward sign propagation discussed in Sec. 3.4). This means that, while the information on the observable σ_z flows from the first to the second qubit, the information on σ_x travels from the second to the first qubit. If we require the information to alway goes from the system to the apparatus, then we must identify the first qubit with the system and the second qubit with the measurement apparatus in (6.162) and *vice versa* in (6.163).

The above ambiguities are resolved if we take into account the interaction of the measurement apparatus with the environment. For instance, we can represent the environment as a third qubit, interacting with the apparatus by means of a CNOT gate (see Fig. 6.20). The system-apparatus density matrix is obtained after tracing over the environmental degrees of freedom. In this example we obtain

$$\rho_{SA} = \begin{bmatrix} |\alpha|^2 & 0 & 0 & 0 \\ 0 & 0 & 0 & 0 \\ 0 & 0 & 0 & 0 \\ 0 & 0 & 0 & |\beta|^2 \end{bmatrix}. \tag{6.164}$$

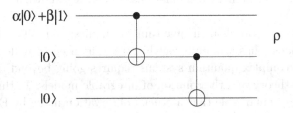

Fig. 6.20 A single-qubit premeasurement followed by a system–environment interaction process. The lines represent the system (top), the measurement apparatus (middle) and the environment (bottom).

In this density matrix, quantum correlations have disappeared, and we may interpret the density matrix (6.164) as follows. There exist *classical correlations* between the states $|0\rangle_S$ and $|0\rangle_A$ as well as between the states

$|1\rangle_S$ and $|1\rangle_A$: if the apparatus is in the state $|0\rangle_A$, we know that the system is in the state $|0\rangle_S$; on the contrary, if the apparatus is in the state $|1\rangle_A$, the system is in the state $|1\rangle_S$. We emphasize that the density matrix (6.164) is diagonal in a *preferential basis* whose states (known as *pointer states*) are determined by the form of the apparatus–environment interaction. It is the CNOT interaction of the apparatus with the environment that determines the preferential basis in which the density matrix (6.164) is diagonal. A different apparatus–environment interaction would determine a different preferential basis. Therefore, according to the pointer-state theory, it is the interaction with the environment that determines which observable is measured by the apparatus (see Zurek, 2003).

6.5   * Quantum chaos

The discovery of classical dynamical chaos is now recognized as a major scientific achievement of the past century. More than three hundred years after the establishment of Newton's laws, we finally attained a reasonable understanding of the qualitative features of the solutions of the equations of motion for non-linear systems. Indeed, apart from the pioneering contribution of scientists such as Henri Poincaré and a few others, it was only after the work of Kolmogorov and Fermi–Pasta–Ulam in the fifties that the scientific community gradually recognized the important role of what are now known as *non-linear dynamical systems* or *complex systems*. It has to be noted that in the growth of this field a major role is being played by computers, which allow numerical simulations (sometimes known as "numerical experiments").

The analogous problem in quantum mechanics, sometimes known as "quantum chaos", has a more recent history. The need to understand the behaviour of complex quantum systems requires going beyond traditional perturbation theory and the solution of integrable models. In this connection it is quite remarkable that a crucial observation made by Einstein in 1917 had passed unnoticed for more than half a century. Einstein noted that the quantization rules introduced by Niels Bohr in 1913 were only applicable to systems for which there existed invariant tori in classical mechanics. However, invariant tori do not exist for most systems and therefore Bohr's quantization rules cannot be applied. Indeed, quantum mechanics has been developed on the basis of integrable systems such as the harmonic oscillator and the hydrogen atom. On the other hand such systems are the exception rather than the rule and the typical behaviour of classical dynamical systems is instead chaotic. Therefore, the understanding of the quantum behaviour of systems that are chaotic in the classical limit is very important and is the subject of the present section. The problem of the quantum to classical transition is naturally posed within this context. A related question is: how accurate can the description of natural phenomena based on classical mechanics be?

6.5.1 * Dynamical chaos in classical mechanics

As once stated by Sommerfeld, for a better understanding of a quantum problem it is advisable to study its classical counterpart. This is particularly true for quantum chaos. Therefore, in this section we briefly discuss dynamical chaos in classical mechanics. Dynamical chaos destroys the deterministic image of classical physics and shows that, typically, the trajectories of the deterministic equations of motion are, in some sense, *random and unpredictable*. Such classical behaviour is rooted in the *exponential local instability of motion*. For Hamiltonian (non-dissipative) systems the local instability is described by the linearized equations of motion

$$
\begin{cases}
\dot{\xi} = \left(\dfrac{\partial^2 H}{\partial q \partial p}\right)_r \xi + \left(\dfrac{\partial^2 H}{\partial p^2}\right)_r \eta, \\[4mm]
\dot{\eta} = -\left(\dfrac{\partial^2 H}{\partial q^2}\right)_r \xi - \left(\dfrac{\partial^2 H}{\partial q \partial p}\right)_r \eta,
\end{cases}
\tag{6.165}
$$

where $H = H(q, p, t)$ is the Hamiltonian, $(q, p) \equiv (q_1, \ldots, q_f, p_1, \ldots, p_f)$ are the coordinates of the $2f$-dimensional phase space and $\xi = dq$, $\eta = dp$ are f-dimensional vectors in the tangent space (f is the number of degrees of freedom). The coefficients $(\cdots)_r$ of the linear equations (6.165) are taken along the reference trajectory and therefore depend on time.

An important quantity, characterizing the stability of the motion along the reference trajectory, is the so-called (*maximum*) *Lyapunov exponent* λ, which is defined as the limit

$$
\lambda = \lim_{|t| \to \infty} \frac{1}{|t|} \ln\left(\frac{d(t)}{d(0)}\right),
\tag{6.166}
$$

with $d^2(t) = \xi^2(t) + \eta^2(t)$ the length of the tangent vector at time t. Positivity of the Lyapunov exponent ($\lambda > 0$) implies exponential instability of the motion: two nearby trajectories separate exponentially, with a rate given by λ. Note that, in Hamiltonian systems the instability does not depend on the direction of time and is therefore reversible, as is the chaotic motion.

The reason why the exponentially unstable motion is said to be chaotic is that almost all orbits, though deterministic, are unpredictable (see also Sec. 1.4), in the framework of fixed finite precision for specifying initial data and performing calculations. Indeed, according to the Alekseev–Brudno theorem (see Alekseev and Jacobson, 1981), in the algorithmic theory of dynamical systems the *information* $I(t)$ associated with a segment of trajectory of length t is equal, asymptotically, to

$$\lim_{|t| \to \infty} \frac{I(t)}{|t|} = h, \qquad (6.167)$$

where h is the Kolmogorov–Sinai entropy (see Sec. 5.14.3), which is positive when $\lambda > 0$.[3] This means that in order to predict a new segment of a chaotic trajectory one needs an amount of additional information proportional to the length of the segment and independent of the previous length of the trajectory. In such a situation information cannot be extracted from the observation of the past history of the motion. If, on the other hand, the instability is not exponential but follows a power-law, the information required per unit time is *inversely* proportional to the previous length of the trajectory and, asymptotically, prediction becomes possible. Of course, for a sufficiently short time interval prediction is possible even for a chaotic system and can be characterized by the randomness parameter

$$r = \frac{h|t|}{|\ln \mu|}. \qquad (6.168)$$

Here μ is the accuracy of trajectory recording, so that $|\ln \mu|$ gives the amount of information (number of digits) necessary to specify the state of the system at a given time. The prediction is possible in the *finite* interval of "temporary determinism" ($r < 1$), while $r \gg 1$ corresponds to the *infinite* region of asymptotic randomness. Note that in chaotic systems prediction is possible up to a time that scales only *logarithmically* with the accuracy μ: we have $r < 1$ when $|t| < \frac{1}{h}|\ln \mu|$.

Exponential instability implies a *continuous spectrum of motion*.[4] The continuous spectrum, in turn, implies *correlation decay*. This property, which is known as *mixing* in ergodic theory, may be seen as the basis of

---

[3] As stated in Eq. (5.262), for conservative systems the Kolmogorov–Sinai entropy is given by the sum of all positive Lyapunov exponents.

[4] The spectrum of motion is also known as the *power spectrum*. If $x(t)$ is a dynamical variable, such as position $q(t)$ or momentum $p(t)$, the power spectrum $S(\omega)$ is defined

a *statistical description* of dynamical systems. The point is that mixing provides the statistical independence of different parts of a dynamical trajectory, sufficiently separated in time. This is the main condition for the application of probability theory, which allows the calculation of statistical characteristics such as diffusion or relaxation to equilibrium.[5]

Dynamical chaos represents a limiting case of classical motion. The opposite limiting case is given by completely *integrable* f-freedom systems. By definition, an integrable Hamiltonian system is defined as one having as many single-valued, analytic constants of motion $\Phi_i(q, p, t) = C_i$ as degrees of freedom in involution (that is, all pairwise Poisson brackets $[\Phi_i, \Phi_j] = 0$). For such systems there exists a single-valued, analytic, canonical transformation bringing the Hamiltonian $H(q, p, t)$ into the form $H(I) \equiv H(I_1, \ldots, I_f)$, that is, a function of the new momenta (known as *actions*) alone. The equations of motion are then trivial to integrate and lead to trajectories winding around an f-torus with f discrete frequencies. The resulting motion is, in general, *quasi-periodic* and a cloud of points draws out a bundle of trajectories such that *close-by points only separate linearly.*[6]

---

as the Fourier transform of the *autocorrelation function* $C(\tau)$ of $x(t)$:

$$C(\tau) = \lim_{T \to \infty} \frac{1}{T} \int_0^T dt\, x(t) x(t + \tau),$$

$$S(\omega) = \frac{1}{2\pi} \int_{-\infty}^{+\infty} d\tau\, C(\tau) e^{-i\omega\tau} = \frac{1}{2\pi} \left| \lim_{T \to \infty} \frac{1}{T} \int_0^T dt\, x(t) e^{-i\omega t} \right|^2.$$

For an integrable system the solution $x(t)$ can be written as a multiply periodic function:

$$x(t) = \sum_{m_1, \ldots, m_f} x_{m_1, \ldots, m_f} e^{i(m_1 \omega_1 + \cdots + m_f \omega_f)t},$$

where the m_i are integers. Notice that, since $x(t)$ is real, we have $x_{-m_1, \ldots, -m_f} = x^\star_{m_1, \ldots, m_f}$. Inserting the above expression into the previous, we obtain

$$S(\omega) = \sum_{m_1, \ldots, m_f} |x_{m_1, \ldots, m_f}|^2 \delta(\omega - m_1 \omega_1 - \ldots - m_f \omega_f),$$

that is, a set of sharp lines at the fundamental frequencies $\omega_1, \ldots, \omega_f$ and their linear combinations with integer coefficients $m_1, \ldots, m_f$. For a chaotic system the function $S(\omega)$ is instead continuous.

[5]On the other hand, we point out that exponential instability is not necessary for a meaningful statistical description: examples of systems with linear instability have been found, which exhibit normal diffusion (see Li *et al.*, 2004).

[6]Clearly, the motion is not ergodic on the $(2f - 1)$-dimensional energy surface. Yet it may be ergodic on the f-torus, provided the f frequencies are incommensurate. We recall that ergodicity means that almost all trajectories cover the energy surface homogeneously; that is, the sojourn time of a trajectory in any region of the energy surface is proportional to the invariant measure of that region. This property implies that temporal

The property of complete integrability is very delicate and atypical as it is, in general, destroyed by an arbitrarily weak perturbation that converts a completely integrable system into a KAM-integrable system (after Kolmogorov, Arnold and Moser).[7] We stress that, unlike integrable motion, chaotic motion is typically very robust: it is *structurally stable*; that is, small perturbations do not qualitatively alter the system behaviour.[8]

Instead of using trajectories (in what might be called the "Newton picture") classical dynamics can be described in terms of distribution functions in phase space (the "Liouville picture"). Distribution functions obey the *linear* Liouville equation. Therefore, the condition of non-linearity for dynamical chaos to occur refers to the description in terms of trajectories. In terms of distribution functions the property of mixing implies the approach, on average (that is, after *coarse graining*, as explained below), of any initially smooth distribution to a steady state. This process is known as *statistical relaxation*. The time-reversibility of the distribution function is related to its very complicated structure, which becomes more and more "scarred" as the relaxation proceeds. In the case of exponential instability of motion the spatial scale of the oscillations of the distribution function decreases exponentially with time. It is in these fine spatial oscillations that the memory of the initial state is retained. To remove these complicated structures, the distribution function must be coarse-grained; that is, averaged over some domain. The evolution of the coarse-grained function is described by a kinetic equation, *e.g.*, a diffusion equation. The coarse-grained function converges to a smooth steady state.

In closing this subsection, we wish to stress the two crucial properties of classical mechanics necessary for dynamical chaos to occur: (i) a continuous spectrum of motion and (ii) a continuous phase space.

6.5.2   * *Quantum chaos and the correspondence principle*

As we saw in the previous subsection, a well developed ergodic theory exists for classical systems with a finite number f of degrees of freedom (see also

---

and ensemble averages converge to the same mean but it is not sufficient for a statistical description of motion since relaxation of a given distribution to some statistical steady state is not guaranteed. To this end the mixing property is necessary.

[7]The structure of KAM motion is very intricate: the motion is confined to invariant tori for most initial conditions yet a single, connected, chaotic motion component (for $f > 2$) of exponentially small measure (with respect to the perturbation) arises, which is nevertheless everywhere dense (see Arnold, 1997).

[8]Strictly speaking, this property is only valid for the so-called Anosov flows, see Lichtenberg and Lieberman (1992).

Fig. 6.21). This theory allows a fairly good understanding of the statistical properties of such systems.

A corresponding theory is completely lacking for the following important classes of systems:

(i) Classical systems with an infinite number of degrees of freedom, such as matter interacting with an electromagnetic field (*e.g.*, the problem of black-body radiation). The main difficulty here is that the two limits $|t| \to \infty$ and $f \to \infty$ do not commute.

(ii) Quantum systems with a finite number f of degrees of freedom. The difficulty here is that the two limits $|t| \to \infty$ and $\hbar_{\text{eff}} \to 0$ do not commute (see Fig. 6.21), where $\hbar_{\text{eff}} = \hbar/I$ is the effective Planck constant of the system, I being a characteristic value of the action variable. The classical limit is obtained when $\hbar_{\text{eff}} \to 0$.

(iii) Quantum systems with infinite degrees of freedom.

In the following we shall only consider systems of class (ii); namely, we shall present a short introduction to so-called "quantum chaos"

The problem of quantum chaos grew out of attempts to understand the very peculiar phenomenon of classical dynamical chaos in terms of quantum mechanics. The distinction between regular and chaotic motion survives quantization, even though the distinction criteria change. In particular, the alternative of exponential or power-law divergence of trajectories disappears in quantum mechanics, Heisenberg's uncertainty principle forbidding the notion of trajectories. Conversely, as we shall discuss below, the quantum mechanics of systems that are chaotic in the classical limit is characterized by genuine quantum phenomena, such as *quantum dynamical localization* and *level repulsion*, as in random-matrix theory.

The essential conditions for classical chaos, discussed in the previous subsection, are violated in quantum mechanics. Indeed, the energy and the frequency spectrum of any quantum motion, bounded in phase space, are always *discrete*. As a result, the motion is always almost periodic and, according to the existing theory of dynamical systems, corresponds to the limiting case of *regular motion*. The ultimate origin of this fundamental quantum property is the *discreteness of phase space* itself or, in modern mathematical language, a non-commutative geometry of the latter: The uncertainty principle implies a finite size of an elementary phase-space cell: $\Delta q \Delta p \geq \hbar$. On the other hand the *correspondence principle* requires the transition from quantum to classical mechanics for all phenomena, including dynamical chaos. How can the correspondence principle be reconciled with

GENERAL THEORY OF DYNAMICAL SYSTEMS

Finite number f of freedoms

$$H(\theta, I, t) = H_0(I) + \epsilon V(\theta, I, t)$$

Asymptotic ergodic theory $|t| \to \infty$

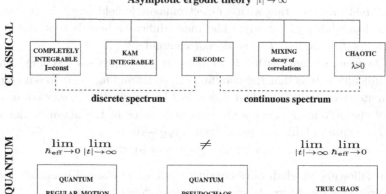

Fig. 6.21 Classical ergodic theory and the place of quantum chaos: I, θ are action-angle variables, λ the Lyapunov exponent, $\hbar_{\text{eff}}$ the dimensionless Planck constant. Notice that ergodicity is compatible with both discrete or continuous spectrum. Quantum pseudochaos takes place for finite f, t and $\hbar_{\text{eff}}$.

a discrete quantum energy spectrum when the limit is to be chaotic and thus characterized by a frequency continuum? The answer to this question must lie in the existence of $\hbar_{\text{eff}}$-*dependent time scales* ($\hbar_{\text{eff}}$ being the effective Planck constant) and, equivalently, energy scales. For quantum features to become manifest one must resolve discrete energy levels (whose spacings vanish as $\hbar_{\text{eff}} \to 0$); that is, sustain observation times that diverge in the limit $\hbar_{\text{eff}} \to 0$. Quantum chaos possesses all the properties of classical dynamical chaos but only on finite time scales that grow indefinitely in the classical limit. In other words, the quantum to classical correspondence for chaotic phenomena can be understood from the observation that the distinction between continuous and discrete spectra only becomes sharp in the limit $|t| \to \infty$.[9]

---

[9]It is interesting to note that the same mechanism of transient chaos works in the case of any (*e.g.*, classical) linear wave or even in the case of completely integrable dynamical systems. In this sense, quantum chaos can be viewed as a particular case of a phenomenon known as *pseudochaos* (or *finite-time dynamical chaos*) and is different from the "true" dynamical chaos defined in existing ergodic theory. It is important to

Note that the entire problem of quantum mechanics is in general divided into two qualitatively different aspects:

(i) proper quantum dynamics, as described by a specific dynamical variable, the wave function $|\psi(t)\rangle$;

(ii) quantum measurement, including the recording of the result and hence the collapse of the wave function.

The first aspect is described by a *deterministic* equation such as the Schrödinger equation and naturally belongs to the general theory of dynamical systems. In the following, only the first aspect will be discussed. At any rate, we wish to point out that the absence of classical-like chaos in quantum dynamics is true for the first aspect only. As far as the result is concerned, quantum measurement is fundamentally a random process.

6.5.3 * Time scales of quantum chaos

In order to understand the main features of quantum chaos it is convenient to consider simple models that nevertheless exhibit the rich variety and complexity typical of general non-linear systems. Particularly useful are area-preserving maps, as these maps may be very conveniently handled in computer simulations. To illustrate the time scales of quantum chaos, we shall consider the so-called *kicked rotator*, also known as the *Chirikov standard map*. The relevant Hamiltonian reads

$$H(I,\theta,\tau) = \tfrac{1}{2}I^2 + k\cos\theta \sum_{m=-\infty}^{+\infty} \delta(\tau - mT), \qquad (6.169)$$

where (I,θ) are conjugate momentum–angle variables. The expression "kicked rotator" refers to a particular physical interpretation of this model as a rotator with angular momentum I, driven by a series of periodic pulses (the kicks). The corresponding equations of motion reduce to the standard map

$$\begin{cases} \bar{I} = I + k\sin\theta, \\ \bar{\theta} = \theta + T\bar{I}, \end{cases} \qquad (6.170)$$

which gives the new variables $(\bar{I},\bar{\theta})$ after one period T of the perturbation. The standard-map model may be studied on either the infinite cylinder (unbounded motion, $-\infty < I < +\infty$) or a finite torus (bounded motion)

---

point out that, in contrast to classical linear waves, the linearity of quantum evolution is not an approximation but a fundamental physical property.

of length $2\pi L$ (along the rescaled variable $J = TI$), with L integer to avoid discontinuities being introduced into (6.170).

The quantized standard map is obtained by means of the usual quantization rules: $\theta \to \theta$ and $I \to I = -i\partial/\partial\theta$ (we set $\hbar = 1$). The quantum evolution in one map iteration is described by a unitary operator U, known as the Floquet operator, acting on the wave function ψ:

$$\bar{\psi} = U\,\psi = e^{-iTI^2/2}\,e^{-ik\cos\theta}\,\psi. \qquad (6.171)$$

Since $[\theta, J] = [\theta, TI] = iT$, the effective Planck constant is given by $\hbar_{\text{eff}} = \hbar/k$ and the classical limit corresponds to $k \to \infty$ and $T \to 0$ while keeping $K = kT$ constant.

The kicked-rotator model on the torus can also be considered as the Poincaré surface-of-section map for a *conservative* system with two degrees of freedom. What makes the standard map almost universal is the local (in momentum) approximation it provides for a broad class of more complicated physical models. For example, as we shall discuss in Sec. 6.5.5, the excitation and ionization of a hydrogen atom under a microwave field can be approximated, in the quasiclassical regime, by map (6.170). Finally, the quantum kicked-rotator model is studied experimentally with cold atoms in a pulsed optical lattice, created by laser fields (the first experiment was performed by Moore *et al.*, 1995).

Note that the kicked rotator belongs to the class of periodically driven dynamical systems described in Sec. 3.15.3 and therefore its classical dynamics depends only on the parameter $K = kT$. For $K \gg 1$ the classical motion may be considered ergodic, mixing and exponentially unstable with Lyapunov exponent $\lambda \approx \ln(K/2)$, negligibly small stability islands apart. In particular, the rescaled momentum variable $J = TI$ displays a random walk type of motion and, for K larger than the chaos border $K_c \approx 1$, exhibits normal diffusion:

$$\langle (\Delta J)^2 \rangle = \langle (J - \langle J \rangle)^2 \rangle \approx D(K)\,t, \qquad (6.172)$$

where $t = \tau/T$ measures the time in units of map iterations and the diffusion coefficient $D(K) = C(K)\frac{K^2}{2}$. Here the function $C(K)$ accounts for dynamical correlations. In particular, $C(K) \to 0$ when $K \to K_c$ and $C(K) \to 1$ for $K \gg 1$ (random phase approximation). As discussed in Sec. 3.15.3, the evolution of a distribution function $f(J, t)$ at $K \gg K_c$ is governed by the Fokker–Planck equation. Therefore, starting from a distribution initially peaked at $J = J_0$ (*i.e.*, $f(J, 0) = \delta(J - J_0)$), we obtain a Gaussian distribution at time t, whose width is given by $\sqrt{D(K)t}$ (see Eq. (3.204)). These conclusions are valid for the standard map taken on the infinite cylinder. When the motion is bounded on a torus with $0 \leq J < 2\pi L$, the diffusion leads to the statistical relaxation of any non-singular distribution function to the uniform steady state $f_s(J) = \frac{1}{2\pi L}$. The relaxation time scale is estimated by the diffusion time

$$t_d \sim \frac{L^2}{D(K)}. \tag{6.173}$$

In order to understand the existence of different time scales in the quantum motion, we compare the classical and quantum evolution starting from the same initial conditions. According to the *Ehrenfest theorem*, a quantum wave packet follows a beam of classical orbits as long as the packet remains narrow. During this time interval the quantum wave-packet motion is exponentially unstable and random as is the underlying classical trajectory. However, the initial size of the quantum wave packet is bounded from below by the elementary quantum phase-space cell of order $\hbar$. Let us start from an initial minimum-uncertainty wave packet of size $\Delta\theta_0 \Delta J_0 = T\Delta\theta_0 \Delta I_0 \sim T\hbar = \hbar_{\text{eff}}$, with $\Delta\theta_0 \sim \Delta J_0 \sim \sqrt{\hbar_{\text{eff}}}$ (this choice corresponds to the optimal, least-spreading wave packet). Then $\Delta\theta$ grows exponentially due to classical exponential instability: $\Delta\theta(t) \sim \Delta\theta_0 \exp(\lambda t) \sim \sqrt{\hbar_{\text{eff}}} \exp(\lambda t)$, with λ the maximum Lyapunov exponent of the system. Therefore, complete spreading over the angle variable θ is obtained after the so-called *Ehrenfest* (or *random*) *time scale*

$$t_E \sim \frac{1}{\lambda} \big| \ln \hbar_{\text{eff}} \big|. \tag{6.174}$$

True dynamical chaos, characterized by exponential instability, is limited in quantum mechanics (for Hamiltonian systems) to the logarithmically short (in $\hbar_{\text{eff}}$) Ehrenfest time scale (see Berman and Zaslavsky, 1978). Note that t_E increases indefinitely as $\hbar_{\text{eff}} \to 0$, in agreement with the correspondence principle.

The second time scale t^* (known as the *Heisenberg* or the *relaxation time scale*), at which the quantum evolution breaks away from the classical diffusion, is related to the phenomenon of *quantum dynamical localization*. For $t > t^*$, while the classical distribution goes on diffusing, the quantum distribution reaches a steady state that *decays exponentially* over the momentum eigenbasis (see Figs. 6.22 and 6.23):

$$W_m \equiv |\langle m|\psi\rangle|^2 \approx \frac{1}{\ell}\exp\left(-\frac{2|m-m_0|}{\ell}\right), \qquad (6.175)$$

where the index m denotes the eigenstates of I ($I|m\rangle = m|m\rangle$), m_0 is the initial value of the momentum and the *localization length* ℓ gives the width of the localized distribution.

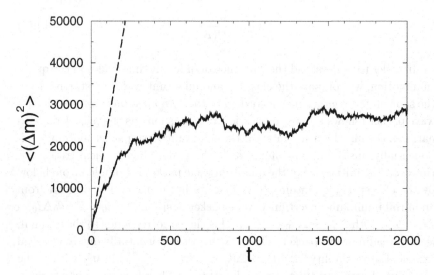

Fig. 6.22   Quantum localization (solid curve) of classical diffusion (dashed curve) for the kicked-rotator model. Classical and quantum evolution are obtained for parameter values $K = 5$, $L = 300$, $N = 2^{13} = 8192$ levels ($\hbar_{\text{eff}} \approx 0.23$). Classical evolution is computed by iterating the classical Chirikov standard map starting at time $t = 0$ with an ensemble of 10^4 orbits chosen in the interval $(\theta, I) \in [2-0.5, 2+0.5] \times [-2, 2]$. Quantum evolution is obtained by iterating the quantum kicked-rotator map starting from the initial least-spreading Gaussian wave packet of size $\Delta\theta = (\Delta I)^{-1} \sim \sqrt{\hbar_{\text{eff}}} \approx 0.5$, centred on the initial classical density of points.

An estimate of $t^\star$ and ℓ can be obtained by means of the following argument (see Chirikov *et al.*, 1981). During the initial stage the quantum motion mimics classical diffusion, so that the number of unperturbed levels significantly involved in the dynamics increases with time as $\Delta m \approx \sqrt{D_m t}$, where $D_m = D(K)/\hbar_{\text{eff}}^2 \approx k^2/2 \sim 1/\hbar_{\text{eff}}^2$ is the classical diffusion coefficient, measured in number of levels. Since the number of levels involved grows diffusively $\propto \sqrt{t}$ and, due to the Heisenberg principle, the discreteness of levels is resolved down to an energy spacing $\propto 1/t$, then the discreteness of

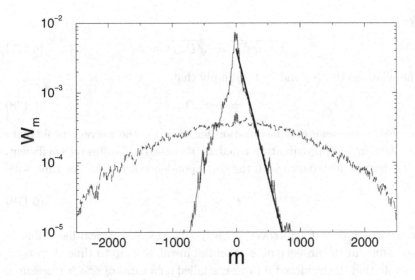

Fig. 6.23 Classical (dashed curve) versus quantum (solid curve) probability distributions over the momentum basis for the kicked-rotator model, at time $t = 5000 \gg t^\star \sim D_m \sim \frac{1}{2} \left(\frac{K}{\hbar_{\text{eff}}} \right)^2 \sim 2.4 \times 10^2$. The dashed curve gives the Gaussian classical distribution, while the solid curve is exponentially localized. The straight line shows the decay predicted by (6.175), with $\ell = D_m$. Parameter values and initial conditions are the same as in Fig. 6.22.

spectrum eventually dominates. This is the fundamental reason for which localization takes place. The localization length ℓ can then be estimated as follows. The localized wave packet has significant projection over approximately ℓ basis states, both in the basis of the momentum eigenstates and in the basis of the eigenstates of the Floquet operator U, which gives the evolution of the wave packet from time t to $t+1$ ($|\psi(t+1)\rangle = U|\psi(t)\rangle$). This operator is unitary and therefore its eigenvalues can be written as $\exp(i\lambda_i)$, and the so-called *quasi-energies* λ_i are in the interval $[0, 2\pi[$. Thus, the mean level spacing between "significant" quasi-energy eigenstates is $\Delta E \approx 2\pi/\ell$. The Heisenberg principle tells us that the minimum time required for the dynamics to resolve this energy spacing is given by

$$t^\star \approx 1/\Delta E \approx \ell. \tag{6.176}$$

This is the *break time*, after which the quantum features of the dynamics become apparent. Diffusion up to time $t^\star$ involves a number of levels given by

$$\sqrt{\langle (\Delta m)^2 \rangle} \approx \sqrt{D_m t^\star} \approx \ell. \qquad (6.177)$$

The relations (6.176) and (6.177) imply that

$$t^\star \approx \ell \approx D_m. \qquad (6.178)$$

Therefore, the *quantum* localization length ℓ for the average probability distribution is approximately equal to the *classical* diffusion coefficient. Note that, in accordance with the correspondence principle, the time scale

$$t^\star \sim \frac{1}{\hbar_{\text{eff}}^2} \qquad (6.179)$$

diverges as $\hbar_{\text{eff}} \to 0$. Moreover, $t^\star \gg t_E$. Therefore, classical-like diffusion is possible, in the absence of exponential instability, up to time $t^\star \gg t_E$.

Note that, if the kicked rotator is studied on a torus of size N (measured in number of levels), then localization cannot take place if $\ell > N$. In this case the diffusion time t_d (given by Eq. (6.173)) is shorter than the localization time $t^\star$ and the Heisenberg time t_H, after which the discreteness of the energy spectrum becomes manifest, is simply given by the inverse mean level spacing: $t_H \sim N$.

In Fig. 6.24 we compare the quantum evolution of an initially narrow wave packet with the classical evolution of an ensemble of trajectories, demonstrating both the Ehrenfest and the Heisenberg time scales. In order to allow direct comparison between the "quantum phase-space distribution" and the classical phase-space distribution, we plot the quantum Husimi function.[10] Due to exponential instability, the initial wave packet as well as the classical trajectories are spread over the entire interval $\theta \in [0, 2\pi[$ in approximately a couple of kicks (see the plots of Fig. 6.24 at times $t = 0$, $t = 1$, and $t = 3$). This corresponds to the Ehrenfest time. After that time the initial wave packet is "destroyed", in that it splits into many new small packets. Nevertheless, the quantum distribution follows the distribution of classical orbits (see Fig. 6.24, middle). However, while the distribution of classical points spreads indefinitely according to the diffusion process, the quantum distribution saturates at some maximum width (see Fig. 6.24, bottom). This determines the Heisenberg time scale.

---

[10]The Husimi function at a given point (θ, I) is obtained by the projection of the quantum state on the coherent state centred at that point. This corresponds to the smoothing of the Wigner function on the scale of the Planck constant (see Chang and Shi, 1986).

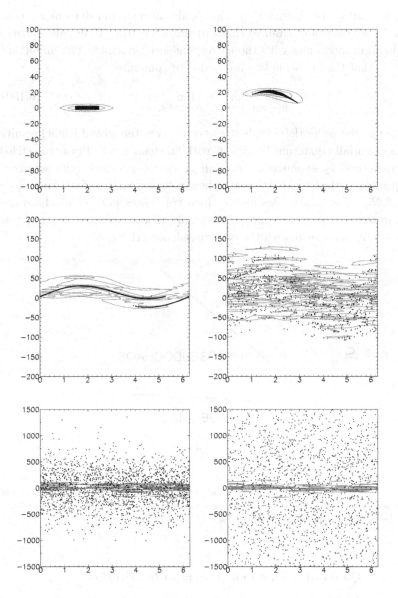

Fig. 6.24 Classical and quantum evolution for the kicked-rotator model in momentum–angle variables I (vertical axis) and θ (horizontal axis), with parameter values and initial conditions as in Fig. 6.22. The dots represent classical trajectories while in the quantum case the Husimi function is plotted (contour plots). The figure shows snapshots at different times: $t = 0$ (top left), $t = 1$ (top right), $t = 3$ (middle left), $t = 10$ (middle right), $t = 500$ (bottom left) and $t = 5000$ (bottom right).

The concept of characteristic time scales of quantum dynamics reconciles the absence of dynamical chaos (defined according to ergodic theory) in quantum mechanics with the correspondence principle. The important point is that the following two limits do not commute:

$$\lim_{|t|\to\infty}\lim_{\hbar_{\mathrm{eff}}\to 0} \neq \lim_{\hbar_{\mathrm{eff}}\to 0}\lim_{|t|\to\infty}. \tag{6.180}$$

While the first order (left) leads to classical chaos, the second (right) results in an essentially quantum behaviour with no chaos at all. Both true chaos (characterized by exponential instability) and pseudochaos (characterized by quantum diffusion) are transient phenomena confined to finite times (see Fig. 6.25). The time scales up to which true chaos and pseudochaos are seen in the quantum dynamics of classically chaotic systems diverge when $\hbar_{\mathrm{eff}} \to 0$, in agreement with the correspondence principle.

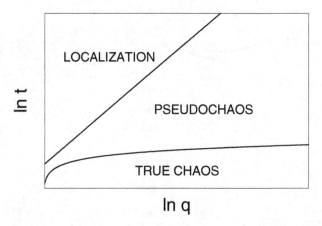

Fig. 6.25 A schematic drawing, for the kicked-rotator model, of the quantum-chaos times scales as a function of the quasiclassical parameter $q = 1/\hbar_{\mathrm{eff}}$: Ehrenfest time scale (lower curve) and Heisenberg time scale (upper curve).

6.5.4 * *Quantum chaos and Anderson localization*

Dynamical localization has profound analogies with Anderson localization of electronic transport in *disordered* solids. For the latter problem, of particular interest are the so-called tight-binding models, which are lattice discretization of the Schrödinger equation. These models play an important role in the investigation of the transport properties of solids at low temperature, where the electron wave function becomes very sensitive to local impurities and imperfections of the crystal lattice.

The *Anderson model* in one dimension is described by the eigenvalue equation

$$(Hu)_n = W_n u_n + t(u_{n-1} + u_{n+1}) = E u_n, \qquad (6.181)$$

where u_n is the electronic wave function at site n, t the strength of the hopping terms between nearest-neighbour lattice sites (kinetic energy) and the random site energies W_n are independently and homogeneously distributed in the interval $[-W/2, W/2]$, where W determines the disorder strength. The one-dimensional Anderson model exhibits exponentially localized eigenfunctions, no matter how small the disorder strength W is (see, *e.g.*, Lee and Ramakrishnan, 1985). For a given eigenfunction,

$$u_n \propto \exp\left(-\frac{1}{\ell}|n - n_0|\right), \qquad (6.182)$$

where ℓ is the localization length and the eigenfunction is centred at some site n_0.

In the Anderson model, as well as in the kicked rotator, quantum interference effects forbid unbounded diffusion, in real space in the first case, in momentum in the latter. This analogy can be formalized by means of the following transformation, which maps the kicked rotator into a disordered tight-binding model. The eigenvalue equation for the Floquet operator of the kicked rotator reads

$$U|\psi\rangle = \exp(-iH_0 T)\exp(-iV)|\psi\rangle = \exp(-i\epsilon T)|\psi\rangle, \qquad (6.183)$$

where $H_0 = \frac{I^2}{2}$ is the unperturbed Hamiltonian and (for the kicked rotator) $V(\theta) = k\cos\theta$. After introducing the operators W and t, defined by

$$\exp[-i(H_0 - \epsilon)T] = \frac{1 + iW}{1 - iW}, \qquad \exp(-iV) = \frac{1 + it}{1 - it}, \qquad (6.184)$$

and after defining

$$|\phi\rangle = (1 - it)^{-1}|\psi\rangle, \qquad (6.185)$$

the eigenvalue equation (6.183) reduces to

$$(t + W)|\phi\rangle = 0. \qquad (6.186)$$

We now expand $|\phi\rangle$ over the momentum basis: $|\phi\rangle = \sum_{m=-\infty}^{+\infty} u_m |m\rangle$, where $\langle\theta|m\rangle = \frac{1}{\sqrt{2\pi}}\exp(im\theta)$. We then end up with the eigenvalue equation

$$W_n u_n + \sum_{\substack{l=-\infty \\ (l\neq 0)}}^{+\infty} t_l u_{n+l} = E u_n, \qquad (6.187)$$

where we have defined

$$W_n = \langle n|W\rangle = \langle n|\tan\left(\tfrac{1}{2}(\epsilon - H_0)T\right)\rangle = \tan\left(\tfrac{1}{2}\epsilon T - \tfrac{1}{4}n^2 T\right), \qquad (6.188)$$

$$t_l = \langle l|t\rangle = -\langle l|\tan\left(\tfrac{V}{2}\right)\rangle = -\frac{1}{2\pi}\int_0^{2\pi} d\theta\,\exp(il\theta)\tan\left(\tfrac{V(\theta)}{2}\right), \qquad (6.189)$$

with $E \equiv -t_0$. The tight-binding model (6.187) can be seen as a generalization of the Anderson model (6.181) and describes the motion of an electron in a crystal with site energies W_n and hopping matrix elements t_l. Note that, in contrast to the Anderson model, the hopping terms are not limited to nearest-neighbour lattice sites.[11]

Note that in (6.187) the site energies, in contrast to the Anderson model (6.181), are not random but determined by the dynamics. Nevertheless, if $T/(4\pi)$ is irrational, then the sequence W_n turns out to be "pseudo-random" and leads to the same localization phenomenon as the random sequence. This has been confirmed by several numerical computations, showing the localization of eigenfunctions in the case in which T is an irrational multiple of 4π.

When $T/4\pi$ is rational the site energies W_n become periodic and therefore the electron is described by Bloch waves and moves freely in the crystal. This situation corresponds to the so-called *quantum resonances* in the kicked rotator. Let us observe that quantum evolution is endowed with two different periods: the first (T) is explicitly specified by the perturbation, the second is 4π and follows from the free-evolution peculiarity of having a spectrum given by integers, so that the Floquet operator is unchanged when $T \to T + 4\pi$. Naively speaking, the free rotator has energy levels $E_m = m^2/2$ and the photon energy is $2\pi/T$; the resonance condition is met whenever an integer number of photons matches the transition between unperturbed levels. This condition corresponds to rationality of the ratio

---

[11]In the case of the kicked-rotator, the transformation $t = -\tan\left(\frac{V}{2}\right) = -\tan\left(\frac{k\cos\theta}{2}\right)$ must satisfy the bound $|k\cos\theta| < \pi$, to avoid singularities. This restriction can be overcome by means of a more complicated mapping, as shown by Shepelyansky (1986).

$T/4\pi$ between the two periods. In this case the energy of the rotator grows quadratically in time, a phenomenon without a classical analogue.

It should be stressed that in the dynamical case (i) localization is related to quasi-energy eigenfunctions and occurs in momentum instead of config-uration space and (ii) no external random element is introduced since the perturbation is periodic. For this reason localization due to classical chaos is known as "dynamical localization", to distinguish it from Anderson lo-calization due the presence of disorder in the Hamiltonian.

6.5.5 * *The hydrogen atom in a microwave field*

One of the most significant cases in which classical and quantum chaos confronted each other was the explanation of the experiments on the mi-crowave ionization of highly excited hydrogen atoms, first performed by Bayfield and Koch (1974). Hydrogen atoms prepared in very elongated states with a high principal quantum number $n_0 \approx 63 - 69$ were injected into a microwave cavity and the ionization rate was measured. The mi-crowave frequency was $\omega = 9.9\,\mathrm{GHz}$, corresponding to a photon energy approximately a hundred times lower than the ionization energy of level 66 and even lower than the energy required for the transition from level 66 to 67. Much surprise therefore followed the discovery that very efficient ionization occurred when the electric field intensity exceeded a threshold value $\epsilon \approx 20\,\mathrm{V/cm}$, much lower than the value for Stark ionization in a static field. Numerical simulations (Leopold and Percival, 1978) showed that classical mechanics could reproduce the experimental data quite well, as suggested by the correspondence principle, in view of the high quantum numbers involved. Subsequent analysis (Jensen, 1982, Delone *et al.*, 1983), still in classical terms, explained the threshold intensity as the critical value for the onset of chaotic diffusion in the action variable n. However, the hy-drogen atom is a quantum object and it is possible to find experimentally appropriate parameters such that, similarly to the kicked-rotator model, the classical chaotic diffusion halts due to quantum interference effects.

The Hamiltonian for a hydrogen atom interacting with a time-periodic linearly polarized electric field is, in the dipole approximation,

$$H = \frac{p^2}{2} - \frac{1}{r} + \epsilon z \cos(\omega t). \qquad (6.190)$$

Here ϵ and ω are the strength and the frequency of the electric field, directed along z. Note that atomic units are used.[12] When the electron is in a state that is very extended along the direction of the field, the one-dimensional model, described by the Hamiltonian

$$H_{1d} = \frac{p^2}{2} - \frac{1}{z} + \epsilon z \cos(\omega t), \quad z > 0, \qquad (6.191)$$

is a good approximation to the real three-dimensional motion. The unperturbed Hamiltonian $H_{1d}^{(0)} = \frac{p^2}{2} - \frac{1}{z}$ describes both bounded (with negative energy, $E < 0$) and unbounded motion ($E > 0$); as far as we are interested in exploring the dynamics that precedes ionization, we are confined to negative energies and, accordingly, we introduce action-angle variables (n, θ), thus obtaining

$$H_{1d} = -\frac{1}{2n^2} + \epsilon z(n, \theta) \cos(\omega t). \qquad (6.192)$$

We remark that classical dynamics depends only on the rescaled quantities $\epsilon_0 \equiv \epsilon n_0^4$ and $\omega_0 = \omega n_0^4$, where n_0 is the initial value of the action variable (in the quantum case, the initial value of the principal quantum number). To prove this scaling, it is sufficient to operate the transformations

$$z \to \frac{1}{n_0^2} z, \quad t \to \frac{1}{n_0^3} t, \quad \epsilon \to n_0^4 \epsilon, \quad \omega \to n_0^3 \omega. \qquad (6.193)$$

They imply $p = \frac{dz}{dt} \to n_0 p$ and $H_{1d} \to n_0^2 H_{1d}$. Since the Hamiltonian is only multiplied by a constant, the equations of motion do not change. We simply rescale the size of the trajectories. If the initial value of the unperturbed energy is $E = -\frac{1}{2n_0^2}$, then (6.193) allows us to scale E to the energy of the first Bohr level, $E \to n_0^2 E = -\frac{1}{2}$. Note that ϵ_0 and ω_0 are the ratios of ϵ and ω with the strength of the Coulomb field ($\epsilon_C = z^{-2} \approx n_0^{-4}$) and with the frequency of the unperturbed electronic Kepler motion ($\omega_K = n_0^{-3}$). The rescaled time is measured, up to a factor 2π, in number of periods of the unperturbed Kepler motion: $\frac{1}{n_0^3} t = 2\pi \frac{t}{T_K}$, where $T_K = 2\pi n_0^3$.

Exercise 6.16   Prove that the scaling (6.193) is no longer valid in quantum mechanics.

If the field frequency exceeds the electron frequency ($\omega_0 > 1$), it can be shown (see Casati *et al.*, 1988) that the motion of system (6.192) is

---

[12]We pass from the Gauss system to atomic units (a.u.) by setting $m = 1$, $e = 1$, $\hbar = 1$, where m and e are the mass and charge of the electron. We have ϵ (V/cm) = $5.14 \times 10^9 \epsilon$ (a.u.), ν (GHz) = $6.58 \times 10^6 \omega$ (a.u.), where $\nu = \omega/(2\pi)$.

approximately described by a map over one Kepler period of the electron, the so-called *Kepler map*:

$$\begin{cases} \bar{N} = N + k\sin\phi, \\ \bar{\phi} = \phi + \dfrac{\pi}{\sqrt{2\omega}}\left(-\bar{N}\right)^{-3/2}, \end{cases} \qquad (6.194)$$

where $N = E/\omega$, ϕ is the field phase at the perihelion and

$$k \approx 2.6\,\frac{\epsilon}{\omega^{5/3}} \qquad (6.195)$$

is the perturbation parameter.[13] Note that the Kepler map is defined only for states such that $\bar{N} < 0$: if, on iterating this map, we end up in the continuum ($\bar{N} > 0$, that is $E > 0$) then the electron must be considered as ionized. The linearization of the second equation in (6.194) around the initial value $N_0 = -\frac{1}{2n_0^2\omega}$ reduces the Kepler map to the standard map:

$$\begin{cases} \bar{N}_\phi = N_\phi + k\sin\phi, \\ \bar{\phi} = \phi + T\bar{N}_\phi, \end{cases} \qquad (6.196)$$

where $N_\phi \equiv N - N_0$ has, in the quantum case, the meaning of the number of photons absorbed by the atoms, $T = 6\pi\omega^2 n_0^5$ and an irrelevant constant phase shift has been neglected in the second equation.

As we saw in Sec. 6.5.3, the border for the transition to chaos is given by $K = kT \approx 1$, which gives

$$\epsilon_c \approx \frac{1}{49 n_0^5 \omega^{1/3}}, \qquad (6.197)$$

that is, in rescaled units,

$$\epsilon_{oc} \approx \frac{1}{49\omega_0^{1/3}}. \qquad (6.198)$$

This is the classical *chaos border* above which, according to classical mechanics, the electron diffuses until ionization.

On the other hand, we know from the quantum kicked-rotator model that localization takes place over a length (here measured in number of photons)

$$\ell_\phi \approx \frac{k^2}{2} \approx 3.3\,\frac{\epsilon^2}{\omega^{10/3}}. \qquad (6.199)$$

---

[13] A further condition for the validity of the Kepler map is that $\epsilon \ll 5\omega^{4/3}$, see Shepelyansky (1994).

Therefore, the ionization process in the *quantum* hydrogen atom under a microwave field is governed by two relevant lengths:

(i) the localization length (measured in number of photons) ℓ_ϕ,
(ii) the "*sample size*" N_I, namely the number of photons required to reach the continuum starting from the initial state with principal quantum number n_0. This number is given by

$$N_I = \frac{1}{2n_0^2 \omega}. \tag{6.200}$$

If $\ell_\phi \ll N_I$, then quantum localization takes place and no ionization is possible. On the other hand, if $\ell > N_I$ then the process of chaotic diffusion goes on until ionization. The condition $\ell_\phi = N_I$, that is

$$\epsilon_q \approx 0.4 \frac{\omega^{7/6}}{n_0}, \tag{6.201}$$

gives the so-called *quantum delocalization border*. In order to have chaotic ionization, $\epsilon > \epsilon_q$ is required.

The above predictions have been confirmed by experimental results on the microwave ionization of hydrogen atoms (Galvez *et al.*, 1988; Bayfield *et al.*, 1989), thus providing experimental evidence of the quantum suppression of classically chaotic diffusion due to the localization phenomenon.

A comparison of the localization theory with the experimental data obtained by Bayfield *et al.* (1989) is shown in Fig. 6.26. The empty circles represent the experimentally observed threshold values of the microwave peak intensity for ionization of 10% of the atoms.[14] The dotted curve is the classical chaos border and the dashed curve is the prediction of the localization theory for 10% ionization. The numerical data (filled circles) are obtained from the integration of the quantized Kepler map. In such simulations the interaction time, including the switching on and off of the microwave field, was chosen to be the same as in actual experiments. The agreement between experimental and numerical data is quite remarkable, in that the Kepler map is only a crude approximation for the actual quantum dynamics.[15] Moreover, the localization theory gives a satisfactory *average*

---

[14]Both in the experiment and in numerical computations the ionization probability is obtained as the total probability above a cutoff level n_c.

[15]Furthermore, though the numerical model is one-dimensional, in actual experiments the initially excited state corresponds to a microcanonical distribution over the shell with a given principal quantum number. The classical counterpart for this would be a microcanonical ensemble of orbits. Nevertheless, the experimental data agree fairly well with the predictions of the one-dimensional quantum Kepler map. The reason for this

picture.

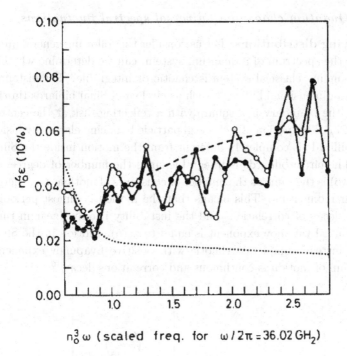

$$n_0^3 \omega \text{ (scaled freq. for } \omega / 2\pi = 36.02 \, GH_2)$$

Fig. 6.26 The scaled 10% threshold field from experimental results (empty circles, taken from Bayfield *et al.*, 1989) and from numerical integration of the quantum Kepler map (filled circles). The dashed curve is the quantum theoretical prediction according to localization theory, the dotted curve is the classical chaos border (figure taken from Casati *et al.*, 1990).

It can be seen from Fig. 6.26 that both the numerical and experimental data exhibit sensible deviations from the average prediction. These *fluctuations* are analogous to the conductance fluctuations observed in finite solid-state samples (see Casati *et al.*, 1990). In the microwave ionization of hydrogen atoms there is a finite "photonic lattice" since a finite number of photons is required to ionize the atom. However, while in the solid-state case (described by the Anderson model) the fluctuations can be traced back to the randomness of the potential, in the hydrogen atom the source

agreement is the following: due to the existence of an approximate integral of motion, the main contribution to excitation turns out to be given by orbits that are extended along the direction of the linearly polarized external field (see Casati *et al.*, 1988). For such orbits, the use of the one-dimensional model is fully justified.

of fluctuations is dynamical chaos.

6.5.6 * Quantum chaos and universal spectral fluctuations

Level-spacing distributions. Let us consider the following general question: given the spectrum of a quantum system, can we determine whether the corresponding classical system is chaotic or integrable? For instance, consider two-dimensional billiards, such as circle and Sinai billiards (in the latter case, the boundary is a square with a reflecting disk at the centre) (see Fig. 6.27). The motion of a classical particle bouncing elastically inside the circle billiards is completely different from the motion inside the Sinai billiards. The circle billiards is integrable since the number of degrees of freedom (two) is the same as the number of constants of motion, *i.e.*, energy and angular momentum. This means that the motion is almost periodic, there is no decay of correlations and the instability is only linear in time (the maximum Lyapunov exponent is equal to zero). Motion in the Sinai billiards is instead completely chaotic with positive Lyapunov exponent. The spectrum of motion is continuous and correlations decay.[16]

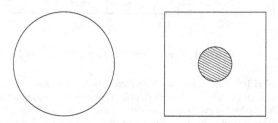

Fig. 6.27   The circle billiards (left) and Sinai billiards (right).

If we quantize these two systems, then we know that in both cases we have a discrete sequence of eigenvalues $E_n = \frac{k_n^2}{2m}$ and eigenfunctions ψ_n of the Schrödinger equation $(\nabla^2 + k_n^2)\psi_n = 0$ with Dirichlet boundary conditions (the wave function vanishes at the boundary of the billiards). For the circle, the sequence of eigenvalues E_n can be computed analytically and is given by the zeros of the Bessel functions of the first kind. For Sinai billiards the sequence of levels E_n' can only be computed numerically, yet we know it forms a discrete sequence. Is there any qualitative difference between the two sequences E_n and E_n'? Certainly, the structure of the eigenvalues and

---

[16]Note that the origin of the chaotic behaviour of the Sinai billiards is the defocusing effect of the disk, due to its negative curvature.

eigenfunctions must possess some peculiarity in such a manner that, when taking the classical limit, in one case (circle billiards) the motion becomes regular, while in the other case (Sinai billiards) it becomes wildly erratic and chaotic. Can we identify such properties? We should expect that they will reside in different statistical properties of the eigenvalues.

A first quantity of interest is the smoothed level density $\bar{\rho}(E)$. For a two-dimensional billiards it is given by the famous Weyl formula (obtained by imposing the Dirichlet boundary conditions):

$$\bar{\rho}(E) \approx \frac{m}{2\pi\hbar^2}A - \frac{L}{4\pi}\sqrt{\frac{2m}{\hbar^2 E}}, \qquad (6.202)$$

where A is the area of the billiards and the second term, which contains the length L of the perimeter of the billiards, vanishes for large E.[17] Therefore, by increasing the energy E, the density tends to a constant that only contains the area of the billiards and does not depend on its shape. It is now clear that the density $\bar{\rho}(E)$ cannot carry information on the chaotic or integrable nature of the billiards (which clearly depends on its shape).

---

[17]The Weyl formula (6.202) can be derived from a semiclassical computation of the density of states. We assume that the number of states having energy smaller than E is equal to the number of cells of size $(2\pi\hbar)^2$ contained in the phase-space volume $\Omega(E)$ corresponding to energy smaller than E:

$$\Omega(E) = \int_{H(q,p)<E} dqdp,$$

where (q,p) are the phase-space coordinates and $H(q,p)$ the system Hamiltonian. For a particle with energy $E = \frac{|p|^2}{2m}$, moving in a two-dimensional billiards of area A, the phase-space volume $\Omega(E) = \pi|p|^2 A = 2\pi mEA$. Therefore, the mean number of levels with energy smaller than E is

$$\bar{N}(E) = \frac{\Omega(E)}{(2\pi\hbar)^2} = \frac{m}{2\pi\hbar^2}AE,$$

which implies

$$\bar{\rho}(E) = \frac{d\bar{N}(E)}{dE} = \frac{m}{2\pi\hbar^2}A.$$

This gives the first term in (6.202). The correction term in (6.202) can be obtained by observing that the wave function ψ must vanish at the boundary of the billiards, so that the effective area A is smaller by a factor δA. The width of the strip where $\psi \approx 0$ is of the order of the de Broglie wavelength λ. Hence, $\delta A \approx \lambda L = \frac{\hbar}{\sqrt{2mE}}L$, where L is the length of the billiards perimeter. Thus,

$$\bar{\rho}(E) = \frac{m}{2\pi\hbar^2}(A - \delta A),$$

which directly leads to the Weyl formula (6.202).

In particular its consideration cannot answer the famous question posed by Mark Kac [*Amer. Math. Monthly* **73**, 1 (1966)]: *"Can one hear the shape of a drum?"*. More precisely, can one deduce the shape of a plane region by knowing the frequencies at which it resonates?

In order to discriminate between regular and chaotic billiards one needs to consider *fluctuation* properties; that is, how levels are distributed around the average density $\bar{\rho}(E)$.

As spectra corresponding to different systems or different spectral regions (corresponding to different values of the energy) of the same system have in general different average densities $\bar{\rho}$, we must ensure uniformity of the average spacings before comparing fluctuations. This is achieved by a local (in energy) renormalization of the unit of energy. Such a procedure is known as *unfolding* and is accomplished by defining

$$e_k = \overline{N}(E_k), \tag{6.203}$$

where $\overline{N}(E)$ is the average number of levels with energy smaller than E.[18] In this manner, we separate the smooth part $\bar{\rho}(E) = \frac{d\overline{N}(E)}{dE}$ of the energy density from its fluctuations. We now substitute the original set $\{E_k\}$ of eigenvalues with the new set of numbers $\{e_k\}$. By construction, these numbers have mean level spacing $\langle e_{k+1} - e_k \rangle_k$ equal to unity ($\langle \ldots \rangle_k$ denoting the average over k). From now on, we shall call E_k the unfolded levels e_k.

The simplest quantity of interest in describing level fluctuations is the *level-spacing distribution* $P(s)$, where $P(s)ds$ denotes the probability of finding two adjacent levels with energy separation in the interval $[s, s+ds]$. That is, if $s_i = E_{i+1} - E_i$, then the probability that $s \leq s_i \leq s + ds$ is given by $P(s)ds$. Note that $\int_{s=0}^{+\infty} ds P(s) = 1$ and, due to the unfolding procedure, $\bar{s} = \int_0^{+\infty} ds s P(s) = 1$.

The energy levels of the harmonic oscillator are perfectly correlated and equally spaced, so that $P(s)$ reduces to a δ function centred at $\bar{s} = 1$, namely $P(s) = \delta(s - \bar{s}) = \delta(s - 1)$. In the opposite limit, in which the energy levels are completely uncorrelated (randomly distributed), it is possible to prove

---

[18] At large energies we can generalize what explained above for two-dimensional billiards and compute $\overline{N}(E)$ as

$$\overline{N}(E) = \frac{1}{(2\pi\hbar)^d} \int\limits_{H(q,p)<E} dq\,dp,$$

where d is the number of degrees of freedom for the system under examination.

(see exercise 6.17) that $P(s)$ is given by the Poisson distribution:

$$P(s) = P_P(s) = \exp(-s). \tag{6.204}$$

Exercise 6.17   Derive the Poisson distribution (6.204) for a random sequence of energy levels.

Another measure of the deviation of the levels from the average spacing is given by the *spectral rigidity* $\Delta_3(L)$, which is defined as the mean square deviation of the best local straight-line fit to the staircase cumulative spectral density $N(E)$ over the scale L. Here the staircase function $N(E)$ gives the number of levels in the interval $(-\infty, E]$, and we assume that the spectrum has been unfolded, so that the local average level density is independent of E. We then define

$$\Delta_3(L, \alpha) = \frac{1}{L} \min_{A,B} \int_\alpha^{\alpha+L} dE \, [N(E) - AE - B]^2. \tag{6.205}$$

Usually, one is interested in the quantity $\Delta_3(L) \equiv \langle \Delta_3(L, \alpha) \rangle_\alpha$, obtained after averaging $\Delta_3(L, \alpha)$ over α, *i.e.*, over different parts of the energy spectrum (if we assume that the spectrum is translationally invariant, then Δ_3 does not depend on α). For the harmonic oscillator $\Delta_3(L) = \frac{1}{12}$, independently of L. When instead the energy levels are uncorrelated we obtain

$$\Delta_3(L) = \frac{L}{15}, \tag{6.206}$$

independently of α. Note that, in contrast to the nearest-neighbour spacing distribution $P(s)$, the spectral rigidity $\Delta_3(L)$ measures long-range correlations in the energy spectrum.[19]

Random-matrix theory. There exists a well developed theory, known as random-matrix theory (RMT), which describes the statistical properties of complex quantum systems. RMT was introduced and developed at the beginning of the 1950's on the basis of the observation that, for very "complicated" systems such as those with a large number of degrees of freedom, it is too difficult and practically meaningless to integrate the equations of motion or to diagonalize huge (Hamiltonian) matrices. In analogy with

---

[19] An alternative measure of the stiffness of the spectrum often used in the literature is the *level number variance* Σ_2, defined as

$$\Sigma_2(L) = \langle [n(L, \alpha) - \langle n(L, \alpha) \rangle_\alpha]^2 \rangle_\alpha,$$

where (after unfolding) $n(L, \alpha)$ counts the number of levels in the interval $[\alpha, \alpha + L]$.

classical statistical mechanics, one is not interested in the exact determination of the individual energy levels but only in their statistical properties. Indeed, in complex systems such as heavy nuclei or excited molecules, the knowledge of exact states is meaningless in the same sense as the precise knowledge of positions and velocities in systems with a large number of degrees of freedom is meaningless. For instance, when considering a gas of $O(10^{23})$ molecules enclosed in a vessel, one is not interested in the positions and velocities of the individual molecules but in global thermodynamic quantities such as pressure or entropy.

As stated by Dyson [*J. Math. Phys.* **3**, 140 (1962)]: *"... there must come a point beyond which such analyses of individual levels cannot usefully go. For example, observations of levels of heavy nuclei in the neutron-capture region give precise information concerning a stretch of levels from number N to number $(N+n)$, where n is an integer of the order of 100 while N is of the order of 10^6. It is improbable that level assignments based on shell structure and collective or individual-particle quantum numbers can ever be pushed as far as the millionth level. It is therefore reasonable to inquire whether the highly excited states may be understood from the diametrically opposite point of view, assuming as a working hypothesis that all shell structure is washed out and that no quantum number other than spin and parity remain good. The result of such inquiry will be a statistical theory of energy levels. The statistical theory will not predict the detailed sequence of levels in any one nucleus, but it will describe the general appearance and the degree of irregularity of the level structure that is expected to occur in any nucleus which is too complicated to be understood in detail ... What is here required is a new kind of statistical mechanics, in which we renounce exact knowledge not of the state of a system but of the nature of the system itself. We picture a complex nucleus as a "black box" in which a large number of particles are interacting according to unknown laws. The problem then is to define in a mathematically precise way an ensemble of systems in which all possible laws of interaction are equally probable".*

This program, initiated by Wigner, led to the development of random-matrix theory. The main idea is that the statistical properties of complex systems are the same as those of an appropriate ensemble of random matrices. The *space-time symmetries* obeyed by the system impose certain conditions on the admissible matrix ensemble. Here we limit ourselves to the consideration of the following two main cases, corresponding to different RMT ensembles:

(i) If the system is invariant under time-reversal and rotations, the Hamiltonian matrices can be chosen as real symmetric. In this case, the appropriate ensemble of random matrices is the Gaussian orthogonal ensemble (GOE), defined in the space of $N \times N$ real symmetric matrices (where N is a large integer) by two requirements:

(a) The ensemble is *invariant* under every orthogonal transformation

$$H \rightarrow H' = W^T H W, \qquad (6.207)$$

where W is any real orthogonal matrix. This means that the probability $p_N(H)dH$ that a matrix H belongs to the volume element $dH = \prod_{i \leq j} dH_{ij}$ (dH_{ij} being the differential increment of the matrix element H_{ij}) is invariant under orthogonal transformations; that is, $p_N(H)dH = p_N(H')dH'$.

(b) The various matrix elements H_{ij}, $i \leq j$, are statistically *independent* random variables. Therefore, the probability density function $p_N(H)$ is a product of functions $p_{ij}(H_{ij})$ depending on a single matrix element: $p_N(H) = \prod_{i \leq j} p_{ij}(H_{ij})$.

It can be shown that these two requirements uniquely determine the ensemble (see Mehta, 1991). We can then write the function $p_N(H)$ as

$$p_N(H) = C_N \exp\left[-\frac{\mathrm{Tr}(H^2)}{4\sigma^2}\right], \qquad (6.208)$$

where

$$\mathrm{Tr}(H^2) = \sum_{1 \leq i \leq N} H_{ii}^2 + 2 \sum_{1 \leq i < j \leq N} H_{ij}^2 \qquad (6.209)$$

and the constants C_N and σ are fixed by the normalization and choice of the unit of energy. Therefore, for the GOE the matrix elements are Gaussian distributed with zero mean and the same variance σ^2, except for the diagonal elements, for which the variance is $2\sigma^2$.

(ii) If time-reversal invariance is violated (this is the case, for instance, of a charged particle in a magnetic field), then the Hamiltonian matrices are complex Hermitian. In this case, the appropriate ensemble is the Gaussian unitary ensemble (GUE), defined in the space of Hermitian matrices by the following properties:

(a) The ensemble is invariant under every unitary transformation

$$H \rightarrow H' = U^{-1} H U, \qquad (6.210)$$

where U is any unitary matrix. In this case the probability $p_N(H)dH$ that a matrix H belongs to the volume element $dH = \prod_{i \leq j} d[\text{Re}(H_{ij})] \prod_{i<j} d[\text{Im}(H_{ij})]$ is invariant under unitary transformations: $p_N(H)dH = p_N(H')dH'$.

(b) The matrix elements $\text{Re}(H_{ij})$ $(i \leq j)$ and $\text{Im}(H_{ij})$ $(i < j)$ are statistically independent random variables.

Note that Eq. (6.208) is still valid but now

$$\text{Tr}(H^2) = \sum_{1 \leq i \leq N} H_{ii}^2 + 2 \sum_{1 \leq i < j \leq N} \left\{ [\text{Re}(H_{ij})]^2 + [\text{Im}(H_{ij})]^2 \right\}, \qquad (6.211)$$

so that both $\text{Re}(H_{ij})$ and $\text{Im}(H_{ij})$ are Gaussian distributed.

A detailed theory has been developed for the random-matrix ensembles GOE and GUE. A main result refers to the level-spacing statistics $P(s)$. For the GOE case, $P(s)$ is well approximated by the famous Wigner surmise:

$$P_W^O(s) \approx \frac{\pi}{2} s \exp\left(-\frac{\pi}{4}s^2\right), \qquad (6.212)$$

while for the GUE ensemble

$$P_W^U(s) \approx \frac{32}{\pi^2} s^2 \exp\left(-\frac{4}{\pi}s^2\right). \qquad (6.213)$$

We note that the Wigner surmise is exact for 2×2 matrices (see exercise 6.18).

Exercise 6.18 Derive the Wigner surmise (6.212) for an ensemble of 2×2 real symmetric matrices with independent random matrix elements H_{11}, H_{22}, H_{12}.

It is also interesting that the spectral rigidity $\Delta_3(L) \propto \ln L$ both in the GOE and in the GUE case. The logarithmic dependence of $\Delta_3(L)$ indicates a strong rigidity of the spectrum, which has to be compared with $\Delta_3(L) = \frac{L}{15}$ for a random sequence of eigenvalues and with the maximum rigidity $\Delta_3(L) = \frac{1}{12}$ for a regular sequence of equally spaced levels.

The predictions of RMT agree very well with experimental data. This statement is demonstrated in Fig. 6.28, which shows the distribution $P(s)$ of nuclear-level spacings (as usual, s is measured in units of the mean level spacing). Data are obtained from neutron resonance spectroscopy and high-resolution proton scattering and refer to quasi-bound states of the compound nucleus far from the ground state region. The agreement with the predictions of RMT is impressive.[20] This is especially true since the theory has no adjustable parameter.

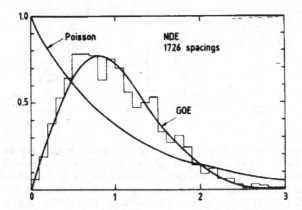

Fig. 6.28 The nearest neighbour spacing distribution for the "nuclear data ensemble" (NDE), constructed from 1726 spacings of levels of the same spin and parity, corresponding to 36 sequences of 32 different nuclei. For comparison, the RMT prediction (GOE ensemble) and the Poisson distribution are also shown (solid curves). The figure is reprinted with permission from Bohigas *et al.* (1983). Copyright (1983) by the American Physical Society.

We remark that this satisfactory agreement is related, for many-body interacting systems, to the complexity of the systems considered and not to the underlying interaction, which may be nuclear or electromagnetic. One therefore expects that spectra of highly excited atoms or complex molecules should also be described by GOE. This expectation is confirmed in Fig. 6.29, obtained by using atomic energy levels of neutral and ionized atoms in the rare-earth region (left) and the spectrum of a polyatomic molecule such as NO_2 (right).

The following point should be stressed: we are interested here in the

---

[20]Note that the energy-level spacing distribution for states with the same spin J and parity π agree with the Wigner surmise. If instead we consider all states, corresponding to different values of the quantum numbers J and π, then the Poisson distribution follows. This result can be explained as follows. When there are good quantum numbers corresponding to exact integrals of motion, such as angular momentum and parity, and when the basis states are labelled by these quantum numbers, then the Hamiltonian matrix splits into independent blocks (the matrix elements connecting these blocks vanish). Therefore, energy levels coming from different blocks are perfectly uncorrelated. If several such independent sequences are analyzed together, then the Poisson distribution follows (if two adjacent levels correspond to different values of the good quantum numbers, then they are uncorrelated).

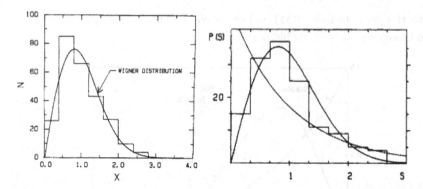

Fig. 6.29   A histogram of level spacings for 140 vibrational levels taken from level sequences of atoms in the rare-earth region (left) and for 140 energy levels of NO_2 (right).
In both cases the Wigner distribution (6.212) is shown (in the right-hand plot the Poisson distribution is also drawn). The figures are reprinted with permission from Camarda
and Georgopulos (1983) (left) and from Zimmermann *et al.* (1988) (right). Copyright
(1983, 1988) by the American Physical Society.

properties of a single system and we are attempting to describe them by
averaging over an ensemble of Hamiltonians. This is only meaningful if the
properties in which we are interested are the same for almost all systems of
the ensemble. More precisely, what is required here is a technical property,
known as *self-averaging*, which is an ergodic-like property, according to
which in some appropriate limit (*e.g.*, as $N \to \infty$, where N is the dimension
of the matrices in the ensemble) the dispersion of the relevant quantities
over the ensemble tends to zero. Under this condition, "typical" values
of these quantities are very close to the average values: the latter are in
general easier to compute and this constitutes the main advantage of the
ensemble method. The level-spacing distribution for example turns out to
be the same when computed along several levels of a given nucleus or when
computed by averaging over an ensemble of nuclei.

Quantum chaos and level statistics. The main idea that led to the introduction of RMT was the notion of "complexity", which was at that time
quite vague and mainly related to the large number of degrees of freedom
involved in a many-body problem. Nowadays we have a well defined notion
of complexity usually referred to as "chaos". As we know, also systems with
a very small number of degrees of freedom (*e.g.*, two-dimensional billiards)
can exhibit dynamical chaos and therefore their motion is very complicated.
It is then quite reasonable to expect that the level statistics of such systems
are described by RMT. This expectation was confirmed by many numer-

ical simulations. In Fig. 6.30, the nearest-neighbour-spacing distributions for circular and Sinai billiards are shown. The first model corresponds to integrable classical dynamics, the second to a fully chaotic classical system. In contrast to the case of circular billiards, for which the smallest spacings are more frequent, the results for Sinai billiards are fully consistent with the GOE predictions.[21,22] Note that the nearest-neighbour-spacing distribution for the circular billiards is in agreement with analytical (Berry and Tabor, 1977) and numerical results, showing that (apart from exceptions such as the harmonic oscillator), if the corresponding classical dynamics is completely integrable, then $P(s)$ is equal to the Poisson distribution, $P(s) = \exp(-s)$.

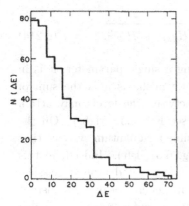
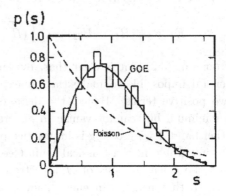

Fig. 6.30 The nearest-neighbour-spacing distributions $P(s)$ for the circular (left) and Sinai billiards (right). Note that in the left-hand figure the energy is not normalized to the mean energy and the histogram is not normalized to total unit area. The figures are reprinted with permission from McDonald and Kaufman (1979) (left) and Bohigas *et al.* (1984) (right). Copyright (1979, 1984) by the American Physical Society.

The main difference between the level-spacing statistics corresponding to classically integrable or chaotic systems is that in the first case (Poisson

---

[21] These results have been reproduced in microwave cavities constructed in the shape of integrable or chaotic billiards. The spacings between microwave normal mode frequencies of the cavity follow the Poisson distribution in the integrable case and the Wigner distribution in the chaotic case (see Stöckmann, 1999). These results are interesting as they show that RMT can be applied not only to quantum mechanics but also to classical electromagnetic waves.

[22] Note that the successful application of RMT to classically chaotic systems is not confined to toy models such as the Sinai billiards. For instance, highly excited levels of the hydrogen atom in a strong magnetic field exhibit spectral fluctuations in agreement with RMT (see Friedrich and Wintgen, 1989).

distribution) there is a high probability of finding small spacings (*level clustering*), while in the latter case (Wigner distribution) negligible probability is assigned to spacings that are very small compared to the mean spacing. Therefore, one can say that levels repel each other and this phenomenon is known as *level repulsion*.

A qualitative justification of level repulsion is the following. Consider two levels for the Hamiltonian $H(\lambda)$ that upon variation of the parameter λ undergo a close encounter. Around the crossing point it is then possible to describe these two levels by means of nearly degenerate perturbation theory. Hence, it is possible to restrict the analysis to that of a two-level subspace; that is, to a 2×2 Hermitian matrix with matrix elements H_{ij} $(i, j = 1, 2)$. Its eigenvalues are given by

$$E_{\pm} = \tfrac{1}{2} (H_{11} + H_{22}) \pm \sqrt{\tfrac{1}{4} (H_{11} - H_{22})^2 + |H_{12}|^2}. \qquad (6.214)$$

From Eq. (6.214) it is clear that, by varying a single parameter, it is in general impossible to make the square root vanish, as it is the sum of two positive terms. Hence, the distance between the levels may attain a minimum but cannot vanish in general (see Fig. 6.31, right). On the other hand, level crossing is in practice possible for quantum systems that are integrable in the classical limit (see Fig. 6.31, left). Indeed, in this case the eigenfunctions of H are strongly concentrated around classical tori, so that nearby (in energy) levels are in general peaked on distant tori. Thus, the superposition between these two eigenfunctions is negligible and $H_{12}(\lambda) \approx 0$. In this case, it is sufficient that the single condition $(H_{11} - H_{22})(\lambda) = 0$ be fulfilled to obtain the level crossing $E_{+}(\lambda) = E_{-}(\lambda)$.

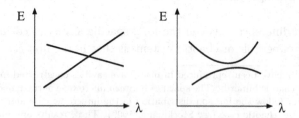

Fig. 6.31   A schematic drawing of level crossing (left) and avoided crossing (right).

The fact that the spectra of systems whose classical analogue is chaotic show the same *universal fluctuations* as predicted by RMT is a conjecture mainly supported by numerical data.[23] Today there is overwhelming evidence that, apart from exceptions corresponding to particular cases (see Bohigas, 1991), fluctuations are described by the Poisson distribution for integrable systems and are in agreement with RMT for chaotic systems. As a consequence, the spectral statistics of quantum systems is usually taken as a sort of characterization of quantum chaos, even in systems where the classical limit is hard to define. Thus, the transition from integrability to quantum chaos when some parameter of the system Hamiltonian is varied is associated with the transition from the Poisson to the Wigner distribution in the level-spacing statistics (see the model described in Sec. 6.5.7).

We shall not discuss in detail the properties of quantum eigenfunctions in relation to the integrable or chaotic properties of a dynamical system. We just quote a result due to Shnirelman, which states that for chaotic billiards the squared n-th eigenfunction $u_n^2(x)$ tends weakly to the uniform distribution as $n \to \infty$. This indicates that the eigenfunctions of quantum chaotic systems are delocalized in phase space over the energy surface (*ergodic eigenstates*).[24] For integrable systems the eigenstates are non-ergodic: they are instead concentrated around classical tori.

Note that classically chaotic systems that exhibit quantum localization, such as the kicked-rotator model, conform to RMT predictions only when the localization length is larger than the system size. In the opposite limit of small localization length, eigenstates are very far from being ergodic and the spectrum exhibits Poissonian statistics (see Izrailev, 1990).

Finally, it is worth mentioning that there is a conjecture that relates the statistical behaviour of the complex zeros of the Riemann zeta function to the statistical behaviour of the eigenvalues of large random matrices. The Riemann zeta function is defined as

$$\zeta(z) = \sum_{n=1}^{\infty} \frac{1}{n^z} = \prod_p \frac{1}{1 - p^{-z}}, \quad \text{Re} z > 1, \qquad (6.215)$$

where the product is over primes. The function $\zeta(z)$ can be extended by analytic continuation elsewhere in the complex z plane, except for a simple pole at $z = 1$. The Riemann hypothesis states that the complex zeros of $\zeta(z)$ all are along the line $\text{Re}(z) = \frac{1}{2}$. If E_n denotes the imaginary part of the n-th complex zero of

---

[23] Universal fluctuations in dynamical systems with classical chaos are beginning to be understood in terms of Gutzwiller's semiclassical periodic orbit theory (Müller *et al.*, 2004).

[24] Note, however, that for finite n there exist *scars*; that is, non-ergodic eigenstates that have prominent density near classical periodic orbits (see Heller, 1991).

ζ; that is, $\zeta(\frac{1}{2} + iE_n) = 0$, $E_{n+1} > E_n$, then there is evidence that the level-spacing distribution for the quantities E_n follow the GUE spectral statistics. It has been suggested that the E_n are eigenvalues of a quantum Hamiltonian obtained by quantizing a classical chaotic system without time-reversal symmetry (Berry, 1985).

6.5.7 ⋆ *The chaos border for the quantum computer hardware*

As an example of the transition to chaos, in this section we consider a model of n spin-$\frac{1}{2}$ particles (qubits) placed on a two-dimensional lattice in the presence of an external magnetic field directed along z. Nearest neighbour spins interact via an Ising coupling with random strength. The Hamiltonian of the system is

$$H = \sum_i \Gamma_i \sigma_i^z + \sum_{<i,j>} J_{ij} \sigma_i^x \sigma_j^x, \qquad (6.216)$$

where the operators σ_i are the standard Pauli matrices acting on the i-th qubit, the second sum in the Hamiltonian only runs over nearest-neighbour spins (at the borders of the lattice periodic boundary conditions are imposed), Γ_i corresponds to the energy separation between the states of the qubit and J_{ij} is the interaction strength between qubit i and qubit j. The parameters Γ_i and J_{ij} are randomly and uniformly distributed in the intervals $[\Delta_0 - \delta/2, \Delta_0 + \delta/2]$ and $[-J, J]$, respectively. Hamiltonian (6.216) was proposed (Georgeot and Shepelyansky, 2000) as a simple model of the quantum computer hardware, in which system imperfections generate undesired interqubit couplings J_{ij} and energy fluctuations δ_i.[25]

For $J = \delta = 0$, the spectrum of the Hamiltonian is composed of $n + 1$ degenerate levels, and the interlevel spacing is $2\Delta_0$, which corresponds to the energy required to flip a single qubit. We study the case $\delta, J \ll \Delta_0$, in which the degeneracies are resolved and the spectrum is composed of $n + 1$ bands, each band corresponding to states with given numbers of spins up and down. Since $\delta, J \ll \Delta_0$, the interband coupling in (6.216)

---

[25]For $\Gamma_i = \Gamma$ and $J_{ij} = J$ independently of the site index, that is, without randomness, Eq. (6.216) reduces to the Ising model in a transverse field. In the thermodynamic limit $n \to \infty$ this model exhibits a quantum phase transition at the critical value $\lambda = J/\Gamma = 1$. The magnetization $\langle \sigma_x \rangle$ of the ground state is different from zero for $\lambda > 1$ and vanishes at the transition. We note that the Ising model is paradigmatic for the study of the relationship between the entanglement structure of the ground state and quantum phase transitions, see Osborne and Nielsen (2002) and Osterloh *et al.* (2002).

may be neglected and each single band can be studied separately. The number of states in the j-th band $(j = 0, \ldots, n)$ is equal to the binomial coefficient $\binom{n}{j}$. Since the δ_i fluctuate randomly in an interval of size δ, each band at $J = 0$ (except the extremes) has a Gaussian shape of width $E_B \sim \sqrt{n}\delta$.[26] The average number of states inside a band N_B is of the order of $N/n = 2^n/n$, so that the energy spacing between adjacent states inside one band is $\Delta_n = E_B/N_B \sim n^{3/2}2^{-n}\delta$, which becomes exponentially small when n increases. The highest density of states is obtained for the central energy band (with equal numbers of spins up and down) and we therefore expect quantum chaos to show up there first. Hence, we shall focus on the central band with zero magnetization ($\sum_i \sigma_i^z = 0$).

The first term in Hamiltonian (6.216) describes independent particles and therefore cannot lead to quantum chaos. Indeed, the energy spectrum of a non-interacting many-body system is given by the sum of independent single-particle energies and thus RMT spectral statistics does not apply. As we shall see in what follows, the second (interaction) term in (6.216) may lead to quantum chaos behaviour.

The transition to ergodic eigenstates and to quantum chaos behaviour can be detected from the change in the spectral statistics of the system. In particular, the level-spacing statistics $P(s)$ is a very convenient quantity for numerical studies. Figure 6.32 (left) shows the $P(s)$ distribution for a lattice with $n = 16$ spins and three different values of the coupling strength J. The transition is clearly seen in the spectral statistics from a Poisson distribution (6.204) at small J to a Wigner (GOE) distribution (6.212), characteristic of RMT for large enough J. To probe this transition in more detail, it is useful to define a parameter η that varies continuously from $\eta = 1$ (Poisson) to $\eta = 0$ (Wigner). We thus define

$$\eta = \frac{\int_0^{s_0} \left[P(s) - P_W^O(s)\right]ds}{\int_0^{s_0} \left[P_P(s) - P_W^O(s)\right]ds}, \qquad (6.217)$$

where $P_P(s)$ and $P_W^O(s)$ are the Poisson and Wigner level-spacing distributions and $s_0 = 0.4729\ldots$ is their first intersection point. Figure 6.32 (right) gives the dependence of the parameter η on the scaled coupling Jn/δ (see below for the physical motivation of this scaling). The Poisson to Wigner crossover becomes sharper when n increases, suggesting a sharp transition in the thermodynamic limit $n \to \infty$. The minimum spreading of curves is

---

[26]The majority of states are inside this interval, while the total band width is $\approx n\delta/2$. This is due to rare events in the sum of n random numbers.

for $\eta(J_c) \approx 0.2$, corresponding to the *chaos border* $J_c n/\delta \approx 3.7$

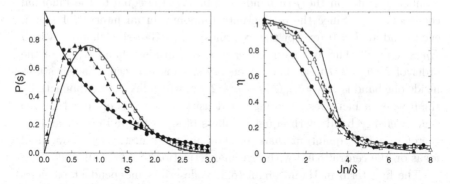

Fig. 6.32   Left: Level-spacing statistics for $n = 16$ spins, $J = 0.05\delta$ (circles, $\eta = 1.01$), $J = 0.2\delta$ (triangles, $\eta = 0.32$), and $J = 0.4\delta$ (squares, $\eta = 0.05$). The full curves show Poisson and Wigner (GOE) distributions. The statistics is obtained from the states in the middle of the central energy band ($\pm 5\%$ of states around the centre). Right: Dependence of η on the scaled coupling Jn/δ, for $n = 9$ qubits (circles, $N_D = 10^4$ random realizations of δ_i, J_{ij}), $n = 12$ (squares, $N_D = 10^3$), $n = 15$ (diamonds, $N_D = 45$), $n = 16$ (empty triangles, $N_D = 23$), and $n = 18$ (filled triangles, $N_D = 3$). The figures are taken from Benenti *et al.* (2001a).

Let us now explain the border for the transition to chaos by means of a qualitative physical argument. It is convenient to refer to the "quantum register states"; that is, to the basis of the eigenstates $|i_{n-1}, i_{n-2}, \ldots, i_1, i_0\rangle$ of (6.216) at $J = 0$ (where $|i_k\rangle$ is an eigenstate of σ_k^z and $i_k = 0$ or 1). The interaction term in (6.216) only couples quantum register states that differ by the value of σ_z for exactly two spins. As a consequence, the Hamiltonian matrix in the basis of quantum register states is very *sparse*; that is, only a few matrix elements are non-zero. Indeed, a quantum register state is only coupled to $2n$ other states among the 2^n available. We stress that this sparsity is a very general property of many-body systems and is due to the fact that the interaction in the natural world is two-body. Hence, a relevant energy scale for the model is the level spacing Δ_c between *directly coupled* multi-particle states. This energy scale can be evaluated as follows. A quantum register state is coupled to $\sim n$ states in an energy interval of size $\sim \delta$: out of the $2n$ non-zero couplings mentioned above we consider only those inside the energy band; that is, we exclude the two-qubit transitions $|00\rangle \rightarrow |11\rangle$ and $|11\rangle \rightarrow |00\rangle$ whose energy cost is $2\Delta_0 \gg J, \delta$. Thus, $\Delta_c \sim \delta/n$. Note that there are two other relevant energy scales in the problem: the average level spacing Δ_0 between the two states of a qubit (one-particle spacing) and the mean level spacing $\Delta_n \sim n^{3/2}2^{-n}\delta$ between multi-qubit states. We have $\Delta_n \ll \Delta_c \ll \Delta_0$. The important point is that Δ_c and Δ_n vary in extremely different manners with respect to n: Δ_n drops exponentially with n, Δ_c in a polynomial manner. The three energy scales that are relevant to the problem are drawn in Fig. 6.33. At small J the many-body problem (6.216) can be solved using perturbation theory. This approach breaks down when the typical interaction matrix element J between directly coupled states is of the order of their energy separation $\Delta_c \sim \delta/n$. We can therefore estimate the chaos border J_c from the condition

$$J_c \sim \Delta_c \sim \delta/n. \tag{6.218}$$

The quantum chaos regime corresponds to $J > J_c$. This expectation is in agreement with the numerical results shown in Fig. 6.32 (right). We stress that the chaos border is exponentially larger than the multi-qubit level spacing Δ_n.

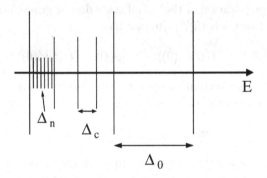

Fig. 6.33 A schematic drawing of the different relevant energy scales in (6.216): one-qubit level spacing Δ_0, level spacing Δ_c between multi-qubit states directly coupled by the two-body interaction and energy spacing Δ_n between multi-qubit states.

The transition in the level-spacing statistics reflects a qualitative change in the structure of the eigenstates of Hamiltonian (6.216). While for $J \ll J_c$ these eigenstates are very close to the quantum register states, for $J > J_c$ they become a superposition of an exponentially large number of non-interacting eigenstates. The mixing of the non-interacting eigenstates takes place inside a Breit–Wigner energy width Γ given by the Fermi golden rule: $\Gamma \sim J^2/\Delta_c \sim J^2 n/\delta$. As a result, the residual interaction spreads a quantum register state over an exponentially large number of states after a chaotic time scale

$$\tau_\chi = \frac{1}{\Gamma} \sim \frac{\delta}{J^2 n}. \tag{6.219}$$

This sets an upper time limit to the stability of a generic superposition of states coded in the quantum computer wave function. In addition, it is clear that a necessary requirement for quantum computer operability is the possibility to operate many quantum gates within the chaotic time scale.

Exercise 6.19 Adapt the estimate (6.218) to the case in which the interaction term in (6.216) is all to all; that is, $H = \sum_i \Gamma_i \sigma_i^z + \sum_{i \neq j} J_{ij} \sigma_i^x \sigma_j^x$.

6.5.8 * *The quantum Loschmidt echo*

As we have discussed in Sec. 6.5.1, classical chaotic motion is characterized by local exponential instability with respect to initial conditions. Such characterization of dynamical chaos cannot be translated *sic et simpliciter* in the quantum domain since, due to unitarity of quantum evolution, imperfections in the preparation of the initial state do not grow in time. Indeed, given two initial states $|\psi(0)\rangle, |\phi(0)\rangle$ we have

$$|\psi(t)\rangle = U(t)|\psi(0)\rangle, \qquad |\phi(t)\rangle = U(t)|\phi(0)\rangle, \tag{6.220}$$

with $U(t)$ unitary time-evolution operator from time 0 to time t. Therefore, the scalar product does not change in time:

$$\langle \psi(t)|\phi(t)\rangle = \langle \psi(0)|\phi(0)\rangle. \tag{6.221}$$

In other words, the fact that in quantum mechanics there is no notion of trajectories does not allow us to apply the criterion of stability with respect to variation of the initial conditions.

In classical mechanics, one may describe the motion in terms of a phase-space distribution function, whose evolution is unitary and governed by the Liouville equation. However, in this description, instead of slightly changing the initial conditions, one may study the stability properties by introducing small variations of system parameters. It turns out that exponentially unstable systems exhibit the same rate of exponential instability by slightly changing the initial conditions with fixed parameters or the parameters with fixed initial conditions. The advantage of the latter procedure is that it can be directly extended to quantum mechanics, thus allowing a comparison of the stability properties of classical and quantum motion.

Let us therefore consider two slightly different Hamiltonians H and $H_\epsilon = H + \epsilon V$ and take the scalar product

$$f(t) = |\langle \psi(0)|U_\epsilon^\dagger(t)U(t)|\psi(0)\rangle|^2 = |\langle \psi_\epsilon(t)|\psi(t)\rangle|^2, \qquad (6.222)$$

where $U(t)$ and $U_\epsilon(t)$ are the time-evolution operators corresponding to H and H_ϵ, respectively. The quantity $f(t)$ is known as the *fidelity* and measures how accurate the solution remains under small perturbations of the Hamiltonian. Expression (6.222) is also known as the *Loschmidt echo* since it can be seen as a measure of the accuracy at which the initial state $|\psi(0)\rangle$ is recovered by inverting the dynamics at time t and returning to time 0 with the perturbed Hamiltonian H_ϵ.

The name Loschmidt echo derives from the famous *Loschmidt paradox*, which was raised by the Bohemian physicist Joseph Loschmidt against Boltzmann statistical theory. As is well known, in the second half of the nineteenth century Boltzmann derived his famous equation. This equation describes the temporal evolution of the function $f(v,t)$, which gives the probability that a given molecule in a gas has velocity v at time t. Boltzmann's intent was to derive the second law of thermodynamics from dynamical principles. He was able to prove the so-called H-theorem. That is, he introduced the function

$$H(t) = \int dv \, f(v,t) \ln f(v,t) \qquad (6.223)$$

and showed that, as $t \to \infty$, $H(t)$ approaches a minimum value. Correspondingly, the function $f(v,t)$ approaches the Maxwell velocity distribution. The entropy S, which is proportional to $-H$, approaches a maximum value.

Loschmidt's observation was based on the fact that Newton's equations of motion for the molecules in a gas are exactly time reversible. Therefore, by reverting the velocities of the molecules, the system "goes backwards". This implies that for any initial condition for which H decreases (S increases) there is an initial condition for which H increases (S decreases). Hence, Boltzmann's conclusion cannot be correct and these time-reversed evolutions appear to violate the second law of thermodynamics. Boltzmann took Loschmidt's criticism very seriously and this led him to the statistical interpretation of the second law and finally to the well-known expression (5.251) for the entropy.[27]

We now know the solution to the Loschmidt paradox. In an isolated dynamical system there are *fluctuations*, during which entropy may decrease for some time. These fluctuations are less and less frequent as they become stronger and stronger. In any event, if the system possesses the mixing property (see Sec. 6.5.1), then for $t \to +\infty$ (and also for $t \to -\infty$) the phase-space density $\rho(x,v,t)$ always

---

[27]When Loschmidt died in 1895 Boltzmann said in his eulogy that "His work forms a mighty cornerstone that will be visible as long as science exists."

converges, after coarse graining, to the microcanonical ensemble and the entropy $S = -k_B \int \rho \ln \rho$ to its maximum, microcanonical value (5.251).

It is instructive to consider the classical Loschmidt echo f_c, which can be defined in terms of classical probability distributions $\rho(q, p, t)$:

$$f_c(t) = A \int dq dp \, \rho_\epsilon(q, p, t) \rho(q, p, t), \qquad (6.224)$$

where ρ and ρ_ϵ are the classical probability distributions obtained by evolving the same initial distribution $\rho(q, p, 0)$ up to time t under the Hamiltonians H and H_ϵ, respectively. The prefactor A is a normalization constant:

$$A = \frac{1}{\sqrt{\int dq dp \, \rho_\epsilon^2(q, p, t)} \sqrt{\int dq dp \, \rho^2(q, p, t)}}. \qquad (6.225)$$

Note that ρ_ϵ and ρ are normalized to unit total probability:

$$\int dq dp \, \rho(q, p, t) = \int dq dp \, \rho_\epsilon(q, p, t) = 1. \qquad (6.226)$$

For chaotic systems, with positive Lyapunov exponent λ, it turns out that the asymptotic decay of fidelity to its microcanonical value $f_c(\infty)$ is the same as for correlation functions and therefore can be either power-law or exponential. In the latter case, the rate of the exponential decay is governed by the gap in the discretized *Perron–Frobenius operator*, which describes the evolution of the coarse-grained distribution. Note that the asymptotic decay rate is not given, in general, by the Lyapunov exponent. The short-time decay of f_c is instead different from that of correlation functions. Indeed, it is exponential with a decay rate given by the Lyapunov exponent. The Lyapunov decay starts after an initial transient time

$$t_\nu = \frac{1}{\lambda} \ln\left(\frac{\nu}{\epsilon}\right), \qquad (6.227)$$

where ν is the size of the initial distribution and ϵ the perturbation strength. Notice that this is the time required to amplify the perturbation up to the size ν of the initial distribution. The Lyapunov decay takes place in a short time interval $t_\nu < t < t_\epsilon$, where

$$t_\epsilon \sim \frac{1}{\lambda} \ln\left(\frac{2\pi}{\epsilon}\right) \qquad (6.228)$$

is the time taken to amplify the perturbation up to randomization of phases. After this a diffusion regime ($f(t) \propto 1/\sqrt{Dt}$, with D the diffusion constant) follows. This power-law decay goes on until the diffusion time $t_D \sim L^2/D$ (L is the system size) and is followed by relaxation to the microcanonical equilibrium distribution with a rate determined by the gap in the discretized Perron–Frobenius operator. Both the short-time and the asymptotic fidelity decays are shown in Fig. 6.34 for the classical sawtooth map.[28]

Two characteristics of classical fidelity decay are worth mentioning:

1. In both the short-time Lyapunov decay as well as in the asymptotic exponential decay, *the decay rate is independent of the perturbation strength* ϵ (see Fig. 6.35). This fact, which may look quite surprising, is due to the exponential character of the instability, which renders the strength of the perturbation irrelevant. In systems with linear instability fidelity depends on the perturbation strength. Note that in quantum mechanics there is no exponential instability outside the Ehrenfest time scale. Therefore, outside this scale, one expects the decay rate of quantum fidelity to depend on perturbation strength.

2. If instead of a static perturbation we apply stochastic noise, the fidelity decay remains the same. This means that the effect of a noisy environment on the decay of fidelity for a classical chaotic system is similar to that of a generic static Hamiltonian perturbation. Indeed, owing to internal dynamical chaos, the deterministic or noisy character of the perturbation is not important. This raises the question whether for quantum systems that are classically chaotic static errors will have the same effect as stochastic perturbations induced by the environment. An

---

[28] The gap in the discretized Perron–Frobenius operator can be numerically computed using the following method:

(i) the phase-space torus ($0 \leq \theta < 2\pi$, $-\pi L \leq p < \pi L$) is divided into $N \times NL$ square cells;

(ii) the transition matrix elements between cells are determined numerically by iterating for one map step the phase-space distributions given by the characteristic functions of each cell: in this manner we build a finite dimensional approximation to the one-period evolution operator U;

(iii) this truncated evolution matrix $U^{(N)}$ (of size $LN^2 \times LN^2$) is diagonalized: it is no longer unitary, and its eigenvalues $z_i^{(N)}$ are inside the unit circle in the complex plane. The non-unitarity of the coarse-grained evolution is due to the fact that the transfer of probability to finer scale structures in the phase space is cut-off, and this results in effective dissipation;

(iv) the *Ruelle–Pollicott resonances* correspond to "frozen" non-unimodular eigenvalues, namely $z_i^{(N)} \to z_i$ when $N \to \infty$, with $|z_i| < 1$. Convergence of eigenvalues to values inside the unit circle comes from the asymptotic self-similarity of chaotic dynamics. The asymptotic ($t \to \infty$) relaxation of correlations is determined by the resonance with largest moduli, $|\bar{z}| = \max_i |z_i| < 1$, giving a decay rate $\gamma_0 = \ln |\bar{z}|$. Note that there is a gap $1 - |\bar{z}|$ between this resonance and the unit circle.

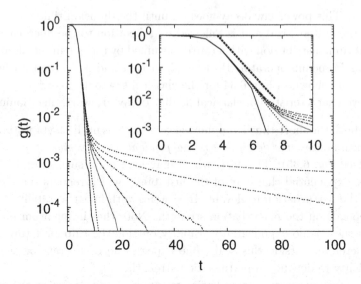

Fig. 6.34 Decay of the classical fidelity $g(t) = [f_c(t) - f_c(\infty)]/[f_c(0) - f_c(\infty)]$ for the sawtooth map (defined by Eq. (3.200)) on the torus $0 \leq \theta < 2\pi$, $-\pi L \leq p < \pi L$, with the parameters $K = (\sqrt{5} + 1)/2$ and $\epsilon = 10^{-3}$ (H and H_ϵ correspond to the sawtooth map with parameter K and $K + \epsilon$, respectively) for different values of $L = 1, 3, 5, 7, 10, 20, \infty$ from the fastest to the slowest decaying curve. The initial phase-space density is chosen to be uniform in the region $\theta \in [0, 2\pi)$, $p \in [-\pi/100, \pi/100]$ and zero elsewhere. Note that for $t_\epsilon < t < t_D \sim L^2/D$, that is, between the Lyapunov decay and the exponential asymptotic decay (with rate determined by the gap in the discretized Perron–Frobenius operator), there is a $\propto 1/\sqrt{Dt}$ decay, as expected from diffusive behaviour. Inset: magnification of the same plot for short times, with the corresponding Lyapunov decay indicated as a thick dashed line. The figure is taken from Benenti *et al.* (2003).

important role here is expected to be played by quantum chaos.

Let us now turn to the quantum case. The behaviour of quantum fidelity has been studied, numerically and analytically, with different tools: semiclassical methods, perturbation theory, random-matrix theory (see the guide to the bibliography at the end of this chapter). For chaotic systems, one may distinguish three main regimes for the fidelity decay:

1. *Lyapunov regime* – For times shorter than the Ehrenfest time (6.174) the quantum fidelity follows the classical behaviour, characterized by a perturbation independent decay rate equal to the Lyapunov exponent λ:

$$f(t) \sim e^{-\lambda t}. \tag{6.229}$$

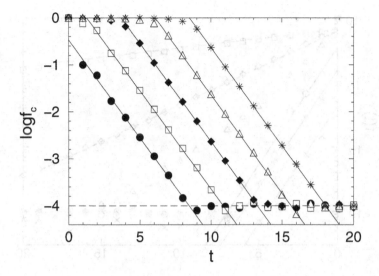

Fig. 6.35 Decay of classical fidelity for the classical sawtooth map with $K = 1$, $L = 1$, initial distribution uniform in the strip $0 \le \theta < 2\pi$, $-\nu/2 \le p < \nu/2$ ($\nu = 2\pi/10^4$) and zero elsewhere, perturbation strength $\epsilon = 10^{-3}$ (circles), 10^{-4} (squares), 10^{-5} (diamonds), 10^{-6} (triangles), and 10^{-7} (stars). The straight lines show the decay $f_c(t) \propto \exp(-\lambda t)$, with Lyapunov exponent $\lambda = \ln[(2 + K + \sqrt{K^2 + 4K})/2] \approx 0.96$. Note that this decay starts after a time $\propto \ln(1/\epsilon)$, in agreement with (6.227). The dashed line indicates saturation to the microcanonical value $f_c(\infty) = \nu/(2\pi L) = 10^{-4}$. The figure is taken from Benenti and Casati (2002b).

For sufficiently strong perturbation strength ϵ, namely $\sigma \equiv \epsilon/\hbar_{\text{eff}} > 1$, the fidelity drops to its saturation value $f(\infty) = \frac{1}{N}$ before the Ehrenfest time is reached. Here $N \sim \hbar_{\text{eff}}^{-d}$ is the dimension of the Hilbert space and d is the number of degrees of freedom.

2. *Fermi golden rule regime* – When the dimensionless parameter $\sigma < 1$, one may apply perturbation theory. The Fermi golden rule regime is characterized by the exponential decay

$$f(t) \sim e^{-\Gamma t}, \qquad (6.230)$$

where $\Gamma = \frac{U^2}{\Delta}$. Here Δ is the average level spacing and U the typical matrix element of the perturbation operator ϵV connecting the eigenstates of H. This regime takes place for perturbation strengths $\sigma < 1$ but such that $U > \Delta$. As an example, in Fig. 6.36 we show the crossover between the Fermi golden rule and the Lyapunov regime for the quantum sawtooth map.

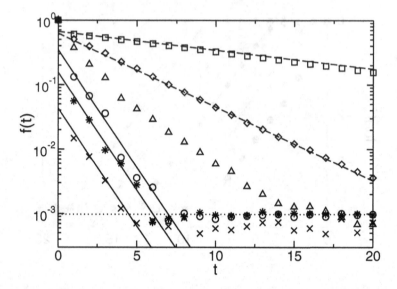

Fig. 6.36   Quantum fidelity decay for the sawtooth map (3.200) in the quantum-chaos regime. Here $N = 2^{10}$ levels, $K = 0.75$, $\epsilon = 10^{-3}$ (squares), 2×10^{-3} (diamonds), 3×10^{-3} (triangles), 10^{-2} (circles), 2×10^{-2} (stars), and 3×10^{-2} (crosses). Dashed lines correspond to Fermi Golden rule decay: $f(t) \sim \exp(-A\sigma^2 t)$, where $A \approx 2.4$. Full lines show Lyapunov decay: $f(t) \sim \exp(-\lambda t)$, with maximum Lyapunov exponent $\lambda \approx 0.84$. Note that the crossover between the Fermi golden rule and Lyapunov regimes takes place at $\epsilon \sim 5 \times 10^{-3}$, corresponding to $\sigma \sim 1$. The dotted line shows the fidelity saturation value $f(\infty) = 1/N$. A momentum eigenstate is chosen as initial wave function and data are obtained after averaging over 100 different initial conditions. Figure taken from Benenti *et al.* (2004).

3. *Perturbative regime* – If the perturbation is small enough, that is, $U < \Delta$, then stationary perturbation theory may be used and we have the Gaussian decay

$$f(t) \sim e^{-U^2 t^2}. \qquad (6.231)$$

It is clear from Eqs. (6.230) and (6.231) that the crossover between the Fermi golden rule and Gaussian regimes takes place at time $\bar{t}$ such that

$$\Gamma \bar{t} \approx U^2 \bar{t}^2. \qquad (6.232)$$

Therefore, $\bar{t} \sim \frac{\Gamma}{U^2} \sim t_H$, where $t_H = \frac{1}{\Delta}$ is the Heisenberg time. If the perturbation is strong enough, then the fidelity completely decays within the Heisenberg time. In this case, exponential decay dominates. On the other hand, if the perturbation is sufficiently small then no significant decay of the fidelity takes place before the Heisenberg time. In this case the decay is Gaussian and occurs after the Heisenberg time. The value of the perturbation that divides these two regimes is given by the condition $\Gamma \bar{t} \approx 1$; that is, $\Gamma \sim \Delta$, which coincides with (6.232). Note that, if the perturbation is memoryless instead of static, then Fermi golden rule decay applies for any $\sigma < 1$.

Notice that quantum fidelity decay is perturbation dependent, except for the Lyapunov regime. The existence of a Lyapunov regime is in agreement with the correspondence principle, according to which within the Ehrenfest time scale a quantum wave packet is exponentially unstable, just as a classical chaotic orbit.

6.5.9 * *Dynamical stability of quantum motion*

Strong numerical evidence (see Shepelyansky, 1983) has been obtained that quantum evolution is very stable, in sharp contrast to the extreme sensitivity to initial conditions and rapid loss of memory that is the very essence of classical chaos. In computer simulations this effect leads to practical irreversibility of classical motion. Indeed, even though the exact equations of motion are reversible, any, however small, imprecision such as *computer round-off errors*, is magnified by the exponential instability of orbits to the extent that any memory of the initial conditions is effaced and reversibility is destroyed. This fact is illustrated in Fig. 6.37, drawn for the kicked-rotator model. Here the momentum is reversed after $t = 100$ kicks. Due to the exponential instability of classical orbits, the time reversibility for orbits with inverted momentum disappears after $t_\epsilon \approx 35$. Note that computer simulations are performed with round-off errors of the order of $\epsilon \sim 10^{-14}$ and

$$
t_\epsilon \approx \frac{1}{\lambda} |\ln \epsilon| \approx \frac{|\ln \epsilon|}{\ln(K/2)} \tag{6.233}
$$

is the time taken to amplify the perturbation and significantly modify the trajectories. In contrast, as shown in Fig. 6.37, in the quantum case almost exact reversion is observed in numerical simulations (in this case velocity reversal is obtained by complex conjugation of the wave function, $|\psi\rangle \rightarrow |\psi^\star\rangle$). Therefore, quantum dynamics, although diffusive (for times shorter than the localization time $t^\star$), lacks dynamical instability. The physical reason for this striking difference between quantum and classical motion is rooted in the discreteness of phase space in quantum mechanics. If we consider the classical motion (governed by the Liouville equation) of some phase-space density, smaller and smaller scales are explored exponentially fast. These fine details of the density distribution are rapidly lost under small perturbations. In quantum mechanics, there is a lower limit to this process, set by the size of the Planck cell.

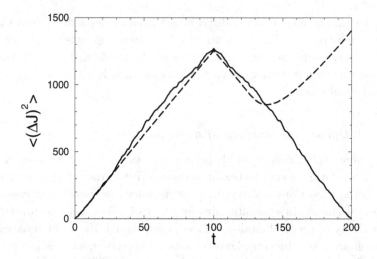

Fig. 6.37 Practical irreversibility of classical motion (dashed curve) and reversibility of quantum motion (solid curve) for the kicked-rotator model in the chaotic regime for $K = 5$, $k = 50$.

The results on fidelity decay give an explanation of the different stability of quantum and classical motion observed in computer simulations. Round-off errors are always so small that $\sigma \ll 1$. If we model round-off errors as memoryless errors of size ϵ, then the quantum fidelity decays as $f(t) \sim \exp(-\sigma^2 t)$. For $\sigma \sim 10^{-15}$, the simulation is then stable up to an enormously long time scale

$$t_f^{(q)} \sim \frac{1}{\sigma^2}. \qquad (6.234)$$

On the other hand, in the classical case fidelity decays after a time

$$t_f^{(c)} \sim \frac{1}{\lambda} + t_\nu \sim \frac{1}{\lambda}\left[1 + \ln\left(\frac{\nu}{\epsilon}\right)\right], \qquad (6.235)$$

where t_ν is the time to start the exponential fidelity decay with rate λ. Note that in this case the scaling is only logarithmic in ϵ, due to exponential instability. Thus, in practice, classical motion is irreversible after the logarithmically short time scale $t_f^{(c)}$.

We also point out that an exponential decay of the fidelity alone does not imply exponential instability. For instance, Fermi golden rule decay takes place after the Ehrenfest time scale in the absence of exponential instability.

6.5.10 * *Dynamical chaos and dephasing: the double-slit experiment*

A word of caution is necessary when discussing quantum fidelity decay and its relation with the corresponding classical decay. Indeed, for a pure state $|\psi(0)\rangle$ one has

$$f_\psi(t) = |a_\psi(t)|^2 = \left|\langle\psi(0)|U_\epsilon^\dagger(t)U(t)|\psi(0)\rangle\right|^2 = \left|\operatorname{Tr}(E(t)\rho(0))\right|^2, \quad (6.236)$$

where $a_\psi(t)$ is the fidelity amplitude, $\rho(0) = |\psi(0)\rangle\langle\psi(0)|$ the initial density matrix and $E(t) = U_\epsilon^\dagger(t)U(t)$ the echo operator. On the other hand, for a mixed state $\rho(0) = \sum_\psi p_\psi|\psi(0)\rangle\langle\psi(0)|$ ($\sum_\psi p_\psi = 1$) one may define fidelity in the following two manners:

1.

$$f(t) = \frac{\operatorname{Tr}[\rho_\epsilon(t)\rho(t)]}{\operatorname{Tr}[\rho(0)]^2} = \frac{\operatorname{Tr}[E^\dagger(t)\rho(0)E(t)\rho(0)]}{\operatorname{Tr}[\rho(0)]^2}, \quad (6.237)$$

2.

$$\mathcal{F}(t) = \left|\sum_\psi p_\psi a_\psi(t)\right|^2 = \sum_\psi p_\psi^2 f_\psi(t) + \sum_{\psi \neq \psi'} p_\psi p_{\psi'}' a_\psi(t) a_{\psi'}^\star(t). \quad (6.238)$$

Definition (6.237) is perhaps the most natural and popular, while definition (6.238) is a straightforward generalization of expression (6.236) to the case of arbitrary mixed initial states $\rho(0)$; note that (6.238) can also be written as $\mathcal{F}(t) = |\operatorname{Tr}(E(t)\rho(0))|^2$. Both expressions reduce to (6.236) for a pure initial state ($[\rho(0)]^2 = \rho(0)$).

Notice that the function (6.237) has a well-defined classical limit that coincides with the classical fidelity (6.224). We may write

$$f(t) = \frac{1}{\operatorname{Tr}[\rho(0)]^2} \sum_{\psi,\psi'} p_\psi p_{\psi'}' W_{\psi\psi'}, \quad (6.239)$$

with transition probabilities

$$W_{\psi\psi'} = \left| \langle \psi | E | \psi' \rangle \right|^2. \tag{6.240}$$

Therefore, the decay of this quantity has nothing to do with quantum dephasing and is just due to the transitions induced by the echo operator E from the initially populated states to all empty states.

Expression $\mathcal{F}(t)$ is instead composed of two terms. The first is a sum of fidelities $f_\psi = |a_\psi|^2$ of individual pure initial states with weights p_ψ^2. The second term depends on the relative phases of the fidelity amplitudes; therefore, fidelity $\mathcal{F}$ *accounts for quantum interference* and is expected to retain quantal features even in the deep semiclassical region.

A question of interest is under what conditions these quantum interference terms decay. It is known that the presence of an environment leads to decoherence and thus to the decay of interference terms. Indeed, external noise induces non-unitary evolution leading to the decay of the off-diagonal elements of the density matrix in the eigenbasis of some physical observable, thus restoring the classical behaviour. On the other hand we know that classical deterministic systems, due to internal dynamical chaos, can exhibit a motion that is indistinguishable from that of systems under the action of an external random perturbation. Analogously one may inquire whether deterministic chaotic evolution of a quantum state can lead to quantum dephasing. It is possible to show that, for classically chaotic systems, internal chaos induces a dephasing, leading to the decay of fidelity $\mathcal{F}$ at a rate that is determined by the decay of an appropriate classical correlation function. Therefore, in this case internal dynamical chaos produces a dephasing effect similar to the decoherence induced by the environment.

A more direct and vivid illustration of the dephasing effect of classical dynamical chaos is provided by the following numerical double-slit experiment. The time-dependent Schrödinger equation $i\hbar \frac{\partial}{\partial t} \psi(x, y, t) = H\psi(x, y, t)$, with $H = \frac{p^2}{2m}$, is solved numerically for a quantum particle that moves freely inside the two-dimensional domain as indicated in Fig. 6.38 (full line). Note that the domain is composed of two regions that are only connected via two narrow slits. We shall refer to the upper bounded region as to the *billiards domain*, and to the lower as the *radiating region*.

The lower, radiating region, should in principle be infinite. Thus, in order to efficiently damp waves at finite boundaries, an absorbing layer is introduced around the radiating region. More precisely, in the region referred to as the absorber, a negative imaginary potential is added to the

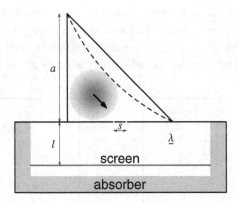

Fig. 6.38   The geometry of the numerical double-slit experiment. All scales are in true proportions. The two slits are placed at a distance s on the lower side of the billiards. We choose scaled units in which the Planck constant $\hbar = 1$, mass $m = 1$ and the base of the triangular billiards has length $a = 1$. The circular arc (dashed curve) has radius $R = 2$. The initial state $|\psi(t = 0)\rangle$ is a Gaussian wave packet (a coherent state) centred at a distance $a/4$ from the lower-left corner of the billiards (in both Cartesian directions) and with velocity v pointing to the midpoint between the slits. The screen is located at a distance $l = 0.4$ from the base of the triangle. The magnitude of velocity v (in our units equal to the wave-number $k = v$) sets the de Broglie wavelength $\lambda = 2\pi/k$. We also take $k = 180$ corresponding to approximately 1600 excited states of the closed quantum billiards. The slit distance has been set to $s = 0.1 \approx 3\lambda$ and the width of the slits is $d = \lambda/4$. The wave-packet is also characterized by the position uncertainty $\sigma_x = \sigma_y = 0.24$, which was chosen as large as possible in the present geometry in order to have a small uncertainty in momentum $\sigma_k = 1/(2\sigma_x)$. The figure is taken from Casati and Prosen (2005).

Hamiltonian: $H \to H - iV(x,y)$, $V \geq 0$, which, according to the time-dependent Schrödinger equation, ensures exponential damping in time.

While the wave-function evolves with time, a small probability current leaks from the billiards and radiates through the slits. The radiation probability is recorded on a horizontal line $y = -l$, referred to as the screen. The experiment stops when the probability that the particle remains in the billiards region becomes vanishingly small. The intensity at the position x on the screen is defined as the perpendicular component of the probability current, integrated in time:

$$I(x) = \int_0^\infty dt \, \mathrm{Im}\left[\psi^*(x,y,t)\frac{\partial}{\partial y}\psi(x,y,t)\right]_{y=-l}. \qquad (6.241)$$

Via the conservation of probability the intensity is normalized, $\int_{-\infty}^{\infty} dx \, I(x) = 1$, and is positive, $I(x) \geq 0$. $I(x)$ is interpreted as the probability density for a particle to arrive at the screen position x. The main result of the numerical simulations is shown in Fig. 6.39. As expected, the intensity $I(x)$ exhibits interference fringes when both slits are open, and is a simple unimodal distribution when only a single slit is open.

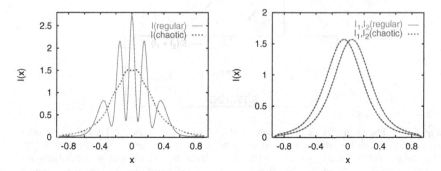

Fig. 6.39 Left: The total intensity after the double-slit experiment as a function of the position on the screen. $I(x)$ is obtained as the perpendicular component of the probability current, integrated in time. The full curve indicates the case of a regular billiards, while the dotted curve indicates the case of chaotic one. The dashed curve indicates the intensity averaged over the two 1-slit experiments (with either one or the other of the two slits closed). Right: The two pairs of curves represent the intensities on the screen for the two 1-slit experiments. The full curves indicate the case of the regular billiards while the dotted curves indicate the case of chaotic billiards, the results being practically the same. The figure is taken from Casati and Prosen (2005).

Let us now consider a simple modification of the above numerical simulation. The hypotenuse of the triangle is replaced by a circular arc (dashed curve in fig. 6.38). This change has a dramatic consequence for the classical ray dynamics inside the billiards; namely, the latter becomes fully chaotic. This has also a dramatic effect on the result of the double slit experiment. The interference fringes almost completely disappear and the intensity can be very accurately reproduced by the sum of intensities $[I_1(x) + I_2(x)]/2$ for the two experiments where only a single slit is open. This means that the result of such an experiment is the same as it would be in terms of classical ray dynamics. Notice, however, that at any given instant there is a definite phase relation between the wave function at both slits. Yet, as time proceeds, this phase relation changes and is lost after averaging over time.

In conclusion, if the billiards problem is classically integrable then interference fringes are observed, as in the case of the usual configuration of the *gedanken* double-slit experiment with plane waves. However, for a classically chaotic billiards, fringes disappear completely and the observed intensity on the screen is the sum of the intensities obtained by opening one slit at a time.

Notice that in the standard treatment of decoherence one starts from a pure state and then takes the trace over the environment. In this manner the state becomes mixed, the off-diagonal matrix elements decay and the system looses its quantal features. In our case, we have unitary evolution, the state is always a pure state and there is no decay of off-diagonal matrix elements. However, provided we are in presence of internal dynamical chaos, the process of integration over time leads to the same result. In this case off-diagonal matrix elements decay, on average. This numerical experiment shows that, by considering a pure quantum state in the absence of any external decoherence mechanism, internal dynamical chaos can provides the required randomization to ensure a quantum to classical transition in the semiclassical region.

In closing this section, we wish to briefly discuss a different definition of fidelity, which provides a measure of the "distance" between two probability distributions or two quantum states (an introduction to this quantity can be found in Nielsen and Chuang, 2000). One starts with the fidelity F of two probability distributions $\{p_x\}$ and $\{q_x\}$, defined as

$$F(p_x, q_x) = \sum_x \sqrt{p_x q_x}. \tag{6.242}$$

Clearly, when the two distributions coincide, $F(p_x, q_x) = \sum_x p_x = 1$.
The fidelity of two quantum states ρ and σ is defined as

$$F(\rho, \sigma) = \text{Tr} \sqrt{\rho^{1/2} \sigma \rho^{1/2}}. \tag{6.243}$$

Note that, if ρ and σ commute, we can find a basis $\{|i\rangle\}$ where

$$\rho = \sum_i r_i |i\rangle\langle i|, \qquad \sigma = \sum_i s_i |i\rangle\langle i|. \tag{6.244}$$

The fidelity can then be written as

$$F(\rho, \sigma) = \text{Tr}\left(\sum_i \sqrt{r_i s_i} |i\rangle\langle i|\right) = \sum_i \sqrt{r_i s_i} = F(r_i, s_i). \tag{6.245}$$

That is, in this case the quantum fidelity $F(\rho, \sigma)$ reduces to the classical fidelity $F(r_i, s_i)$ for the distributions of the eigenvalues r_i and s_i.

It is also interesting to note that the fidelity of a pure state $|\psi\rangle$ and an arbitrary state σ is given by

$$F(|\psi\rangle, \sigma) = \sqrt{\langle\psi|\sigma|\psi\rangle},$$ (6.246)

which is the square root of the overlap between $|\psi\rangle$ and σ. We can compare this definition with that we introduced above, Eq. (6.237): for the present case, in which $\rho = |\psi\rangle\langle\psi|$ and $\rho_\epsilon = \sigma$, it reduces to

$$f(t) = \langle\psi(t)|\sigma|\psi(t)\rangle.$$ (6.247)

Exercise 6.20   Another useful measure of the distance between two quantum states ρ and σ is the *trace distance* $D(\rho, \sigma)$, defined as

$$D(\rho, \sigma) = \tfrac{1}{2} \operatorname{Tr}|\rho - \sigma|,$$ (6.248)

where $|\rho - \sigma| \equiv \sqrt{(\rho - \sigma)^\dagger(\rho - \sigma)}$. Show that, if ρ and σ are single-qubit states, then $D(\rho, \sigma)$ is equal to half their distance on the Bloch ball.

6.5.11   * *Entanglement and chaos*

When a quantum system interacts with the environment, non-classical correlations (entanglement) between the system and the environment are in general established. On the other hand, when tracing over the environmental degrees of freedom, we expect that the entanglement between internal degrees of freedom of the system is reduced or even destroyed. This expectation is confirmed in models in which the environment is represented by a many-body system (for instance, a multimode environment of oscillators in the Caldeira–Leggett model or a spin bath). On the other hand, the following question arises: could the many-body environment be substituted with a closed deterministic system with a small number of degrees of freedom, but chaotic? In other words, can the complexity of the environment arise not from being many-body but from having chaotic dynamics? In this section, we give a positive answer to this question and discuss a simple *fully deterministic* model of *chaotic environment* (see Rossini *et al.*, 2006).

Let us consider the interaction of a two-qubit system with a quantum kicked rotator (the environment). The overall Hamiltonian H reads as follows:

$$H = H_S + H_R + H_{SR},$$

$$H_S = h_1\sigma_x^{(1)} + h_2\sigma_x^{(2)},$$

$$H_R = \tfrac{1}{2}I^2 + k\cos\theta \sum_j \delta(\tau - jT), \qquad (6.249)$$

$$H_{SR} = \epsilon\big(\sigma_z^{(1)} + \sigma_z^{(2)}\big)\cos\theta \sum_j \delta(\tau - jT),$$

where H_S is the system Hamiltonian, H_R describes the kicked rotator (the environment, also known as the reservoir) and H_{SR} the interaction. We consider $T = 2\pi/N$, with $N \gg 1$; that is, the kicked rotator is in the semiclassical regime. It is convenient to introduce a discrete time $t = \tau/T$, measured in units of kicks. The unitary operator describing the evolution of the overall system (qubits plus kicked rotator) in one kick is given by

$$U = \exp\Big\{-i\Big[k + \epsilon\big(\sigma_z^{(1)} + \sigma_z^{(2)}\big)\Big]\cos\theta\Big\}\exp\Big[-\tfrac{1}{2}iTI^2\Big]$$
$$\times \exp\big(-i\delta_1\sigma_x^{(1)}\big)\exp\big(-i\delta_2\sigma_x^{(2)}\big), \quad (6.250)$$

where we have defined $\delta_1 = h_1T$ and $\delta_2 = h_2T$. Let us numerically simulate the evolution of the overall system starting from a separable initial state $|\Psi_0\rangle = |\phi^+\rangle \otimes |\psi_0\rangle$.[29] Note that, given an initially pure state $|\Psi_0\rangle$, also the state $|\Psi(t)\rangle$ at time t is pure since the overall evolution (6.250) is unitary. The reduced density matrix $\rho_S(t)$ describing the two qubits at time t is then obtained after tracing the overall density matrix $\rho(t) = |\Psi(t)\rangle\langle\Psi(t)|$ over the kicked-rotator degree of freedom.

Let us evaluate the *entanglement of formation* $E(t)$ of the state $\rho_S(t)$ following Wootters (1998).[30] First of all we compute the so-called *concurrence*, defined as $C = \max(\lambda_1 - \lambda_2 - \lambda_3 - \lambda_4, 0)$, where the λ_i are the square roots of the eigenvalues of the matrix $R = \rho_S\tilde{\rho}_S$, in decreasing order. Here $\tilde{\rho}_S$ is the spin-flipped matrix of ρ_S, which is defined by

---

[29] The results discussed in this section do not depend on the initial condition $|\psi_0\rangle$, provided the kicked rotator is in the chaotic regime.

[30] The entanglement of formation E is defined as the mean entanglement of the pure states forming ρ_S, minimized over all possible decompositions $\rho_S = \sum_j p_j|\psi_j\rangle\langle\psi_j|$:

$$E(\rho_S) = \inf_{\text{dec}}\sum_j p_j E(|\psi_j\rangle).$$

We remind the reader that, according to Eq. (5.231), $E(|\psi_j\rangle) = S(|\psi_j\rangle\langle\psi_j|)$.

$\tilde{\rho}_S = (\sigma_y \otimes \sigma_y)\rho_S^*(\sigma_y \otimes \sigma_y)$; note that the complex conjugate is taken in the computational basis $\{|00\rangle, |01\rangle, |10\rangle, |11\rangle\}$). Once the concurrence has been computed, entanglement is obtained as $E = h\big((1 + \sqrt{1 - C^2})/2\big)$, where h is the binary entropy function: $h(x) = -x\log_2 x - (1 - x)\log_2(1 - x)$. Note that $0 \le E \le 1$ and that the limiting cases $E = 1$ and $E = 0$ correspond to maximally entangled states and separable states, respectively. We also compute the von Neumann entropy $S(t) = -\operatorname{Tr}[\rho_S(t)\log\rho_S(t)]$ of the reduced density matrix ρ_S. This quantity measures the entanglement between the two qubits and the kicked rotator.

In Fig. 6.40 we show the entanglement $E(t)$ and the reduced von Neumann entropy $S(t)$, for the cases in which the kicked rotator is in the fully chaotic ($K = 100$) or in the KAM-integrable regime ($K = 0.5$). The entanglement $E(t)$ decays in time and, in parallel, the reduced entropy $S(t)$ increases. This shows that the entanglement $E(t)$ between the two qubits drops due to the creation of entanglement (measured by $S(t)$) between the two qubits and the environment, whose role in model (6.249) is played by the kicked rotator. Note that there is a remarkable difference between the integrable and the chaotic case. In particular, in the chaotic case $S(t)$ saturates (up to corrections $O(1/N)$ due to the finite dimensionality of the environment) to the maximum possible value $S = 2$ for a two-qubit system.

It is interesting to compare the results obtained from the above deterministic model with those of a map derived for the two-qubit density matrix within the framework of the Kraus representation formalism discussed in Sec. 5.4. We model the effect of the interaction with the kicked rotator as a phase kick (see Sec. 6.1.9) rotating both qubits through the same angle about the z-axis of the Bloch sphere. This rotation is described by the matrix

$$R_{12}(\theta) = \begin{bmatrix} e^{-i\epsilon\cos\theta} & 0 \\ 0 & e^{i\epsilon\cos\theta} \end{bmatrix} \otimes \begin{bmatrix} e^{-i\epsilon\cos\theta} & 0 \\ 0 & e^{i\epsilon\cos\theta} \end{bmatrix}. \tag{6.251}$$

That is to say, we assume that the angle θ in the interaction Hamiltonian of (6.249) is drawn from a random uniform distribution in $[0, 2\pi[$. This is motivated by the fact that for the kicked-rotator model in the chaotic regime with $K \gg 1$ the phases at consecutive kicks can be considered as uncorrelated (the random-phase approximation). The evolution of the reduced density matrix from $\rho(t)$ to $\rho(t+1)$ is then obtained after averaging

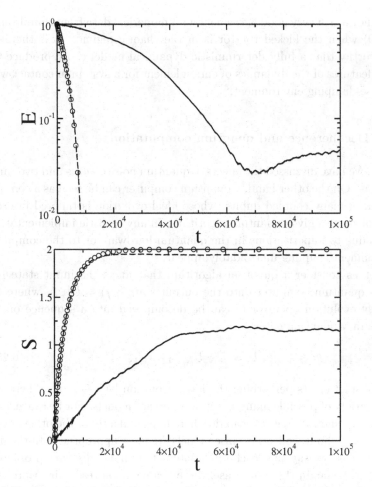

Fig. 6.40 Entanglement E (top) and von Neumann entropy S (bottom) as a function of time t (measured in number of kicks) for the model (6.249), for $N = 2^{13}$, $\delta_1 = 10^{-2}$, $\delta_2 = \sqrt{2}\delta_1$, $\epsilon = 10^{-2}$, $K = kT = 0.5$ (full curve) and $K = 100$ (dashed curve). The initial state of the kicked rotator is $|\psi_0\rangle = |-\frac{N}{2}\rangle$. The circles show the results obtained from the phase-damping model (6.252).

over θ and we end up with the map

$$\rho(t+1) = \frac{1}{2\pi} \int_0^{2\pi} d\theta\, R_{12}(\theta)\, e^{-i\delta\sigma_x^{(2)}} e^{-i\delta\sigma_x^{(1)}} \rho(t)\, e^{i\delta\sigma_x^{(1)}} e^{i\delta\sigma_x^{(2)}} R_{12}^\dagger(\theta).$$

$$(6.252)$$

This map can be iterated, so that we obtain $\rho(t)$ and then $E(t)$ and $S(t)$. The results derived from map (6.252) are shown in Fig. 6.40. It can be seen that they are in very good agreement with numerical data from Hamiltonian (6.249) when the kicked rotator is in the chaotic regime. It is therefore noteworthy that a fully deterministic dynamical model can reproduce the main features of the dynamics of entanglement for a system in contact with a phase-damping environment.[31]

6.6   Decoherence and quantum computation

So far, we have discussed the effects of quantum noise on one- and two-qubit systems. On the other hand, a quantum computer can be seen as a complex system of many coupled qubits, whose ideal evolution is tailored in order to implement a given quantum algorithm. In any realistic implementation, errors due to imperfections in the quantum hardware or to the computer-environment coupling unavoidably appear.

Let us consider a quantum algorithm that maps the input state $|\psi_i\rangle$ of the quantum computer onto the output state $|\psi_f\rangle = U|\psi_i\rangle$, where the unitary evolution operator U can be decomposed into a sequence on N_g elementary quantum gates:

$$U = U_{N_g} U_{N_g-1} \cdots U_1. \qquad (6.253)$$

We consider errors perturbing the ideal evolution $|\psi_i\rangle \rightarrow |\psi_f\rangle$. There is a large variety of possible many-qubit decoherence models. For instance, each qubit in a quantum register can decohere independently of the others or, in the opposite limit, collective noise models are more appropriate. In the first limiting case we say that each qubit interacts individually with a different reservoir, while, in the latter case, there is a common reservoir. Moreover, the noise may be memoryless (Markovian approximation) or correlated over a time scale comparable or larger than the time between two consecutive quantum gates. In the latter case, memory effects should be taken into account.

We discuss the following classes of errors:

(i) *unitary memoryless errors (noisy gates):* the operators U_j change as follows:

$$U_j \rightarrow W_\epsilon(j)\, U_j, \qquad (6.254)$$

---

[31]The phase-damping map (6.251) can also be derived, in the Markovian limit, from the Caldeira–Leggett model, see Palma *et al.* (1996).

with $W_\epsilon(j)$ unitary and changing without any memory from gate to gate. An example is provided by phase-shift gates ($|0\rangle \to |0\rangle$, $|1\rangle \to e^{i\phi}|1\rangle$), with *random phase fluctuations*, $\phi \to \phi + \delta\phi(j)$, with $\delta\phi(j)$ randomly drawn from the interval $[-\epsilon, \epsilon]$.

(ii) *static imperfections:* the error $W_\epsilon(j)$ in (6.254) is the same for all j. For instance, undesired phase rotations or qubit couplings can result from static imperfections in the quantum computer hardware, such as a coupling between two qubits that cannot be exactly switched off to zero after the application of a quantum gate.

(iii) *non-unitary decoherence:* the unavoidable coupling of the quantum computer to the external world in general causes a non-unitary evolution of the quantum computer itself.

The accuracy of quantum computation in the presence of the above errors is typically measured by the *fidelity*

$$f(t) = \text{Tr}[\rho_\epsilon(t)\rho(t)] = \langle\psi(t)|\rho_\epsilon(t)|\psi(t)\rangle, \qquad (6.255)$$

where $\rho(t) = |\psi(t)\rangle\langle\psi(t)|$ is the ideal pure state of the quantum computer at time t and $\rho_\epsilon(t)$ is the density matrix describing the state (in general, a mixed state) of the perturbed quantum computer. Note that, as discussed in Sec. 6.5.8, the fidelity is also an important quantity in the study of the stability of quantum motion in the general theory of dynamical systems.

In the study of the stability of quantum computation under decoherence and imperfection effects, a basic tool is the numerical simulation of noisy many-qubit quantum algorithms. As a first example, we consider the quantum simulation of the one-dimensional harmonic oscillator, whose Hamiltonian reads $H = H_0 + V(x)$, with kinetic energy $H_0 = \frac{p^2}{2}$ and a harmonic potential $V(x) = \frac{1}{2}\omega^2(x - x_0)^2$. In Fig. 6.41 we show the results obtained for $n = 6$ qubits, evolving an initial (Gaussian) wave function $\psi(x, 0)$ by means of the Schrödinger equation. The propagation of the wave function up to time $\bar{t}$ is performed, as described in Sec. 3.15.1, by means of the Trotter decomposition:

$$\psi(x, \bar{t}) \approx \left[e^{-\frac{i}{\hbar}H_0\,dt}e^{-\frac{i}{\hbar}V(x)\,dt}\right]^{\bar{t}/dt} \psi(x, 0), \qquad (6.256)$$

where dt is a time much smaller than the time scales of interest for the system and $\bar{t}/dt$ is the number of time steps used to evolve the system up to time $\bar{t}$. We set $\hbar = 1$. In Fig. 6.41 we consider $\omega = 1$, so that the oscillation period is $T = 2\pi$, and we consider $dt = \frac{1}{20}T \ll T$. The dynamics is integrated up to time $\bar{t} = 2T$, and therefore the number of time steps is $\bar{t}/dt = 40$. As explained in Sec. 3.15.1, we evaluate the operators $U_0(dt) = e^{-\frac{i}{\hbar}H_0 dt}$ and $U_V(dt) = e^{-\frac{i}{\hbar}V dt}$ in the basis in which they are diagonal; that is, in the coordinate (x) and momentum (p) basis, respectively. The quantum Fourier transform F then allows to pass efficiently (using $\frac{1}{2}n(n-1)$ controlled phase-shift and n Hadamard gates) from the x to the p representation. Therefore, we have

$$\psi(x,\bar{t}) \approx [U(dt)]^{\bar{t}/dt}\psi(x,0), \qquad U(dt) = F^{-1}U_0(dt)FU_V(dt). \quad (6.257)$$

We investigate the motion inside the region $x \in [a,b]$, and discretize the wave function by means of a grid of 2^n equally spaced points in the interval $[a,b]$. The $2^n \times 2^n$ diagonal matrices U_0 and U_V can be implemented without ancillary qubits by means of 2^n generalized controlled phase-shift gates, similarly to Sec. 3.5 (see Fig. 3.16).[32] In the case of Fig. 6.41, we have $n = 6$ and therefore the number of quantum gates required to simulate the evolution of the wave function in a time step dt is $n_g = 2^{n+1} + n(n+1) = 170$ (2^{n+1} to implement $H_0(dt)$ and $H_V(dt)$ and $n(n-1)$ to implement F and F^{-1}), while $N_g = (\bar{t}/dt)n_g = 6.8 \times 10^3$ gates are needed to build the wave function $\psi(x,\bar{t})$ starting from $\psi(x,0)$.

Let us consider the effects of decoherence on the above-described quantum simulation. For this purpose, we assume that each qubit interacts independently with a different reservoir. A quantum operation is therefore applied to each qubit after each quantum gate. As we saw in Sec. 6.1.1, quantum noise acting on a single qubit is described by 12 parameters, associated with rotations, deformations and displacements of the Bloch sphere. To give a concrete example, in Fig. 6.41 we show the impact of the phase-flip noise (corresponding to the deformation of the Bloch sphere into an ellipsoid with z as symmetry axis), at noise strengths $\epsilon = 0.01$ and $\epsilon = 0.02$, with $\sin\epsilon = |\gamma|$ in Eq. (6.31).[33] It is interesting that the quantum simulation in Fig. 6.41 is rather robust against significant noise strengths: for $\epsilon = 0.01$ we have quite a high value of fidelity ($f \approx 0.71$) after a number of quantum gates as large as $N_g = 6.8 \times 10^3$.

It is very important to assess how errors scale with the input size n.

---

[32]This implementation is not efficient as it scales exponentially with the number of qubits. Efficient implementations (polynomial in n) are possible for analytic potentials $V(x)$ but require, in general, the use of ancillary qubits. We also point out that usually such efficient implementations outperform the above described inefficient implementation only when the number of qubits n is quite large.

[33]Note that very similar results are obtained for the other single-qubit noise channels introduced in Sec. 6.1.

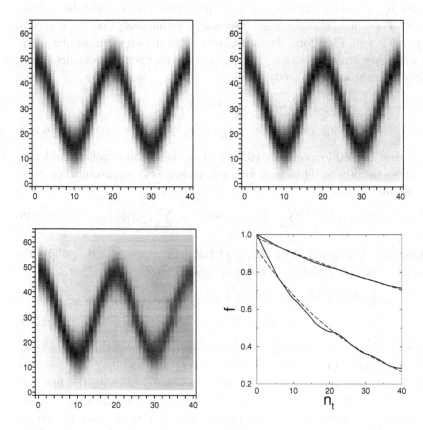

Fig. 6.41 Top left plot: $|\psi(x,t)|^2$ for a harmonic oscillator with potential $V(x) = \frac{1}{2}\omega^2(x - x_0)^2$, for $\omega = 1$, $x_0 = 0$. The interval $-5 \leq x \leq 5$ is discretized by means of a grid of 64 points (vertical axis) and the total integration time $\bar{t} = 2T = \frac{4\pi}{\omega}$ is divided in 40 time steps (horizontal axis). The initial condition is a Gaussian wave function centred at $x = \frac{5}{2}$. The top right and bottom left plots are the same as that top left but with phase-flip noise of strength $\epsilon = 0.01$ and 0.02, respectively. The full curves in the bottom right plot show the fidelity of the quantum simulation at $\epsilon = 0.01$ (above) and 0.02 (below), as a function of the number of time steps $n_t = \frac{t}{dt}$. The dashed lines correspond to the exponential fit $f(n_t) \propto \exp(-C\epsilon^2 n_g n_t)$, with $C(\epsilon = 0.01) \approx 0.49$ and $C(\epsilon = 0.02) \approx 0.45$.

A simple argument can be put forward (see Ekert *et al.*, 2001) under the assumption that each qubit decoheres separately. Let us assume that the qubit-environment interaction is such as

$$|0\rangle|0\rangle_E \rightarrow |0\rangle|e_0\rangle_E, \quad |1\rangle|0\rangle_E \rightarrow |1\rangle|e_1\rangle_E, \qquad (6.258)$$

where $|0\rangle_E$ is the initial state and $|e_0\rangle_E$, $|e_1\rangle_E$ two final states of the environment. These final states are not necessarily orthogonal; that is, in general $_E\langle e_0|e_1\rangle_E \neq 0$. Therefore, the overall unitary evolution of an n-qubit quantum register state $|i\rangle \equiv |i_{n-1}\rangle \cdots |i_1\rangle |i_0\rangle$ together with the environment, initially in the state $|0 \cdots 00\rangle_E \equiv |0\rangle_E \cdots |0\rangle_E |0\rangle_E$, is given by

$$|i\rangle |0 \cdots 00\rangle = |i_{n-1} \cdots i_1 i_0\rangle |0\rangle_E \cdots |0\rangle_E |0\rangle_E$$
$$\rightarrow |i_{n-1} \cdots i_1 i_0\rangle |e_{i_{n-1}}\rangle_E \cdots |e_{i_1}\rangle_E |e_{i_0}\rangle_E \equiv |i\rangle |E_i\rangle_E. \quad (6.259)$$

Note that (6.259) follows from (6.258) when each qubit decoheres independently of the others. Therefore, for a generic state of the quantum computer we obtain

$$|\psi\rangle |0 \cdots 00\rangle_E = \sum_i c_i |i\rangle |0 \cdots 00\rangle_E \rightarrow \sum_i c_i |i\rangle |E_i\rangle_E. \quad (6.260)$$

Thus, the off-diagonal elements ρ_{ij} of the density operator describing the n-qubit quantum register are reduced by a factor

$$|_E\langle E_i|E_j\rangle_E| = |_E\langle e_{i_{n-1}}|e_{j_{n-1}}\rangle_E| \cdots |_E\langle e_{i_1}|e_{j_1}\rangle_E| \, |_E\langle e_{i_0}|e_{j_0}\rangle_E|$$
$$= |_E\langle e_1|e_0\rangle_E|^{d_H(i,j)}, \quad (6.261)$$

where $d_H(i,j)$ is the Hamming distance between i and j; that is, the number of binary digits in which i and j differ. For instance, if $i = 01010$ and $j = 11011$, then $d_H(i,j) = 2$ because i and j differ in the first and last digits. Note that there are off-diagonal terms that drop by a factor $|_E\langle e_1|e_0\rangle_E|^n$. Therefore, the error probability in a generic quantum computation is expected to grow exponentially with n or, equivalently, the fidelity $f \propto \exp(-An)$, where A is some constant. Of course, this naive expectation needs to be checked using more realistic theoretical models, in which the dynamics of the quantum computer operating a given quantum algorithm is taken into account. Moreover, collective-noise models also deserve investigation. These issues will be discussed in Secs. 6.6.1 and 6.7.

6.6.1 * Decoherence and quantum trajectories

In this section we shall describe the quantum-trajectory approach, a theory developed mainly in the field of quantum optics to investigate physical phenomena such as spontaneous emission, resonance fluorescence and Doppler cooling, to name but a few. Here we discuss quantum trajectories as a powerful technique for numerical simulation of quantum information processing in a noisy environment.

As a consequence of the undesired environmental coupling, a quantum processor becomes, in general, entangled with its environment. Therefore, under the assumption that the environment is Markovian, the state is described by a density matrix whose evolution is governed by a master equation. Solving this equation for a state of several qubits is a prohibitive task in terms of memory cost. Indeed, for a system whose Hilbert space has dimension N, one has to store and evolve a density matrix of size $N \times N$. Quantum trajectories allow us instead of doing so, to store only a stochastically evolving state vector of size N. By averaging over many runs we obtain the same probabilities (within statistical errors) as those obtained directly through the density matrix. Therefore, quantum trajectories are the natural approach for simulating equations otherwise very hard to solve.

The GKSL master equation (6.99) can be written as

$$\dot{\rho} = -\frac{i}{\hbar}[H, \rho] - \frac{1}{2}\sum_{\mu}\{L_{\mu}^{\dagger}L_{\mu}, \rho\} + \sum_{\mu}L_{\mu}\rho L_{\mu}^{\dagger}, \qquad (6.262)$$

where L_{μ} are the Lindblad operators ($\mu \in [1, \ldots, \mathcal{M}]$, the number $\mathcal{M}$ depending on the noise model), H is the system Hamiltonian and $\{\,,\}$ denotes the anticommutator. The first two terms of this equation can be regarded as the evolution generated by an effective non-Hermitian Hamiltonian, $H_{\text{eff}} = H + iK$, with $K = -\frac{\hbar}{2}\sum_{\mu}L_{\mu}^{\dagger}L_{\mu}$. In fact, we see that

$$-\frac{i}{\hbar}[H, \rho] - \frac{1}{2}\sum_{\mu}\{L_{\mu}^{\dagger}L_{\mu}, \rho\} = -\frac{i}{\hbar}[H_{\text{eff}}\rho - \rho H_{\text{eff}}^{\dagger}], \qquad (6.263)$$

which reduces to the usual evolution equation for the density matrix in the case when H_{eff} is Hermitian. The last term in (6.262) is responsible for the so-called *quantum jumps*. In this context the Lindblad operators L_{μ} are also called *quantum-jump operators*. If the initial density matrix is in a pure state $\rho(t_0) = |\phi(t_0)\rangle\langle\phi(t_0)|$, after an infinitesimal time dt it evolves to the following statistical mixture:

$$\rho(t_0 + dt) = \rho(t_0) - \frac{i}{\hbar} \left[H_{\text{eff}} \rho(t_0) - \rho(t_0) H_{\text{eff}}^\dagger \right] dt + \sum_\mu L_\mu \rho(t_0) L_\mu^\dagger dt$$

$$\approx \left(I - \frac{i}{\hbar} H_{\text{eff}} dt \right) \rho(t_0) \left(I + \frac{i}{\hbar} H_{\text{eff}}^\dagger dt \right) + \sum_\mu L_\mu \rho(t_0) L_\mu^\dagger dt$$

$$= \left(1 - \sum_\mu dp_\mu \right) |\phi_0\rangle\langle\phi_0| + \sum_\mu dp_\mu |\phi_\mu\rangle\langle\phi_\mu|, \tag{6.264}$$

with the probabilities dp_μ defined by

$$dp_\mu = \langle\phi(t_0)|L_\mu^\dagger L_\mu|\phi(t_0)\rangle dt, \tag{6.265}$$

and the new states by

$$|\phi_0\rangle = \frac{\left(I - \frac{i}{\hbar} H_{\text{eff}} dt \right) |\phi(t_0)\rangle}{\sqrt{1 - \sum_\mu dp_\mu}} \tag{6.266}$$

and

$$|\phi_\mu\rangle = \frac{L_\mu|\phi(t_0)\rangle\sqrt{dt}}{\sqrt{dp_\mu}} = \frac{L_\mu|\phi(t_0)\rangle}{\|L_\mu|\phi(t_0)\rangle\|}. \tag{6.267}$$

The *quantum-jump picture* turns out then to be clear: a jump occurs and the system is prepared in the state $|\phi_\mu\rangle$ with probability dp_μ. With probability $1 - \sum_\mu dp_\mu$ there are no jumps and the system evolves according to the effective Hamiltonian H_{eff} (normalization is also included in this case because the evolution is given by a non-unitary operator).

In order to simulate the master equation one may employ a numerical method usually known as the Monte Carlo wave function approach. We start from a pure state $|\phi(t_0)\rangle$ and at intervals dt, much smaller than the time scales relevant for the evolution of the density matrix, we perform the following evaluation. We choose a random number ϵ from a uniform distribution in the unit interval $[0, 1]$. If $\epsilon < dp$, where $dp = \sum_\mu dp_\mu$, the system jumps to one of the states $|\phi_\mu\rangle$ (to $|\phi_1\rangle$ if $0 \leq \epsilon \leq dp_1$, to $|\phi_2\rangle$ if $dp_1 < \epsilon \leq dp_1 + dp_2$, and so on). On the other hand, if $\epsilon > dp$, evolution with the non-Hermitian Hamiltonian H_{eff} takes place, ending up in the state $|\phi_0\rangle$. In both circumstances we renormalize the state. We repeat this process as many times as $n_{\text{steps}} = \bar{t}/dt$ where $\bar{t}$ is the entire time elapsed during the evolution. Each realization provides a different *quantum trajectory* and a particular set of them (given a choice of the

Lindblad operators) is an "unravelling" of the master equation.[34] It is easy to see that if we average over different runs, we recover the probabilities obtained with the density operator. In fact, given an operator A, we can write the mean value $\langle A \rangle_t = \text{Tr}[A\rho(t)]$ as the average over $\mathcal{N}$ trajectories:

$$\langle A \rangle_t = \lim_{\mathcal{N} \to \infty} \frac{1}{\mathcal{N}} \sum_{i=1}^{\mathcal{N}} \langle \phi_i(t) | A | \phi_i(t) \rangle. \tag{6.268}$$

The advantage of using the quantum-trajectory method is clear since we need to store a vector of length N (N being the dimension of the Hilbert space) rather than an $N \times N$ density matrix. Moreover, there is also an advantage in computation time with respect to direct density-matrix calculations. It is indeed generally found that a reasonably small number of trajectories ($\mathcal{N} \approx 100 - 500$) is needed in order to obtain a satisfactory statistical convergence, so that there is an advantage in computer time provided $N > \mathcal{N}$.[35]

We can say that a quantum trajectory represents a single member of an ensemble whose density operator satisfies the corresponding master equation (6.262). This picture can be formalized by means of the stochastic Schrödinger equation

$$|d\phi\rangle = -iH|\phi\rangle dt - \tfrac{1}{2} \sum_\mu \left(L_\mu^\dagger L_\mu - \langle \phi | L_\mu^\dagger L_\mu | \phi \rangle \right) |\phi\rangle dt$$

$$+ \sum_\mu \left(\frac{L_\mu}{\sqrt{\langle \phi | L_\mu^\dagger L_\mu | \phi \rangle}} - I \right) |\phi\rangle \, dN_\mu, \quad (6.269)$$

where the stochastic differential variables dN_μ are statistically independent and represent measurement outcomes (for instance, in indirect measurement models the environment is measured, see the example from quantum optics below). Their ensemble average is given by $M[dN_\mu] = \langle \phi | L_\mu^\dagger L_\mu | \phi \rangle dt$. The probability that the variable dN_μ is equal to 1 during a given time step

---

[34]Different unravellings are possible since there is always freedom in the choice of the Lindblad operators that induce a given temporal evolution of the density matrix $\rho(t)$ (see, *e.g.*, Brun, 2002). This corresponds to the freedom in the operator-sum representation discussed in Sec. 5.4.

[35]The updating of a density matrix and of a wave vector, performed after each time step dt, require $O(N^3)$ and $O(N^2)$ operations, respectively. In the first case, we must multiply $N \times N$ matrices, in the latter $N \times N$ matrices by a vector of size N. Hence, the cost in computer time for the quantum-trajectory approach is $\propto \mathcal{N}N^2$, to be compared with the cost $\propto N^3$ for the density-matrix calculations.

dt is $\langle\phi|L_\mu^\dagger L_\mu|\phi\rangle dt$. Therefore, most of the time the variables dN_μ are 0 and as a consequence the system evolves continuously by means of the non-Hermitian effective Hamiltonian H_{eff}. However, when a variable dN_μ is equal to 1, the corresponding term in Eq. (6.269) is the most significant. In these cases a quantum jump occurs. Therefore, Eq. (6.269) is a stochastic non-linear differential equation, where the stochasticity is due to the measurement and non-linearity appears as a consequence of the renormalization of the state vector after each measurement process. We point out that, in contrast to the master equation (6.262) for the density operator, Eq. (6.269) represents the evolution of an individual quantum system, as exemplified by a single run of a laboratory experiment.

There is a close connection between the quantum-jump picture and the Kraus-operator formalism. To see this, we write the solution to the master equation (6.262) as a completely positive map:

$$\rho(t_0 + dt) = \mathcal{S}(t_0; t_0 + dt)\rho(t_0) = \sum_{\mu=0}^{\mathcal{M}} E_\mu(dt)\rho(t)E_\mu^\dagger(dt), \qquad (6.270)$$

where, for $\mu = 0$, we have $E_0 = I - iH_{\text{eff}}dt/\hbar$ and, for $\mu > 0$, $E_\mu = L_\mu\sqrt{dt}$ (see Sec. 6.2.2), satisfying $\sum_{\mu=0}^{\mathcal{M}} E_\mu^\dagger E_\mu = I$ to first order in dt. The action of the superoperator $\mathcal{S}$ in (6.270) can be interpreted as ρ being randomly replaced by $E_\mu\rho E_\mu^\dagger / \text{Tr}(E_\mu\rho E_\mu^\dagger)$, with probability $\text{Tr}(E_\mu\rho E_\mu^\dagger)$. Equivalently, the set $\{E_\mu\}$ defines a Positive Operator-Valued Measurement (POVM), with POVM elements $F_\mu = E_\mu^\dagger E_\mu$ satisfying $\sum_{\mu=0}^{\mathcal{M}} F_\mu = I$. The process outlined is equivalent to performing a continuous (weak) measurement on the system, which can be seen as an indirect measurement if the environment is actually measured.

A simple example will help us clarify the general quantum-trajectory theory sketched above. We consider the simplest, zero-temperature instance of the quantum optical master equation (6.100):

$$\dot{\rho} = -\frac{i}{\hbar}[H, \rho] - \frac{\gamma}{2}(\sigma_-\sigma_+\rho + \rho\sigma_-\sigma_+) + \gamma\sigma_+\rho\sigma_-, \qquad (6.271)$$

where the Hamiltonian $H = \frac{1}{2}\hbar\omega_0\sigma_z$ describes the free evolution of a two-level atom[36] and γ is the atom–field coupling constant. In this case there is a single Lindblad operator $L_1 = \sqrt{\gamma}\sigma_+$ and a jump is a transition from

---

[36]The exact expression of the Hamiltonian H is not important here and one could equally well consider the same example for a generic Hamiltonian H, provided the interaction picture is considered (see, *e.g.*, Scully and Zubairy, 1997).

the excited state $|1\rangle$ to the ground state $|0\rangle$ of the atom. Starting from a pure initial state $|\phi(t_0)\rangle = \alpha|0\rangle + \beta|1\rangle$ and evolving it for an infinitesimal time dt, the probability of a jump in a time dt is given by

$$dp = \langle\phi(t_0)|L_1^\dagger L_1|\phi(t_0)\rangle dt = \gamma\langle\phi(t_0)|\sigma_-\sigma_+|\phi(t_0)\rangle dt = \gamma p_e(t_0) dt, \tag{6.272}$$

where $p_e(t_0) = |\beta|^2$ is the population of the excited state $|1\rangle$ at time t_0. If a jump occurs, the new state of the atom is

$$|\phi_1\rangle = \frac{L_1|\phi(t_0)\rangle}{\|L_1|\phi(t_0)\rangle\|} = \frac{\sqrt{\gamma}\,\sigma_+(\alpha|0\rangle + \beta|1\rangle)\sqrt{dt}}{\sqrt{dp}} = \frac{\beta}{|\beta|}|0\rangle. \tag{6.273}$$

In this case, the transition $|1\rangle \to |0\rangle$ takes place and the emitted photon is detected. As a consequence, the atomic state vector collapses onto the ground state $|0\rangle$. If instead there are no jumps, the system evolution is governed by the non-Hermitian effective Hamiltonian $H_{\text{eff}} = H - i\frac{\hbar}{2}L_1^\dagger L_1 = H - i\frac{\hbar}{2}\gamma\sigma_-\sigma_+$, so that the state of the atom at time $t_0 + dt$ is

$$\begin{aligned}
|\phi_0\rangle &= \frac{\left(I - \frac{i}{\hbar}H_{\text{eff}}dt\right)|\phi(t_0)\rangle}{\sqrt{1-dp}} \\
&= \frac{\left(1 - i\frac{\omega_0}{2}dt\right)\alpha|0\rangle + \left(1 + i\frac{\omega_0}{2}dt - \frac{\gamma}{2}\right)\beta|1\rangle}{\sqrt{1 - \gamma|\beta|^2 dt}}.
\end{aligned} \tag{6.274}$$

Note that the normalization factor $1/\sqrt{1-dp}$ is due to the fact that, if no counts are registered by the photodetector, then we consider it more probable that the system is unexcited. To illustrate the fact that the normalization factor leads to the correct physical result, let us consider the evolution without jumps in a finite time interval, from t_0 to $t_0 + t$, and then let $t \to \infty$. If we first write the unnormalized state vector as

$$|\phi_0^{(u)}(t)\rangle = \alpha^{(u)}(t)|0\rangle + \beta^{(u)}(t)|1\rangle, \tag{6.275}$$

we see that the coefficients $\alpha^{(u)}$ and $\beta^{(u)}$ obey the simple equations of motion

$$\dot\alpha^{(u)}(t) = -i\frac{\omega_0}{2}\alpha^{(u)}(t), \qquad \dot\beta^{(u)}(t) = \left[i\frac{\omega_0}{2} - \frac{\gamma}{2}\right]\beta^{(u)}(t), \tag{6.276}$$

which imply

$$\begin{aligned}
\alpha^{(u)}(t_0 + t) &= \exp\left[-i\frac{\omega_0}{2}(t - t_0)\right]\alpha^{(u)}(t_0), \\
\beta^{(u)}(t_0 + t) &= \exp\left[\left(i\frac{\omega_0}{2} - \frac{\gamma}{2}\right)(t - t_0)\right]\beta^{(u)}(t_0).
\end{aligned} \tag{6.277}$$

Therefore, after normalization, the evolution of the state vector conditional on there being no photons detected is

$$|\phi_0(t_0 + t)\rangle = \frac{\alpha \exp\left[-i\frac{\omega_0}{2}(t - t_0)\right]|0\rangle + \beta \exp\left[\left(i\frac{\omega_0}{2} - \frac{\gamma}{2}\right)(t - t_0)\right]|1\rangle}{\sqrt{|\alpha|^2 + |\beta|^2 \exp[-\gamma(t - t_0)]}}.$$

(6.278)

We stress that as $t \to +\infty$ the state $|\phi_0(t)\rangle \to |0\rangle$ (up to an overall phase factor). That is, if after some long time we have never seen a count, then we conclude that we have been in the ground state $|0\rangle$ from the beginning.

Let us now demonstrate the ability of the quantum-trajectory approach to model noisy quantum-information protocols with a large number of qubits. A first issue is the generalization of the single-qubit quantum-noise channels discussed at the beginning of this chapter (amplitude damping, phase shift, ...) to many qubits. Of course, many different generalizations are possible. In what follows we shall take two different viewpoints, illustrated in the example of the amplitude-damping channel. In the first case (generalized amplitude-damping channel) we assume that a single damping probability describes the action of the environment, irrespective of the internal many-body state of the system. In the second approach (simple amplitude-damping channel), we assume that each qubit has its own interaction with the environment, independently of the other qubits. This makes the damping probability grow with the number of qubits that can perform the transition (the jump, in the quantum-trajectory language) $|1\rangle \to |0\rangle$. Both models assume that only one qubit of the system can decay at a time.

In the generalized amplitude-damping channel we assume that an n-qubit state $|i_{n-1} \ldots i_0\rangle$ ($i_l = 0, 1$, with $0 \le l \le n - 1$) decays during the interval dt with a probability $dp = \Gamma dt/\hbar$, where Γ is the system–environment coupling constant. The possible states of the system after the damping process are those in which the transition $|1\rangle \to |0\rangle$ has occurred in one of the qubits, the transition probability being the same for all the qubits. For example, the (pure) states of the computational basis for a two-qubit system are transformed, after a time dt, as follows:

$$|11\rangle\langle 11| \to \left(1 - \frac{\Gamma dt}{\hbar}\right)|11\rangle\langle 11| + \frac{\Gamma dt}{2\hbar}\left(|01\rangle\langle 01| + |10\rangle\langle 10|\right),$$

$$|10\rangle\langle 10| \to \left(1 - \frac{\Gamma dt}{\hbar}\right)|10\rangle\langle 10| + \frac{\Gamma dt}{\hbar}|00\rangle\langle 00|,$$

$$|01\rangle\langle 01| \to \left(1 - \frac{\Gamma dt}{\hbar}\right)|01\rangle\langle 01| + \frac{\Gamma dt}{\hbar}|00\rangle\langle 00|,$$

(6.279)

$$|00\rangle\langle 00| \to |00\rangle\langle 00|.$$

Note that the final states are in general statistical mixtures.

The evolution of the same initial states is different for the simple amplitude-damping model. In this case we have

$$|11\rangle\langle 11| \to \left(1 - \frac{2\Gamma dt}{\hbar}\right)|11\rangle\langle 11| + \frac{\Gamma dt}{\hbar}\left(|01\rangle\langle 01| + |10\rangle\langle 10|\right),$$

$$|10\rangle\langle 10| \to \left(1 - \frac{\Gamma dt}{\hbar}\right)|10\rangle\langle 10| + \frac{\Gamma dt}{\hbar}|00\rangle\langle 00|,$$

$$|01\rangle\langle 01| \to \left(1 - \frac{\Gamma dt}{\hbar}\right)|01\rangle\langle 01| + \frac{\Gamma dt}{\hbar}|00\rangle\langle 00|, \tag{6.280}$$

$$|00\rangle\langle 00| \to |00\rangle\langle 00|.$$

Note that in this model the decay probability for a state of the computational basis is proportional to the number of qubits in the $|1\rangle$ state.

Let us illustrate the application of the quantum-trajectory approach to both models. We study the fidelity of quantum teleportation through a noisy chain of qubits. A schematic drawing of this quantum protocol is shown in Fig. 6.42. We consider a chain of n qubits, and assume that Alice can access the qubits located at one end of the chain, Bob those at the other end. Initially Alice owns an EPR pair (for instance, we take the Bell state $|\phi^+\rangle = \frac{1}{\sqrt{2}}[|00\rangle + |11\rangle])$, while the remaining $n - 2$ qubits are in a pure state. Thus, the global initial state of the chain is given by

$$\sum_{i_{n-1},\dots,i_2} c_{i_{n-1},\dots,i_2}|i_{n-1}\dots i_2\rangle \otimes \frac{1}{\sqrt{2}}\left(|00\rangle + |11\rangle\right), \tag{6.281}$$

where $i_k = 0, 1$ denotes the up or down state of qubit k. In order to deliver one of the qubits of the EPR pair to Bob, we implement a protocol consisting of $n - 2$ SWAP gates that exchange the states of pairs of qubits:

$$\sum_{i_{n-1},\dots,i_2} \frac{c_{i_{n-1},\dots,i_2}}{\sqrt{2}}\left(|i_{n-1}\dots i_2 00\rangle + |i_{n-1}\dots i_2 11\rangle\right) \to$$

$$\sum_{i_{n-1},\dots,i_2} \frac{c_{i_{n-1},\dots,i_2}}{\sqrt{2}}\left(|i_{n-1}\dots 0 i_2 0\rangle + |i_{n-1}\dots 1 i_2 1\rangle\right) \to \dots$$

$$\dots \to \sum_{i_{n-1},\dots,i_2} \frac{c_{i_{n-1},\dots,i_2}}{\sqrt{2}}\left(|0 i_{n-1}\dots i_2 0\rangle + |1 i_{n-1}\dots i_2 1\rangle\right). \tag{6.282}$$

After this, Alice and Bob share an EPR pair, and therefore an unknown state of a qubit ($|\psi\rangle = \alpha|0\rangle + \beta|1\rangle$) can be transferred from Alice to Bob by means of the standard teleportation protocol described in Sec. 4.5. Here we take random coefficients $c_{i_{n-1},\dots,i_2}$; that is, they have amplitudes of the order of $1/\sqrt{2^{n-2}}$ (to assure wave function normalization) and random phases. This ergodic hypothesis models the transmission of a qubit through a chaotic quantum channel. We assume that the quantum protocol is implemented by a sequence of ideally instantaneous and perfect SWAP gates, separated by a time interval τ, during which the quantum noise introduces errors.

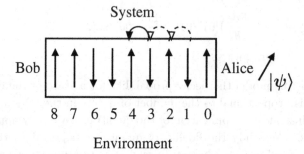

Fig. 6.42 A schematic drawing of the teleportation procedure studied in the text. Alice sends one of the qubits of a Bell state in her possession to Bob. Meanwhile dissipation is induced by the environment. In the figure, there is a chain of $n = 9$ qubits and the third of the $n - 2$ SWAP gates required by this quantum protocol has been applied. The qubit $|\psi\rangle$ is to be teleported.

In the quantum-trajectory method, the fidelity of the teleportation protocol is computed as

$$f = \lim_{\mathcal{N}\to\infty} \frac{1}{\mathcal{N}} \sum_{i=1}^{\mathcal{N}} \langle\psi|(\rho_B)_i|\psi\rangle, \qquad (6.283)$$

where $(\rho_B)_i$ is the reduced density matrix of the teleported qubit (owned by Bob), obtained from the wave vector of the trajectory i at the end of the quantum protocol. The effect of the two amplitude-damping channels described above on the fidelity of the teleportation protocol is illustrated in Fig. 6.43, where we show the results of numerical simulations for the special case in which the state to be teleported is $|\psi\rangle = \frac{1}{\sqrt{2}}\big(|0\rangle + |1\rangle\big)$. In this case, in the limit of infinite chain ($n \to \infty$) or of large damping rate, the density matrix ρ_B describing the state of Bob's qubit becomes $\rho_B = |0\rangle\langle0|$. Thus, the asymptotic value of fidelity is given by $f_\infty = \frac{1}{2}$ and we plot the values of $\bar{f} = f - f_\infty$, for the generalized and simple amplitude-damping channels. For both cases we have checked the accuracy of the quantum trajectory simulations by reproducing the results with direct density matrix calculations (possible only up to approximately $n = 10$ qubits). For the simple amplitude-damping channel the fidelity decays exponentially, in agreement with the theoretical formula

$$f = \tfrac{1}{2} + \tfrac{1}{2}\exp(-\gamma k), \qquad (6.284)$$

where $\gamma = \Gamma\tau/\hbar$ is the dimensionless damping rate and $k = t/\tau = n - 2$ measures the time in units of quantum (SWAP) gates. To derive this theoretical formula, we observe that this quantum-noise model does not generate entanglement between the two qubits of the Bell pair and the other qubits of the chain. Therefore, it is sufficient to study the evolution of the Bell state $|\phi^+\rangle\langle\phi^+|$ under the amplitude-damping noise model to obtain Eq. (6.284). In contrast, the generalized amplitude-damping model entangles these two qubits with the rest of the chain. In this case, it can be seen from Fig. 6.43 that the fidelity decay is slower and not exponential. The most important point here is that the quantum-trajectory approach allows one to simulate a much larger number of qubits than accessible by direct solution of the master equation, which, due to memory restrictions in a classical computer, is only possible for up to $n \approx 10$ qubits.

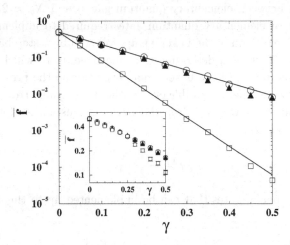

Fig. 6.43 The fidelity $\bar{f} = f - f_\infty$ ($f_\infty = \tfrac{1}{2}$) for the teleportation of the state $|\psi\rangle = \frac{1}{\sqrt{2}}(|0\rangle + |1\rangle)$ as a function of the dimensionless damping rate $\gamma = \Gamma\tau/\hbar$, for the amplitude-damping model. The circles and squares are the results of the quantum-trajectory calculations for chains with $n = 10$ and $n = 20$ qubits, respectively; the triangles give the results of the density matrix calculations at $n = 10$; the straight lines correspond to the theoretical result $f = \tfrac{1}{2} + \tfrac{1}{2}\exp[-\gamma(n - 2)]$. Inset: the same but for the generalized amplitude-damping channel. The figure is taken from Carlo *et al.* (2004).

Besides the simple quantum teleportation protocol, the quantum-trajectory method can also be used for the simulation of complex quantum computations. As an example, let us consider the effect of a noisy environment on the quantum algorithm for the n-qubit quantum baker's map (described in Sec. 3.15.2). More precisely, we consider the phase-flip channel, assuming that the noise strength is the same for each qubit. We take an initial state $|\psi_0\rangle$ with amplitudes of the order of $1/\sqrt{2^n}$ and random phases. The forward evolution of the baker's map is performed up to k map steps, followed by the k-step backward evolution (the forward evolution in one map step is governed by the unitary transformation T defined in Eq. (3.194), the backward evolution by $T^\dagger$). Owing to quantum noise, the initial state $|\psi_0\rangle$ is not exactly recovered and the final state of the system is described by a density matrix ρ_f. The quantum-computation fidelity is given by $f = \langle\psi_0|\rho_f|\psi_0\rangle$. It can be seen in Fig. 6.44 that the fidelity decay induced by phase-flip noise is in agreement with the formula

$$f = \exp(-n\gamma N_g) = \exp(-2\gamma n^3 k), \qquad (6.285)$$

where $\gamma = \Gamma\tau/\hbar$ is the dimensionless noise strength (again, τ denotes the time interval between elementary quantum gates) and $N_g = 2n^2 k$ is the total number of elementary quantum gates required to implement the k steps forward evolution of the baker's map, followed by k step backward.[37] From Eq. (6.285) we can determine the time scale up to which a reliable quantum computation of the baker's map evolution in the presence of the phase-flip noise channel is possible even without quantum error correction. The time scale k_f at which f drops below some constant A (for instance, $A = 0.9$) is given by

$$k_f = -\frac{\ln A}{2\gamma n^3}. \qquad (6.286)$$

The total number of gates that can be implemented up to this time scale is given by

$$(N_g)_f = 2n^2 k_f = -\frac{\ln A}{\gamma n}. \qquad (6.287)$$

Thus, the number of gates $(N_g)_f$ that can be reliably implemented without quantum error correction drops only polynomially with the number of qubits: $(N_g)_f \propto 1/n_q$.

---

[37]The analytic derivation of formula (6.285) is provided in Carlo *et al.* (2004).

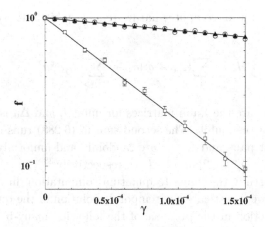

Fig. 6.44 A semilogarithmic plot of the fidelity as a function of the dimensionless decay rate γ, for the baker's map after one map step in the presence of the phase-flip channel. Circles and squares correspond to quantum-trajectory simulations for $n = 10$ and $n = 20$ qubits, respectively. Triangles give the results obtained by direct computation of the density matrix evolution at $n = 10$. Solid lines stand for $f(\gamma) = \exp(-2\gamma n^3)$. The figure is taken from Carlo *et al.* (2004).

6.7 * Quantum computation and quantum chaos

The aim of this section is to discuss the limits to quantum computation due to chaos effects. Let us first point out that, even when a quantum processor is ideally isolated from the environment, *i.e.*, in situations where the decoherence time of the processor is very large compared to the computational time scales, the operability of the quantum computer is not yet guaranteed. Indeed, the presence of *device imperfections* also hinders the implementation of any quantum computation. A quantum computer can be seen as a *complex many-body (-qubit) system*, in which interaction between the qubits composing the quantum registers is needed to produce the multipartite (many-qubit) entanglement necessary in efficient quantum algorithms. Moreover, device imperfections such as small inaccuracies in the coupling constants induce errors. Unwanted mutual interactions between qubits can be a source of essential errors in practical implementations of quantum computation. For instance, in the ion-trap quantum processors magnetic dipole–dipole interactions couple qubits. In NMR quantum computing, undesired residual interactions survive after imperfect spin echoes.

In this section we discuss the impact of *static hardware imperfections* on the stability of quantum algorithms. For this purpose, it is convenient to model the quantum computer hardware as a qubit lattice described by the Hamiltonian

$$H_{\rm s} = \sum_i (\Delta_0 + \delta_i)\sigma_i^z + \sum_{<i,j>} J_{ij}\sigma_i^x\sigma_j^x, \qquad (6.288)$$

where $\sigma_i^x, \sigma_i^y, \sigma_i^z$ are the Pauli matrices for qubit i, and Δ_0 is the average level spacing for one qubit. The second sum in (6.288) runs over nearest-neighbour qubit pairs and δ_i, J_{ij} are randomly and uniformly distributed in the intervals $[-\bar\delta/2, \bar\delta/2]$ and $[-\bar J, \bar J]$, respectively.[38]

In order to study the limits to quantum computation due to hardware imperfections, we investigate the temporal evolution of the quantum computer wave function in the presence of the following many-body Hamiltonian:

$$H(\tau) = H_{\rm s} + H_{\rm g}(\tau), \qquad (6.289)$$

where

$$H_{\rm g}(\tau) = \sum_k \delta(\tau - k\tau_g)h_k. \qquad (6.290)$$

Here h_k realizes the k-th elementary gate of a sequence prescribed to implement a given quantum algorithm. The algorithm is therefore implemented by a sequence of (ideally) instantaneous and perfect one- and two-qubit gates, separated by a time interval τ_g, during which the Hamiltonian (6.288) gives undesired phase rotations and qubit couplings. We assume that the phase accumulation given by Δ_0 is eliminated by standard spin-echo techniques (see Sec. 8.1.4). In this case, the remaining terms in the static Hamiltonian (6.288) can be seen as residual terms after imperfect spin echoes.

To provide a concrete example, we assume in the rest of this section that the sequence $\{h_k\}$ implements the quantum algorithm simulating the quantum sawtooth map

$$\begin{cases} \bar I = I + k(\theta - \pi), \\ \bar\theta = \theta + T\bar I. \end{cases} \qquad (6.291)$$

The dynamics of this map is described in detail in Sec. 3.15.3.

---

[38]The transition to quantum chaos for Hamiltonian (6.288) is discussed in Sec. 6.5.7.

6.7.1 * *Quantum versus classical errors*

It is interesting to compare the effects of *quantum errors*, described, for instance, by the static imperfections model (6.288), with the action of the *round-off errors* typical of classical computation. For this purpose, both the quantum Husimi functions and the classical density plots are shown in Fig. 6.45. These pictures are obtained for $K = kT = -0.1$, so that the motion is stable (maximum Lyapunov exponent $\lambda = 0$), the phase space has a complex structure of elliptic islands down to smaller and smaller scales and anomalous diffusion in the rescaled momentum variable $J = TI$ is observed: $\langle (\Delta J)^2 \rangle \propto t^\alpha$ with $\alpha = 0.57$. We consider $-\pi \leq J < \pi$; the classical limit is obtained by increasing the number of qubits $n = \log N$ (and the number of levels $N = 2^n$), with $T = 2\pi/N$ ($k = K/T$, $J = TI$, $-N/2 \leq I < N/2$). As initial state at time $t = 0$ we consider a momentum eigenstate, $|\psi(0)\rangle = |I_0\rangle$, with $I_0 = [0.38N]$. The dynamics of the sawtooth map reveals the complexity of the phase-space structure, as shown by the Husimi functions of Fig. 6.45, taken after 1000 map iterations. We note that $n = 6$ qubits are sufficient to observe the quantum localization of the anomalous diffusive propagation through hierarchical integrable islands. At $n = 9$ one can see the appearance of integrable islands and at $n = 16$ the quantum Husimi function explores the complex hierarchical structure of the classical phase space down to small scales. The effect of static imperfections, modelled by Eq. (6.288), on the operability of the quantum computer is shown in Fig. 6.45 (right column). The main features of the wave packet dynamics remain evident even in the presence of significant imperfections, characterized by the dimensionless strength $\epsilon = \bar{\delta}\tau_g$. The main manifestation of imperfections is the injection of quantum probability inside integrable islands. This creates characteristic concentric ellipses, which follow classical periodic orbits moving inside integrable islands. These structures become more and more pronounced with increasing n. Thus, quantum errors strongly affect quantum tunnelling inside integrable islands, which in a pure system drops exponentially (proportional to $\exp(-cN)$, with c constant).

It is interesting to stress that the effect of quantum errors is qualitatively different from classical round-off errors, which produce only slow diffusive spreading inside integrable islands (see Fig. 6.45, bottom right). This difference is related to the fact that spin flips in quantum computation can make direct transfer of probability over large distances in phase space; that is, quantum errors are *non-local in phase space*. This is a consequence of the binary encoding of the discretized angle and momentum variables. For instance, we represent the momentum eigenstates $|I\rangle$ $(-N/2 \leq I < N/2)$ in the computational basis as $|\alpha_{n-1} \cdots \alpha_1 \alpha_0\rangle$, where $\alpha_j \in \{0,1\}$ and $I = -N/2 + \sum_{j=0}^{n-1} \alpha_j 2^j$. If we take, say, $n = 6$ qubits $(N = 2^6 = 64)$, the state $|000000\rangle$ corresponds to $|I = -32\rangle$ $(J = -\pi)$, $|000001\rangle$ to $|I = -31\rangle$ $(J = -\pi + 2\pi(1/2^6))$, and so on until $|111111\rangle$, corresponding to $|I = 31\rangle$ $(J = -\pi + 2\pi(63/2^6))$. Let us consider the simplest quantum error, the bit flip: if we flip the least significant qubit $(\alpha_0 = 0 \leftrightarrow 1)$, we exchange $|I\rangle$ with $|I + 1\rangle$, while, if we flip the most significant qubit $(\alpha_{n-1} = 0 \leftrightarrow 1)$, we exchange $|I\rangle$ with $|I + 32\rangle$. It is clear that this latter error transfers probability very far away in phase space.

6.7.2   * *Static imperfections versus noisy gates*

It is interesting to compare the effect of static imperfections generated by the Hamiltonian (6.288) with the case when the static imperfections are absent but the computer operates with *noisy gates*. To model the noisy gates we set $\bar{J} = 0$ and δ_i fluctuating randomly and independently from one gate to another in the interval $[-\bar{\delta}/2, \bar{\delta}/2]$ (similar results are obtained for $\bar{J} \neq 0$).

A quantitative comparison can be performed by computing the *fidelity* of quantum computation, defined by $f(t) = |\langle \psi_\epsilon(t)|\psi_0(t)\rangle|^2$, where $\psi_\epsilon(t)$ is the actual quantum wave function in the presence of imperfections (or noisy gates) and $\psi_0(t)$ is the quantum state for a perfect computation. We show in Fig. 6.46 the fidelity $f(t)$ as a function of time, for different values of the static imperfection strength. There is a clear change of behaviour in the fidelity decay. At short times the fidelity drops exponentially ($f(t) \approx \exp(-At)$, with $A \propto \epsilon^2$), then, at time $\bar{t} \approx 250$, the decay quite abruptly becomes Gaussian ($f(t) \approx \exp(-Bt^2)$, with $B \propto \epsilon^2$). Moreover, it can be shown that the time $\bar{t}$ does not depend on the static imperfection strength. It turns out that $\bar{t} \sim t_H$, where $t_H = N$ is the Heisenberg time, given by the inverse mean level spacing. A qualitative reason for this behaviour is the following. Before t_H the system does not resolve the discreteness of the spectrum. Therefore, the density of states can be treated as continuous and the Fermi golden rule can be applied. In the case of noisy gates the fidelity decay is exponential at all times. This corresponds to the Fermi golden rule regime, where at each gate operation a probability of order ϵ^2 is transferred from the ideal state to other states. Since there are no correlations between consecutive noisy gates, the population of the ideal

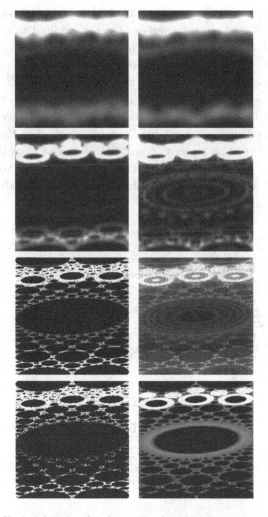

Fig. 6.45 The Husimi function for the sawtooth map in momentum–angle variables (J, θ), with $-\pi \leq J < \pi$ (vertical axis) and $0 \leq \theta < 2\pi$ (horizontal axis), for $K = -0.1$, $T = 2\pi/2^n$, $I_0 = J_0/T = [0.38 \times 2^n]$, averaged over the interval $950 \leq t \leq 1000$. From top to bottom: $n = 6, 9, 16$ and classical density plot, obtained from an ensemble of 10^8 trajectories, with initial momentum $J_0 = 0.38 \times 2\pi$ and random angles. Left- and right-hand columns show the case without and with imperfections: in the quantum case the imperfection strength $\epsilon = \bar{\delta}\tau_g = 2 \times 10^{-3}$ ($n = 6$), 6×10^{-4} ($n = 9$), 10^{-4} ($n = 16$), for $\bar{J} = 0$ (similar results are obtained for $\bar{J} = \bar{\delta}$); in the classical case round-off errors have amplitude 10^{-3}. In the Husimi functions we choose the ratio of the momentum–angle uncertainties $s = \Delta p/\Delta \theta = 1$ ($\Delta p \Delta \theta = T/2$). Black corresponds to the minimum of the probability distribution and white to the maximum. The figure is taken from Benenti *et al.* (2001b).

noiseless state decays exponentially. We can write $f(t) \approx \exp(-C\epsilon^2 N_g)$, where $N_g = n_g t$ is the total number of gates required to evolve t steps of the sawtooth map, $n_g = 3n^2 + n$ being the number of gates per map iteration, and C is a constant, which can be determined numerically. The fidelity time scale for noisy gates is therefore given by $t_f^{(n)} \propto 1/(\epsilon^2 n_g)$. The dependence on ϵ is qualitatively different compared to the Gaussian decay in the static imperfection case at small ϵ, where $t_f^{(s)} \propto 1/\epsilon$ (in this latter case the fidelity decay essentially takes place after the Heisenberg time). The fidelity time scales $t_f^{(n)}$ and $t_f^{(s)}$ (obtained from the condition $f(t_f) = c = 0.9$) are shown in Fig. 6.47 as a function of ϵ. We stress that the static imperfections give shorter time scales t_f and are therefore more dangerous for quantum computation.

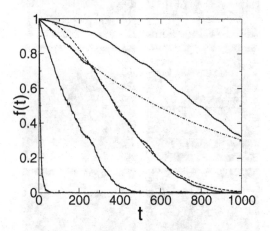

Fig. 6.46   Fidelity decay for $n = 11$ qubits, $\bar{J} = 0$ and, from top to bottom, $\epsilon = \delta\tau_g = 5 \times 10^{-5}$, 10^{-4}, 2×10^{-4}, 10^{-3} (solid lines). Dashed and dot-dashed lines show the fitting functions $\exp(-At)$ and $\exp(-Bt^2)$, with $A \approx 1.2 \times 10^{-3}$ and $B \approx 5 \times 10^{-4}$. The initial condition is a Gaussian wave packet.

The results for static imperfections in Figs. 6.46 and 6.47 can be explained following the random-matrix theory approach developed in Frahm *et al.* (2004). Assuming that the unitary Floquet operator U corresponding to the quantum evolution in one map step can be modelled by a random matrix, one obtains from a random-matrix theory calculation that

$$-\ln f(t) \approx \frac{t}{t_c} + \frac{t^2}{t_c t_H}, \tag{6.292}$$

where $t_c \approx 1/(\epsilon^2 n n_g^2)$ characterizes the inverse effective strength of the

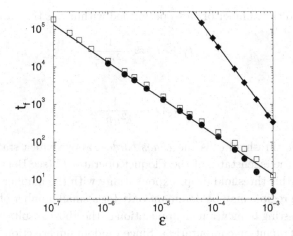

Fig. 6.47 Fidelity time scale t_f as a function of ϵ, for $n = 9$, in the case of static imperfections $[\bar{J} = \bar{\delta}$ (circles) and $\bar{J} = 0$ (squares)] and noisy gates at $\bar{J} = 0$ (diamonds). The straight lines have slopes -1 and -2. The figure is taken from Benenti *et al.* (2001b).

perturbation and $t_H = 2^n$. Relation (6.292) is valid as long as ϵ and t are sufficiently small so that the fidelity remains close to unity. The time scale t_c governs the exponential fidelity decay at times smaller than the Heisenberg time t_H. In the random-matrix theory approach we can distinguish two regimes:

(i) *perturbative regime:* for $\epsilon < \epsilon_c \approx 1/(\sqrt{n 2^n} n_g)$; that is, for $t_c > t_H$, the fidelity decay is dominated by the second term in the right-hand side of Eq. (6.292). The decay essentially takes place only after the Heisenberg time and is Gaussian, $f(t) \approx \exp(-t^2/t_c t_H)$. The fidelity time scale is therefore

$$t_f \sim \sqrt{t_c t_H} \approx \frac{\sqrt{2^n}}{\epsilon \sqrt{n} n_g} \qquad (6.293)$$

and the total number of gates that can be performed within this time is

$$(N_g)_f \sim \frac{\sqrt{2^n}}{\epsilon \sqrt{n}}. \qquad (6.294)$$

(ii) *quantum chaos regime:* for $\epsilon > \epsilon_c$ $(t_c < t_H)$ the fidelity decay is dominated by the t/t_c term in (6.292), so that it is exponential, $f(t) \approx \exp(-t/t_c)$, and occurs before the Heisenberg time. The fidelity time scale and the number of gates performed within this time are given by

$$t_f \sim t_c \approx \frac{1}{\epsilon^2 n n_g^2}, \tag{6.295}$$

$$(N_g)_f \sim \frac{1}{\epsilon^2 n n_g}. \tag{6.296}$$

Note that the threshold ϵ_c is the *chaos border*, above which static imperfections mix the eigenstates of the Floquet operator U (see Benenti *et al.*, 2002). Since this threshold drops exponentially with the number of qubits, the Gaussian regime may be dominant only for a small number of qubits.[39]

It is interesting to mention an application of the above results to improve the fidelity of quantum computation. Since random imperfections changing from gate to gate always lead to an exponential decay of the fidelity, it is tempting to try to randomize the static imperfections to slow down the fidelity decay from Gaussian to exponential. This idea can be formalized if one observes that the fidelity may be expressed in terms of a *correlation function* of the perturbation. Let us briefly illustrate this point. We consider a quantum algorithm that performs a given unitary transformation U by means of a sequence of T quantum gates:

$$U = U(T) U(T-1) \cdots U(1). \tag{6.297}$$

We denote the perturbed evolution by U_ϵ, where

$$U_\epsilon = e^{-i\epsilon V(T)} U(T) e^{-i\epsilon V(T-1)} U(T-1) \cdots e^{-i\epsilon V(1)} U(1). \tag{6.298}$$

Here ϵ denotes the strength of the perturbation, which is generated by the Hermitian operator $V(t)$. The fidelity of the quantum algorithm can be written as

$$f(T) = \left| \frac{1}{N} \text{Tr} \left[U_\epsilon(T, 0) U(T, 0) \right] \right|^2, \tag{6.299}$$

where the trace averages the result over a complete set of initial states (for instance, quantum register states) and $U(t, t') \equiv U(t) U(t-1) \cdots U(t'+1)$ is the evolution operator from t' to $t > t'$, $U_\epsilon(t, t')$ is defined in the same manner for the perturbed evolution. Defining the Heisenberg evolution of

---

[39] In contrast, for a quantum computer at rest (without quantum gates being applied) the chaos border decreases only polynomially with the number of qubits (see Sec. 6.5.7).

the perturbation as $V(t,t') = U^\dagger(t,t')V(t)U(t,t')$ we obtain

$$f(T) = \left| \frac{1}{N} \text{Tr}\left(e^{i\epsilon V(1,0)} e^{i\epsilon V(2,0)} \cdots e^{i\epsilon V(T,0)} \right) \right|^2 . \qquad (6.300)$$

As we are interested in the case in which the fidelity is close to unity, we can expand it up to the second order in ϵ (see Prosen and Žnidarič, 2002):

$$f(T) \approx 1 - \epsilon^2 \sum_{t,t'=1}^{T} C(t,t'), \qquad (6.301)$$

where

$$C(t,t') = \frac{1}{N} \text{Tr}\left[V(t',0)\, V(t,0) \right] \qquad (6.302)$$

is a two-point temporal correlation of the perturbation. It is therefore clear that a quantum algorithm is more stable when the correlation time of the perturbation is smaller. This can be performed by devising a "less regular" sequence of gates that realize the transformation U required by the algorithm (see Prosen and Žnidarič, 2001). For instance, using the Pauli operators one can change the computational basis repeatedly and randomly during a quantum computation. The effectiveness of this method in randomizing static imperfections, thus improving the fidelity of the quantum computation, was demonstrated in Kern *et al.* (2005), see also Viola and Knill (2005).

6.8   A guide to the bibliography

Reviews on decoherence are to be found in Zurek (2003) and Kiefer and Joos (1998) while a simple introduction can be found in Zurek (1991). There are several studies of the effect of decoherence and imperfections on the stability of quantum computation: for instance, Palma *et al.* (1996), Miquel *et al.* (1996), Georgeot and Shepelyansky (2000), Benenti *et al.* (2001b), Strini (2002), Carvalho *et al.* (2004) and Facchi *et al.* (2005).

A discussion of the master equation from the perspective of quantum optics can be found in Gardiner and Zoller (2000). References on dissipative quantum systems are Caldeira and Leggett (1983), Weiss (1999), Dittrich *et al.* (1998) and Prokof'ev and Stamp (2000).

Classical chaotic motion in non-linear dynamical systems is treated in several books, see for instance Lichtenberg and Lieberman (1992). General references on quantum chaos are Casati and Chirikov (1995), Haake (2000) and Stöckmann (1999). Random matrix theories in quantum physics are discussed in Bohigas (1991) and Guhr *et al.* (1998). A review on the quantum Loschmidt echo is provided by Gorin *et al.* (2006). Quantum chaos theory in many-body systems is discussed in Shepelyansky (2001). Quantum chaos experiments with hydrogen atoms in a microwave field and with cold ions in optical lattices are reviewed in Koch and van Leeuwen (1995) and Raizen *et al.* (2000), respectively.

The quantum-trajectory approach to quantum noise is discussed in Carmichael (1993), Gardiner and Zoller (2000), Scully and Zubairy (1997) and Plenio and Knight (1998). This approach can be generalized to treat non-Markovian effects, see for instance Breuer *et al.* (1999). An introduction to quantum trajectories closer to quantum information can be found in Brun (2002). The use of quantum trajectories for the simulation of quantum-information protocols is investigated in Barenco *et al.* (1997), Carlo *et al.* (2003) and Carlo *et al.* (2004).

The limits to quantum computation due to quantum chaos effects are reviewed in Georgeot (2006) and Benenti and Casati (2006).

Chapter 7

Quantum Error Correction

In this chapter we discuss how to protect quantum information from errors. The use of error-correcting codes to fight the effect of noise is a well developed technique in classical information processing. The key ingredient to protect against errors is *redundancy*.

To grasp this point, it is useful to consider the following example. Alice wishes to send Bob a classical bit through a classical communication channel; that is, a channel described by the laws of classical mechanics. The effect of noise in the channel is to flip the bit $(0 \rightarrow 1$ or $1 \rightarrow 0)$ with probability ϵ $(0 \leq \epsilon \leq 1)$, while the bit is transmitted without error with probability $1 - \epsilon$. The simplest manner to protect the bit is to send three copies of it: Alice sends 000 instead of just 0, say, or 111 instead of 1. Bob receives the three bits and applies *majority voting*: if, for instance, he receives 010, he assumes that, most probably, there was a single error affecting the second bit $(0 \rightarrow 1)$. He therefore concludes that the transmitted bit of information was 0.

We should point out that the underlying hypothesis is that the noisy channel is memoryless; namely, noise acts independently on each bit. Therefore, if Alice sends 000, Bob will receive 000 with probability $(1 - \epsilon)^3$. The error-correcting code succeeds if there is a single error; that is, when Bob receives 100, 010, or 001. Each of these messages is received with probability $\epsilon(1 - \epsilon)^2$. The code fails if two or more bits have been flipped. This is the case if Bob receives 011, 101, 110 (with probability $\epsilon^2(1 - \epsilon)$), or 111 (with probability ϵ^3). Therefore, the failure probability of the code is $\epsilon_c = 3\epsilon^2(1 - \epsilon) + \epsilon^3 = 3\epsilon^2 - 2\epsilon^3$. For just a single bit, the error probability was ϵ. Hence, the code improves the probability of successful transmission if $\epsilon_c < \epsilon$; that is, if $\epsilon < 1/2$. The improvement is greater for ϵ smaller since the error probability is reduced by a factor $\approx 3\epsilon$. For instance, for

$\epsilon = 10^{-1}$, $\epsilon_c = 2.8 \times 10^{-2}$, while, for $\epsilon = 10^{-2}$, $\epsilon_c = 2.98 \times 10^{-4}$.

The application of the same redundancy principle to quantum information encounters difficulties directly related to the basic principles of quantum mechanics:

1. Owing to the no-cloning theorem (discussed in Sec. 4.2), it is impossible to make copies of an unknown quantum state. Therefore, we cannot mimic the above-described classical code by sending $|\psi\rangle|\psi\rangle|\psi\rangle$ to protect an unknown quantum state $|\psi\rangle$.
2. In order to operate classical error correction, we observe (measure) the output from the noisy channel. In quantum mechanics, we know that, in general, measurements disturb the quantum state under investigation. For instance, if we receive the state $|\psi\rangle = \alpha|0\rangle + \beta|1\rangle$ and measure its polarization along the z-axis, the state will collapse onto $|0\rangle$ (with probability $|\alpha|^2$) or $|1\rangle$ (with probability $|\beta|^2$). In either case, the coherent superposition of the states $|0\rangle$ and $|1\rangle$ will be destroyed.
3. While the only possible classical error affecting a single bit is the bit flip ($0 \to 1$ and $1 \to 0$), the class of possible quantum errors is much richer. For instance, we can have the phase-flip error: $\alpha|0\rangle + \beta|1\rangle \to \alpha|0\rangle - \beta|1\rangle$. This error has no classical counterpart. Moreover, a continuum of quantum errors may occur in a single qubit. Given the state $|\psi\rangle$, noise may slightly rotate it: $|\psi\rangle \to R|\psi\rangle$, with R a rotation matrix. Such small errors will accumulate in time, eventually leading to incorrect computations (see Sec. 3.6). At first sight, it might thus appear that infinite resources are required to correct such errors since infinite precision is required to determine a rotation angle exactly.

However, we shall see in this chapter that, in spite of the above difficulties, quantum error correction is possible. We shall first discuss some simple examples: the three-qubit bit-flip and phase-flip codes, the nine-qubit Shor code and the five-qubit code. Then, on more general grounds, we shall discuss quantum codes, such as the CSS code, based on results of classical linear error correction. We shall also introduce passive error correction and include a discussion of the quantum Zeno effect. Finally, we shall discuss fault-tolerant quantum computation and show that, under certain hypotheses, if the noise level is below some threshold, then arbitrarily long, but reliable quantum computation is possible. We shall close this chapter by discussing two quantum-communication problems: purification of the quantum information transmitted through a noisy channel and entanglement-enhanced information transmission over a quantum channel with memory.

7.1 The three-qubit bit-flip code

Let us assume that Alice wishes to send a qubit, prepared in a generic state $|\psi\rangle = \alpha|0\rangle + \beta|1\rangle$, to Bob via a noisy quantum channel. The following hypothesis is made: the noise acts on each qubit independently, leaving the state of the qubit unchanged (with probability $1 - \epsilon$) or applying the Pauli operator σ_x (with probability ϵ). We remind the reader that σ_x produces a bit-flip error since $\sigma_x|0\rangle = |1\rangle$ and $\sigma_x|1\rangle = |0\rangle$. To protect the quantum state $|\psi\rangle$, Alice employs the following *encoding*:

$$|0\rangle \rightarrow |0_L\rangle \equiv |000\rangle, \qquad |1\rangle \rightarrow |1_L\rangle \equiv |111\rangle. \qquad (7.1)$$

The subscript L indicates that the states $|0_L\rangle$ and $|1_L\rangle$ are the *logical* $|0\rangle$ and $|1\rangle$ states (also known as *codewords*), encoded by means of three physical qubits. Correspondingly, a generic state is encoded as follows:

$$|\psi\rangle = \alpha|0\rangle + \beta|1\rangle \rightarrow \alpha|0_L\rangle + \beta|1_L\rangle = \alpha|000\rangle + \beta|111\rangle. \qquad (7.2)$$

This encoding is implemented by means of the quantum circuit in Fig. 7.1: the first CNOT gate maps $(\alpha|0\rangle + \beta|1\rangle)|00\rangle$ into $(\alpha|00\rangle + \beta|11\rangle)|0\rangle$ and the second CNOT leads to the encoded state $\alpha|000\rangle + \beta|111\rangle = \alpha|0_L\rangle + \beta|1_L\rangle$. This state is an entangled three-qubit state, known as a GHZ (Greenberger, Horne and Zeilinger) state or *cat state*. We should point out that Alice's encoding does not violate the no-cloning theorem since the encoded state is not the same as three copies of the original unknown state:

$$\alpha|000\rangle + \beta|111\rangle \neq |\psi\rangle|\psi\rangle|\psi\rangle = (\alpha|0\rangle + \beta|1\rangle)(\alpha|0\rangle + \beta|1\rangle)(\alpha|0\rangle + \beta|1\rangle). \qquad (7.3)$$

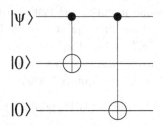

Fig. 7.1 A quantum circuit encoding a single qubit into three.

The three qubits, prepared in the cat state, are sent from Alice to Bob through the noisy channel. As a result, Bob receives one of the following states:

$$
\begin{array}{ll}
\alpha|000\rangle + \beta|111\rangle, & (1-\epsilon)^3, \\
\alpha|100\rangle + \beta|011\rangle, & \epsilon(1-\epsilon)^2, \\
\alpha|010\rangle + \beta|101\rangle, & \epsilon(1-\epsilon)^2, \\
\alpha|001\rangle + \beta|110\rangle, & \epsilon(1-\epsilon)^2, \\
\alpha|110\rangle + \beta|001\rangle, & \epsilon^2(1-\epsilon), \\
\alpha|101\rangle + \beta|010\rangle, & \epsilon^2(1-\epsilon), \\
\alpha|011\rangle + \beta|100\rangle, & \epsilon^2(1-\epsilon), \\
\alpha|111\rangle + \beta|000\rangle, & \epsilon^3,
\end{array}
\tag{7.4}
$$

where in the right-hand column we have written the probabilities of receiving the different states.

In order to correct a single bit-flip error, Bob might be tempted to measure the polarizations $\sigma_z^{(1)}$, $\sigma_z^{(2)}$ and $\sigma_z^{(3)}$ of the three qubits. To give a concrete example, let us assume that he receives the state $\alpha|100\rangle + \beta|011\rangle$. The three-qubit polarization measurement gives outcome 100 (with probability $|\alpha|^2$) or 011 (with probability $|\beta|^2$). In both case, Bob could apply majority voting and would conclude that the first qubit has been flipped. However, the coherent superposition of the states $|0\rangle$ and $|1\rangle$ would then be lost.

The problem may be solved by performing *collective measurements* on two qubits simultaneously. This can be achieved by means of the circuit in Fig. 7.2, which allows Bob to measure $\sigma_z^{(1)}\sigma_z^{(2)}$ and $\sigma_z^{(1)}\sigma_z^{(3)}$. Bob employs two ancillary qubits, both prepared in the state $|0\rangle$. The first two CNOT gates and the measurement of the polarization x_0 of the first ancillary qubit (by means of the detector D_0) tell him the value of $\sigma_z^{(1)}\sigma_z^{(2)}$. Note that $x_0 = 0$ corresponds to $\sigma_z^{(1)}\sigma_z^{(2)} = 1$, while $x_0 = 1$ corresponds to $\sigma_z^{(1)}\sigma_z^{(2)} = -1$. In the same manner, the last two CNOT gates and the measurement of the second ancillary qubit provide him with the value of $\sigma_z^{(1)}\sigma_z^{(3)}$ ($x_1 = 0$ when $\sigma_z^{(1)}\sigma_z^{(3)} = 1$ and $x_1 = 1$ when $\sigma_z^{(1)}\sigma_z^{(3)} = -1$).

As an example, we consider the case in which the first qubit has been flipped. The initial state of the five qubits is then

$$
\big(\alpha|100\rangle + \beta|011\rangle\big)|00\rangle.
\tag{7.5}
$$

It is easy to check that the four CNOT gates map this state into

$$\left(\alpha|100\rangle + \beta|011\rangle\right)|11\rangle. \tag{7.6}$$

The measurement of the two ancillary qubits gives Bob two classical bits of information, x_0 and x_1, known as the *error syndrome*, of value $x_0 = 1$ and $x_1 = 1$. Since $x_0 = 1$ Bob concludes that one of the first two qubits has been flipped. In the same manner, from $x_1 = 1$ Bob concludes that either the first or the third qubit has been flipped. Put together, the information provided by the values of x_0 and x_1 leads Bob to conclude that the first qubit has been flipped. Therefore, he applies a NOT gate (σ_x) to this qubit to recover the encoded state $\alpha|000\rangle + \beta|111\rangle$.

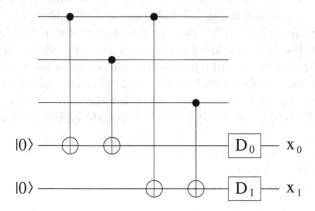

Fig. 7.2   A quantum circuit for extracting the error syndrome in the three-qubit bit-flip code.

In general, the measured syndrome and the action taken by Bob are the following (see Fig. 7.3):

$$
\begin{aligned}
x_0 &= 0,\ x_1 = 0, \quad \text{no action,} \\
x_0 &= 0,\ x_1 = 1, \quad \text{apply NOT to the third qubit,} \\
x_0 &= 1,\ x_1 = 0, \quad \text{apply NOT to the second qubit,} \\
x_0 &= 1,\ x_1 = 1, \quad \text{apply NOT to the first qubit.}
\end{aligned}
\tag{7.7}
$$

After Bob's action, the five-qubit states and their probabilities will be given by

$$(\alpha|000\rangle + \beta|111\rangle)|00\rangle, \qquad (1 - \epsilon)^3,$$
$$(\alpha|000\rangle + \beta|111\rangle)|11\rangle, \qquad \epsilon(1 - \epsilon)^2,$$
$$(\alpha|000\rangle + \beta|111\rangle)|10\rangle, \qquad \epsilon(1 - \epsilon)^2,$$
$$(\alpha|000\rangle + \beta|111\rangle)|01\rangle, \qquad \epsilon(1 - \epsilon)^2,$$
$$(\alpha|111\rangle + \beta|000\rangle)|01\rangle, \qquad \epsilon^2(1 - \epsilon), \qquad (7.8)$$
$$(\alpha|111\rangle + \beta|000\rangle)|10\rangle, \qquad \epsilon^2(1 - \epsilon),$$
$$(\alpha|111\rangle + \beta|000\rangle)|11\rangle, \qquad \epsilon^2(1 - \epsilon),$$
$$(\alpha|111\rangle + \beta|000\rangle)|00\rangle, \qquad \epsilon^3.$$

From now on, we may neglect the ancillary qubits. Finally, to extract the qubit sent by Alice, Bob applies the inverse of the encoding procedure. This *decoding* is shown in Fig. 7.3 and leads the three qubits sent by Alice to the state $(\alpha|0\rangle + \beta|1\rangle)|00\rangle$ (for the first four states in Eq. 7.8) or to $(\alpha|1\rangle + \beta|0\rangle)|00\rangle$ (for the last four states in Eq. 7.8). Hence, the three-qubit bit-flip code is successful if no more than one qubit has been flipped. This is the most likely possibility if $\epsilon \ll 1$. The code fails if more than two qubits have been corrupted by the noisy channel. This takes place with probability $\epsilon_c = 3\epsilon^2(1 - \epsilon) + \epsilon^3$. Therefore, the encoding improves the transmission of quantum information provided $\epsilon_c < \epsilon$; that is, $\epsilon < \frac{1}{2}$. This requirement is the same as in the classical three-bit code discussed at the beginning of this chapter.

A few comments are in order:

1. From the syndrome measurement Bob does not learn anything about the quantum state (the values of α and β). Hence, quantum coherence is not destroyed. This is possible because a qubit of information is encoded in a many-qubit entangled state and we only measure collective properties of this state.

2. If we repeat quantum-error correction in the case of several uses of a quantum noisy channel (for instance, if we wish to stabilize the state of a quantum computer, namely the *quantum memory*, against environmental noise), every time we must supply new ancillary qubits or erase them to the $|0\rangle$ state. This process requires the expenditure of power since, according to Landauer's principle, erasure of information dissipates energy.

Exercise 7.1 Design a circuit to measure the error syndrome in the

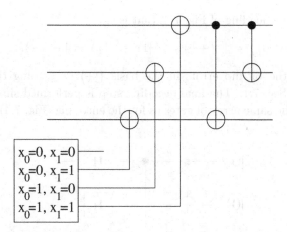

Fig. 7.3 Error correction and decoding in the three-qubit bit-flip code. The values of the classical bits x_0 and x_1 control the application of the NOT gates. The two CNOT gates decode the single-qubit message.

three-qubit bit-flip code without using any ancillary qubits.

Exercise 7.2 Compute the fidelity of a generic pure state sent from Alice to Bob through a bit-flip noisy channel. Compare with the result obtained when the three-qubit bit-flip error-correcting code is applied.

7.2 The three-qubit phase-flip code

In this section, we shall show that it is also possible to correct phase errors. These are quantum errors with no classical analogue. The phase-flip error affects the states of the computational basis as follows:

$$|0\rangle \rightarrow \sigma_z|0\rangle = |0\rangle, \qquad |1\rangle \rightarrow \sigma_z|1\rangle = -|1\rangle. \qquad (7.9)$$

Thus, a generic state $|\psi\rangle = \alpha|0\rangle + \beta|1\rangle$ is mapped into $\sigma_z|\psi\rangle = \alpha|0\rangle - \beta|1\rangle$. The method developed in Sec. 7.1 cannot correct phase errors. However, we observe that a phase-flip error in the computational basis $\{|0\rangle, |1\rangle\}$ becomes a bit-flip error in the basis $\{|+\rangle, |-\rangle\}$, where

$$|+\rangle = \frac{1}{\sqrt{2}}(|0\rangle + |1\rangle), \qquad |-\rangle = \frac{1}{\sqrt{2}}(|0\rangle - |1\rangle). \qquad (7.10)$$

Indeed, we have $\sigma_z|+\rangle = |-\rangle$ and $\sigma_z|-\rangle = |+\rangle$. We may transform the vectors of the computational basis into the new basis vectors (and *vice versa*) by means of the Hadamard gate. Therefore, to correct phase errors we exploit the encoding of Fig. 7.4; that is,

$$|0\rangle \rightarrow |0_L\rangle = |+++\rangle, \qquad |1\rangle \rightarrow |1_L\rangle = |---\rangle, \qquad (7.11)$$

and correct the bit-flip errors in the basis $\{|+\rangle, |-\rangle\}$ using the method described in Sec. 7.1. The final decoding step is performed simply by implementing the same array of gates as for the encoding (Fig. 7.4) but in reverse order.

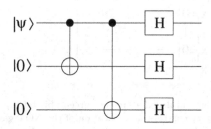

Fig. 7.4 A quantum circuit encoding a single qubit into three for the phase-flip code.

7.3 The nine-qubit Shor code

The nine-qubit Shor code corrects the most general possible noise acting on a single qubit. We employ the following encoding:

$$|0\rangle \rightarrow |0_L\rangle \equiv \frac{1}{\sqrt{8}} \big(|000\rangle + |111\rangle\big)\big(|000\rangle + |111\rangle\big)\big(|000\rangle + |111\rangle\big),$$
$$|1\rangle \rightarrow |1_L\rangle \equiv \frac{1}{\sqrt{8}} \big(|000\rangle - |111\rangle\big)\big(|000\rangle - |111\rangle\big)\big(|000\rangle - |111\rangle\big), \qquad (7.12)$$

so that a generic quantum state $|\psi\rangle = \alpha|0\rangle + \beta|1\rangle \rightarrow \alpha|0_L\rangle + \beta|1_L\rangle$. The quantum circuit implementing this encoding is shown in Fig. 7.5. The first two CNOT and the Hadamard gates of this circuit implement the three-qubit phase-flip encoding as in Fig. 7.4,

$$|0\rangle \rightarrow |+++\rangle, \qquad |1\rangle \rightarrow |---\rangle. \qquad (7.13)$$

Then, the last CNOT gates encode each of these three qubits into a block of three, by means of the three-qubit bit-flip encoding of Fig. 7.1

$$|+\rangle = \tfrac{1}{\sqrt{2}}\big(|0\rangle + |1\rangle\big) \rightarrow \tfrac{1}{\sqrt{2}}\big(|000\rangle + |111\rangle\big),$$
$$|-\rangle = \tfrac{1}{\sqrt{2}}\big(|0\rangle - |1\rangle\big) \rightarrow \tfrac{1}{\sqrt{2}}\big(|000\rangle - |111\rangle\big). \qquad (7.14)$$

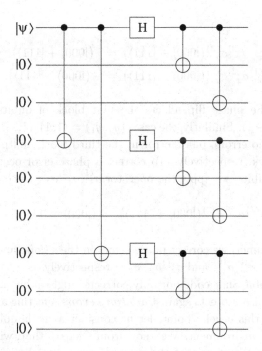

Fig. 7.5 A quantum circuit encoding a single qubit into nine.

This code can correct both bit and phase-flip errors. The quantum circuit extracting the error syndrome is shown in Fig. 7.6. In each three-qubit block a single bit-flip error can be detected and corrected following the method described in Sec. 7.1. Moreover, we can deal with phase errors affecting a single qubit. Let us assume that the phase error occurs in the first qubit. As a consequence, the state of the first block of qubits is modified as follows (neglecting the wave function normalization):

$$|000\rangle + |111\rangle \rightarrow |000\rangle - |111\rangle,$$
$$|000\rangle - |111\rangle \rightarrow |000\rangle + |111\rangle. \tag{7.15}$$

In order to detect this phase-flip error without disturbing the encoded quantum state $|\psi\rangle = \alpha|0\rangle + \beta|1\rangle$, we must perform collective measurements. More precisely, we measure

$$y_0 = \sigma_x^{(1)}\sigma_x^{(2)}\sigma_x^{(3)}\sigma_x^{(4)}\sigma_x^{(5)}\sigma_x^{(6)},$$
$$y_1 = \sigma_x^{(1)}\sigma_x^{(2)}\sigma_x^{(3)}\sigma_x^{(7)}\sigma_x^{(8)}\sigma_x^{(9)}. \tag{7.16}$$

We have

$$\sigma_x^{(1)}\sigma_x^{(2)}\sigma_x^{(3)}\big(|000\rangle + |111\rangle\big) = \big(|000\rangle + |111\rangle\big),$$
$$\sigma_x^{(1)}\sigma_x^{(2)}\sigma_x^{(3)}\big(|000\rangle - |111\rangle\big) = -\big(|000\rangle - |111\rangle\big). \tag{7.17}$$

Therefore, if the phase flip affects the first block of qubits, we obtain $(y_0, y_1) = (-1, -1)$. Similarly, the cases $(y_0, y_1) = (1, 1)$, $(1, -1)$ and $(-1, 1)$ correspond to no errors, phase error in the third block and phase error in the second block, respectively. To correct a phase error occurring in the first block of qubits, we apply the operator $\sigma_z^{(1)}\sigma_z^{(2)}\sigma_z^{(3)}$ since

$$\sigma_z^{(1)}\sigma_z^{(2)}\sigma_z^{(3)}\big(|000\rangle \pm |111\rangle\big) = \big(|000\rangle \mp |111\rangle\big). \tag{7.18}$$

In the same manner, we correct phase errors in the second and third block by applying $\sigma_z^{(4)}\sigma_z^{(5)}\sigma_z^{(6)}$ and $\sigma_z^{(7)}\sigma_z^{(8)}\sigma_z^{(9)}$, respectively.

The nine-qubit Shor code not only corrects single-qubit bit and phase-flip errors, but also protects against *arbitrary errors* affecting a single qubit. To understand this crucial point, let us consider a single qubit which interacts with its environment. We know from Chap. 5 that, without loss of generality, we can assume that the environment is initially in a pure state, which we call $|0\rangle_E$. The most general unitary evolution U of the qubit and its environment may be written as

$$U|0\rangle|0\rangle_E = |0\rangle|e_0\rangle_E + |1\rangle|e_1\rangle_E,$$
$$U|1\rangle|0\rangle_E = |0\rangle|e_2\rangle_E + |1\rangle|e_3\rangle_E, \tag{7.19}$$

where $|e_0\rangle_E$, $|e_1\rangle_E$, $|e_2\rangle_E$ and $|e_3\rangle_E$ are states of the environment, not necessarily normalized or mutually orthogonal. For a generic initial state of the system, $|\psi\rangle = \alpha|0\rangle + \beta|1\rangle$, we have

$$U\big(\alpha|0\rangle + \beta|1\rangle\big)|0\rangle_E$$

$$= \alpha\big(|0\rangle|e_0\rangle_E + |1\rangle|e_1\rangle_E\big) + \beta\big(|0\rangle|e_2\rangle_E + |1\rangle|e_3\rangle_E\big)$$

$$= \big(\alpha|0\rangle + \beta|1\rangle\big)\tfrac{1}{2}\big(|e_0\rangle_E + |e_3\rangle_E\big) + \big(\alpha|0\rangle - \beta|1\rangle\big)\tfrac{1}{2}\big(|e_0\rangle_E - |e_3\rangle_E\big)$$
$$+ \big(\alpha|1\rangle + \beta|0\rangle\big)\tfrac{1}{2}\big(|e_1\rangle_E + |e_2\rangle_E\big) + \big(\alpha|1\rangle - \beta|0\rangle\big)\tfrac{1}{2}\big(|e_1\rangle_E - |e_2\rangle_E\big)$$

$$= I|\psi\rangle|e_I\rangle_E + \sigma_z|\psi\rangle|e_z\rangle_E + \sigma_x|\psi\rangle|e_x\rangle_E + \sigma_x\sigma_z|\psi\rangle|e_{xz}\rangle_E, \tag{7.20}$$

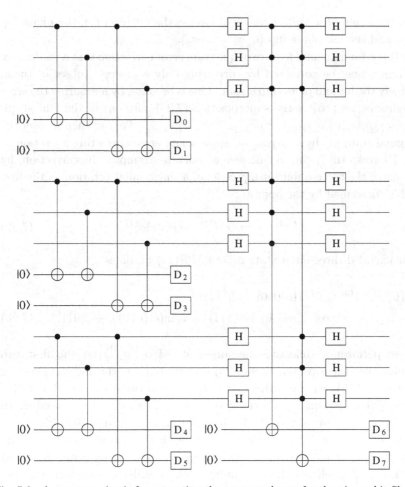

Fig. 7.6 A quantum circuit for extracting the error syndrome for the nine-qubit Shor code. The symbols D_i ($i = 0, \ldots, 7$) denote detectors measuring single-qubit polarizations.

where

$$|e_I\rangle_E \equiv \tfrac{1}{2}\big(|e_0\rangle_E + |e_3\rangle_E\big), \qquad |e_z\rangle_E \equiv \tfrac{1}{2}\big(|e_0\rangle_E - |e_3\rangle_E\big),$$

$$|e_x\rangle_E \equiv \tfrac{1}{2}\big(|e_1\rangle_E + |e_2\rangle_E\big), \qquad |e_{xz}\rangle_E \equiv \tfrac{1}{2}\big(|e_1\rangle_E - |e_2\rangle_E\big). \tag{7.21}$$

Therefore, the action of U can be expanded over the *discrete* set of operators $\{I, \sigma_x, \sigma_y = i\sigma_x\sigma_z, \sigma_z\}$. This is because, as can be readily checked, these operators are a basis for the Hilbert space of 2×2 matrices. This expansion embodies the fact that arbitrary single-qubit errors can be expressed as a weighted sum of a *finite* number of errors: the bit flip (σ_x), the phase flip (σ_z) and the bit–phase flip $(\sigma_x\sigma_z = -i\sigma_y)$.

It is a fundamental feature of quantum error correction that a *continuum* of errors may be corrected by correcting only a *discrete* subset of them, namely the bit and phase-flip errors. This is because, by measuring the error syndrome, we project the superposition (7.20) onto one of the four states $I|\psi\rangle|e_I\rangle_E$, $\sigma_z|\psi\rangle|e_z\rangle_E$, $\sigma_x|\psi\rangle|e_x\rangle_E$, $\sigma_x\sigma_z|\psi\rangle|e_{xz}\rangle_E$. We can then recover the original state $|\psi\rangle$ by applying an appropriate error-correcting operation.

To grasp this point, let us give a concrete example: the correction, by means of the three-qubit bit-flip code, of a single-qubit rotation on the first qubit[1] described by the operator

$$U_\epsilon^{(1)} = \cos(\epsilon)I^{(1)} + i\sin(\epsilon)\sigma_x^{(1)}. \tag{7.22}$$

The encoded three-qubit state $\alpha|000\rangle + \beta|111\rangle$ becomes

$$\begin{aligned}
\left(U_\epsilon^{(1)} \otimes I^{(2)} \otimes I^{(3)}\right)&\left(\alpha|000\rangle + \beta|111\rangle\right) \\
&= \cos(\epsilon)\left(\alpha|000\rangle + \beta|111\rangle\right) + i\sin(\epsilon)\left(\alpha|100\rangle + \beta|011\rangle\right).
\end{aligned} \tag{7.23}$$

If we perform a collective measurement (of $\sigma_z^{(1)}\sigma_z^{(2)}$) on the first two qubits, then the wave function (7.23) is projected over the undamaged state $\alpha|000\rangle + \beta|111\rangle$ with probability $\cos^2(\epsilon)$ or over the state $\alpha|100\rangle + \beta|011\rangle$ with probability $\sin^2(\epsilon)$. In the first case, no further action is needed. In the latter case, we correct the bit-flip error as explained in Sec. 7.1.

Finally, we wish to discuss in more depth the role of encoding in the Shor code. Let us consider the case in which an arbitrary error, described by Eq. (7.20), effects the first qubit. We consider the evolution of the codewords $|0_L\rangle$ and $|1_L\rangle$ separately. It is sufficient to write the evolution of only the first three-qubit block since the other blocks are unchanged. We have

$$\begin{aligned}
\big(|000\rangle + |111\rangle\big)&|0\rangle_E \\
\rightarrow\ & |000\rangle|e_0\rangle_E + |100\rangle|e_1\rangle_E + |011\rangle|e_2\rangle_E + |111\rangle|e_3\rangle_E \\
=\ & \big(|000\rangle + |111\rangle\big)\tfrac{1}{2}\big(|e_0\rangle_E + |e_3\rangle_E\big) + \big(|000\rangle - |111\rangle\big)\tfrac{1}{2}\big(|e_0\rangle_E - |e_3\rangle_E\big) \\
& + \big(|100\rangle + |011\rangle\big)\tfrac{1}{2}\big(|e_1\rangle_E + |e_2\rangle_E\big) + \big(|100\rangle - |011\rangle\big)\tfrac{1}{2}\big(|e_1\rangle_E - |e_2\rangle_E\big).
\end{aligned} \tag{7.24}$$

---

[1] Of course, the same error can also be corrected by the Shor code.

Similarly, we obtain

$$\big(|000\rangle - |111\rangle\big)|0\rangle_E$$

$$\rightarrow |000\rangle|e_0\rangle_E + |100\rangle|e_1\rangle_E - |011\rangle|e_2\rangle_E - |111\rangle|e_3\rangle_E$$

$$= \big(|000\rangle - |111\rangle\big)\tfrac{1}{2}\big(|e_0\rangle_E + |e_3\rangle_E\big) + \big(|000\rangle + |111\rangle\big)\tfrac{1}{2}\big(|e_0\rangle_E - |e_3\rangle_E\big)$$

$$+ \big(|100\rangle - |011\rangle\big)\tfrac{1}{2}\big(|e_1\rangle_E + |e_2\rangle_E\big) + \big(|100\rangle + |011\rangle\big)\tfrac{1}{2}\big(|e_1\rangle_E - |e_2\rangle_E\big).$$

$$(7.25)$$

This implies that the final state of the environment is the same if the system is initially either in the encoded state $|0_L\rangle$ or in $|1_L\rangle$ (see exercise 7.3). The deep reason for this result is that the states $|0_L\rangle$ and $|1_L\rangle$ are entangled and it is impossible to tell them apart by observing just a single qubit (the state of a single qubit is equal to $\tfrac{1}{2}I$ for both $|0_L\rangle$ and $|1_L\rangle$). Therefore, given an arbitrary state $\alpha|0_L\rangle + \beta|1_L\rangle$, the environment cannot learn anything about α and β through interaction with a single qubit (inducing single-qubit errors). Since quantum information is not destroyed by this interaction, error recovery is possible.

Exercise 7.3 Compute the final state of the environment when the initial state of system plus environment is described by $|0_L\rangle|0\rangle_E$ or $|1_L\rangle|0\rangle_E$ and a generic single-qubit error occurs ($|0_L\rangle$ and $|1_L\rangle$ are the codewords of the nine-qubit Shor code).

7.4 General properties of quantum error correction

So far, we have described quantum error correction in the case of single-qubit errors. We now discuss how to implement quantum error correction when more general errors occur. First of all, note that errors affecting n qubits can be expanded over a set of 4^n operators $\{E_k\}$, constructed as tensor products of the single-qubit operators I, σ_x, σ_y, and σ_z. An example, for $n = 5$, is given by $I^{(1)} \otimes \sigma_y^{(2)} \otimes \sigma_x^{(3)} \otimes I^{(4)} \otimes \sigma_z^{(5)}$. The action of an arbitrary unitary operator U on the n-qubit system plus the environment is

$$U|\psi\rangle|0\rangle_E = \sum_{k=0}^{4^n-1} E_k|\psi\rangle|e_k\rangle_E, \tag{7.26}$$

where $|\psi\rangle$ is the initial n-qubit state. We call $\mathcal{E} = \{E_0, \ldots, E_{4^n-1}\}$ the set of all possible errors affecting n qubits and $\mathcal{E}_c$ the subset of errors that can be corrected by a code. Let us discuss what conditions should be satisfied to allow error correction. First of all, correctable errors should map two different codewords $|i_L\rangle$ and $|j_L\rangle$ into orthogonal states:

$$\langle i_L|E_a^\dagger E_b|j_L\rangle = 0 \quad \text{for } i \neq j, \tag{7.27}$$

where E_a and $E_b \in \mathcal{E}_c$. If this condition were not satisfied, then the states $E_a|i_L\rangle$ and $E_b|j_L\rangle$ could not be distinguished with certainty and therefore perfect error correction would be impossible. The second condition is that, for any correctable errors E_a and E_b,

$$\langle i_L|E_a^\dagger E_b|i_L\rangle = C_{ab}, \tag{7.28}$$

where C_{ab} does not depend on the state $|i_L\rangle$. If this were not the case, we would obtain some information on the encoded state from the measurement of the error syndrome. Therefore, we would inevitably disturb the quantum state. Note that $C_{ab} = C_{ba}^*$.

Conditions (7.27) and (7.28) can be put together and it is possible to prove that error correction is possible if and only if

$$\langle i_L|E_a^\dagger E_b|j_L\rangle = C_{ab}\delta_{ij}, \tag{7.29}$$

where E_a and E_b belong to the set $\mathcal{E}_c$ of correctable errors and the matrix C_{ab} is Hermitian (for a proof see, *e.g.*, Preskill, 1998a). If $C_{ab} = \delta_{ab}$, the code is known as *non-degenerate*, In this case, it is possible to identify with certainty which error occurred. In contrast, if $C_{ab} \neq \delta_{ab}$, we call the code *degenerate*.

Exercise 7.4 Show that condition (7.29) is fulfilled by the three-qubit bit-flip code.

Exercise 7.5 Show that the three-qubit bit-flip code is non-degenerate while the nine-qubit Shor code is degenerate.

It is instructive to describe the error recovery procedure in the simple case of non-degenerate codes. Provided the system has been subjected to correctable errors, the most general system plus environment state is given by

$$\sum_{E_k \in \mathcal{E}_c} E_k|\psi\rangle|e_k\rangle_E. \tag{7.30}$$

To measure the error syndrome, we can attach ancillary qubits, initially in a well known state $|0\rangle_A$, to the system, and operate the unitary transformation

$$\sum_{E_k \in \mathcal{E}_c} E_k |\psi\rangle |e_k\rangle_E |0\rangle_A \;\rightarrow\; \sum_{E_k \in \mathcal{E}_c} E_k |\psi\rangle |e_k\rangle_E |a_k\rangle_A. \tag{7.31}$$

A projective measurement of the ancillary qubits will then collapse this sum to a single term

$$E_{\bar{k}} |\psi\rangle |e_{\bar{k}}\rangle_E |a_{\bar{k}}\rangle_A. \tag{7.32}$$

Note that the system is now disentangled from the environment and from the ancillary qubits. Since the operators E_k are unitary (they are constructed as tensor products of the Pauli matrices, which are unitary), it is sufficient to apply the unitary operator $E_{\bar{k}}^{\dagger} = E_{\bar{k}}$ to the system to recover the original state $|\psi\rangle$.

7.4.1 The quantum Hamming bound

The quantum Hamming bound only applies to non-degenerate codes. It tells us the minimum number n of physical qubits required to encode k logical qubits, in such a manner that errors affecting at most t qubits can be corrected. If j errors occur, there are $\binom{n}{j}$ possible locations for these errors. For instance, if $n = 3$ and $j = 2$, the $\binom{3}{2}$ possibilities are: (i) errors in the first and second qubit, (ii) in the first and third qubit, and (iii) in the second and third qubit. Each qubit may be subjected to three possible errors (bit flip σ_x, phase flip σ_z, and bit–phase flip $\sigma_x \sigma_z = -i\sigma_y$). Hence, there are 3^j possible errors for each error location. The total number of possible errors affecting t or less qubits is therefore given by

$$\sum_{j=0}^{t} \binom{n}{j} 3^j. \tag{7.33}$$

Note that the sum over j starts from zero to include the error-free case too. To encode k qubits by means of a non-degenerate code, each of these errors must correspond to a 2^k-dimensional subspace. These subspaces must be mutually orthogonal and belong to the 2^n-dimensional Hilbert space for n qubits. Therefore, we can write the quantum Hamming bound

$$\sum_{j=0}^{t} \binom{n}{j} 3^j 2^k \leq 2^n. \tag{7.34}$$

For non-degenerate codes correcting a single error ($t = 1$), the quantum Hamming bound reduces to $(1 + 3n) 2^k \leq 2^n$. Let us call $n_{\min}$ the smallest value of n satisfying this bound. For codes encoding a single qubit ($k = 1$) and correcting arbitrary single-qubit errors, $n_{\min} = 5$ qubits.[2] Note that the ratio $n_{\min}/k$ decreases with k. For example, $n_{\min} = 12$ for $k = 6$. Therefore, the encoding of quantum information is more efficient for large k. The price to pay is a greater complexity of the corresponding quantum error-correcting codes.

7.5   * The five-qubit code

In this section, we describe a quantum error-correcting code which protects a qubit of information against arbitrary single-qubit errors. To accomplish this, we encode a single logical qubit into five physical qubits, the minimum number required for this task. The encoding is given by

$$|0\rangle \rightarrow |0_L\rangle \equiv \tfrac{1}{\sqrt{8}}\big(|00000\rangle - |01111\rangle - |10011\rangle + |11100\rangle$$
$$+ |00110\rangle + |01001\rangle + |10101\rangle + |11010\rangle\big),$$
$$\text{(7.35)}$$
$$|1\rangle \rightarrow |1_L\rangle \equiv \tfrac{1}{\sqrt{8}}\big(|11111\rangle - |10000\rangle + |01100\rangle - |00011\rangle$$
$$+ |11001\rangle + |10110\rangle - |01010\rangle - |00101\rangle\big),$$

and can be implemented by the circuit in Fig. 7.7.

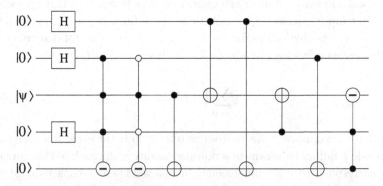

Fig. 7.7   A quantum circuit encoding a single qubit into five. The circles with a minus sign correspond to a phase shift of π. The qubit $|\psi\rangle = \alpha|0\rangle + \beta|1\rangle$ is encoded into the five-qubit state $\alpha|0_L\rangle + \beta|1_L\rangle$.

---

[2]It is possible to prove that $n_{\min} = 5$ also in the case of degenerate codes, see Knill and Laflamme (1997).

A remarkable feature of the five-qubit code is that the circuit for detecting the error syndrome, drawn in Fig. 7.8, is exactly the same as that for encoding, but run backwards. There are 15 possible single-qubit errors, 3 for each of the five qubits (bit flip, phase flip and bit–phase flip). The four measurements in Fig. 7.8 provide the 4 classical bits a, b, c and d allowing us to distinguish the 15 possible errors plus the case without errors. Table 7.1 exhibits all possibilities. For instance, the case $a = b = c = d = 0$ corresponds to no errors. If instead the outcomes of the measurements are $a = 1$ and $b = c = d = 0$, then the bit-flip error affected the first qubit. Different outcomes (error syndromes) are associated with different errors, as shown in Table 7.1. The post-measurement state $|\psi'\rangle$ of the qubit carrying the quantum information is shown in the same table. It is easy to see that the original state $|\psi\rangle = \alpha|0\rangle + \beta|1\rangle$ is recovered by a unitary transformation U that depends upon the results a, b, c and d of the measurements. For examples, if $a = b = c = 0$ and $d = 1$, then $|\psi'\rangle = \alpha|0\rangle - \beta|1\rangle$ and we restore $|\psi\rangle$ by means of $U = \sigma_z$. Indeed, $\sigma_z|\psi'\rangle = |\psi\rangle$.

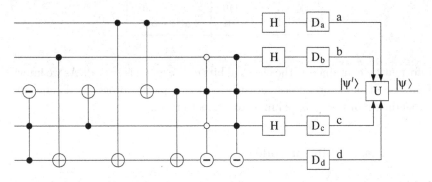

Fig. 7.8 A quantum circuit extracting the error syndrome and recovering the correct state $|\psi\rangle$ in the five-qubit code. The four detectors D_a, D_b, D_c and D_d measure single qubit polarizations. The resulting classical bits a, b, c and d drive the unitary operator U, which maps $|\psi'\rangle$ onto the original state $|\psi\rangle$.

Exercise 7.6 Verify Table 7.1.

We point out that the five-qubit code does not require any ancillary qubits. In any case, the code is dissipative: to apply the code again we must first of all encode the state $|\psi\rangle$ onto the five-qubit state $\alpha|0_L\rangle + \beta|1_L\rangle$. For this purpose, we must supply four new ancillary qubits prepared in the state $|0\rangle$. Alternatively, if we wish to recycle the ancillary qubits, we must first map their state $|abcd\rangle$ into $|0000\rangle$. This means that the

Table 7.1 Error, syndrome and resulting
state in the five-qubit code.

Error	$abcd$	$|\psi'\rangle$
None	0000	$\alpha|0\rangle + \beta|1\rangle$
$\sigma_x^{(3)}\sigma_z^{(3)}$	1101	$-\alpha|1\rangle + \beta|0\rangle$
$\sigma_x^{(5)}\sigma_z^{(5)}$	1111	$-\alpha|0\rangle + \beta|1\rangle$
$\sigma_x^{(2)}$	0001	$\alpha|0\rangle - \beta|1\rangle$
$\sigma_z^{(3)}$	1010	$\alpha|0\rangle - \beta|1\rangle$
$\sigma_z^{(5)}$	1100	$\alpha|0\rangle - \beta|1\rangle$
$\sigma_x^{(2)}\sigma_z^{(2)}$	0101	$\alpha|0\rangle - \beta|1\rangle$
$\sigma_x^{(5)}$	0011	$-\alpha|0\rangle - \beta|1\rangle$
$\sigma_z^{(1)}$	1000	$-\alpha|0\rangle - \beta|1\rangle$
$\sigma_z^{(2)}$	0100	$-\alpha|0\rangle - \beta|1\rangle$
$\sigma_z^{(4)}$	0010	$-\alpha|0\rangle - \beta|1\rangle$
$\sigma_x^{(1)}$	0110	$-\alpha|1\rangle + \beta|0\rangle$
$\sigma_x^{(3)}$	0111	$-\alpha|1\rangle + \beta|0\rangle$
$\sigma_x^{(4)}$	1011	$-\alpha|1\rangle + \beta|0\rangle$
$\sigma_x^{(1)}\sigma_z^{(1)}$	1110	$-\alpha|1\rangle + \beta|0\rangle$
$\sigma_x^{(4)}\sigma_z^{(4)}$	1001	$-\alpha|1\rangle + \beta|0\rangle$

information contained in the classical bits a, b, c and d is erased. As we know from Landauer's principles, erasure is a dissipative process. Therefore, expenditure of power is required to correct errors.

7.6 * Classical linear codes

In this section, we shall discuss a few elements of the theory of classical error correction. We shall focus on concepts that have proved useful for the development of quantum error-correcting codes.

A classical error-correcting code C is defined by a set of codewords $\mathcal{C}$ and a set of correctable errors $\mathcal{E}$ such that, for any $u, v \in \mathcal{C}$, with $u \neq v$, and for any $e, f \in \mathcal{E}$, we have

$$u + e \neq v + f. \tag{7.36}$$

This implies that correctable errors cannot map two different codewords into the same string of bits. This is a necessary condition to unambiguously recover the original codeword.

A code C is called an $[n, k, d]$ code if we encode a k-bit message into an n-bit message and codewords differ from each other in at least d bits. The *Hamming distance* $d_H(u, v)$ of two codewords u, v is defined as the number of bits in which u and v differ. For example, given $u = 00011010$ and $v = 01011000$, we have $d_H(u, v) = 2$ since the two strings u and v differ in the second and the penultimate bits. We define

$$d \equiv \min_{\substack{u,v \in C \\ u \neq v}} d_H(u, v). \tag{7.37}$$

It is clear that the code C allows the correction of errors that affect at most $[(d - 1)/2]$ bits, where $[\ ]$ means the integer part. Indeed, we must ensure that the action of any two correctable errors e, f cannot lead to the same n-bit message. In other words, we require that $u + e \neq v + f$.

As an example, we consider the majority voting described at the beginning of this chapter. This is a $[3, 1, 3]$ code since we encode a single bit ($k = 1$) into a block of three ($0 \rightarrow 000$ and $1 \rightarrow 111$) and the two codewords 000 and 111 differ in $d = 3$ bits. Since $[(d - 1)/2] = 1$, this code is able to correct errors only affecting a single bit.

A *linear code* encoding k bits of information into an n-bit message is determined by an n by k matrix G, called the *generator matrix*. The entries of G are zeros and ones and a k-bit message x is encoded into an n-bit string y as follows:

$$y = Gx. \tag{7.38}$$

Note that here x and y must be treated as column vectors and that, in computing Gx, additions are taken modulo 2. In the rest of this section, all arithmetic operations must be understood modulo 2. The addition (modulo 2) of two codewords is again a codeword. Moreover, an arbitrary codeword can be expressed as a linear combination of the columns of G. The message is uniquely encoded if the columns of G are linearly independent. This implies that the 2^k codewords (of length n) constitute a k-dimensional subspace in the n-dimensional space of the 2^n binary strings of length n.

Another very important matrix is the $(n - k)$ by n *parity-check matrix* H, defined by

$$HG = 0. \tag{7.39}$$

For any codeword $y = Gx$, we have $Hy = H(Gx) = (HG)x = 0$. This means that the codewords are the kernel of H. In order to have 2^k linearly independent codewords, the kernel of H must have dimension k. Thus, the $(n - k)$ rows of H must be linearly independent.

If a codeword y is corrupted by an error e, that is

$$y \rightarrow y' = y + e, \tag{7.40}$$

then

$$Hy' = H(y + e) = Hy + He = HGx + He = He. \tag{7.41}$$

Therefore, application of the parity matrix H gives an *error syndrome* He that depends only on the vector error e and not on the codeword y. Error correction is possible if we can unambiguously derive the error e from the error syndrome He.

We point out that a code can be defined either by the generator matrix G or by the parity-check matrix H. Given H, we can construct G as follows. We take a basis $\{y_1, \ldots, y_k\}$ for the kernel of H. These vectors are the columns of

$$G \equiv [y_1, y_2, \ldots, y_k]. \tag{7.42}$$

On the other hand, given G we can construct H. We take $(n - k)$ linearly independent vectors $z_1, \ldots, z_{n-k}$ orthogonal to the columns of G. We then define

$$H \equiv \begin{bmatrix} z_1^T \\ z_2^T \\ \vdots \\ z_{n-k}^T \end{bmatrix}. \tag{7.43}$$

7.6.1   * The Hamming codes

In order to illustrate the working of a classical linear codes, it is useful to introduce two codes, which we shall call C_1 and C_2.[3] As we shall see in the next section, these codes are important for quantum error correction.

---

[3]The code C_1 belongs to the class of the so-called Hamming codes.

Code C_1 is a $[7, 4, 3]$ code, defined by the generator matrix

$$
G(C_1) = \begin{bmatrix} 1 & 0 & 0 & 0 \\ 0 & 1 & 0 & 0 \\ 0 & 0 & 1 & 0 \\ 0 & 0 & 0 & 1 \\ 0 & 1 & 1 & 1 \\ 1 & 0 & 1 & 1 \\ 1 & 1 & 0 & 1 \end{bmatrix}. \tag{7.44}
$$

If we label the 4-bit messages as

$$
x_0 = \begin{bmatrix} 0 \\ 0 \\ 0 \\ 0 \end{bmatrix}, \quad x_1 = \begin{bmatrix} 0 \\ 0 \\ 0 \\ 1 \end{bmatrix}, \quad x_2 = \begin{bmatrix} 0 \\ 0 \\ 1 \\ 0 \end{bmatrix}, \quad \ldots, \quad x_{15} = \begin{bmatrix} 1 \\ 1 \\ 1 \\ 1 \end{bmatrix}, \tag{7.45}
$$

then we obtain the 16 codewords $y_i = G x_i$ $(i = 0, 1, \ldots, 15)$. For example, we have

$$
y_7 = G(C_1) \, x_7 = \begin{bmatrix} 1 & 0 & 0 & 0 \\ 0 & 1 & 0 & 0 \\ 0 & 0 & 1 & 0 \\ 0 & 0 & 0 & 1 \\ 0 & 1 & 1 & 1 \\ 1 & 0 & 1 & 1 \\ 1 & 1 & 0 & 1 \end{bmatrix} \begin{bmatrix} 0 \\ 1 \\ 1 \\ 1 \end{bmatrix} = \begin{bmatrix} 0 \\ 1 \\ 1 \\ 1 \\ 1 \\ 0 \\ 0 \end{bmatrix}. \tag{7.46}
$$

The 16 codewords $y_0, \ldots, y_{15}$ are given in Table 7.2. Note that they are a subset of the $2^7 = 128$ possible 7-bit messages. It is easy to check that the 16 codewords differ in at least 3 bits, so that only single-bit errors can be corrected ($[(d - 1)/2] = 1$).

Table 7.2 Codewords of the $[7, 4, 3]$ Hamming code.

y_0	y_1	y_2	y_3	y_4	y_5	y_6	y_7	y_8	y_9	y_{10}	y_{11}	y_{12}	y_{13}	y_{14}	y_{15}
0	0	0	0	0	0	0	0	1	1	1	1	1	1	1	1
0	0	0	0	1	1	1	1	0	0	0	0	1	1	1	1
0	0	1	1	0	0	1	1	0	0	1	1	0	0	1	1
0	1	0	1	0	1	0	1	0	1	0	1	0	1	0	1
0	1	1	0	1	0	0	1	0	1	1	0	1	0	0	1
0	1	1	0	0	1	1	0	1	0	0	1	1	0	0	1
0	1	0	1	1	0	1	0	1	0	1	0	0	1	0	1

The rows of the parity-check matrix $H(C_1)$ must be linearly independent and orthogonal to the columns of $G(C_1)$. These conditions are fulfilled by

$$H(C_1) = \begin{bmatrix} 0 & 0 & 0 & 1 & 1 & 1 & 1 \\ 0 & 1 & 1 & 0 & 0 & 1 & 1 \\ 1 & 0 & 1 & 0 & 1 & 0 & 1 \end{bmatrix}. \tag{7.47}$$

Let us assume that a correctable error e_k occurs, so that $y_i \rightarrow y_i' = y_i + e_k$. We know from Eq. (7.41) that $H(C_1)y_i' = H(C_1)e_k$. As previously discussed, only single-bit errors can be corrected by this code. Let us call e_i the error affecting the i-th bit; that is, $e_1 = (1, 0, 0, 0, 0, 0, 0)$, $e_2 = (0, 1, 0, 0, 0, 0, 0)$, $\ldots$, $e_7 = (0, 0, 0, 0, 0, 0, 1)$. It is easy to check that the error syndrome $H(C_1)e_k$ is just the binary representation of k. For instance, for $k = 3$ we have

$$H(C_1)\, e_3 = \begin{bmatrix} 0 & 0 & 0 & 1 & 1 & 1 & 1 \\ 0 & 1 & 1 & 0 & 0 & 1 & 1 \\ 1 & 0 & 1 & 0 & 1 & 0 & 1 \end{bmatrix} \begin{bmatrix} 0 \\ 0 \\ 1 \\ 0 \\ 0 \\ 0 \\ 0 \end{bmatrix} = \begin{bmatrix} 0 \\ 1 \\ 1 \end{bmatrix}, \tag{7.48}$$

and 011 is the binary representation of $k = 3$. It is therefore sufficient to flip the k-th bit to correct the error.

Let us now describe the code C_2. We first of all note that, in general, if $HG = 0$, also $G^T H^T = 0$. Therefore, we can interchange the role of the matrices $\{G, H\}$ and $\{H^T, G^T\}$. Given a code $C[n, k]$ (encoding k logical bits into n physical bits), we can define another code $C^{\perp}[n, n-k]$ (encoding $n - k$ logical bits into n physical bits), known as the *dual code* of C. The dual code has generator matrix H^T and parity-check matrix G^T. We define $C_2 \equiv C_1^{\perp}$. Hence,

$$G(C_2) = [H(C_1)]^T = \begin{bmatrix} 0 & 0 & 1 \\ 0 & 1 & 0 \\ 0 & 1 & 1 \\ 1 & 0 & 0 \\ 1 & 0 & 1 \\ 1 & 1 & 0 \\ 1 & 1 & 1 \end{bmatrix}, \tag{7.49}$$

$$H(C_2) = [G(C_1)]^T = \begin{bmatrix} 1 & 0 & 0 & 0 & 0 & 1 & 1 \\ 0 & 1 & 0 & 0 & 1 & 0 & 1 \\ 0 & 0 & 1 & 0 & 1 & 1 & 0 \\ 0 & 0 & 0 & 1 & 1 & 1 & 1 \end{bmatrix}. \tag{7.50}$$

The $2^{n-k} = 2^{7-4} = 8$ codewords $\tilde{y}_i$ of C_2 are given in Table 7.3. They are obtained from $\tilde{y}_i = G(C_2)\tilde{x}_i$, with

$$\tilde{x}_0 = \begin{bmatrix} 0 \\ 0 \\ 0 \end{bmatrix}, \quad \tilde{x}_1 = \begin{bmatrix} 0 \\ 0 \\ 1 \end{bmatrix}, \quad \tilde{x}_2 = \begin{bmatrix} 0 \\ 1 \\ 0 \end{bmatrix}, \quad \dots, \quad \tilde{x}_7 = \begin{bmatrix} 1 \\ 1 \\ 1 \end{bmatrix}. \tag{7.51}$$

Note that the $\tilde{y}_i$ are a subset of the codewords y_i of C_1 This can be checked by direct inspection or simply by noting that the columns of $G(C_2)$ are linear superpositions of the columns of $G(C_1)$.

Table 7.3   Codewords of the C_2 code.

$\tilde{y}_0$	$\tilde{y}_1$	$\tilde{y}_2$	$\tilde{y}_3$	$\tilde{y}_4$	$\tilde{y}_5$	$\tilde{y}_6$	$\tilde{y}_7$
0	1	0	1	0	1	0	1
0	0	1	1	0	0	1	1
0	1	1	0	0	1	1	0
0	0	0	0	1	1	1	1
0	1	0	1	1	0	1	0
0	0	1	1	1	1	0	0
0	1	1	0	1	0	0	1

Exercise 7.7   Construct the generator matrix and a parity-check matrix for the majority-voting code described at the beginning of this chapter. Compute the error syndromes for correctable errors.

Exercise 7.8   Check with examples that the $[7, 4, 3]$ Hamming code cannot correct errors affecting more than one bit.

7.7   * CSS codes

The Calderbank–Shor–Steane (CSS) quantum codes are defined as follows. Let C_1 and C_2 be classical error-correcting codes. We assume that C_1 is a $[n, k_1]$ code (k_1 logical bits encoded into n physical bits) and C_2 a $[n, k_2]$ code, with $k_2 < k_1$. Moreover, we assume that the codewords of C_2 are a subset of the codewords of C_1. Therefore, C_2 induces an equivalence relation in C_1: by definition, two codewords $v_1, v_2 \in C_1$ are equivalent if there is a $w \in C_2$ such that $v_1 = v_2 + w$. The equivalence classes are known as the *cosets* of C_2 in C_1.

The CSS quantum error-correcting code associates a codeword $|\tilde{v}\rangle$ with each equivalence class. We define

$$|\tilde{v}\rangle = \frac{1}{\sqrt{2^{k_2}}} \sum_{w \in C_2} |v + w\rangle. \tag{7.52}$$

Note that here and in the rest of this section additions are bitwise modulo 2. It is easy to check that, if v_1 and v_2 belong to the same coset, then $|\tilde{v}_1\rangle = |\tilde{v}_2\rangle$. On the other hand, if v_1 and v_2 belong to different cosets, then $\langle \tilde{v}_1 | \tilde{v}_2 \rangle = 0$. Therefore, there are $2^{k_1 - k_2}$ linearly independent codewords, corresponding to the $2^{k_1 - k_2}$ cosets. Hence, the CSS code encodes $k_1 - k_2$ logical qubits into n physical qubits.

As an example, we consider the $[n = 7, k_1 = 4]$ Hamming code C_1 and the $[n = 7, k_2 = 3]$ code $C_2 = C_1^\perp$, introduced in the previous section. In this case, we have a CSS code encoding $k_2 - k_1 = 1$ logical qubit into $n = 7$ physical qubits. Using Eq. (7.52), we construct $|0_L\rangle$ as follows:

$$
\begin{aligned}
|0_L\rangle &= \tfrac{1}{\sqrt{8}} \sum_{i=0}^{7} |y_0 + \tilde{y}_i\rangle \\
&= \tfrac{1}{\sqrt{8}} (|0000000\rangle + |1010101\rangle + |0110011\rangle + |1100110\rangle \\
&\quad + |0001111\rangle + |1011010\rangle + |0111100\rangle + |1101001\rangle),
\end{aligned}
\tag{7.53}
$$

where $y_0 = 0000000$ is a codeword of C_1 (see Table 7.2) and the $\tilde{y}_i$ are the codewords of C_2. The other codeword $|1_L\rangle$ is associated with the other coset of C_2 in C_1. That is, we need to find a codeword of C_1 which is not in the coset determining $|0_L\rangle$. For instance, we can consider $y_{15} = 1111111$, so that

$$
\begin{aligned}
|1_L\rangle &= \tfrac{1}{\sqrt{8}} \sum_{i=0}^{7} |y_{15} + \tilde{y}_i\rangle \\
&= \tfrac{1}{\sqrt{8}} (|1111111\rangle + |0101010\rangle + |1001100\rangle + |0011001\rangle \\
&\quad + |1110000\rangle + |0100101\rangle + |1000011\rangle + |0010110\rangle).
\end{aligned}
\tag{7.54}
$$

Let us show that, if the classical codes C_1 and $C_2^\perp$ can correct errors affecting up to t bits, then the quantum CSS code can also correct up to t-qubits. As we have seen in Sec. 7.3, it is sufficient to correct bit and phase-flip errors to correct arbitrary errors. Owing to these two type of errors, the state (7.52) becomes

$$|\tilde{v}\rangle_{ap} = \frac{1}{\sqrt{2^{k_2}}} \sum_{w \in C_2} (-1)^{(v+w) \cdot e_p} |v + w + e_a\rangle, \qquad (7.55)$$

where the dot denotes the bitwise scalar product and the n-bit vector e_a (e_p) describes amplitude (phase) errors. The j-th bit of the vector e_a (e_p) is equal to 1 if a bit flip (phase flip) corrupted the j-th qubit. The error e_a (e_p) is correctable by the CSS code if the number of 1's is not greater than t (we remind the reader that both C_1 and $C_2^{\perp}$ can correct t-bit errors).

In order to detect the amplitude error e_a we introduce a number of ancillary qubits sufficient to store the error syndrome $|H_1 e_a\rangle$ (H_1 is the $(n - k_1) \times n$ parity-check matrix associated with the classical code C_1). We map the state

$$|\tilde{v}\rangle_{ap} |0\rangle \qquad (7.56)$$

(the ancillary qubits are prepared in the $|0\rangle$ state) into

$$\frac{1}{\sqrt{2^{k_2}}} \sum_{w \in C_2} (-1)^{(v+w) \cdot e_p} |v + w + e_a\rangle |H_1 e_a\rangle. \qquad (7.57)$$

We then measure the ancillary qubits to obtain the error syndrome $H_1 e_a$. This tells us the qubits for which a bit flip occurred. These errors can be corrected by applying a NOT gate to each of these qubits. The resulting state is

$$|\tilde{v}\rangle_p = \frac{1}{\sqrt{2^{k_2}}} \sum_{w \in C_2} (-1)^{(v+w) \cdot e_p} |v + w\rangle, \qquad (7.58)$$

where we have neglected the ancillary qubits, which are factorized and do not concern us any more.

Exercise 7.9 Construct the quantum circuit that maps the state (7.56) into (7.57), for the case in which C_1 is the $[7, 4, 3]$ Hamming code and $C_2 = C_1^{\perp}$.

To correct phase errors, we first apply the Hadamard gate to each qubit, obtaining the state

$$\frac{1}{\sqrt{2^{k_2}}} \frac{1}{\sqrt{2^n}} \sum_{w \in C_2} \sum_{z=0}^{2^n - 1} (-1)^{(v+w) \cdot (e_p + z)} |z\rangle. \qquad (7.59)$$

If we define $z' = z + e_p$, this state may be rewritten as

$$\frac{1}{\sqrt{2^{k_2}}} \frac{1}{\sqrt{2^n}} \sum_{w \in C_2} \sum_{z'=0}^{2^n - 1} (-1)^{(v+w) \cdot z'} |z' + e_p\rangle, \qquad (7.60)$$

and, using the result of exercise 7.10, as

$$\frac{1}{\sqrt{2^{n-k_2}}} \sum_{z' \in C_2^\perp} (-1)^{v \cdot z'} |z' + e_p\rangle. \qquad (7.61)$$

The error e_p now behaves as an amplitude error and can therefore be corrected as described above: we introduce ancillary qubits in the state $|0\rangle$ and apply the parity-check matrix $H_2^\perp$ for $C_2^\perp$ (note that, in the special case in which $C_2 = C_1^\perp$, then $H_2^\perp = H_1$). Thus, we map $|z' + e_p\rangle|0\rangle$ into $|z' + e_p\rangle|H_2^\perp e_p\rangle$. We then correct the error e_p to obtain the state

$$\frac{1}{\sqrt{2^{n-k_2}}} \sum_{z' \in C_2^\perp} (-1)^{v \cdot z'} |z'\rangle. \qquad (7.62)$$

Finally, we apply the Hadamard gate to each qubit to recover the original uncorrupted state $|\tilde{v}\rangle$.

Exercise 7.10   Prove that, for a linear code $C[n, k]$, $\sum_{w \in C} (-1)^{w \cdot z} = 2^k$ if $z \in C^\perp$ and 0 otherwise.

7.8   Decoherence-free subspaces

In this section, we shall discuss *passive* quantum error-avoiding codes, in which no measurements or recovery operations are performed to detect and correct errors. The basic idea of passive codes is to encode the information in decoherence-free subspaces. This is possible if the system–environment interaction has certain *symmetries*.

An example will help us clarify this concept. Let us assume that a system of n qubits is coupled to the environment in a symmetric manner and undergoes a dephasing process, defined as

$$|0\rangle_j \rightarrow |0\rangle_j, \qquad |1\rangle_j \rightarrow e^{i\phi} |1\rangle_j, \qquad (j = 1, \ldots, n). \qquad (7.63)$$

In this model, we suppose that the phase ϕ has no dependence on the qubit j; that is, the dephasing process in invariant under qubit permutations. This is an example of *collective decoherence*: several qubits couple identically to the environment. A concrete example of collective decoherence is obtained when an n-qubit register is implemented by a solid state system and the main source of errors is the coupling of each qubit with phonons whose wavelength is much larger than the distance between the qubits.

It is instructive to consider what happens for $n = 2$ qubits. We have

$$\begin{aligned}
|00\rangle &\rightarrow |00\rangle, \\
|01\rangle &\rightarrow e^{i\phi}|01\rangle, \\
|10\rangle &\rightarrow e^{i\phi}|10\rangle, \\
|11\rangle &\rightarrow e^{2i\phi}|11\rangle.
\end{aligned} \tag{7.64}$$

Since the states $|01\rangle$ and $|10\rangle$ acquire the same phase, a simple encoding allows us to avoid phase errors:

$$|0_L\rangle \equiv |01\rangle, \qquad |1_L\rangle \equiv |10\rangle. \tag{7.65}$$

Then the state $|\psi_L\rangle = \alpha|0_L\rangle + \beta|1_L\rangle$ evolves under the dephasing process as follows:

$$|\psi_L\rangle \rightarrow \alpha e^{i\phi}|01\rangle + \beta e^{i\phi}|10\rangle = e^{i\phi}|\psi_L\rangle. \tag{7.66}$$

The overall phase factor $e^{i\phi}$ acquired due to the dephasing process has no physical significance. Therefore, the two-dimensional subspace spanned by the states $|01\rangle$ and $|10\rangle$ is a decoherence-free subspace, which can be used to encode a single qubit. It is interesting that, if the dephasing angles are different for the two qubits ($|1\rangle_1 \rightarrow e^{i\phi_1}|1\rangle_1$ and $|1\rangle_2 \rightarrow e^{i\phi_2}|1\rangle_2$), then the dephasing angle affecting the state $|\psi_L\rangle$ is $\phi_1 - \phi_2$. Indeed, we have

$$|\psi_L\rangle \rightarrow e^{i\phi_2}\left(\alpha|01\rangle + \beta e^{i(\phi_1 - \phi_2)}|10\rangle\right). \tag{7.67}$$

For $n = 3$, it is easy to check that the subspaces spanned by $\{|000\rangle\}$, $\{|100\rangle, |010\rangle, |001\rangle\}$, $\{|011\rangle, |101\rangle, |110\rangle\}$ and $\{|111\rangle\}$ are decoherence-free. Indeed, the states residing in these subspaces acquire global phases 1, $e^{i\phi}$, $e^{2i\phi}$ and $e^{3i\phi}$, respectively.

More generally, in the n-qubit case any subspace spanned by the states of the computational basis with an equal number of 1's and 0's (say, k 1's and $n - k$ 0's) is decoherence-free. These subspaces have dimension $d_k = \binom{n}{k}$ and may be used to encode $\log_2 d_k$ qubits.

Exercise 7.11 Is it possible to find decoherence-free subspaces in the case in which amplitude (bit-flip) errors act identically on every qubit of a quantum register?

7.8.1   * *Conditions for decoherence-free dynamics*

Let us establish the conditions for decoherence-free dynamical evolution. By definition, a subspace $\tilde{\mathcal{H}}$ of a Hilbert space $\mathcal{H}$ is decoherent-free if the evolution inside $\tilde{\mathcal{H}}$ is unitary. We point out that this definition of decoherence-free subspace does not rule out the possible presence of unitary errors in quantum computation. Such errors may result from inaccurate implementations of quantum logic gates. For instance, in ion-trap quantum processors laser pulses are used to implement sequences of quantum gates and fluctuations in the duration of each pulse induce unitary errors, which accumulate during a quantum computation.

We first formulate the conditions for decoherence-free dynamics in terms of the Hamiltonian description for a system in interaction with a reservoir. Following Sec. 6.2.1, we can write the most general Hamiltonian describing such a situation as

$$
\begin{aligned}
H &= H_S \otimes I_R + I_S \otimes H_R + H_{SR} \\
&= H_S \otimes I_R + I_S \otimes H_R + \sum_i \sigma_i \otimes B_i,
\end{aligned}
\tag{7.68}
$$

where H_S and H_R describe the system and the reservoir and the operators σ_i and B_i act on the system and on the reservoir, respectively.

A decoherence-free subspace is found by assuming that there exists a set of eigenvectors $\{|\tilde{k}\rangle\}$ of the operators σ_i such that

$$
\sigma_i|\tilde{k}\rangle = c_i|\tilde{k}\rangle,
\tag{7.69}
$$

for any $i, |\tilde{k}\rangle$. Note that the eigenvalues c_i are *degenerate* since they depend only on the index i of the operator σ_i and not on $\tilde{k}$. If we limit ourselves to considering the subspace $\tilde{\mathcal{H}}$ spanned by the states $\{|\tilde{k}\rangle\}$, we can write the Hamiltonian as

$$
\tilde{H} = H_S \otimes I_R + I_S \otimes \left[H_R + \sum_i c_i B_i \right],
\tag{7.70}
$$

where $\tilde{H}$ is the restriction of H to $\tilde{\mathcal{H}}$. If we assume that the system Hamiltonian H_S leaves the Hilbert subspace $\tilde{\mathcal{H}}$ invariant and if the initial state resides in $\tilde{\mathcal{H}}$, then the evolution of the system is decoherence-free.

To show this, we assume that at time $t = 0$ the system and the environment are not entangled. Then, the initial system plus environment state may be written as

$$
\rho_{SR}(0) = \rho_S(0) \otimes \rho_R(0),
\tag{7.71}
$$

where $\rho_S(0)$ and $\rho_R(0)$ are the system and environment density matrices at time $t = 0$. Moreover, we assume that the initial density matrix describes a state residing in $\tilde{\mathcal{H}}$; that is,

$$\rho_S(0) = \sum_{\tilde{i},\tilde{j}} s_{\tilde{i}\tilde{j}} |\tilde{i}\rangle\langle\tilde{j}|, \tag{7.72}$$

with $|\tilde{i}\rangle, |\tilde{j}\rangle \in \tilde{\mathcal{H}}$. We can also write

$$\rho_R(0) = \sum_{\mu,\nu} r_{\mu\nu} |\mu\rangle\langle\nu|, \tag{7.73}$$

where $\{|\mu\rangle\}$ is a basis for the Hilbert space of the environment. It is easy to see that temporal evolution does not take the system out of the subspace $\tilde{\mathcal{H}}$. Indeed, we have

$$U_{SR}(t)(|\tilde{i}\rangle \otimes |\mu\rangle) = U_S(t)|\tilde{i}\rangle \otimes U_R(t)|\mu\rangle, \tag{7.74}$$

where $U_{SR}(t) = \exp(-i\tilde{H}t/\hbar)$, $U_S(t) = \exp(-iH_S t/\hbar)$ and $U_R(t) = \exp[-i(H_R + \sum_j c_j B_j)t/\hbar]$. Hence, the evolution of the state (7.71) is given by

$$\rho_{SR}(t) = \sum_{\tilde{i},\tilde{j}} s_{\tilde{i}\tilde{j}} U_S(t)|\tilde{i}\rangle\langle\tilde{j}|U_S^\dagger(t) \otimes \sum_{\mu,\nu} r_{\mu\nu} U_R(t)|\mu\rangle\langle\nu|U_R^\dagger(t). \tag{7.75}$$

It follows that

$$\rho_S(t) = \text{Tr}_R[\rho_{SR}(t)] = U_S(t)\rho_S(0)U_S^\dagger(t), \tag{7.76}$$

and therefore the evolution of the system is unitary.

The conditions for decoherence-free dynamics can also be expressed in the framework of the Kraus representation. As we saw in Sec. 5.4, the Kraus operator E_μ is defined as $E_\mu = {}_R\langle\mu|U_{SR}|0\rangle_R$. The matrix representation of E_μ in the basis in which the first states span $\tilde{\mathcal{H}}$ is given by

$$E_\mu = \begin{bmatrix} g_\mu \tilde{U}_S & 0 \\ 0 & C_\mu \end{bmatrix}, \tag{7.77}$$

where $g_\mu = {}_R\langle\mu|U_R|0\rangle_R$, $\tilde{U}_s$ is the restriction of U_S to $\tilde{\mathcal{H}}$ and C_μ is a block matrix acting on the subspace $\tilde{\mathcal{H}}^\perp$ orthogonal to $\tilde{\mathcal{H}}$. Therefore, all Kraus operators E_μ, when restricted to a decoherence-free subspace $\tilde{\mathcal{H}}$, have an identical unitary representation $\propto \tilde{U}_S$, up to a multiplicative constant g_μ. The normalization constraint $\sum_\mu E_\mu^\dagger E_\mu = I_S$ implies $\sum_\mu |g_\mu|^2 = 1$. If the initial state ρ_S resides in the subspace $\tilde{\mathcal{H}}$; that is,

$$\rho_S = \begin{bmatrix} \tilde{\rho}_S & 0 \\ 0 & 0 \end{bmatrix}, \tag{7.78}$$

then the final state ρ_S' also resides in $\tilde{\mathcal{H}}$, and the system's evolution is unitary since

$$\rho_S' = \begin{bmatrix} \sum_\mu |g_\mu|^2 \tilde{U}_S \tilde{\rho}_S \tilde{U}_S^\dagger & 0 \\ 0 & 0 \end{bmatrix} = \begin{bmatrix} \tilde{U}_S \tilde{\rho}_S \tilde{U}_S^\dagger & 0 \\ 0 & 0 \end{bmatrix}. \tag{7.79}$$

7.8.2   * *The spin-boson model*

A nice example of decoherence-free dynamics is the spin-boson model, which describes n spin-$\frac{1}{2}$ particles (the system) interacting with a bosonic field (the reservoir). The interaction Hamiltonian is

$$H_{SR} = \sum_{i=1}^{n} \sum_{k} \left[g_{ik}^+ \sigma_i^+ \otimes b_k + g_{ik}^- \sigma_i^- \otimes b_k^\dagger + g_{ik}^z \sigma_i^z \otimes (b_k + b_k^\dagger) \right], \tag{7.80}$$

where $\sigma_i^\pm = \sigma_x \pm i\sigma_y$ and σ_i^z are Pauli operators acting on the i-th spin, b_k ($b_k^\dagger$) is the annihilation (creation) operator for the k-th mode of the bosonic field and $g_{ik}^\pm$, g_{ik}^z are coupling constants (note that the requirement of Hermitian H_{SR} implies that $(g_{ik}^+)^\star = g_{ik}^-$). This model describes the interaction between a system of qubits (spins) and a bosonic environment, including both dissipative coupling (the terms $\sigma_i^+ \otimes b_k$ and $\sigma_i^- \otimes b_k^\dagger$ describe energy exchanges between the system and the environment) and phase-damping processes (through the $\sigma_i^z \otimes (b_k + b_k^\dagger)$ term).

Let us assume that the coupling constants are independent of the qubit index; that is, $g_{ik}^\pm \equiv g_k^\pm$ and $g_{ik}^z \equiv g_k^z$. This collective decoherence situation is relevant in solid-state systems, provided the coupling to a phononic bath is the dominant source of decoherence and that the wavelength of the relevant phonon modes is much larger than the qubit spacing.[4]

---

[4]Another physical situation in which the spin-boson model is relevant is the coupling of n identical two-level atoms to a single mode of the electromagnetic field (for instance, we can consider n ions in a trap coupled to a laser field). In this case, provided the wavelength of the radiation field is much longer than the distance between the atoms, Hamiltonian (7.80) reduces to

$$H_{SR} = \sum_{i=1}^{n} (g^+ \sigma_i^+ \otimes b + g^- \sigma_i^- \otimes b^\dagger).$$

Given the collective decoherence assumption, a decoherence-free subspace exists. Indeed, we can define the total spin operators

$$S_\alpha \equiv \sum_{i=1}^{n} \sigma_i^\alpha, \tag{7.81}$$

with $\alpha = +, -, z$, so that the coupling Hamiltonian becomes

$$H_{SR} = \sum_{\alpha=+,-,z} S_\alpha \otimes B_\alpha, \tag{7.82}$$

where $B_+ \equiv \sum_k g_k^+ b_k$, $B_- \equiv B_+^\dagger$ and $B_z \equiv \sum_k g_k^z (b_k + b_k^\dagger)$. The condition (7.69) for decoherence-free dynamics is fulfilled if we encode the quantum information in *singlet states* $(S = 0)$; that is, in states $|\tilde{k}\rangle$ satisfying

$$S_\alpha |\tilde{k}\rangle = 0, \tag{7.83}$$

for $\alpha = +, -, z$.

For the case $n = 2$, the only singlet state is

$$\frac{1}{\sqrt{2}} \big(|01\rangle - |10\rangle \big). \tag{7.84}$$

For $n = 4$, the (singlet) decoherence-free subspace has dimension two and is spanned by the states[5]

$$
\begin{aligned}
|0_L\rangle &= \tfrac{1}{2}\big(|0101\rangle + |1010\rangle - |0110\rangle - |1001\rangle\big), \\
|1_L\rangle &= \tfrac{1}{\sqrt{12}}\big(2|0011\rangle + 2|1100\rangle - |0101\rangle - |1010\rangle - |0110\rangle - |1001\rangle\big).
\end{aligned} \tag{7.85}
$$

---

[5] These states can be computed using standard methods for the addition of angular momenta. In general, given two angular momenta $\boldsymbol{j}_1$ and $\boldsymbol{j}_2$ and the total angular momentum $\boldsymbol{J} = \boldsymbol{j}_1 + \boldsymbol{j}_2$, we have

$$|j_1 j_2; JM\rangle = \sum_{m_1, m_2} |j_1 m_1; j_2 m_2\rangle \langle j_1 m_1; j_2 m_2 | j_1 j_2; JM\rangle,$$

where $|j_1 j_2; JM\rangle$ and $|j_1 m_1; j_2 m_2\rangle$ are eigenstates of j_1^2, j_2^2, J^2, J_z and $j_1^2, j_{1z}, j_2^2, j_{2z}$, while the matrix elements $\langle j_1 m_1; j_2 m_2 | j_1 j_2; JM\rangle$, usually denoted as $\langle j_1 j_2 m_1 m_2 | JM\rangle$, are known as the Clebsch–Gordan coefficients. Note that the conditions $M = m_1 + m_2$, $|M| \le J$, $|j_1 - j_2| \le J \le j_1 + j_2$ must be fulfilled. The four-qubit singlet states of (7.85) are obtained as combination of the two-qubit singlet ($|s\rangle \equiv |j = 0, m = 0\rangle = \frac{1}{\sqrt{2}}(|01\rangle - |10\rangle)$) and triplet ($|t_+\rangle \equiv |j = 1, m = 1\rangle = |00\rangle$, $|t_0\rangle \equiv |j = 1, m = 0\rangle = \frac{1}{\sqrt{2}}(|01\rangle + |10\rangle)$, $|t_-\rangle \equiv |j = 1, m = -1\rangle = |11\rangle$) states, with the correct Clebsch–Gordan coefficients: $|0_L\rangle \equiv |j_1 = 0, j_2 = 0; J = 0, M = 0\rangle = |s\rangle_{12} \otimes |s\rangle_{34}$ and $|1_L\rangle \equiv |j_1 = 1, j_2 = 1; J = 0, M = 0\rangle = \frac{1}{\sqrt{3}}(|t_+\rangle_{12} \otimes |t_-\rangle_{34} - |t_0\rangle_{12} \otimes |t_0\rangle_{34} + |t_-\rangle_{12} \otimes |t_+\rangle_{34})$.

Hence, this subspace can be used to encode a single-qubit state. In this manner we can construct the singlet states for progressively higher numbers of qubits. Group theory tells us (see, *e.g.*, Lidar *et al.*, 2000) that the dimension of the (singlet) decoherence-free subspace for n qubits is

$$\dim[\mathrm{DFS}(n)] = \frac{n!}{(n/2 + 1)! \, (n/2)!}. \tag{7.86}$$

This subspace can be used to encode n_L logical qubits, with

$$n_L = \log_2\{\dim[\mathrm{DFS}(n)]\} \approx n - \tfrac{3}{2}\log_2 n, \tag{7.87}$$

where the right-hand expression is obtained after application of Stirling's formula $n! \sim \sqrt{2\pi}\, n^{(n+1/2)} e^{-n}$ for large n. This means that the *encoding efficiency* $\epsilon \equiv n_L/n$, defined as the number of logical qubits n_L per number of physical qubits n, tends to unity as $n \to \infty$.

7.9   * The Zeno effect

The Zeno effect, in its simplest instance, refers to the freezing of the evolution of a quantum state due to frequent measurements. However, as we shall discuss below, the Zeno effect also takes place in systems in which a strong disturbance dominates the temporal evolution of the quantum systems. In general, there is no need to invoke the collapse of the wave function. Even more importantly from the viewpoint of quantum computation, the Zeno effect does not necessarily freeze the dynamics. The system can evolve away from its initial state, although it remains in a "decoherence-free" subspace, which can in principle be appropriately engineered. These issues are discussed in the present section, following the presentation of Facchi and Pascazio (2003).

We first consider a simple example where the Zeno phenomenon is induced by frequent projective measurements. Let H be the total, time-independent Hamiltonian of a quantum system and $|\psi(0)\rangle = |a\rangle$ its initial state at time $t = 0$. The survival probability $p(t)$; that is, the probability to find the system in the same state $|a\rangle$ at time t, is given by

$$p(t) = \left| \langle a | \psi(t) \rangle \right|^2 = \left| \langle a | \exp\left(-\frac{i}{\hbar} H t \right) | a \rangle \right|^2. \tag{7.88}$$

A short-time expansion yields a quadratic behaviour:

$$p(t) \sim \left| \langle a | \left(I - \frac{i}{\hbar} H t - \frac{1}{2\hbar^2} H^2 t^2 \right) | a \rangle \right|^2$$

$$\sim 1 - \frac{1}{\hbar^2} \left(\langle a | H^2 | a \rangle - \langle a | H | a \rangle^2 \right) t^2 = 1 - \frac{t^2}{t_Z^2}, \qquad (7.89)$$

where

$$t_Z = \frac{\hbar}{\sqrt{\langle a | H^2 | a \rangle - \langle a | H | a \rangle^2}} \qquad (7.90)$$

is the so-called Zeno time.[6]

If N projective measurements are performed at time intervals $\tau = \frac{t}{N}$,[7] then the survival probability at time t is

$$p(t) = \left[|\langle a | \psi(\tau) \rangle|^2 \right]^N = \left[p(\tau) \right]^N = \left[p\left(\frac{t}{N} \right) \right]^N$$

$$\sim \left(1 - \frac{t^2}{N^2 t_Z^2} \right)^N \sim \exp\left(-\frac{t^2}{N t_Z^2} \right). \qquad (7.91)$$

If $N \to \infty$, then $p(t) \to 1$; namely, the evolution is completely frozen. Note that the decay in time of the survival probability (7.91) is exponential: for a given τ and $t = N\tau$ (N integer),

$$p(t) \sim \exp\left[-\gamma(\tau) t \right], \qquad (7.92)$$

with the decay rate $\gamma(\tau) \sim \tau / t_Z^2$.

We remark that the quantum Zeno effect is a direct consequence of the following mathematical property of the Schrödinger equation (sketched in Fig. 7.9): in a short time $\delta\tau = t/N = O(1/N)$, the phase of the wave function evolves as $O(\delta\tau)$, while the probability changes by $O((\delta\tau)^2)$, so that $p(t) \sim [1 - O(1/N^2)]^N \to 1$ when $N \to \infty$.

Exercise 7.12 Discuss the Zeno effect for a two-level system driven by the Hamiltonian $H = H_0 + H_{\text{int}}$, with $H_0 = \frac{1}{2}\hbar\omega\sigma_z$ and $H_{\text{int}} = \frac{1}{2}\hbar\Omega\sigma_x$, the initial state of the system being $|0\rangle$.

---

[6]Note that, if the Hamiltonian H is divided into free and interaction parts, $H = H_0 + H_{\text{int}}$, with the initial state $|a\rangle$ eigenstate of the free Hamiltonian ($H_0|a\rangle = \omega_a|a\rangle$) and the interaction part off-diagonal in the basis of the eigenstates of H_0, so that $\langle a|H_{\text{int}}|a\rangle = 0$, then the Zeno time is given by $t_Z = \hbar/\sqrt{\langle a|H_{\text{int}}^2|a\rangle}$ and only depends on the interaction Hamiltonian.

[7]In this example we assume that the measurement is *selective*: we select only the survived component ($|\psi(\tau)\rangle \to |a\rangle\langle a|\psi(\tau)\rangle$) and stop the others. Note that, as we shall discuss below, the Zeno effect also takes place for non-selective measurements.

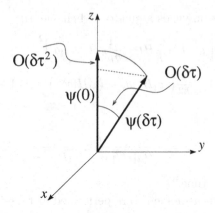

Fig. 7.9   A schematic drawing of the short-time evolution of phase and probability for a wave function whose evolution is governed by the Schrödinger equation.

Note that the collapse of the wave function (an inherently non-unitary and irreversible process) is not necessarily required for the quantum Zeno effect. In order to illustrate this concept we consider a three-level system governed by the Hamiltonian

$$H = \Omega_1\big(|0\rangle\langle 1| + |1\rangle\langle 0|\big) + \Omega_2\big(|1\rangle\langle 2| + |2\rangle\langle 1|\big). \qquad (7.93)$$

In the basis $\{|0\rangle, |1\rangle, |2\rangle\}$ this Hamiltonian reads

$$H = \begin{bmatrix} 0 & \Omega_1 & 0 \\ \Omega_1 & 0 & \Omega_2 \\ 0 & \Omega_2 & 0 \end{bmatrix}. \qquad (7.94)$$

If the system is prepared at time $t = 0$ in the state $|\psi(0)\rangle = |0\rangle$, then the survival probability is given by (see exercise 7.13)

$$p(t) = |\langle 0|\psi(t)\rangle|^2 = \frac{1}{(\Omega_1^2 + \Omega_2^2)^2}\left[\Omega_2^2 + \Omega_1^2 \cos\left(\frac{\sqrt{\Omega_1^2 + \Omega_2^2}\,t}{\hbar}\right)\right]^2. \qquad (7.95)$$

Note that for large values of the ratio Ω_2/Ω_1 the system is in practice frozen in the level $|0\rangle$. In this case, as soon as the system makes a transition from $|0\rangle$ to $|1\rangle$ it undergoes a very fast Rabi oscillation to level $|2\rangle$. Therefore, we can say that level $|2\rangle$ acts as a measuring apparatus: when the ratio Ω_2/Ω_1 is large, then a better observation of the state of the system is performed, thus hindering the transition $|0\rangle \rightarrow |1\rangle$. We stress that the measurement performed by level $|2\rangle$ is continuous (there is no wave-function collapse) and Hermitian (the model is purely Hamiltonian).

Exercise 7.13   Prove Eq. (7.95).

The following theorem provides a very general formulation of the quantum Zeno effect. Let us consider a quantum system whose states reside in the Hilbert space $\mathcal{H}$. The evolution of the system density matrix ρ is described by the superoperator

$$\rho(t) = \mathcal{U}_t\,\rho(0) = U(t)\,\rho(0)\,U^\dagger(t), \qquad U(t) = \exp\left(-\frac{i}{\hbar}Ht\right), \qquad (7.96)$$

where H is a time-independent bounded Hamiltonian. We also introduce a set of projectors P_i such that $P_iP_j = \delta_{ij}P_i$ and $\sum_i P_i = I$. The subspaces relative to the operators P_i are denoted by $\mathcal{H}_i = P_i\mathcal{H}$ ($\mathcal{H} = \oplus_i\mathcal{H}_i$). We consider a non-selective measurement (the measuring apparatus does not select the different outcomes) described by the superoperator

$$\mathcal{P}\rho = \sum_n P_n\rho P_n. \qquad (7.97)$$

The evolution of the system after N such measurements performed in a time t is determined by the superoperator

$$\mathcal{S}_t^{(N)} = \left(\mathcal{P}\mathcal{U}_{t/N}\right)^N\mathcal{P}, \qquad (7.98)$$

where we have also operated a first measurement at time $t = 0$, which prepares the state $\mathcal{P}\rho(0) = \sum_i P_i\rho(0)P_i$. The evolution of the system density matrix reads

$$\rho(t) = \sum_i W_i(t)\rho_0 W_i^\dagger(t), \qquad (7.99)$$

with $W_i^\dagger(t)W_i(t) = P_i$. Moreover, the probability that the system is found in the subspace $\mathcal{H}_i$ is

$$p_i(t) = \text{Tr}[\rho(t)P_i] = \text{Tr}[W_i(t)\rho_0 W_i^\dagger(t)] = \text{Tr}[\rho_0 P_i] = p_i(0). \qquad (7.100)$$

It is clear from Eqs. (7.99) and (7.100) that any interference term between the different subspaces $\mathcal{H}_i$ is destroyed and that the probability is conserved in each subspace. Each operator $W_i(t)$ is unitary within the subspace $\mathcal{H}_i$ and has the form

$$W_i(t) = P_i\exp\left(-\frac{i}{\hbar}P_iHP_it\right). \qquad (7.101)$$

The above theorem is interesting from the viewpoint of quantum computation, as it suggests strategies to contrast decoherence. A simple example will help clarify this concept. We consider again Hamiltonian (7.94) and the projectors

$$
P_1 = \begin{bmatrix} 1 & 0 & 0 \\ 0 & 1 & 0 \\ 0 & 0 & 0 \end{bmatrix}, \quad P_2 = \begin{bmatrix} 0 & 0 & 0 \\ 0 & 0 & 0 \\ 0 & 0 & 1 \end{bmatrix}, \tag{7.102}
$$

satisfying $P_1 + P_2 = I$. The subspace $\mathcal{H}_1 = P_1 \mathcal{H}$ is the two-dimensional subspace (qubit) of interest for quantum computation, while the coupling Ω_2 mimics decoherence. In the limit $N \to \infty$ the operators (7.101) become

$$
\begin{aligned}
W_1(t) &= P_1 \exp\left(-\frac{i}{\hbar} P_1 H P_1 t\right) = P_1 \exp\left\{-\frac{i}{\hbar} \begin{bmatrix} 0 & \Omega_1 t & 0 \\ \Omega_1 t & 0 & 0 \\ 0 & 0 & 0 \end{bmatrix}\right\} \\
&= \begin{bmatrix} \cos\left(\frac{\Omega_1 t}{\hbar}\right) & -i\sin\left(\frac{\Omega_1 t}{\hbar}\right) & 0 \\ -i\sin\left(\frac{\Omega_1 t}{\hbar}\right) & \cos\left(\frac{\Omega_1 t}{\hbar}\right) & 0 \\ 0 & 0 & 0 \end{bmatrix},
\end{aligned} \tag{7.103}
$$

$$
W_2(t) = P_2 \exp\left(-\frac{i}{\hbar} P_2 H P_2 t\right) = P_2 = \begin{bmatrix} 0 & 0 & 0 \\ 0 & 0 & 0 \\ 0 & 0 & 1 \end{bmatrix}
$$

and the qubit evolves according to the Hamiltonian

$$
P_1 H P_1 = \begin{bmatrix} 0 & \Omega_1 & 0 \\ \Omega_1 & 0 & 0 \\ 0 & 0 & 0 \end{bmatrix}. \tag{7.104}
$$

The subspaces $\mathcal{H}_1$ and $\mathcal{H}_2$ decouple; that is, the evolution of the qubit becomes decoherence-free. Finally, we point out that other Zeno strategies based on unitary disturbances (instead of projective measurements) of the system that we wish to protect are also possible (see Facchi and Pascazio, 2003).

7.10   Fault-tolerant quantum computation

So far, our discussion of quantum error correction has assumed that encoding, decoding of quantum information and error recovery operations can be achieved perfectly. However, these are complex quantum computations subject to errors. Moreover, quantum logic gates performed in quantum information processing may propagate errors in the quantum computer. In spite of these difficulties, we shall show that, under certain assumptions, arbitrarily long quantum computation can, in principle, be performed reliably, provided the noise in individual quantum gates is below a critical threshold. A quantum computer that performs reliably even in the presence of imperfections is said to be *fault-tolerant*. Sophisticated techniques have been developed for the construction of fault-tolerant quantum circuits (for a review see, *e.g.*, Preskill, 1998b). In the following, we shall limit ourselves to illustrate the basic principles of fault-tolerant quantum computation.

7.10.1 *Avoidance of error propagation*

If an error affects one qubit and this qubit interacts with another in order to perform a two-qubit gate, then the error is likely to propagate to the second qubit. To grasp this point, it is sufficient to consider the CNOT gate. If a bit-flip error affects the control qubit, then the error also spreads to the target qubit. For instance, we consider $CNOT(|0\rangle|0\rangle) = |0\rangle|0\rangle$. If there is a bit-flip error affecting the control qubit ($|0\rangle \leftrightarrow |1\rangle$), then $CNOT(|1\rangle|0\rangle) = |1\rangle|1\rangle$, so that both the control and the target qubit are flipped. A more subtle, purely quantum effect, is the *backward sign propagation*, discussed in Sec. 3.4 (see exercise 3.11): a phase error affecting the target qubit is also transferred, after application of the CNOT gate, to the control qubit.

The backward sign propagation problem spoils the efficiency of the error-correcting quantum circuits shown earlier in this chapter. If we assume that the probabilities of errors affecting one and two qubits are $O(\epsilon)$ and $O(\epsilon^2)$, respectively[8], then a single-qubit error-correcting code is useful when it lowers the error probability from $O(\epsilon)$ to $O(\epsilon^2)$. This is not the case, for example, for the circuit drawn in Fig. 7.6 once phase errors affecting the ancillary qubits are taken into consideration. The problem is that we use a single ancillary qubit for more than one CNOT gate. This is clear from Fig. 7.10, which contains the basic building block for error extraction. If, with probability $O(\epsilon)$, a phase error effects the bottom qubit in the left-hand circuit of Fig. 7.10 before the application of the two CNOT gates, then this error spreads to two of the qubits used to encode the quantum information. Therefore, a code able to correct a single-qubit error (such as Shor's nine-qubit code) fails with $O(\epsilon)$ probability. This problem is avoided by the right-hand circuit in Fig. 7.10, which employs each ancillary qubit

---

[8]This is the case, for instance, when errors affecting different qubits are completely uncorrelated with one another.

only once. Therefore, a phase error affecting a single ancillary qubit is propagated only to a single qubit. Then, the probability of having two phase errors transferred from the ancillary qubits to the qubits used for encoding is $O(\epsilon^2)$. We say that the right-hand circuit in Fig. 7.10 is fault-tolerant, while the left-hand circuit is not. More generally, a quantum code correcting up to t errors is said to be fault-tolerant if its failure probability is $O(\epsilon^{t+1})$.

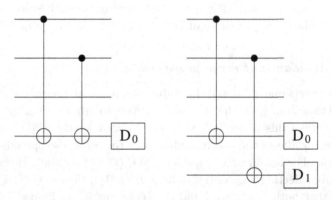

Fig. 7.10   Quantum circuits for extracting the error syndrome. The left-hand circuit employs the same ancillary qubit twice and is therefore not fault-tolerant. In contrast, the right-hand circuit is fault-tolerant.

We point out that the ancillary qubits must be prepared in an appropriate initial state. If we prepare them in the usual state $|00\rangle$, consider a generic encoded initial state $\alpha|000\rangle + \beta|111\rangle$, and assume that a bit-flip error has corrupted the first qubit, then the fault-tolerant circuit in Fig. 7.10 maps the initial state

$$\left(\alpha|100\rangle + \beta|011\rangle\right)|00\rangle \tag{7.105}$$

onto

$$\alpha|10010\rangle + \beta|01101\rangle. \tag{7.106}$$

Therefore, the measurements of the two ancillary qubits projects the five-qubit state onto $|10010\rangle$ (with probability $|\alpha|^2$) or $|01101\rangle$ (with probability $|\beta|^2$). In both case, since one of the two ancillary qubits changed its state from $|0\rangle$ to $|1\rangle$, we may conclude that a bit-flip error affected the first or the second qubit. However, this procedure is not adequate, as we have destroyed the quantum information encoded in the superposition of the states $|000\rangle$ and $|111\rangle$.

To solve this problem, we prepare the ancillary qubits in the equally weighted superposition of the states $|00\rangle$ and $|11\rangle$. Therefore, the initial state

$$\left(\alpha|100\rangle + \beta|011\rangle\right)\tfrac{1}{\sqrt{2}}\left(|00\rangle + |11\rangle\right) \tag{7.107}$$

is mapped by the right-hand circuit in Fig. 7.10 onto

$$\left(\alpha|100\rangle + \beta|011\rangle\right)\frac{1}{\sqrt{2}}\left(|01\rangle + |10\rangle\right). \tag{7.108}$$

The measurement of the ancillary qubits gives us, with equal probabilities, outcome 01 or 10. In both cases, we can conclude that a bit-flip error affected the first or the second qubit, without destroying the quantum superposition $\alpha|100\rangle + \beta|011\rangle$.

Exercise 7.14 Design a fault-tolerant syndrome measurement for the CSS code described in Sec. 7.7.

7.10.2 *Fault-tolerant quantum gates*

In order to implement a reliable quantum computation, we must apply fault-tolerant quantum gates. This is possible if we perform quantum logic operations directly on encoded states.

A fault-tolerant CNOT gate is shown in Fig. 7.11. In this quantum circuit, the first three physical qubits encode the control and the last three physical qubits the target, according to the rule $|0_L\rangle = |000\rangle$ and $|1_L\rangle = |111\rangle$. It is easy to show that, if the CNOT gates are applied *transversally* (that is, bitwise), as shown in Fig. 7.11, then the truth table of the CNOT gate is verified for the logical qubits. Indeed, starting from the six-qubit state $|x_L\rangle|y_L\rangle$, with $x_L, y_L = 0, 1$, we obtain at the end of the circuit $|x_L\rangle|x_L \oplus y_L\rangle$. We point out that the CNOT gate is implemented faulttolerantly, because each qubit in each code block is involved in a single quantum gate. Therefore, errors in one block can propagate at most to one qubit in the other block, not inside the same block and this construction of the CNOT gate is thus fault-tolerant.

We note that it is possible to find a universal set of fault-tolerant quantum gates, in terms of which any quantum computation may be expressed.

7.10.3 *The noise threshold for quantum computation*

The threshold theorem for quantum computation tells us that, given certain assumptions about the noise model (in the simplest case, we consider random and uncorrelated errors) and provided the noise affecting individual quantum gates is below a certain threshold, then it is in principle possible to efficiently implement arbitrarily long quantum computations.

The key ingredient for this result is the use of *concatenated codes*. To

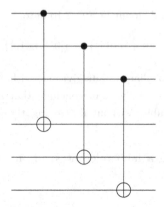

Fig. 7.11   A quantum circuit implementing a transversal CNOT gate between two logical qubits encoded in three-qubit blocks.

understand this concept, let us consider the CSS code described in Sec. 7.7, which encodes a single logical qubit in a block of $n = 7$ qubits. In a concatenated code each qubit of the block is itself a 7-qubit block, and so on (see Fig. 7.12). If there are L levels of concatenation, then a single logical qubit is encoded into $n^L = 7^L$ physical qubits.

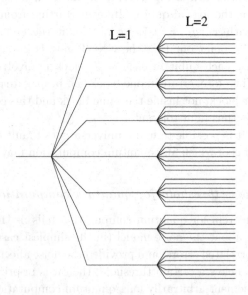

Fig. 7.12   Concatenation of a 7-qubit code up to the $L = 2$ level.

Let us call ϵ the error probability per qubit per appropriate unit of time (for instance, the time required to implement a single elementary quantum gate) and α the number of locations in the quantum circuit where an error can affect a single qubit before that error correction is applied. Typically, for quantum gates such as the fault-tolerant CNOT and for codes correcting a single error, assuming that error correction is applied after each fault-tolerant quantum gate, we obtain $\alpha \sim 10^2$. Error correction at the first level of encoding ($L = 1$) fails if at least two qubits have been corrupted. Therefore, the failure probability is

$$p_1 \approx c\epsilon^2 \approx \alpha^2 \epsilon^2, \qquad (7.109)$$

where $c \approx \alpha^2$ is the number of ways in which a fault-tolerant circuit can introduce at least two errors. At the second level of encoding ($L = 2$), we employ n^2 qubits and error correction fails if at least two of the subblocks of size n fail. Thus, the failure probability is

$$p_2 \approx cp_1^2 \approx \alpha^2(\alpha^2\epsilon^2)^2. \qquad (7.110)$$

We can iterate this procedure. The failure probability at level-L concatenation is

$$p_L \approx cp_{L-1}^2 \approx \frac{(\alpha^2\epsilon)^{2^L}}{\alpha^2}. \qquad (7.111)$$

If we wish to implement a computation of length T (T denotes the number of logic quantum gates) with accuracy ϵ_0, then the error probability per logic gate must be $\leq \epsilon_0/T$. Thus, we must concatenate our code a number of times L such that

$$p_L \approx \frac{(\alpha^2\epsilon)^{2^L}}{\alpha^2} \leq \frac{\epsilon_0}{T}. \qquad (7.112)$$

Provided $\epsilon < \epsilon_{\text{th}} \equiv 1/\alpha^2$, this inequality is fulfilled for

$$L > \bar{L} \approx \log\left[\frac{\log(T/\alpha^2\epsilon_0)}{\log(1/\alpha^2\epsilon)}\right]. \qquad (7.113)$$

The number of physical qubits $\bar{n}_{\text{tot}} = n^{\bar{L}}$ required to achieve this level of accuracy is

$$\bar{n}_{\text{tot}} \approx \left[\frac{\log(T/\alpha^2\epsilon_0)}{\log(1/\alpha^2\epsilon)}\right]^{\log n}. \qquad (7.114)$$

Note that $\bar{n}_{\text{tot}}$ grows only polylogarithmically with T and $1/\epsilon$.

We stress that the above results assume that the quantum computer hardware is such that many quantum gates can be executed in parallel in a single time step. Otherwise, errors in concatenated codes would accumulate too quickly to allow successful error correction.

Finally, we note that, for $\alpha \sim 10^2$, the noise threshold is $\epsilon_{th} \sim 10^{-4}$. Various sophisticated calculations found in the literature give different results $\epsilon_{th} \sim 10^{-6} - 10^{-4}$. The numerical value of the noise threshold depends on the assumed characteristics of the quantum computer hardware.

7.11 * Quantum cryptography over noisy channels

A central problem of quantum communication is how to reliably transmit information through a noisy quantum channel. The carriers of information (the qubits) unavoidably interact with the external world, leading to the phenomenon of decoherence. The problem of noise in communication channels plays a crucial role in quantum cryptography: In this case, noise can be attributed, in the worst case, to the measurements performed on the qubits by an eavesdropper. Since quantum communication protocols can be seen as special instances of quantum computation, the quantum error-correction codes discussed in this chapter could be used to deal with the problem of noisy channels. One should, however, take into account the fact that the qubits belonging to the two communicating parties can be very far away from each other. Therefore, any error correction procedure must be based only on the so-called LOCC, that is on local quantum operations (performed by Alice and Bob on their own qubits), possibly supplemented by classical communication. We shall show below that techniques for the purification of mixed entangled states may be crucial for quantum communication protocols.

Let us consider the E91 cryptographic protocol (see Sec. 4.3.2). We assume that Alice has a source of EPR pairs at her disposal and sends a member of each pair to Bob. The eavesdropper Eve attacks the qubits sent by Alice by means of the quantum copying machine of Bužek and Hillery (see Sec. 5.1.3). As a result, Alice and Bob share partially entangled pairs. Each pair is now entangled with the environment (Eve's qubits) and described by a density operator. In this section, we shall describe an iterative procedure (known as *quantum privacy amplification*) to purify entanglement and, as a consequence, reduce the entanglement with any outside system to arbitrarily low values (note that a maximally entangled

EPR pair is a pure state automatically disentangled from the outside world). We shall also discuss the effect of noisy apparatus. This is important since, under realistic conditions, also the local operations (quantum gates and measurements) that constitute a purification protocol are never perfect and thus introduce a certain amount of noise.

Eve's attack is represented in Fig. 7.13. Note that the unitary transformation W stands for the copying part of the Bužek–Hillery machine (drawn in Fig. 5.1). The two bottom qubits in Fig. 7.13 are prepared in the state

$$|\Phi\rangle = \alpha|00\rangle + \beta|01\rangle + \gamma|10\rangle + \delta|11\rangle \qquad (7.115)$$

and we assume that α, β, γ, δ are real parameters. Let us call ρ_B and ρ_E the density matrices describing the final states of Bob and Eve's qubits. We assume isotropy; that is, if we call (x, y, z) the coordinates of the qubit sent by Alice to Bob before eavesdropping, then the Bloch coordinates (x_B, y_B, z_B) and (x_E, y_E, z_E) associated with ρ_B and ρ_E are such that $x_B/x = y_B/y = z_B/z \equiv R_B$ and $x_E/x = y_E/y = z_E/z \equiv R_E$. These conditions are fulfilled for

$$\beta = \tfrac{1}{2}\alpha - \sqrt{\tfrac{1}{2} - \tfrac{3}{4}\alpha^2}, \qquad \gamma = 0, \qquad \delta = \tfrac{1}{2}\alpha + \sqrt{\tfrac{1}{2} - \tfrac{3}{4}\alpha^2}. \qquad (7.116)$$

It can be checked by direct computation (see exercise 7.15) that in this case $(x_B, y_B, z_B) = 2\alpha\delta(x, y, z)$ and $(x_E, y_E, z_E) = 2\alpha\beta(x, y, z)$. Since the Bloch-sphere coordinates must be real and non-negative, we obtain $\frac{1}{\sqrt{2}} \leq \alpha \leq \frac{2}{\sqrt{6}}$. The ratios R_B and R_E are shown in Fig. 7.14. It can be seen that the two limiting cases $\alpha = \frac{1}{\sqrt{2}}$ and $\alpha = \frac{2}{\sqrt{6}}$ correspond to no intrusion ($x_B = x$, $y_B = y$, $z_B = z$) and maximum intrusion ($x_E = x_B$, $y_E = y_B$, $z_E = z_B$), respectively. Note that in the latter case we recover the symmetric Bužek–Hillery machine ($\rho_E = \rho_B$) described in Sec. 5.1.3 (in this case the state vector (7.115) reduces to the state (5.64), obtained at the end of the preparation stage in the quantum copying network of Fig. 5.1). The degree of Eve's intrusion is conveniently measured by the parameter

$$f_\alpha = \frac{\alpha - \frac{1}{\sqrt{2}}}{\frac{2}{\sqrt{6}} - \frac{1}{\sqrt{2}}}, \qquad (7.117)$$

with $0 \leq f_\alpha \leq 1$.

Exercise 7.15 Show that isotropic cloning is obtained by means of the Bužek–Hillery copying machine if the parameters α, β, γ, δ in the initial state (7.115) are chosen as in Eq. (7.116).

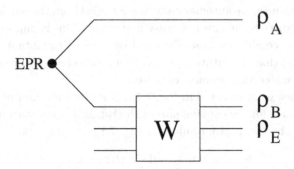

Fig. 7.13　A quantum circuit representing the intrusion of the eavesdropper Eve in the E91 protocol. The unitary transformation W is the copying part of the circuit drawn in Fig. 5.1. The density matrices ρ_A, ρ_B and ρ_E represent the states of Alice's, Bob's and Eve's qubits after tracing over all other qubits.

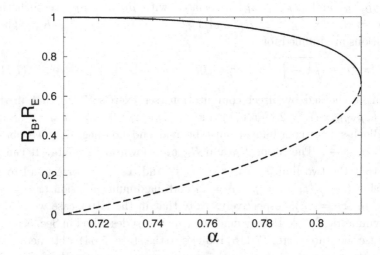

Fig. 7.14　Ratios R_B (solid line) and R_E (dashed line) for the isotropic Bužek–Hillery copying machine as a function of the parameter α.

Alice and Bob purify entanglement by means of the DEJMPS (see Deutsch, Ekert, Jozsa, Macchiavello, Popescu and Sanpera, 1996) protocol. At each step, the EPR pairs are combined in groups of two. The following steps are then taken for each group (see Fig. 7.15):

- To her qubits Alice applies a $\frac{\pi}{2}$ rotation about the x-axis, described by the unitary matrix

$$U = R_x\left(\frac{\pi}{2}\right) = \frac{1}{\sqrt{2}}\begin{bmatrix} 1 & -i \\ -i & 1 \end{bmatrix}. \qquad (7.118)$$

- To his qubits Bob applies the inverse operation

$$V = R_x\left(-\frac{\pi}{2}\right) = U^{-1} = \frac{1}{\sqrt{2}}\begin{bmatrix} 1 & i \\ i & 1 \end{bmatrix}. \qquad (7.119)$$

- Both Alice and Bob perform a CNOT gate using their members of the two EPR pairs.
- They measure the z-components of the two target qubits.
- Finally, Alice and Bob compare the measurement outcomes by means of a public classical communication channel. If the outcomes coincide, the control pair is kept for the next iteration and the target pair discarded. Otherwise, both pairs are discarded.

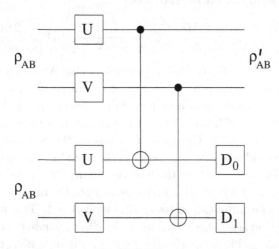

Fig. 7.15   A schematic drawing of the DEJMPS entanglement purification scheme. Note that the density matrix ρ'_{AB} describes the two top qubits only when the detectors D_0 and D_1 give the same outcome.

To see the effect of the DEJMPS procedure, let us consider the special case in which the initial mixed pairs are described by the density matrix ρ_{AB} obtained from the ideal EPR state $|\phi^+\rangle = \frac{1}{\sqrt{2}}\left(|00\rangle + |11\rangle\right)$ after application of the Bužek–Hillery copying machine with intrusion parameter f_α. We obtain

$$\rho_{AB} = \frac{1}{2} \begin{bmatrix} \alpha^2 + \delta^2 & 0 & 0 & 2\alpha\delta \\ 0 & \beta^2 + \gamma^2 & 2\beta\gamma & 0 \\ 0 & 2\beta\gamma & \beta^2 + \gamma^2 & 0 \\ 2\alpha\delta & 0 & 0 & \alpha^2 + \delta^2 \end{bmatrix}. \tag{7.120}$$

We note that this state is diagonal in the Bell basis $\{|\phi^{\pm}\rangle = \frac{1}{\sqrt{2}}(|00\rangle \pm |11\rangle),$ $|\psi^{\pm}\rangle = \frac{1}{\sqrt{2}}(|01\rangle \pm |10\rangle)\}$. Indeed, we have

$$\rho_{AB} = A|\phi^+\rangle\langle\phi^+| + B|\phi^-\rangle\langle\phi^-| + C|\psi^+\rangle\langle\psi^+| + D|\psi^-\rangle\langle\psi^-|, \tag{7.121}$$

where $A = \frac{1}{2}(\alpha + \delta)^2$, $B = \frac{1}{2}(\alpha - \delta)^2$, $C = \frac{1}{2}(\beta + \gamma)^2$ and $D = \frac{1}{2}(\beta - \gamma)^2$. The quantum circuit in Fig. 7.15 maps the state ρ_{AB} of the control pair, in the case in which it is not discarded, onto another state ρ'_{AB} diagonal in the Bell basis. Namely, ρ'_{AB} can be expressed in the form (7.121), provided new coefficients (A', B', C', D') are used instead of (A, B, C, D). A lengthy but straightforward calculation shows that

$$A' = \frac{A^2 + D^2}{N}, \quad B' = \frac{2AD}{N}, \quad C' = \frac{B^2 + C^2}{N}, \quad D' = \frac{2BC}{N}, \tag{7.122}$$

where $N = (A+D)^2 + (B+C)^2$ is the probability that Alice and Bob obtain coinciding outcomes in the measurement of the target qubits. Note that map (7.122) is non-linear as a consequence of the strong nonlinearity of the measurement process. The fidelity after the purification procedure is given by $f = A' = \langle\phi^+|\rho'_{AB}|\phi^+\rangle$. This quantity measures the probability that the control qubits would pass a test for being in the state $|\phi^+\rangle$. Map (7.122) can be iterated and we wish to drive the fidelity to unity. It is possible to prove (see Macchiavello, 1998) that this map converges to the target point $A = 1$, $B = C = D = 0$ for all initial states (7.121) with $A > \frac{1}{2}$. This means that, when this condition is satisfied and a sufficiently large number of initial pairs is available, Alice and Bob can distill asymptotically pure EPR pairs. Note that the quantum privacy amplification procedure is rather wasteful since at least half of the pairs (the target pairs) are lost at every iteration. This means that to extract one pair close to the ideal EPR state after n steps, we need at least 2^n mixed pairs at the beginning. However, this number can be significantly larger since pairs must be discarded when Alice and Bob obtain different measurement outcomes. We therefore compute the survival probability $P(n)$, measuring the probability that an n-step QPA protocol is successful. More precisely, if p_i is the probability that Alice and Bob obtain coinciding outcomes at step i, we have

$$P(n) = \prod_{i=1}^{n} p_i. \tag{7.123}$$

The efficiency $\xi(n)$ of the algorithm is given by the number of pure EPR pairs obtained divided by the number of initial impure EPR pairs. We have

$$\xi(n) = \frac{P(n)}{2^n}. \tag{7.124}$$

Both the fidelity and the survival probability are shown in Fig. 7.16. The different curves of this figure correspond to values of the intrusion parameter from $f_\alpha = 0.05$ (weak intrusion) to $f_\alpha = 0.95$ (strong intrusion). It can be seen that the convergence of the QPA protocol is fast: the fidelity deviates from the ideal case $f = 1$ by less than 10^{-7} in no more than $n = 6$ map iterations. Moreover, the survival probability is quite high: it saturates to $P_\infty \equiv \lim_{n\to\infty} P(n) = 0.60$ for $f_\alpha = 0.95$, $P_\infty = 0.94$ for $f_\alpha = 0.5$ and $P_\infty = 0.9995$ for $f_\alpha = 0.05$.

It is interesting to consider the effects of noisy apparatus on the efficiency of the quantum privacy amplification protocol. For the sake of simplicity we limit ourselves to consider errors affecting only a single qubit. As we have seen in Sec. 6.1.1, we need 12 parameters to characterize a generic quantum operation acting on a two-level system. Each parameter describes a particular noise channel (such as bit flip, phase flip, amplitude damping, ...). It is interesting to point out that the sensitivity of the quantum privacy protocol strongly depends on the kind of noise. To give a concrete example, we show in Fig. 7.17 the deviation $1 - f$ of the fidelity from the ideal value $f = 1$ as a function of the noise strength ϵ. Data are obtained after $n = 5$ iterations of the quantum privacy amplification protocol for the case of strong intrusion by Eve ($f_\alpha = 0.95$). We consider the bit-flip channel (with $\sin \epsilon = |\alpha|$ in Eq. (6.28)), the phase-flip channel ($\sin \epsilon = |\gamma|$ in Eq. (6.31)) and the amplitude-damping channels ($\sin \epsilon = \sqrt{p}$ in Eq. (6.44)). Figure 7.17 is obtained assuming that noise acts on the top qubit of Fig. 7.15 after the U-gate (note, however, that similar curves are obtained when noise acts instead on one of the three remaining qubits). In the noiseless case we start from $1 - f = 1.57 \times 10^{-1}$ and improve the fidelity to $1 - f = 8.20 \times 10^{-6}$ after $n = 5$ iterations of the quantum privacy amplification protocol. Even though all noise channels degrade the performance of the protocol, the level of noise that can be safely tolerated strongly de-

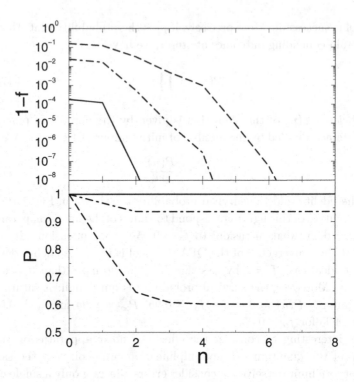

Fig. 7.16   The deviation $1 - f$ of the fidelity f from the ideal case $f = 1$ (top) and survival probability P (bottom) as a function of the number of iterations n of map (7.122). The different curves correspond to the intrusion parameter $f_\alpha = 0.95$ (dashed line), 0.5 (dot-dashed line) and 0.05 (solid line).

pends on the specific channel. For instance, it is clear from Fig. 7.17 that the protocol is much more resilient to bit-flip and amplitude-damping errors than to phase-flip errors.

7.12   * Quantum channels with memory

It is interesting to consider the transmission of information through quantum channels with memory; that is, channels in which correlated noise acts on consecutive uses. This situation occurs in real physical quantum channels, provided the noise is correlated on a time scale larger than the time separation between consecutive uses of the channel. In quantum computation, time correlated noise is important in situations, such as solid state

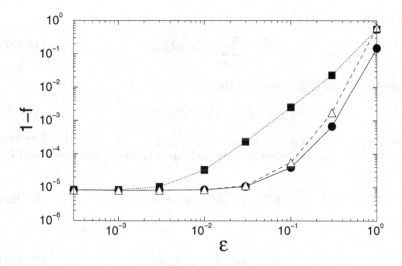

Fig. 7.17 The deviation $1 - f$ of the fidelity f from the ideal case $f = 1$ as a function of the noise strength ϵ, after $n = 5$ steps of the quantum privacy amplification protocol, for $f_\alpha = 0.95$, bit-flip (circles), phase-flip (squares) and amplitude-damping (triangles) channels. The figure is taken from Benenti *et al.* (2006).

qubits, in which noise has components at frequencies much smaller than the time scales of interest for the system dynamics (we say that in this case the environment is non-Markovian).

In this section we consider the case of two consecutive uses of a channel with partial memory, following a model introduced by Macchiavello and Palma (2002). Each use of the channel corresponds to the transmission of a qubit and the action of the channel is described by the Kraus operators E_k, satisfying $\sum_k E_k^\dagger E_k = I$. In particular, we assume that

$$E_k = \sqrt{p_k}\sigma_k, \qquad (7.125)$$

with $i = 0, x, y, z$, $\sigma_0 = I$ and $\sum_k p_k = 1$. If the state ρ is sent by Alice through the channel, then Bob receives the state

$$\rho' = \sum_k E_k \rho E_k^\dagger = p_0\rho + p_x\sigma_x\rho\sigma_x + p_y\sigma_y\rho\sigma_y + p_z\sigma_z\rho\sigma_z. \qquad (7.126)$$

Noise has therefore induced a rotation through an angle π about axis x, y, z of the Bloch sphere with probability p_x, p_y, p_z or left the state unchanged, with probability p_0. In the case of two uses of the channel, we assume that the initial two-qubit density matrix ρ is mapped onto

$$\rho' = \sum_{k_1,k_2} E_{k_1 k_2} \rho E_{k_1 k_2}^\dagger, \tag{7.127}$$

where the Kraus operators have the form

$$E_{k_1 k_2} = \sqrt{p_{k_1 k_2}} \sigma_{k_1} \sigma_{k_2}, \tag{7.128}$$

with $\sum_{k_1,k_2} E_{k_1 k_2}^\dagger E_{k_1 k_2} = I \otimes I$, implying $\sum_{k_1,k_2} p_{k_1 k_2} = 1$. The two limiting cases of memoryless and perfectly correlated channels are described by

$$E_{k_1 k_2}^{(u)} = \sqrt{p_{k_1}} \sqrt{p_{k_2}} \sigma_{k_1} \sigma_{k_2} \tag{7.129a}$$

and

$$E_{k_1 k_2}^{(c)} = \sqrt{p_{k_1}} \sigma_{k_1} \sigma_{k_2} \delta_{k_1 k_2}, \tag{7.129b}$$

respectively. An intermediate case is described by the Kraus operators

$$E_{k_1 k_2}^{(i)} = \sqrt{p_{k_1} [(1-\mu) p_{k_2} + \mu \delta_{k_1 k_2}]} \, \sigma_{k_1} \sigma_{k_2}. \tag{7.130}$$

This corresponds to $p_{k_1 k_2} = p_{k_1} p_{k_2|k_1}$, with $p_{k_2|k_1} = (1-\mu) p_{k_2} + \mu \delta_{k_1 k_2}$. This means that the channels has partial memory: with probability μ the same rotation is applied to both qubits, whereas with probability $1 - \mu$ the two rotations are uncorrelated. Hence, the parameter μ describes the degree of correlation of the channel.

In the following we shall consider the depolarizing channel ($p_0 = 1 - p$, $p_x = p_y = p_z = \frac{p}{3}$) and assume that Alice sends Bob pure orthogonal quantum states drawn with equal *a priori* probabilities $\pi_i = \frac{1}{4}$ ($i = 1, \ldots, 4$) from the ensemble $\{|\psi_1\rangle, |\psi_2\rangle, |\psi_3\rangle, |\psi_4\rangle\}$, where

$$\begin{aligned}
|\psi_1\rangle &= \cos\theta |00\rangle + \sin\theta |11\rangle, & |\psi_2\rangle &= \sin\theta |00\rangle - \cos\theta |11\rangle, \\
|\psi_3\rangle &= \cos\theta |01\rangle + \sin\theta |10\rangle, & |\psi_4\rangle &= \sin\theta |01\rangle - \cos\theta |10\rangle.
\end{aligned} \tag{7.131}$$

Note that these states range from separable ($\theta = 0$) to maximally entangled ($\theta = \frac{\pi}{4}$). We shall maximize, as a function of θ, the Holevo information (see Sec. 5.11.1) $\chi = S(\rho') - \sum_i \pi_i S(\rho_i')$, where $\rho' = \sum_i \pi_i \rho_i'$, $\rho_i' = \sum_{k_1,k_2} E_{k_1 k_2} \rho_i E_{k_1 k_2}^\dagger$ and $\rho_i = |\psi_i\rangle\langle\psi_i|$. We shall show that there exists a memory threshold μ_t above which the Holevo information is maximal when maximally entangled (Bell) states are transmitted. This demonstrates that the transmission of classical information may be enhanced by sending entangled states.

For this purpose, it is useful to write the input two-qubit as follows:

$$\rho_i = \tfrac{1}{4}\Big(I \otimes I + I \otimes \sum_k \alpha_k^{(i)} \sigma_k + \sum_k \beta_k^{(i)} \sigma_k \otimes I + \sum_{kl} \gamma_{kl}^{(i)} \sigma_k \otimes \sigma_l\Big), \quad (7.132)$$

where $\alpha_k^{(i)} = \mathrm{Tr}[\rho_i(I \otimes \sigma_k)]$, $\beta_k^{(i)} = \mathrm{Tr}[\rho_i(\sigma_k \otimes I)]$ and $\gamma_{kl}^{(i)} = \mathrm{Tr}[\rho_i(\sigma_k \otimes \sigma_l)]$. The input state ρ_1 reads

$$\rho_1 = \tfrac{1}{4}\big[I \otimes I + \cos 2\theta(I \otimes \sigma_z + \sigma_z \otimes I) + \sigma_z \otimes \sigma_z$$
$$+ \sin 2\theta(\sigma_x \otimes \sigma_x - \sigma_y \otimes \sigma_y)\big]. \quad (7.133)$$

It can be checked by direct computation that

$$\sum_{k_1,k_2} E_{k_1 k_2}^{(u)}(I \otimes I) E_{k_1 k_2}^{(u)\dagger} = I \otimes I,$$

$$\sum_{k_1,k_2} E_{k_1 k_2}^{(u)}(I \otimes \sigma_i) E_{k_1 k_2}^{(u)\dagger} = \eta I \otimes \sigma_i,$$

$$\sum_{k_1,k_2} E_{k_1 k_2}^{(u)}(\sigma_i \otimes I) E_{k_1 k_2}^{(u)\dagger} = \eta \sigma_i \otimes I, \quad (7.134)$$

$$\sum_{k_1,k_2} E_{k_1 k_2}^{(u)}(\sigma_i \otimes \sigma_j) E_{k_1 k_2}^{(u)\dagger} = \eta^2 \sigma_i \otimes \sigma_j,$$

where $\eta = 1 - \tfrac{4}{3}p$ is the so-called shrinking factor (see Eq. (6.41)). We also obtain

$$\sum_{k_1,k_2} E_{k_1 k_2}^{(c)}(I \otimes I) E_{k_1 k_2}^{(c)\dagger} = I \otimes I,$$

$$\sum_{k_1,k_2} E_{k_1 k_2}^{(c)}(I \otimes \sigma_i) E_{k_1 k_2}^{(c)\dagger} = \eta I \otimes \sigma_i,$$

$$\sum_{k_1,k_2} E_{k_1 k_2}^{(c)}(\sigma_i \otimes I) E_{k_1 k_2}^{(c)\dagger} = \eta \sigma_i \otimes I, \quad (7.135)$$

$$\sum_{k_1,k_2} E_{k_1 k_2}^{(c)}(\sigma_i \otimes \sigma_j) E_{k_1 k_2}^{(c)\dagger} = \delta_{ij}\sigma_i \otimes \sigma_j + (1 - \delta_{ij})\eta \sigma_i \otimes \sigma_j.$$

Taking into account (7.134) and (7.135), we can see that the state ρ_1 is transformed by the depolarizing channel with partial memory (7.130) into the output state

$$\rho_1' = \tfrac{1}{4}\Big\{I \otimes I + \eta \cos 2\theta(I \otimes \sigma_z + \sigma_z \otimes I)$$
$$+ \big[\mu + (1 - \mu)\eta^2\big]\big[\sigma_z \otimes \sigma_z + \sin 2\theta(\sigma_x \otimes \sigma_x - \sigma_y \otimes \sigma_y)\big]\Big\}. \quad (7.136)$$

The eigenvalues of this density matrix are

$$\lambda_{1,2} = \tfrac{1}{4}(1 - \mu)(1 - \eta^2)$$

$$\lambda_{3,4} = \tfrac{1}{4}\Big\{1 + \mu + \eta^2(1 - \mu) \tag{7.137}$$

$$\pm 2\sqrt{\eta^2 \cos^2(2\theta) + [\eta^2(1 - \mu) + \mu]^2 \sin^2(2\theta)}\Big\}.$$

The same eigenvalues are obtained for the other output states ρ_2', ρ_3', ρ_4'. As λ_1 and λ_2 do not depend on θ, the von Neumann entropy $S(\rho_i)$ ($i = 1, \ldots, 4$) is minimized (as a function of θ) when the term under the square root in the expression for λ_3 and λ_4 is maximum. Moreover, we have $\rho' = \tfrac{1}{4}(\rho_1' + \rho_2' + \rho_3' + \rho_4') = \tfrac{1}{4}(I \otimes I)$, so that $S(\rho') = 2$. Therefore, the Holevo information $\chi = S(\rho') - \tfrac{1}{4}\sum_i S(\rho_i') = 2 - S(\rho_i')$ is maximal for separable states ($\theta = 0$) when $\eta^2 > [\eta^2(1 - \mu) + \mu]^2$ and for Bell states ($\theta = \tfrac{\pi}{4}$) when $\eta^2 < [\eta^2(1 - \mu) + \mu]^2$. This latter condition can be equivalently written as

$$\mu > \mu_t = \frac{\eta}{1 + \eta}. \tag{7.138}$$

Therefore, for states of the form (7.131) the Holevo information is maximal for separable states when $\mu < \mu_t$ and for Bell states when $\mu > \mu_t$. At the threshold value $\mu = \mu_t$, the same Holevo information is obtained for any value of θ in (7.131). The Holevo information is shown in Fig. 7.18 as a function of θ, for different values of the parameter μ. The different behaviour below and above the threshold μ_t is evident. Note that, for a perfectly correlated noise channel ($\mu = 1$) we have $\xi = 2$ for Bell states. Indeed, in this case noise does not affect the Bell states: $\rho_i' = \rho_i$, and therefore $S(\rho_i') = S(\rho_i) = 0$.

7.13   A guide to the bibliography

Quantum error correction was invented by Shor (1995) and Steane (1996a). The five-qubit code is discussed in Laflamme *et al.* (1996) and Bennett *et al.* (1996c). The CSS codes were developed by Calderbank and Shor (1996) and Steane (1996b). Tutorials on quantum error correction are Gottesman (2000), Knill *et al.* (2002) and Steane (2006). A very readable introduction is Preskill (1999).

A review on decoherence-free subspaces is Lidar and Whaley (2003). Other useful references that can be used to enter the literature are Palma *et al.* (1996), Zanardi and Rasetti (1998), Lidar *et al.* (1998), Lidar *et al.*

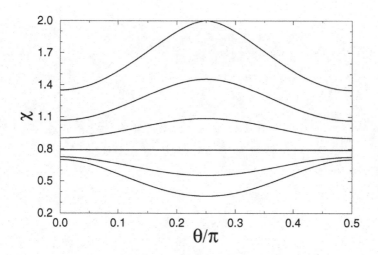

Fig. 7.18   Holevo information χ as a function of the parameter θ, for $\eta = \frac{2}{3}$ and, from bottom to top, $\mu = 0, 0.2, 0.4, 0.6, 0.8, 1$. The function $\chi(\theta)$ has periodicity $\pi/2$. At $\mu = \mu_t = 0.4$, χ is independent of θ.

(2000), Beige *et al.* (2000) and Alber *et al.* (2001b). A useful reference on the link between the quantum Zeno and decoherence-free subspaces is Facchi *et al.* (2004). Both passive quantum error-avoiding codes and active quantum error correction can be treated in a unified picture based on the so-called noiseless subsystems, see Knill *et al.* (2000) and Viola *et al.* (2001).

A review on fault-tolerant quantum computation is Preskill (1998b).

Chapter 8

First Experimental Implementations

The great challenge of quantum computation is to experimentally realize a large scale quantum computer. The requirements that must be fulfilled to achieve this imposing objective are summarized in DiVincenzo (2000):

1. A scalable system with well characterized qubits.
2. The ability to initialize ('reset') the state of the qubits to a fiducial state (such as $|0 \cdots 0\rangle$).
3. Long significant decoherence times, much longer than the gate operation time.
4. An experimentally feasible universal set of quantum gates.
5. A high-fidelity readout method.

It should be remarked that these requirements are to some extent conflicting: we desire the quantum computer to be well isolated from the environment to preserve its coherence and at the same time we must interact with it strongly to prepare the initial state, realize the desired unitary evolution and measure the final state. The problem here is that external control operations typically introduce noise into the computer, thus disturbing the programmed coherent evolution. An important question is how large should the ratio be between the decoherence time τ_d and the "clock time" of the quantum computer; that is, the time τ_g for the execution of a quantum gate. The answer is that the ratio τ_d/τ_g should be large enough to allow quantum error correction. As discussed in Sec. 7.10.3, the threshold value for fault-tolerant quantum computation depends on the characteristics of the quantum hardware. However, optimistic estimates require $\tau_d/\tau_g > 10^4$, an extremely demanding requirement, corresponding to less than one error in 10^4 quantum gate operations.

The five requirements above are sufficient for quantum computation. However, we are also interested in the implementation of quantum communication protocols. For this purpose, two more items must be added to the list of requirements:

1. The ability to interconvert "stationary" and "flying" qubits.
2. Faithful transmission of flying qubits between specified locations.

Using the terms "stationary" and "flying" qubits we emphasize the fact that the physical systems (in practice, photons) used to transmit qubits from place to place are very different from the qubits used for reliable local computation (for instance, two-level atoms or ions). The development of interfaces between quantum information carriers and quantum information storage and processors is an important objective in the development of quantum technologies. On the other hand, the last requirement alone is sufficient for quantum cryptography, which deals with one qubit (or one Bell state) at a time and not with complex many-qubit systems as in the case of quantum computation.

In this chapter, we shall discuss the first experimental realizations of few-qubit quantum computers in different physical systems (nuclear magnetic resonance quantum processors, cavity quantum electrodynamics, trapped ions and solid-state qubits) as well as quantum teleportation and quantum cryptography with photons. We shall not dwell on the technical aspects of the implementations but present instead the basic physical ideas underlying the development of these first quantum machines.

8.1   NMR quantum computation

In the field of nuclear magnetic resonance (NMR) over the last few decades sophisticated techniques have been developed to manipulate and detect nuclear spin states using both static and oscillating magnetic fields simultaneously. These techniques have been used, for instance, to study structural properties of molecules and even biological samples. In the liquid state NMR quantum computation controlled quantum logic operations are performed over a system of spin-$\frac{1}{2}$ nuclei (the qubits) of molecules in solution. This kind of quantum computation is very different from the other implementations, which we shall discuss later in this chapter, for the following main reasons:

1. Nuclear magnetic moments are small and therefore we must average over a large number of molecules to detect a (magnetization) signal (typically $\sim 10^{18}$ molecules in solution are used in NMR quantum computation).
2. We work (at room temperature) on highly mixed (thermal) states, not on pure states.
3. The interaction between qubits (the nuclear spins) is never switched off, so that the evolution due to undesired couplings must be removed by means of appropriate "refocusing" techniques.

As we shall see in this section, the fact that we are working with mixed states implies that the signal-to-noise ratio drops exponentially with the number of qubits. Therefore, liquid-state NMR quantum computation is not scalable. Nevertheless, such an implementation is very important for at least two reasons:

(i) It has allowed the experimental demonstration of quantum algorithms with, so far, up to seven qubits and a number of quantum gates (up to $O(10^2)$) still out of reach of other implementations of quantum computers.
(ii) Many sophisticated quantum control techniques developed in NMR (such as refocusing and composite pulses) are being used by other implementations.

8.1.1 *The system Hamiltonian*

A spin-$\frac{1}{2}$ nucleus in a static magnetic field $\boldsymbol{B}_0$ evolves according to the Hamiltonian $H_0 = -\boldsymbol{\mu} \cdot \boldsymbol{B}_0 = -\gamma \boldsymbol{S} \cdot \boldsymbol{B}_0$, where γ is the gyromagnetic ratio of the nucleus and $\boldsymbol{\mu} = \gamma \boldsymbol{S}$ its magnetic moment, $\boldsymbol{S} = \frac{1}{2}\hbar\boldsymbol{\sigma}$ being the spin operator. The energy difference between the two spin states (the eigenstates $|0\rangle$ and $|1\rangle$ of σ_z) is $\hbar\omega_0$, where $\omega_0 = -\gamma B_0$ is the Larmor frequency. For instance, a proton in a magnetic field of 12 T has a Larmor frequency of approximately 500 MHz; that is, we are in the radiofrequency (RF) range. We can manipulate the state of this spin-$\frac{1}{2}$ nucleus by means of an electromagnetic field $\boldsymbol{B}_1$ which oscillates in the plane perpendicular to the direction of $\boldsymbol{B}_0$ at resonant (radio)frequency $\omega = \omega_0$ (see Sec. 3.16.1). This system provides a physical representation of a qubit.

A quantum register is made of several (spin-$\frac{1}{2}$) atomic nuclei (the qubits) in a molecule. The qubits can be addressed separately by radiofrequency fields if the resonant frequencies are different. This is the case when the qubits are nuclei of different chemical elements (heteronuclear spins) as they have distinct values of gyromagnetic ratio. In a molecule atomic nuclei of the same chemical element (homonuclear spins) can also have different Larmor frequencies since the partial shielding of the magnetic field B_0 by the electrons surrounding the nuclei depends on the electronic environment of each nucleus. Therefore, asymmetries in the molecular structure also lead to different Larmor frequencies for homonuclear spins. We can write the non-interacting Hamiltonian for an n-qubit molecule in a static magnetic field directed along z as follows:

$$H_0 = -\sum_{i=1}^{n} \left(1 - \alpha^{(i)}\right)\gamma^{(i)} B_0 S_z^{(i)} = \sum_{i=1}^{n} \omega_0^{(i)} S_z^{(i)} = \sum_{i=1}^{n} \tfrac{1}{2}\hbar\omega_0^{(i)}\sigma_z^{(i)}, \quad (8.1)$$

where the superscripts (i) label the qubits in the quantum register and $\alpha^{(i)}$, known as the chemical shift, measures the partial shielding of the magnetic field acting on the i-th qubit.[1]

Let us now consider the interactions between nuclear spins in molecules. There are two distinct coupling mechanisms: the direct dipole–dipole interaction and the indirect (electron-mediated) scalar coupling.

The magnetic dipole–dipole interaction between nuclei i and j depends on the internuclear vector r_{ij} connecting the two nuclei and is described by the Hamiltonian

$$(H_d)_{ij} = \frac{\mu_0 \gamma^{(i)}\gamma^{(j)}}{4\pi\hbar|r_{ij}|^3} \left[S^{(i)} \cdot S^{(j)} - \frac{3}{|r_{ij}|^2}\left(S^{(i)} \cdot r_{ij}\right)\left(S^{(j)} \cdot r_{ij}\right) \right], \quad (8.2)$$

where μ_0 is the magnetic permeability of free space. The dipolar interaction is rapidly averaged away in liquid state NMR, due to the fast chaotic motion of molecules.

In the scalar coupling the interaction is mediated by the electrons shared in the chemical bond between the atoms: a nucleus interacts with another nucleus through the overlap of the electronic wave function with the two nuclei. Note that, in contrast to the dipole–dipole interaction, the scalar coupling is an intramolecular coupling (between spins in the same molecule). The Hamiltonian describing the scalar coupling (also known as J-coupling) is

$$H_J = \frac{2\pi}{\hbar}\sum_{i<j} J_{ij} S^{(i)} \cdot S^{(j)} = \frac{2\pi}{\hbar}\sum_{i<j} J_{ij}\left(S_x^{(i)} S_x^{(j)} + S_y^{(i)} S_y^{(j)} + S_z^{(i)} S_z^{(j)}\right),$$

$$(8.3)$$

---

[1]Note that the chemical shift is a very useful phenomenon for the study of the properties of molecules by means of NMR spectroscopy.

where J_{ij} is the coupling strength between nuclei i and j and decreases rapidly with the number of chemical bonds separating these nuclei. When the coupling strength $2\pi J_{ij}$ is much smaller than $|\omega^{(i)} - \omega^{(j)}|$, then the scalar coupling reduces to

$$H_J \approx \frac{2\pi}{\hbar} \sum_{i<j} J_{ij}\, S_z^{(i)} S_z^{(j)}. \tag{8.4}$$

The terms proportional to $S_x^{(i)} S_x^{(j)}$ and $S_y^{(i)} S_y^{(j)}$ have been neglected since they flip the state of both spins and therefore involve an energy cost $\hbar|\omega^{(i)} - \omega^{(j)}|$ much larger than the coupling energy $2\pi\hbar J_{ij}$.

The simplest Hamiltonian for a system of coupled nuclear spins in a molecule is therefore

$$H_{\text{sys}} = H_0 + H_J = \sum_i \omega_0^{(i)} S_z^{(i)} + \frac{2\pi}{\hbar} \sum_{i<j} J_{ij}\, S_z^{(i)} S_z^{(j)}. \tag{8.5}$$

One- and two-qubit quantum gates can then be implemented by means of oscillating magnetic fields, as described in Sec. 3.16.1 (see also exercise 8.1). Note that, if the frequencies $\omega_0^{(i)}$ are all different, we can address single qubits by means of resonant pulses. The time required to resolve frequency $\omega_0^{(i)}$ from frequency $\omega_0^{(j)}$ becomes larger when the frequency separation $|\omega_0^{(i)} - \omega_0^{(j)}|$ is smaller. This sets limits to the speed of quantum logic gates and, therefore, on the number of gates that can be implemented within the decoherence time scale. It is also clear that, in the case of homonuclear molecules, strong chemical shifts are desired.

Exercise 8.1   Consider a single nuclear spin exposed to both a static and a time-dependent magnetic field. Such a system is described by the Hamiltonian

$$H = \omega_0 S_z + \omega_1 \big[\cos(\omega t + \phi) S_x + \sin(\omega t + \phi) S_y \big], \tag{8.6}$$

where the static field is directed along z and the oscillating field rotates in the (x, y) plane at frequency ω. Study the motion of the spin in a coordinate system rotating about the z-axis at frequency ω (*rotating frame*) (note that the wave function $|\psi\rangle_r$ in the rotating frame is related to the wave function $|\psi\rangle$ in the laboratory frame as follows: $|\psi\rangle_r = \exp\big(\frac{i}{\hbar}\omega t S_z\big)|\psi\rangle$).

8.1.2 *The physical apparatus*

The basic structure of an apparatus for NMR quantum computation is shown in Fig. 8.1. A magnet (usually superconducting) creates a strong homogeneous magnetic field B_0 ($10 - 15\,\mathrm{T}$) directed along the z-axis and uniform to within one part in 10^9 over a region of the order of $1\,\mathrm{cm}^3$ where the liquid sample is placed. Two coils, with axes lying in the (x, y)-plane, allow the generation of small radiofrequency fields along the x and y directions. One- and two-qubit gates are implemented by means of radiofrequency (RF) fields of appropriate amplitude, frequency, duration and shape. Sequences of hundreds of RF pulses can be applied, the upper limit on the time of a quantum computation being set by decoherence (typically RF pulses are order of milliseconds long[2] and reliable quantum computation is possible up to several hundred ms).

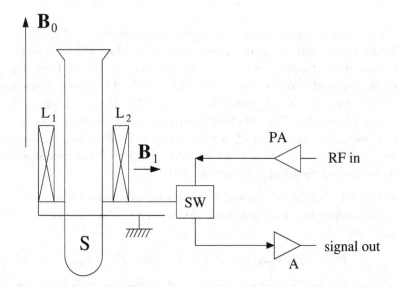

Fig. 8.1 A schematic drawing of an NMR apparatus: B_0 is the static polarizing magnetic field, B_1 the RF magnetic field, S the sample, L_1 and L_2 the coils used both to generate the RF pulses and for magnetization measurements, SW the excitation/measurement switch, PA the power RF amplifier, A the signal preamplifier.

After completion of the pulse sequence, the signal generated by processing nuclei is an induction voltage (known as *free induction decay*) which

---

[2] Pulses that are selective for a single qubit are typically 200-500 μs long, while non-selective RF pulses are typically 10 μs long.

generates an alternating current in the same coils. The signal is then amplified and analyzed. Note that the readout mechanism in NMR quantum information processing is very different from the projective von Neumann measurements described in the previous chapters. Indeed, the system is continuously read out and the voltage induced (by the nucleus labelled as k in the molecules) in the two coils is given by

$$V_{x,y}^{(k)}(t) = V_0 \operatorname{Tr}\left[\rho(t)\sigma_{x,y}^{(k)}\right], \tag{8.7}$$

where $\rho(t)$ is the density matrix describing the state of the nucleus, the Pauli matrices operate only on the k-th spin, and V_0 is a constant depending on the properties of the read-out apparatus. Note that this measurement does not induce a collapse of the wave function, unlike the projective von Neumann measurements. This is due to the fact that the coupling between the magnetization of the spins and the readout apparatus is weak. A detectable signal is nevertheless obtained since it is averaged over the large number of molecules ($O(10^{18})$) in the sample.

8.1.3 Quantum ensemble computation

NMR quantum computation is realized at room temperature. Therefore, the quantum state of a nuclear spin in a molecule is highly mixed, not pure as we have assumed so far when discussing quantum algorithms. Let us explain how this problem can be overcome by the introduction of an appropriate *pseudo-pure state*.

The initial state is at thermal equilibrium and is therefore described by the density matrix

$$\rho = \frac{\exp(-\beta H)}{Z}, \tag{8.8}$$

where $\beta = \frac{1}{k_B T}$ (k_B is Boltzmann's constant and T the temperature) and $Z = \operatorname{Tr}[\exp(-\beta H)]$ is the partition function (a normalization factor introduced in order to satisfy $\operatorname{Tr}(\rho) = 1$).

For a single spin we have $H = \frac{1}{2}\hbar\omega_0\sigma_z$, so that the density matrix in the $\{|0\rangle, |1\rangle\}$ basis reads

$$\rho = \frac{1}{\exp\left(-\frac{1}{2}\beta\hbar\omega_0\right) + \exp\left(\frac{1}{2}\beta\hbar\omega_0\right)} \begin{bmatrix} \exp\left(-\frac{1}{2}\beta\hbar\omega_0\right) & 0 \\ 0 & \exp\left(\frac{1}{2}\beta\hbar\omega_0\right) \end{bmatrix}. \tag{8.9}$$

Note that the off-diagonal matrix elements of ρ (known as coherences) are equal to zero while they are non-vanishing for a generic pure quantum state $|\psi\rangle = c_0|0\rangle + c_1|1\rangle$ when $c_0, c_1 \neq 0$. The diagonal terms of ρ give the probabilities of finding the spin in the state $|0\rangle$ and $|1\rangle$, as for a density matrix in classical statistical mechanics. In typical situations $\beta\hbar\omega_0$ is very small. For example, for a spin-$\frac{1}{2}$ nucleus at room temperature and a static magnetic field inducing a precession frequency $\omega_0/2\pi \approx 500\,\mathrm{MHz}$, we have $\beta\hbar\omega_0 \sim 10^{-5}$. Therefore, we can approximate the single-spin density matrix as

$$\rho \approx \tfrac{1}{2}I - \tfrac{1}{2}\beta\hbar\omega_0\sigma_z. \tag{8.10}$$

For an n-qubit molecule, we can approximate (8.8) as follows:

$$\rho \approx \tfrac{1}{2^n}(I - \beta H). \tag{8.11}$$

The pseudo-pure state used for quantum computation is determined by the deviation ρ_{dev} of the density matrix from the identity (divided by 2^n for normalization reason):

$$\rho_{\mathrm{dev}} = \rho - \tfrac{I}{2^n} \approx -\tfrac{1}{2^n}\beta H. \tag{8.12}$$

For instance, for two non-interacting qubits, whose motion is governed by the Hamiltonian $H = \tfrac{1}{2}\hbar(\omega_0^{(1)}\sigma_z^{(1)} \otimes I^{(2)} + \omega_0^{(2)}I^{(1)} \otimes \sigma_z^{(2)})$, we obtain

$$\rho_{\mathrm{dev}} \approx -\tfrac{1}{4}\beta \left(\tfrac{1}{2}\hbar\omega_0^{(1)}\sigma_z^{(1)} \otimes I^{(2)} + \tfrac{1}{2}\hbar\omega_0^{(2)}I^{(1)} \otimes \sigma_z^{(2)} \right)$$

$$= -\tfrac{1}{8}\beta\hbar \begin{bmatrix} \omega_0^{(1)} + \omega_0^{(2)} & 0 & 0 & 0 \\ 0 & \omega_0^{(1)} - \omega_0^{(2)} & 0 & 0 \\ 0 & 0 & -(\omega_0^{(1)} - \omega_0^{(2)}) & 0 \\ 0 & 0 & 0 & -(\omega_0^{(1)} + \omega_0^{(2)}) \end{bmatrix}. \tag{8.13}$$

In order to work with pseudo-pure states, we exploit two important facts:

1. The quantum mechanical evolution of a system is *linear*, so that we can execute several experiments and combine the results;

2. The observables measured in NMR (spin polarizations) are traceless and therefore not sensitive to the component $\tfrac{1}{2^n}I$ of the density matrix ($\mathrm{Tr}(I\sigma_\alpha^{(k)}) = 0$ for $\alpha = x, y, z$). We also note that this component does not change under temporal evolution: $U(t)IU^\dagger(t) = I$.

There are several techniques to take advantage of the above two facts. In the following, we limit ourselves to describe the so-called *temporal labelling* (or *temporal averaging*). Let us consider, for instance, a two-qubit system, initially prepared in a state described by the diagonal density matrix

$$\rho_1 = \begin{bmatrix} a & 0 & 0 & 0 \\ 0 & b & 0 & 0 \\ 0 & 0 & c & 0 \\ 0 & 0 & 0 & d \end{bmatrix}, \tag{8.14}$$

where a, b, c, d are real non-negative numbers such that $a+b+c+d = 1$ (this guarantees that $\text{Tr}(\rho_1) = 1$). By an appropriate combination of generalized CNOT gates (see Sec. 3.4) we can permute the populations in ρ_1 and obtain the initial states

$$\rho_2 = \begin{bmatrix} a & 0 & 0 & 0 \\ 0 & c & 0 & 0 \\ 0 & 0 & d & 0 \\ 0 & 0 & 0 & b \end{bmatrix}, \quad \rho_3 = \begin{bmatrix} a & 0 & 0 & 0 \\ 0 & d & 0 & 0 \\ 0 & 0 & b & 0 \\ 0 & 0 & 0 & c \end{bmatrix}. \tag{8.15}$$

We then perform three separate experiments starting from the initial (highly mixed) states ρ_1, ρ_2 and ρ_3 and obtain the final states $\rho_i(t) = U(t)\rho_i U^\dagger(t)$ ($i = 1, 2, 3$). Finally, due to the linearity of quantum mechanics, we have

$$\rho_1(t) + \rho_2(t) + \rho_3(t) = U(t) \left\{ (1-a)I + (4a-1) \begin{bmatrix} 1 & 0 & 0 & 0 \\ 0 & 0 & 0 & 0 \\ 0 & 0 & 0 & 0 \\ 0 & 0 & 0 & 0 \end{bmatrix} \right\} U^\dagger(t)$$

$$= (1-a)I + (4a-1)U(t)|00\rangle\langle 00|U^\dagger(t). \tag{8.16}$$

In this equation, the term $(4a-1)U(t)|00\rangle\langle 00|U^\dagger(t)$ is the pseudo-pure state at time t. Since $\sum_{j=1}^{3} \text{Tr}\left(\rho_j(t)\sigma_\alpha^{(k)}\right) = (4a-1)\text{Tr}\left(U(t)|00\rangle\langle 00|U^\dagger(t)\sigma_\alpha^{(k)}\right)$, it is clear that the final outcome of the experiment is proportional to what we would have obtained starting from the pure state $|00\rangle\langle 00|$.

The main drawback of NMR quantum ensemble computation is evident from the previous analysis: assuming that ρ_1 is the density matrix (8.13) describing two qubits at thermal equilibrium[3], we obtain $(4a-1) \approx -2\frac{\beta\hbar}{2^n}(\omega_0^{(1)} + \omega_0^{(2)})$. In general, the pseudo-pure states that we can derive differ from true pure states by a proportionality factor $\propto \frac{1}{2^n}$. This means

---

[3]Qubit–qubit interaction terms are not relevant for the present argument.

that in quantum ensemble computation the signal drops exponentially with
the number of qubits.

8.1.4   Refocusing

An important feature of NMR quantum computation is that the spin–spin
interaction used to implement two-qubit gates is always present. It is therefore necessary to employ appropriate techniques in order to control the
effect of this interaction and to remove undesired evolution (drift term)
generated by the coupling between spins. An interesting method to achieve
this purpose, known as *refocusing* (or *spin echo*), is described in this section. Note that many other physical systems must cope with a drift term
and spin echo techniques borrowed from NMR have already proved very
useful for this purpose (see, *e.g.*, Collin *et al.*, 2004) .

To illustrate refocusing, it is sufficient to consider two qubits in a
molecule, with the qubit–qubit interaction described by the Hamiltonian

$$H_I = \alpha \sigma_z^{(1)} \otimes \sigma_z^{(2)}, \tag{8.17}$$

where the coupling strength α depends on the structure of the molecule.
Let $R_{x_1}(\pi)$ denote the action of a RF pulse (known as π pulse) rotating the
nuclear spin 1 through an angle π about the x-axis of the Bloch sphere:

$$R_{x_1}(\pi) = \exp\left(-i\frac{\pi}{2}\sigma_x^{(1)}\right) = -i\sigma_x^{(1)}. \tag{8.18}$$

The evolution in a time t due to the interaction term is described by the
unitary operator

$$U_I(t) = \exp\left(-i\frac{\alpha t}{\hbar}\sigma_z^{(1)} \otimes \sigma_z^{(2)}\right) = \cos\left(\frac{\alpha t}{\hbar}\right) I - i\sin\left(\frac{\alpha t}{\hbar}\right) \sigma_z^{(1)} \otimes \sigma_z^{(2)}, \tag{8.19}$$

where the last equality follows from Eq. (3.28). The evolution $U_I(t)$ can
be removed if two refocusing pulses $R_{x_1}(\pi)$ are applied at time 0 and $\frac{t}{2}$.
Indeed, we have (see exercise 8.2)

$$U_I\left(\frac{t}{2}\right) R_{x_1}(\pi) U_I\left(\frac{t}{2}\right) R_{x_1}(\pi) = I \tag{8.20}$$

Exercise 8.2   Check Eq. (8.20).

Exercise 8.3   Find an appropriate spin-echo sequence to eliminate the
dynamical evolution due to the single qubit Hamiltonian $H = \frac{1}{2}\omega_0\sigma_z$.

Exercise 8.4 Find an appropriate sequence of pulses for the implementation of a CNOT gate in a two-qubit system whose evolution is governed by the Hamiltonian

$$H = \tfrac{1}{2}\hbar\left(\omega_0^{(1)}\sigma_z^{(1)} + \omega_0^{(2)}\sigma_z^{(2)}\right) + \tfrac{1}{2}\pi\hbar J_{12}\sigma_z^{(1)} \otimes \sigma_z^{(2)}. \qquad (8.21)$$

8.1.5 Demonstration of quantum algorithms

Many quantum algorithms have been implemented using room temperature liquid state NMR techniques with 3–7 qubits, from Grover's to Deutsch–Jozsa and Shor's algorithms, including quantum Fourier transform, teleportation, adiabatic quantum optimization, quantum error correction, concatenated codes, decoherence-free subspaces, noiseless subsystems and quantum simulation. Control methods on a 12-qubit system have been recently reported in Negrevergne $et\ al.$ (2006). So far, no other implementation of a quantum processor has been able to produce similar results. In this section, we briefly describe three relevant experiments.

Quantum Fourier transform. This has been implemented (Weinstein $et\ al.$, 2001) via NMR using the three ^{13}C nuclei of alanine molecules as qubits (see Fig. 8.2).[4] It is clearly seen from Fig. 8.2 that the three carbon atoms have very different local chemical environments. The resulting chemical shifts ($|\omega_0^{(i)} - \omega_0^{(j)}|$ between 2.6 and 12 kHz) thus permit resonant addressing of single qubits.

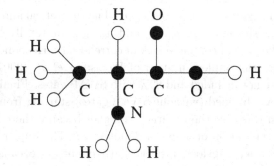

Fig. 8.2 The structure of the alanine molecule used in the implementation of the quantum Fourier transform. The three qubits are the carbon nuclei.

---

[4]A complete characterization of the experimentally determined superoperator in this three-qubit quantum Fourier transform is provided in Weinstein $et\ al.$ (2004).

The experiment looks for the periodicity of the state

$$|\psi\rangle = \tfrac{1}{2}\left(|000\rangle + |010\rangle + |100\rangle + |110\rangle\right) = \tfrac{1}{2}\left(|0\rangle + |2\rangle + |4\rangle + |6\rangle\right)$$
$$= \tfrac{1}{\sqrt{2}}\left(|0\rangle + |1\rangle\right) \otimes \tfrac{1}{\sqrt{2}}\left(|0\rangle + |1\rangle\right) \otimes |0\rangle, \qquad (8.22)$$

prepared by means of two Hadamard gates H applied to the initial (pseudo-pure) state $|000\rangle$ ($|\psi\rangle = (H \otimes H \otimes I)|000\rangle$). Therefore, this experiment demonstrates not only the possibility to implement the quantum Fourier transform but also the capability of NMR quantum computation to prepare non-trivial initial states. The quantum Fourier transform is then implemented by means of three Hadamard and three controlled phase-shift gates (see Sec. 3.11). Finally, one can obtain the correct order for the qubits by means of three SWAP gates or simply by relabelling the qubits.

In order to measure the accuracy with which the QFT had been performed, the following quantity was evaluated:

$$C = \frac{\mathrm{Tr}\left(\rho_{\mathrm{dev,th}}\rho_{\mathrm{dev,exp}}\right)}{\sqrt{\mathrm{Tr}\left(\rho_{\mathrm{dev,th}}^2\right)}\sqrt{\mathrm{Tr}\left(\rho_{\mathrm{dev,exp}}^2\right)}}\sqrt{\frac{\mathrm{Tr}\left(\rho_{\mathrm{dev,exp}}^2\right)}{\mathrm{Tr}\left(\rho_{\mathrm{dev,in}}^2\right)}}, \qquad (8.23)$$

where $\rho_{\mathrm{dev,th}}$ and $\rho_{\mathrm{dev,exp}}$ represent he deviation of the theoretically expected and experimentally measured density matrices from the identity term (more precisely, from $\frac{I}{2^n}$). Note that $\rho_{\mathrm{dev,th}} = U_{\mathrm{th}}\,\rho_{\mathrm{dev,in}}\,U_{\mathrm{th}}^{\dagger}$, where U_{th} is the theoretical evolution operator for the quantum Fourier transform algorithm, and $\rho_{\mathrm{dev,in}}$ is the experimentally obtained initial pseudo-pure state. Note that the first term in (8.23) measures the correlation between $\rho_{\mathrm{dev,exp}}$ and $\rho_{\mathrm{dev,th}}$, while the term under square root weights the reduction in signal over the course of the experiment. The theoretical and experimental deviation density matrices $\rho_{\mathrm{dev,th}}$ and $\rho_{\mathrm{dev,exp}}$ are totally correlated if $C = 1$, while the case of complete lack of correlation corresponds to $C = 0$. The accuracy of the implementation of Weinstein *et al.* (2001) is 80% if SWAP gates are not included and 62% with SWAP gates. Finally, an accuracy of 87% was obtained (without SWAP gates) starting from a thermal initial state. In this case the accuracy is higher because there is no initial stage for the preparation of a special input state. The implementation of the quantum Fourier transform is particularly important because it is a key ingredient of exponentially efficient quantum algorithms, such as factoring and quantum simulations.

Shor's algorithm. The most complex quantum algorithm realized to date is the demonstration of Shor's algorithm using a seven-qubit molecule manipulated with NMR techniques (Vandersypen *et al.*, 2001). The simplest instance of Shor's algorithm was reported: factorization of $N = 15$ into its prime factors 3 and 5. The structure of the molecule used for this experiment is shown in Fig. 8.3. It contains five ^{19}F and two ^{13}C spin-$\frac{1}{2}$ nuclei as qubits. Note that this molecule was specially synthesized in such a manner that the seven resonant frequencies $\omega_0^{(i)}$ were very well separated (a static magnetic field of 11.7 T was applied). This assures that each qubit can be addressed independently. The seven spins interact pairwise via the scalar J-coupling, described by the Hamiltonian $H_I = \frac{2\pi}{\hbar} \sum_{i<j} J_{ij} S_z^{(i)} S_z^{(j)}$.

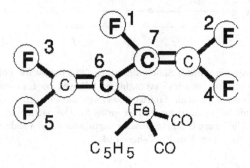

Fig. 8.3 The structure of the molecule used in the implementation of Shor's algorithm. The seven qubits are labelled from 1 to 7. The figure is reprinted with permission from Vandersypen *et al.* (2001). Copyright (2001) by Macmillan Publishers Ltd.

The quantum circuit for Shor's algorithm is shown in Fig. 8.4. The room temperature seven-qubits statistical mixture describing the initial state is converted into a seven-qubit pseudo-pure state by means of the temporal averaging technique. The sequence of quantum gates is realized by applying approximately 300 radiofrequency pulses, at seven different frequencies (from $\omega_0^{(1)}$ to operate single-qubit gates on spin 1 up to $\omega_0^{(7)}$ to operate on spin 7). Pulses are separated by time intervals of free evolution under the system Hamiltonian (8.5). This Hamiltonian includes interaction terms, which are necessary in order to implement Shor's quantum algorithm (two-qubit quantum gates are needed in both the modular exponentiation and the inverse quantum Fourier transform). After completion of the sequence of RF pulses the state of the first three qubits is measured using NMR spectroscopy. As a result, the prime factors 3 and 5 are unambiguously derived from the output spectrum.

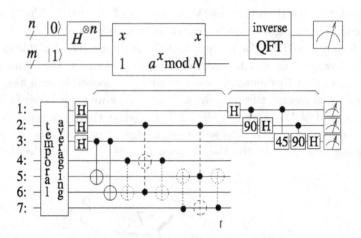

Fig. 8.4   A quantum circuit for Shor's algorithm. The preparation of the initial pseudo-pure state by means of temporal averaging is followed by Hadamard gates, modular exponentiation, inverse quantum Fourier transform and measurement. The gates shown in dotted lines can be replaced by simpler gates for the particular initial state used in the experiment. The figure is reprinted with permission from Vandersypen *et al.* (2001). Copyright (2001) by Macmillan Publishers Ltd.

The quantum sawtooth map. The implementation of the quantum saw-tooth map (see Sec. 3.15.3) on a three-qubit liquid state NMR quantum processor has recently been reported (Henry *et al.*, 2005). The quantum hardware is a solution of tris(trimethylsilyl)silane-acetylene molecules, the three qubits being a hydrogen nucleus and two ^{13}C nuclei. Experiments were performed in a 9.4 T magnetic field, where the resonant frequencies for the carbon qubits were separated by 1.201 kHz due to the chemical shift. The scalar couplings J_{ij} were of the order of 100 Hz. Pulse sequences were optimized accounting for inhomogeneities in the RF field. The quantum simulation of an iteration of the quantum sawtooth map took approximately 0.1 s and up to 4 map steps were simulated in the regime of quantum localization.

The results plotted in Fig. 8.5 show the experimentally measured probability distribution (over the momentum basis) together with the ideal, numerically computed distribution after one and two map iterations. It can be seen that the region around the central (localization) peak does not broaden significantly while the tails are more sensitive to the effects of imperfections and decoherence (which completely destroy localization in approximately four map iterations). The results are interesting because quantum localization is a very fragile quantum interference effect and therefore its observation demonstrates once more that a very sophisticated degree of coherent quantum control has been achieved in NMR quantum processors.

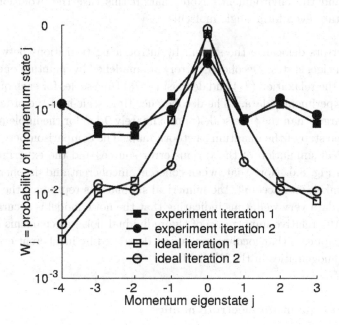

Fig. 8.5 Experimentally measured and ideal probability distributions after one and two iterations of the quantum sawtooth map. The figure is taken from Henry *et al.* (2005).

Furthermore, the degree of stability of the localization effect can be used as a test bed both for measuring the degree of quantum control and for studying the importance of the different noise mechanisms affecting the system. Errors can be classified into three main categories:

1. *Coherent errors:* the evolution is unitary but governed by a unitary operator U_ϵ which differs from the ideal operator U. For instance, undesired spin–spin couplings as well as the action of the system Hamiltonian during a RF pulse generate coherent errors.
2. *Incoherent errors:* the system Hamiltonian varies across the sample (for instance, due to spatial inhomogeneities of the RF field), so that the various members of the ensemble (the molecules) evolve differently from each other. Even though the single molecule evolves unitarily (no entangle-

ment with the environment is generated), the evolution of the ensemble averaged density matrix is not unitary. This kind of error is typical of quantum ensemble computation.

3. *Decoherent errors:* these errors arise from the coupling between the qubits and the environment. Note that in this case the evolution is non-unitary even for a single molecule.

Coherent errors delocalize the system by introducing transitions between momentum eigenstates. Decoherent errors are modelled by quantum operations, with the relaxation (T_1) and dephasing (T_2) time scales for each qubit based on experimental data.[5] The decoherence time scale is of the order of $1\,\mathrm{s}$, much larger than the time scale (approximately $0.1\,\mathrm{s}$) for the implementation of one step of the quantum sawtooth map. The comparison between the numerical simulation of the various error sources and the experiment is shown in Fig. 8.6. Note that, when coherent, incoherent and decoherent errors are taken into account, the numerical simulations reproduce the experimental data very well, thus indicating that the noise model is accurate. Moreover, the relative importance of the individual noise mechanisms can be seen. It appears that localization is first destroyed by incoherent errors due to inhomogeneities in the RF field.

8.2   Cavity quantum electrodynamics

The wording cavity quantum electrodynamics (CQED) denotes a set of techniques allowing the interaction of single atoms and single photons inside a resonating cavity. Here we focus on experiments performed with *Rydberg atoms*; that is, atoms whose valence electrons are in states with a very large principal quantum number n. More precisely, we consider alkali atoms, which have a single valence electron which is highly excited up to $n \sim 20 - 50$. In such conditions, the valence electron is very far from the atomic nucleus and therefore its electric dipole moment is very high (see Table 8.1). As a consequence, the interaction with an applied electromagnetic field is very high. It is therefore possible to achieve the so-called *strong-coupling regime* in which the coherent evolution of a single atom coupled to a single photon stored in a high-quality cavity overwhelms the

---

[5]Of course, as explained in Sec. 5.4, a complete characterization of quantum noise for a three-qubit system would in principle require the determination of $N^4 - N^2 = 4032$ parameters, where $N = 8$ is the dimension of the Hilbert space.

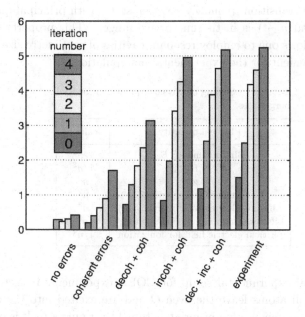

Fig. 8.6   The second moment $\langle(\Delta I)^2\rangle$ of the probability distribution in the ideal case, taking into account coherent, incoherent and decoherent noise models, and in the actual experiment. The figure is taken from Henry *et al.* (2005).

incoherent dissipative processes.[6] This allows for atom–photon entanglement to be produced before decoherence dominates. Moreover, the energy separation $E_n - E_{n-1}$ between two consecutive atomic levels is very low (corresponding to a frequency $\sim 10 - 50\,\mathrm{GHz}$, to be compared with optical frequencies $O(10^{15}\,\mathrm{Hz})$ relevant when $n \sim 1$). This entails two important consequences:

1. these (radio)frequencies are available in laboratories, so that resonant cavities can be excited and then used to manipulate the atoms;
2. the lifetime of Rydberg atoms is very long (much longer, as shown in Table 8.1 than for atoms at $n \sim 1$).

To give some relevant figures, let us note that the lifetime of a Rydberg atom with $n \sim 50$ and high angular momentum $l \sim n$ can be as large as

---

[6]The quality factor Q is a measure of the rate at which a vibrating system dissipates its energy (a higher Q indicates a lower rate of energy dissipation). By definition, Q is 2π times the ratio of the energy stored divided by the energy lost per cycle. For instance, a quality factor $Q \sim 3 \times 10^8$ is reported in Raimond *et al.* (2001).

30 ms and the transition frequency between states with principal quantum numbers n and $n - 1$ is in the microwave range.[7] This property is very useful as it allows one to employ resonant cavities of centimetre size, which are very convenient for the experimental manipulation.

Table 8.1   Scaling of physical quantities for Rydberg states.

Physical quantity	Scales as
Binding energy E_n	$1/n^2$
$E_n - E_{n-1}$	$1/n^3$
Size	n^2
Electric dipole moment	n^2
Lifetime (low angular momentum)	n^3
Lifetime (high angular momentum)	n^5
Critical electric field for ionization	$1/n^4$

The typical experimental setup for CQED experiments is sketched in Fig. 8.7. Alkali atoms leave the oven O and are excited into the desired Rydberg state by means of appropriately tuned laser pulses L. It is possible to select atoms having a well defined velocity using the Doppler effect. Even though the source emits atoms randomly, pulsed lasers allow one to select the incoming atoms and to know the preparation time for the circular Rydberg states within $O(\mu s)$ interval. The position of each atom flying inside the apparatus is then known with $O(mm)$ precision. It is therefore possible to address and control individual atoms. The prepared Rydberg atom crosses one or more cavities (usually, microwave superconducting cavities) R_1, C, and R_2, resonant with the transition between two atomic levels $|g\rangle$ and $|e\rangle$. The two cavities R_1 and R_2 implement microwave Rabi pulses and are used to prepare the initial state in the desired superposition of the states $|g\rangle$ and $|e\rangle$ and to analyze the final state, respectively. Note that the atom can be treated as a two-level system (qubit) since it is prepared in R_1 in a superposition $\alpha|g\rangle + \beta|e\rangle$ and the cavities are resonant with the $|g\rangle \leftrightarrow |e\rangle$ transition. The relevant Hilbert space for the atom is therefore spanned by the $\{|g\rangle, |e\rangle\}$ basis. The cavity C can be prepared in the vacuum state $|0\rangle$ with no photons (the mean photon number can be reduced to 0.1) and can evolve to the one-photon state $|1\rangle$ after interaction with the atom. Note that the photon storage time is $O(ms)$, much larger than the atom–cavity

---

[7]Note that the states with maximum angular momentum $l = n - 1$ (known as circular Rydberg states) are the quantum counterpart of classical circular orbits.

interaction time, a few tens of μs, thus allowing the coherent manipulation of entangled atom–photon states. The up/down ($|g\rangle/|e\rangle$) state of the atom is finally measured using the two detectors D_g and D_e: the atom is ionized by means of a static electric-field ($O(10^2)\,$V/cm) and the resulting electron is counted. This procedure is very effective as the static electric field threshold for ionization strongly depends on the principal quantum number n, as shown in Table 8.1. The detectors D_g and D_e are state selective: if the atom is in the state $|g\rangle$ it is ionized by the static field in D_g, if instead it is in $|e\rangle$ it is ionized by the field in D_e. As an example of this technique, the circular Rydberg states for rubidium atoms with $n = 49$, $n = 50$ and $n = 51$ can be distinguished.

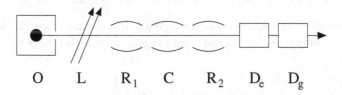

O   L R_1 C R_2 D_e D_g

Fig. 8.7   A sketch of a cavity quantum electrodynamics apparatus: the atoms leaving the oven O are excited into the desired Rydberg state by pulsed lasers L, enter the cavities R_1, C and R_2 and are finally detected using state selective field ionization in D_e and D_g.

We point out that the experimental apparatus sketched in Fig. 8.7 can be seen as the actual implementation of the theoretical procedure described in Sec. 6.1.2 for the measurement of the quantum operation acting on a qubit. The preparation of the initial density matrix ρ involves O, L and R_1 in Fig. 8.7, while the density matrix ρ' obtained after interaction with the cavity C is analyzed through R_2, D_e and D_g. It is therefore possible to measure the 12 parameters determining the mapping (quantum operation) of the single-qubit density matrix ρ into ρ'.

Note that the fields applied in R_1 and R_2 have relaxation times $O(\text{ns})$ and therefore do not produce any entanglement between the atom and the microwave radiation field. Indeed, the time required to induce $|g\rangle \leftrightarrow |e\rangle$ Rabi oscillations is of the order of $10\,\mu$s, much longer than the relaxation time. Hence, we describe the electromagnetic fields in R_1 and R_2 as classical fields. It can be shown (see exercise 8.5) that the action of such a field on a two-level atoms is described, in the $\{|g\rangle, |e\rangle\}$ basis, by the unitary matrix

$$
U = \begin{bmatrix} \cos\frac{\theta}{2} & -ie^{i\phi}\sin\frac{\theta}{2} \\ -ie^{-i\phi}\sin\frac{\theta}{2} & \cos\frac{\theta}{2} \end{bmatrix},
\tag{8.24}
$$

where θ is proportional to the amplitude of the radiation field and to the atom–field interaction time while ϕ is the phase of the field. Since $U = \cos\frac{\theta}{2}I - i\sin\frac{\theta}{2}(\boldsymbol{n}\cdot\boldsymbol{\sigma})$, where the unit vector $\boldsymbol{n} = (\cos\phi, -\sin\phi, 0)$ and $\boldsymbol{\sigma} = (\sigma_x, \sigma_y, \sigma_z)$, then U represents a rotation of the Bloch sphere through an angle θ about the axis directed along $\boldsymbol{n}$ (see Eq. (3.38)). This axis lies in the (x, y) plane of the Bloch sphere and forms an angle $-\phi$ with the x-axis. Starting from a given initial state, say $|\psi_0\rangle = |g\rangle$, we can obtain the generic state of a qubit, $|\psi\rangle = U|\psi_0\rangle = \cos\frac{\theta}{2}|g\rangle - ie^{-i\phi}\sin\frac{\theta}{2}|e\rangle$. Note that, when the atom interacts with a classical field, its state remains pure.

Exercise 8.5   The evolution of a single alkali atom in a classical electromagnetic field is governed, in the dipole approximation, by the Schrödinger equation

$$
i\hbar\frac{d}{dt}|\psi(t)\rangle = \big(H_0 + H_I\big)|\psi(t)\rangle,
$$

$$
H_0 = \frac{p^2}{2m} + V(r),
\tag{8.25}
$$

$$
H_I = -ezE(t),
$$

where the first term in H_0 is the kinetic energy of the valence electron of the atom and $V(r)$ the effective potential acting on such electron, generated by the atomic nucleus and the other electrons, and H_I is due to the interaction of the electron with the electric field generated by a wave linearly polarized along the z-axis. Solve the Schrödinger equation when only two atomic levels are relevant and the electric field is given by

$$
E(t) = E_0\cos(\omega t + \phi).
\tag{8.26}
$$

In particular, derive (8.24).

The Jaynes–Cummings model.   The electromagnetic field in the cavity C must be considered as a quantum object. The interaction between a two-level atom and a single mode of the quantized electromagnetic field is modelled by the *Jaynes–Cummings Hamiltonian*

$$
H = \tfrac{1}{2}\hbar\omega_a\sigma_z + \hbar\omega\big(a^\dagger a + \tfrac{1}{2}\big) + \lambda\sigma_+ a + \lambda^*\sigma_- a^\dagger,
\tag{8.27}
$$

where the Pauli matrices are written in $\{|e\rangle, |g\rangle\}$ basis spanning the Hilbert space associated with the two-level atom, $\sigma_+ = \frac{1}{2}(\sigma_x + i\sigma_y)$ and $\sigma_- = \frac{1}{2}(\sigma_x - i\sigma_y) = \sigma_+^\dagger$, $\hbar\omega_a = E_e - E_g$ is the difference between the energies of the atomic levels $|g\rangle$ and $|e\rangle$, $\hbar\omega$ the single-photon energy, $\hbar\omega\left(a^\dagger a + \frac{1}{2}\right)$ the Hamiltonian describing a single mode of the field, including the zero point energy $\frac{1}{2}\hbar\omega$, $a^\dagger$ and a the photon creation and annihilation operators. Note that the raising and lowering operators σ_+ and σ_- are such that $\sigma_+|g\rangle = |e\rangle$, $\sigma_+|e\rangle = 0$, $\sigma_-|g\rangle = 0$, $\sigma_-|e\rangle = |g\rangle$. On the other hand, the operators $a^\dagger$ and a create and annihilate a photon: $a^\dagger|n\rangle = \sqrt{n+1}|n+1\rangle$, $a|n\rangle = \sqrt{n}|n-1\rangle$ (see exercise 8.6), n being the number of photons in the cavity. Therefore, the operator $\sigma_- a^\dagger$ de-excites the atom ($|e\rangle \rightarrow |g\rangle$) and creates a photon by means of the operator $a^\dagger$ while $\sigma_+ a$ represents the excitation of the atom by absorption of a photon.[8] A very important point is that at resonance ($\omega = \omega_a$) there are coherent Rabi oscillations between the atom–cavity states $|g, n\rangle$ (*i.e.*, atomic state $|g\rangle$ and n photons in the cavity) and $|e, n-1\rangle$. The frequency of these oscillations is proportional to the coupling constant $|\lambda|$ and to $\sqrt{n}$ (see exercise 8.7).

Exercise 8.6   Let us consider the harmonic oscillator, whose Hamiltonian reads

$$H = \frac{p^2}{2m} + \frac{m\omega^2 x^2}{2}, \qquad (8.28)$$

where the operator $p = -i\hbar(d/dx)$ and $[x, p] = i\hbar$. The stationary states of the harmonic oscillator are the eigenfunctions $|n\rangle$ of the Hamiltonian operator (8.28); that is,

$$H|n\rangle = E_n|n\rangle. \qquad (8.29)$$

As shown in quantum mechanics textbooks, the eigenvalues read

$$E_n = \hbar\omega\left(n + \frac{1}{2}\right) \qquad (8.30)$$

and the corresponding stationary states are given by

$$\phi_n(x) \equiv \langle x|n\rangle = \left(\frac{m\omega}{\pi\hbar}\right)^{1/4} \frac{1}{2^{n/2}\sqrt{n!}} H_n\left(\sqrt{\frac{m\omega}{\hbar}}x\right) \exp\left(-\frac{1}{2}\frac{m\omega}{\hbar}x^2\right), \qquad (8.31)$$

---

[8]The Jaynes–Cummings model (8.27) is written in the rotating frame approximation since terms proportional to $\sigma_- a$ and $\sigma_+ a^\dagger$ are not included. See exercise 8.5 on the significance of the rotating wave approximation.

where H_n denotes the n-th Hermite polynomial.[9] Note that the energy of the ground state, $E_0 = \frac{1}{2}\hbar\omega$, is known as the zero-point energy and can be seen as a consequence of the Heisenberg principle. Indeed, it can be seen analogously to exercise 6.15 that the product of the uncertainties Δx and Δp is equal to $\frac{\hbar}{2}$, namely the minimum uncertainty permitted by the Heisenberg principle.

Hamiltonian (8.28) can also be written as

$$H = \hbar\omega \left(a^\dagger a + \tfrac{1}{2} \right),$$ (8.32)

where

$$a = \frac{1}{\sqrt{2m\hbar\omega}}(m\omega x + ip), \qquad a^\dagger = \frac{1}{\sqrt{2m\hbar\omega}}(m\omega x - ip).$$ (8.33)

Note that $[a, a^\dagger] = 1$. Show that the action of a and $a^\dagger$ on the stationary state $|n\rangle$ is as follows:

$$a|n\rangle = \sqrt{n}\,|n-1\rangle, \qquad a^\dagger|n\rangle = \sqrt{n+1}\,|n+1\rangle.$$ (8.34)

It follows that the *number operator* $N = a^\dagger a$ has the property $a^\dagger a|n\rangle = n|n\rangle$; that is, N has the same eigenstates as H.

Exercise 8.7 Solve the Schrödinger equation for the Jaynes–Cummings model.[10]

Exercise 8.8 Discuss the temporal evolution of the atom–field entanglement for the Jaynes–Cummings model at resonance ($\omega = \omega_a$), when the initial state is $|\psi_0\rangle = |g, n\rangle$.

Exercise 8.9 The *coherent states* of the harmonic oscillator are the eigenstates $|\alpha\rangle$ of the annihilation operator a; that is,

$$a|\alpha\rangle = \alpha|\alpha\rangle, \qquad \alpha \in \mathbb{C}.$$ (8.35)

---

[9]The Hermite polynomials satisfy the recurrence relation

$$H_{n+1}(\xi) = 2\xi H_n(\xi) - 2nH_{n-1}(\xi)$$

and the first few Hermite polynomials are

$$H_0(\xi) = 1, \quad H_1(\xi) = 2\xi, \quad H_2(\xi) = 4\xi^2 - 2, \quad H_3(\xi) = 8\xi^3 - 12\xi, \quad \cdots$$

Another useful relation is

$$\frac{d}{d\xi}H_n(\xi) = 2nH_{n-1}(\xi).$$

[10]Note that the Jaynes–Cummings model is one of the few exactly solvable models in quantum field theory.

(i) Show that the representation of a coherent state in the Fock basis (that is, in the basis of the eigenstates of the number operator) is given by

$$|\alpha\rangle = e^{-\frac{1}{2}|\alpha|^2} \sum_{n=0}^{\infty} \frac{\alpha^n}{\sqrt{n!}} |n\rangle. \qquad (8.36)$$

(ii) Show that the mean number of photons in the coherent state $|\alpha\rangle$ is given by $\bar{n} = \langle\alpha|N|\alpha\rangle = \langle\alpha|a^\dagger a|\alpha\rangle = |\alpha|^2$ while the root mean square deviation in the photon number $\Delta n = \sqrt{\bar{n}}$.

(iii) Discuss the temporal evolution of a coherent state. In particular, show that at all times it remains a minimum uncertainty wave packet and that the temporal evolution of the mean values of position and momentum are the same as for a classical particle.

Exercise 8.10 Discuss the temporal evolution of the state of a two-level atom interacting with a single mode of the electromagnetic field according to the resonant Jaynes–Cummings model, when the initial state is $|\psi_0\rangle = |g, \alpha\rangle$, with $|\alpha\rangle$ coherent state corresponding to a large average number of photons, $\bar{n} = |\alpha|^2 \gg 1$. In particular, study the temporal evolution of the Bloch-sphere coordinates and of the von Neumann entropy of the atomic state.

8.2.1 *Rabi oscillations*

As we know, Rabi oscillations consist in the variation of the population of levels (the eigenstates of the system Hamiltonian) induced by an external field, which can be either classical or quantized. The theory of Rabi oscillations is developed in exercises 8.5 and 8.7. Here we are interested in the case in which the electromagnetic field is quantized, so that quantum information can be transferred from the atom to the field and *vice versa*. This is possible in CQED experiments where a two-level atom interacts with a cavity field prepared with a given small number of photons. Such states $|n\rangle$ with fixed photon number are known as *Fock states* or *number states*. In particular, the ground state $|0\rangle$ of the quantum field is the so-called *vacuum state*, in which the photon number is equal to zero.

Let us describe the experiment reported in Varcoe *et al.* (2000). The setup is sketched in Fig. 8.8. A rubidium oven provides two collimated atomic beams. The first (main beam) is sent to the microwave cavity while the second (reference beam) is used as a frequency reference to tune the laser to an atomic resonance. Rubidium (^{85}Rb) atoms are excited from the

$^5S_{1/2}(F=3)$ ground state to the $^{63}P_{3/2}$ Rydberg state. Velocity selection is obtained using the Doppler effect: the laser is at an angle of 11 degrees with respect to the direction normal to the main atomic beam. The laser field is locked to the $^5S_{1/2}(F=3) - {}^{63}P_{3/2}$ transition of the reference beam, the transition frequency being tuned by means of a static electric field. This field changes the energy of the atomic levels (the Stark effect). When the laser frequency is tuned, different atomic velocities are selected in the main beam. It is important to select atoms of different velocities since in this case the atom–cavity interaction time can be varied. Thus, the angle θ of Rabi oscillations (see Eq. (8.24) and exercise 8.7) can be changed, even though the cavity has a well defined photon number.

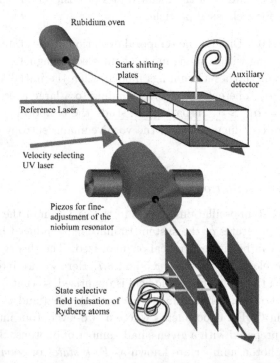

Fig. 8.8　A sketch of the experimental setup in Varcoe *et al.* (2000). The atoms leave the rubidium oven and are excited to a well defined Rydberg state by means of a laser. The atoms of the main beam interact with a superconducting niobium cavity (tuned using two piezo translators) and are finally detected by selective field ionization. The reference beam is used to stabilize the laser frequency to a Stark-shifted atomic resonance. The figure is taken from Varcoe *et al.* (2000).

To generate a state of the cavity with a determined number of photons, the atoms, prepared in the excited state $|e\rangle$ (as usual, $|g\rangle$ and $|e\rangle$ denote the two relevant atomic states), are injected one after the other into the cavity, prepared in the vacuum state $|0\rangle$. The outgoing atoms are then measured. If an atom is now observed in the ground state $|g\rangle$, we deduce that it has emitted a photon in the cavity. If n atoms are measured in the state $|g\rangle$, we conclude that the n-photon Fock state $|n\rangle$ of the cavity field has been generated. Of course, the entire process must require a time much shorter than the photon lifetime (0.35 s in Varcoe *et al.*, 2000). We point out that the above-described generation of a Fock state can be seen as an experimental verification of the collapse of the wave function after the measurement. Indeed, if an excited atom interacts with the cavity field, initially in the state $|n\rangle$, then in general the atom becomes entangled with the cavity field. As shown in exercise 8.8, the atom–field state after interaction is given by

$$-ie^{i\phi_{n+1}}\sin(|\Omega_{n+1}|t)|g, n+1\rangle + \cos(|\Omega_{n+1}|t)|e, n\rangle, \qquad (8.37)$$

where $|\Omega_{n+1}|$ is the frequency of Rabi oscillations, ϕ_{n+1} the field phase and t the interaction time. The state selective field ionization measures the atom in the ground state with probability $p_g = \sin^2(|\Omega_{n+1}|t)$ or in the excited state with probability $p_e = \cos^2(|\Omega_{n+1}|t)$. The postulate of the collapse of the wave function tells us that after the measurement the field is left in either the state $|n+1\rangle$ or $|n\rangle$. Note that in both cases the atom and the field are no longer entangled.

Finally, a new atom prepared in $|e\rangle$ can be sent to the cavity to probe the Fock state $|n\rangle$ by detecting Rabi oscillations, whose expected frequency $|\Omega_{n+1}| \propto \sqrt{n+1}$ (see exercise 8.7). The experimental results are shown in Figs. 8.9 and 8.10. In Fig. 8.9 (left) the atomic inversion $I = p_g - p_e$ is measured as a function of the interaction time t (which can be varied by the above described velocity selection technique). The oscillations in I show that photon emission is a *reversible process* in the strong coupling CQED regime. Ordinary photon emission occurs in free space and is irreversible since the emitted photon escapes and is lost. On the contrary, in CQED experiments the emitted photon remains trapped in the cavity and can be absorbed again by the atom. If the field is not in a Fock state, then

$$I(t) = C\sum_n p_n \left[\sin^2(|\Omega_{n+1}|t) - \cos^2(|\Omega_{n+1}|t)\right]$$

$$= -C\sum_n p_n \cos(2|\Omega_{n+1}|t), \qquad (8.38)$$

where p_n is the probability of finding n photons in the cavity and the factor C accounts for the signal reduction due to dark counts. A fit of the experimental data of Fig. 8.9 (left) according to Eq (8.38) gives the populations p_0, p_1 and p_2 shown in Fig. 8.9 (right). A clear maximum is shown, in correspondence to the expected Fock state. Finally, the dependence of the Rabi frequency on the photon number n is shown in Fig. 8.10, together with the theoretical dependence $T_n \equiv \frac{2\pi}{|\Omega_{n+1}|} \propto \frac{1}{\sqrt{n+1}}$.

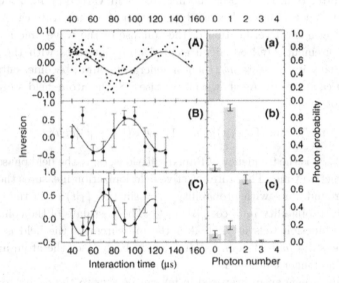

Fig. 8.9   Rabi oscillations for the Fock states (left) and coefficients p_n from the fit (8.38) (right). From top to bottom: $n = 0, 1$ and 2. The figure is taken from Varcoe *et al.* (2000).

The results of Varcoe *et al.* (2000) show that it is possible to prepare Fock states with good accuracy and to observe the interaction of single atoms with such states. Finally, we point out that these experimental results cannot be explained by assuming an interaction between the atoms and a classical field. This shows that it is possible to prepare quantum states of the electromagnetic field in macroscopic resonant cavities.

8.2.2   *Entanglement generation*

Let us describe a CQED experiment (Hagley *et al.*, 1997), in which two initially independent atoms are prepared in an entangled state. The experimental apparatus is sketched in Fig. 8.11. It fulfills very demanding

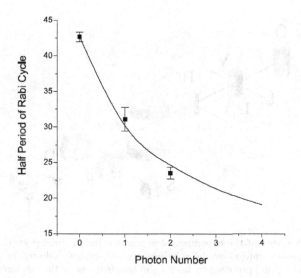

Fig. 8.10   Dependence of the Rabi frequency, obtained from the data of Fig. 8.9, on the photon number. The curve shows the theoretical dependence with the Rabi period $T_n \propto 1/\sqrt{n+1}$. The figure is taken from Varcoe *et al.* (2000).

requirements:

1. The position of each atom along its trajectory is known at any time during the experiment with an error less than 1 mm. This is obtained by exploiting the Doppler effect for velocity selection and using pulsed lasers to prepare the circular Rydberg states at a well defined time.
2. The angle θ of Rabi oscillations in (8.24) can be adjusted so that Rabi pulses with $\theta = \pi/2$ or π can be applied when the atom crosses the cavity. The atomic velocity can be selected so that $\theta = 2\pi$ for a full crossing of the cavity. By applying an electric field across the cavity at an appropriate time, the $|g\rangle \leftrightarrow |e\rangle$ transition is abruptly tuned off resonance, freezing the Rabi oscillations from that time on. Hence, angles $\theta < 2\pi$ and in particular $\pi/2$ and π pulses can be obtained.

Atom–atom entanglement is obtained by sending two atoms, one after the other, through the cavity. The two atoms and the cavity are initially prepared in the state

$$|\psi_i\rangle = |e_1, g_2, 0\rangle, \tag{8.39}$$

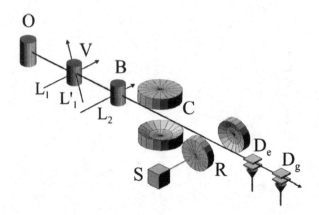

Fig. 8.11   A sketch of the experimental apparatus used in Hagley *et al.* (1997). Rubidium atoms are emitted by the rubidium oven O, velocity selected in zone V using lasers L_1 and L'_1 and prepared by laser L_2 in box B in one of the two circular Rydberg states with principal quantum numbers 50 (state $|g\rangle$) or 51 (state $|e\rangle$). After leaving the superconducting microwave cavity C, the atoms cross the analyzing cavity R, in which classical Rabi field pulses are applied by a source S. Finally, the state of the atom is measured by the detectors D_e and D_g. The figure is reprinted with permission from Hagley *et al.* (1997). Copyright (1997) by the American Physical Society.

where the index 1 refers to the first atom, the index 2 to the second and the third quantum number gives the number of photons in the cavity, initially in the vacuum state $|0\rangle$. The interaction of the first atom with the cavity corresponds to a $\theta = \pi/2$ Rabi pulse. Therefore, with probability $\frac{1}{2}$ the atom emits a photon and evolves into the state $|g\rangle$, whereas with probability $\frac{1}{2}$ it remains in $|e\rangle$. In the first case, the cavity is left in the state $|1\rangle$, in the latter it stays in $|0\rangle$. Therefore, the combined state of the two atoms and the cavity is given by

$$|\psi'\rangle = \tfrac{1}{\sqrt{2}}\big(|e_1, g_2, 0\rangle - |g_1, g_2, 1\rangle\big). \qquad (8.40)$$

Note that the first atom is now maximally entangled with the cavity field, while there is no entanglement with the second atom. The second atom then enters the cavity and the angle θ is set equal to π, corresponding to a complete population reversal $(|g_2, 1\rangle \rightarrow |e_2, 0\rangle)$. If instead the cavity is in the vacuum state, the second atom stays in $|g_2\rangle$ without affecting the cavity field. In both cases the cavity ends up in the vacuum state and the overall state is now given by

$$|\psi_f\rangle = \tfrac{1}{\sqrt{2}}\big(|e_1, g_2\rangle - |g_1, e_2\rangle\big)|0\rangle. \qquad (8.41)$$

Therefore, the two atoms are in a maximally entangled state, while the cavity state $|0\rangle$ is factorized; that is, no atom–cavity entanglement remains.

A $\pi/2$ pulse is then applied to both atoms and finally the two detectors D_e and D_g measure the state of the two atoms. Since the EPR state $\frac{1}{\sqrt{2}}(|e_1, g_2\rangle - |g_1, e_2\rangle)$ is rotationally invariant, it can be written in the same manner also in the basis rotated by the $\pi/2$ pulse. Therefore, the joint probability p_{ge} of finding the first state in $|g\rangle$ and the second in $|e\rangle$ is $\frac{1}{2}$. Similarly, $p_{eg} = 1/2$ and $p_{gg} = p_{ee} = 0$; that is, perfect anticorrelation is expected in the outcomes of the measurements. Experimental imperfections such as photon losses lead to the actual probabilities $p_{eg} = 0.44$, $p_{ge} = 0.27$, $p_{gg} = 0.23$, $p_{ee} = 0.06$.

Convincing evidence of atom–atom entanglement is obtained by detuning the frequency of the analyzing cavity R from the atomic resonance. Since the two atoms cross the cavity R at different times, they experience different phases, ϕ_1 and ϕ_2, of the microwave field. As a consequence (see exercise 8.11) the joint probabilities oscillate as a function of the phase difference $\phi_1 - \phi_2$. We obtain

$$p_{eg} = p_{ge} = \tfrac{1}{4}[1 + \cos(\phi_2 - \phi_1)], \quad p_{gg} = p_{ee} = \tfrac{1}{4}[1 - \cos(\phi_2 - \phi_1)]. \quad (8.42)$$

The phase difference $\phi_2 - \phi_1$ accumulated between the microwave source and the atom is given by the ΔT, where $\Delta = \omega - \omega_a$ is the detuning; that is, the difference between the field frequency ω and the Bohr frequency ω_a associated with the $|g\rangle \leftrightarrow |e\rangle$ transition, while $T = 42\,\mu s$ is the interval separating the times at which the two atoms reach the cavity R. The conditional probabilities $P_c(e_2/g_1)$ and $P_c(e_2/e_1)$ of detecting the second atom in $|e\rangle$, provided the first was measured in $|g\rangle$ or $|e\rangle$, are shown in Fig. 8.12. Since the joint probabilities $p_{ge} = p(g_1)P(e_2/g_1)$, $p_{ee} = p(e_1)P(e_2/e_1)$ and $p(e_1) = p(g_1) = \frac{1}{2}$, the theoretical expectation is

$$P_c(e_2/g_1) = \tfrac{1}{2}[1 + \cos(\phi_2 - \phi_1)], \quad P_c(e_2/e_1) = \tfrac{1}{2}[1 - \cos(\phi_2 - \phi_1)]. \quad (8.43)$$

Oscillations in phase opposition of the conditional probabilities $P_c(e_2/g_1)$ and $P_c(e_2/e_1)$ with period close to $\frac{1}{T}$ are indeed observed, even though the visibility of the interference fringes is only 25% instead of the ideal value of 100%. Such visibility is too low to observe a violation of Bell's inequalities.

Exercise 8.11 Derive Eq. (8.42).

Finally, we point out that entanglement has been established between two atoms separated by a *macroscopic* distance of the order of 1 cm.

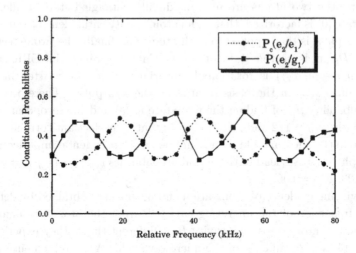

Fig. 8.12   Conditional probabilities for detecting the second atom in the state $|e\rangle$, provided the first was detected in $|e\rangle$ or $|g\rangle$. The figure is reprinted with permission from Hagley *et al.* (1997). Copyright (1997) by the American Physical Society.

8.2.3   The quantum phase gate

In this section, we discuss implementation of the controlled phase-shift gate CPHASE(ϕ) by means of the CQED experiment reported in Rauschenbeutel *et al.* (1999). This two-qubit gate, defined by Eq. (3.47), applies a phase shift of angle ϕ only when both qubits are in their $|1\rangle$ state. We have CPHASE(ϕ)$|x_1, x_0\rangle = \exp(i\phi x_1 x_0)|x_1, x_0\rangle$, with $x_1, x_0 = 0, 1$.

A sketch of the experimental apparatus used in Rauschenbeutel *et al.* (1999) is drawn in Fig. 8.13. Relevant system parameters are: (i) atomic lifetime ≈ 30 ms, (ii) lifetime of the field in the cavity ≈ 1 ms, (iii) atom–cavity interaction time $\approx 20\,\mu$s, (iv) atomic position known within ± 1 mm, (v) setup cooled to 1.3 K.

The two qubits are realized by a single atom (the atomic levels $|i\rangle$ and $|g\rangle$ stand for the $|0\rangle$ and $|1\rangle$ states of the qubit) and by the cavity field ($|0\rangle$ and $|1\rangle$ Fock states). If the atom is in the state $|i\rangle$ or if the cavity is in the vacuum state the atom–field state is unchanged. On the other hand, at resonance a 2π pulse transforms $|g, 1\rangle$ into $-|g, 1\rangle$. Note that the auxiliary level $|e\rangle$ is used to implement this transformation since the Rabi oscillation in the cavity is between the $|g, 1\rangle$ and $|e, 0\rangle$ states. Therefore, a CPHASE($\phi = \pi$) gate is applied. As shown in Rauschenbeutel *et al.* (1999), the controlled phase shift ϕ can varied in the interval $[0, 2\pi)$ by de-

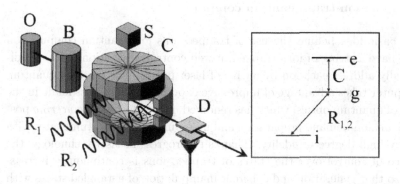

Fig. 8.13 A sketch of the experimental apparatus used in Rauschenbeutel *et al.* (1999) to implement the controlled phase-shift gate (left) and relevant energy levels (right). The atoms are emitted from the oven O and prepared in B in circular Rydberg states with principal quantum number 50 (state $|g\rangle$) or 51 (state $|e\rangle$). The classical fields R_1 and R_2 induce Rabi oscillations between the states $|g\rangle$ and $|i\rangle$ (this latter state has principal quantum number 49). The cavity C is resonant or nearly resonant with the transition $|g\rangle \leftrightarrow |e\rangle$ at 51.1 GHz, S is a classical source and the detector D discriminates between the states $|i\rangle$, $|g\rangle$ and $|e\rangle$. The figure is reprinted with permission from Rauschenbeutel *et al.* (1999). Copyright (1999) by the American Physical Society.

tuning the cavity mode from the $|g\rangle \leftrightarrow |e\rangle$ transition frequency. The action of the CPHASE(ϕ) gate has been demonstrated with the atom prepared in a superposition of the $|i\rangle$ and $|g\rangle$ states by means of the Rabi pulse R_1 and/or with the cavity prepared in a superposition of the $|0\rangle$ and $|1\rangle$ Fock states by injecting a small coherent field in C.

Note that the controlled phase-shift gate, combined with single-qubit gates, can be used to realize any unitary transformation in the Hilbert space of a many-qubit system. Therefore, a universal set of quantum gates can be realized with CQED experiments, even though the scaling of such experiments to systems with a large number of qubits is problematic. The most complex experiments performed so far have engineered the entanglement of three-qubit systems. In particular, a maximally entangled GHZ state of two atoms and the cavity (and of three atoms, used for the readout) has been prepared (see Raimond *et al.*, 2001). Finally, we point out that CQED experiments have been very successful in studying the emergence of classical behaviour due to decoherence effects. In particular, the loss of quantum coherence of a superposition of field states ("Schrödinger's cat") due to entanglement with the environment was experimentally measured (see Raimond *et al.*, 2001).

8.3   The ion-trap quantum computer

The basic idea behind the use of trapped ions for quantum computation is to have a string of ions trapped in well controlled positions and to individually address each ion by means of laser pulses. The ion-trap quantum computer takes advantage of impressive experimental progress made in the field of quantum optics, which has rendered *quantum state engineering* possible, *i.e.*, on-demand preparation and manipulation of quantum states with a very high degree of fidelity. Thanks to progress in laser technology, the degree of control over the states of trapped ions is continuously increasing, so that generation and coherent manipulation of entangled states with several qubits (up to eight) has been achieved. In this section, we shall first describe the main ingredients of ion-trap quantum computation, from the Paul trap mechanism to laser cooling. After this, we shall discuss the operations required to realize a universal set of one- and two-ion quantum gates. Finally, we shall review experimental results showing the potential of trapped ions in the field of quantum computation.

8.3.1   *The Paul trap*

In the *Paul trap*, ions are confined by a spatially varying time-dependent radiofrequency (RF) field. We are interested in the case in which the trapped ions line up along the trap axis (z). This is obtained by means of an oscillating field with a quadrupole geometry in two dimensions, providing confinement along the radial direction ($r = \sqrt{x^2 + y^2}$), while trapping along the z-axis is provided by a static electric field (see figure 8.14 and exercise 8.12).

Let us first consider a single ion in a trap. By averaging over the fast oscillatory motion at (radio)frequency ω_{RF}, an effective harmonic potential is obtained, with frequencies $\omega_x, \omega_y, \omega_z$ along the three principal axes of the trap. Note that the trapping frequency $\omega_t \equiv \omega_z \ll \omega_x, \omega_y$, so that we can limit our considerations to motion along the z-axis. Typical experimental parameters are trap size of approximately $1\,\mathrm{mm}$, applied voltages of $100 - 500\,\mathrm{V}$ and RF field of a few tens of MHz leading to harmonic motion of the trapped ion in the z direction with frequency $\frac{\omega_t}{2\pi} \sim 1 - 5\,\mathrm{MHz}$.

Exercise 8.12   An ion with charge q and mass M is confined in a linear

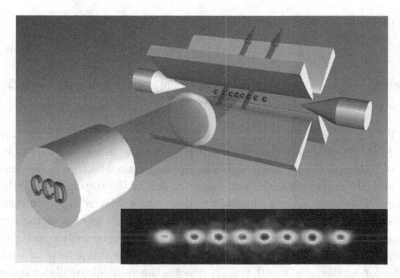

Fig. 8.14   Main figure: a schematic drawing of a linear ion trap setup with a trapped ion string. The four blades are at high voltage (neighbouring blades with opposite potential), oscillating at radio frequency, thus providing confinement in the radial directions. The tip electrodes are at positive high voltage and trap the ions axially (the z direction according to the notations used in the text). A laser addresses the ions individually and manipulates their quantum state. The resonance fluorescence of the ions is imaged onto a CCD (charge-coupled device) camera. Inset: the CCD image of a string of eight ions is shown. The distance between the outer ions is approximately $70\,\mu\mathrm{m}$. Drawing courtesy of Rainer Blatt, Innsbruck.

trap by the quadrupolar electric potential

$$\Phi(x, y, z; t) = \frac{1}{2} \frac{U_0}{R^2} (x^2 - y^2) \cos(\omega_{RF}\, t) + \frac{1}{2} \frac{V_0}{R^2} \left[z^2 - \epsilon x^2 - (1 - \epsilon) y^2 \right], \tag{8.44}$$

with R and ϵ geometric factors.

(i) Show that the equations of motion for the ion lead to harmonic confinement along z, while for both $\xi = x$ and $\xi = y$ we have a Mathieu differential equation, of the form

$$\frac{d^2\xi}{d\tau^2} + \left[a_\xi + 2q_\xi \cos(2\tau)\right] \xi = 0, \tag{8.45}$$

where $\tau = \omega_{RF} t / 2$. Find numerically the region of stability of this equation for parameter values a_ξ and q_ξ around zero.

(ii) Compare the exact numerical solution of the Mathieu equation with the approximate analytic solution

$$\xi = \xi_0 \cos(\beta_\xi \tau) \left[1 + \tfrac{1}{2} q_\xi \cos(2\tau)\right], \tag{8.46}$$

where $\beta_\xi = \sqrt{a_\xi + \tfrac{1}{2} q_\xi^2}$ and the initial conditions $\xi(t = 0) = \xi_0(1 + q_\xi/2)$, $\dot{\xi}(t = 0) = 0$ are assumed.

We are interested in both the vibrational motion of the ion in the trap and in the internal electronic motion. The electronic motion has frequencies $O(10^{15})$ Hz and the motion relative to the hyperfine structure frequencies in the GHz range, while the motion of the ion in the trap is in the MHz range. Therefore, we can employ the Born–Oppenheimer approximation and separate the fast electronic motion from the slow motion of the ion. States relevant to quantum information processing can be written as $|i\rangle|n\rangle$, where $|i\rangle$ refers to the electronic levels $|g\rangle$ and $|e\rangle$ (the computational basis states for a qubit) and $n = 0, 1, 2, \ldots$ denotes the harmonic oscillator states of the vibrational motion of the ion.

Let us now consider a string of N trapped ions (qubits). In this case, there are $3N$ normal modes of vibration ($2N$ radial and N axial modes). We are only interested here in the two lowest frequency axial modes, the centre-of-mass mode, where all ions oscillate together along z as a rigid body, and the stretch mode, where the oscillation amplitude of each ion is proportional to its distance from the centre of the trap. The frequencies of the centre-of-mass and stretch modes are $\omega_c = \omega_t$ and $\omega_s = \sqrt{3}\,\omega_c$, where ω_t is the frequency of the motion along z for a single ion. Note that the frequencies ω_c and ω_s are independent of the number N of ions in the trap (see exercise 8.13 for $N = 2$ and $N = 3$). The vibrational modes at higher frequencies are essentially "frozen" during quantum information processing experiments and therefore we ignore them.

Exercise 8.13   The Hamiltonian governing the motion of N ions in a harmonic linear trap is

$$H = \sum_{i=1}^{N} \frac{p_i^2}{2M} + \sum_{i=1}^{N} \frac{1}{2} M \omega_z^2 z_i^2 + \sum_{i=1}^{N-1} \sum_{j>i} \frac{q^2}{4\pi\epsilon_0 |z_j - z_i|}, \tag{8.47}$$

where q and M are the charge and mass of each ion, ϵ_0 is the electric permittivity of free space and the last term in (8.47) represents the Coulomb repulsion between the ions. Compute the equilibrium positions and the normal modes of vibration for $N = 2$ and $N = 3$.

8.3.2 Laser pulses

Resonant interaction with laser light is used in all stages of ion-trap quantum computations, from state preparation by means of laser cooling techniques to controlled qubit manipulation to state measurement by the quantum-jump technique. The Hilbert space for N ions in a trap is spanned by the states $|i_1, \ldots, i_N; n\rangle$, where $i_1, \ldots, i_N = g, e$ refer to the internal states of the ions, while n determines the collective vibrational motion of the ions. We assume that only one vibrational mode is relevant, say the centre-of-mass mode at frequency ω_t.[11] In $|n\rangle$, the string is in the n-th excited state for the (harmonic oscillator) motion at frequency ω_t and we say that n phonons are excited. Let us consider a laser beam addressing the ion j ($1 \le j \le N$), with the laser frequency ω tuned in such a manner that $\omega = \omega_a + (n' - n)\omega_t$, where $\hbar\omega_a = E_e - E_g$ is the energy difference between the ground state $|g\rangle$ and the excited state $|e\rangle$. A resonant transition between the states $|i_j = 0, n\rangle$ and $|i_j = 1, n'\rangle$ is induced (we do not write the states of the other ions in the trap since they are not modified by the laser). As shown in Fig. 8.15, it is possible to combine two laser pulses in order to change only the vibrational state of the string and not the internal state of the ions. It is evident that, with an appropriate combination of laser pulses, we can build the generic motional superposition state $\sum_n c_n |n\rangle$ starting from the ground state $|0\rangle$ (see exercise 8.14). We can also build a generic superposition $\alpha|g\rangle + \beta|e\rangle$ for the internal state of each ion. Indeed, a classical resonant field with $\omega = \omega_a$ induces Rabi oscillations given by Eq. (8.24). It is then clear that a generic single-ion (-qubit) state can be obtained from the ground state $|g\rangle$ by applying a resonant laser pulse of appropriate duration and phase.

The following three resonant interactions are of special importance for ion-trap quantum computation.

1. *Carrier resonance:* we have $\omega = \omega_a$ and, keeping only the resonant terms, the Hamiltonian describing the trapped ion-laser interaction is given by

$$H_c = \tfrac{1}{2}\hbar\Omega\big(\sigma_+ e^{-i\phi} + \sigma_- e^{i\phi}\big), \qquad (8.48)$$

where Ω is the Rabi frequency measuring the strength of the ion-laser coupling, ϕ is the phase of the laser and the operators $\sigma_+ = |e\rangle\langle g|$, $\sigma_- = |g\rangle\langle e|$. This Hamiltonian gives rise to transitions of the type

---

[11] In some experiments the stretch mode at frequency $\sqrt{3}\omega_t$ is also used.

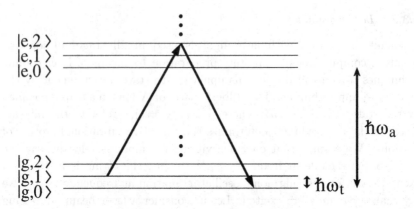

Fig. 8.15   Energy levels of a trapped ion. The global effect of the two transitions shown in the figure ($|g,1\rangle \to |e,2\rangle$ and $|e,2\rangle \to |g,0\rangle$) is to induce a transition between the quantized levels of the harmonic trapping potential, leaving unchanged the electronic state of the ion.

$|g,n\rangle \leftrightarrow |e,n\rangle$. Indeed, the temporal evolution governed by Hamiltonian (8.48) in a time interval t leads to the unitary evolution operator

$$R_c(\theta,\phi) = e^{-\frac{i}{\hbar}H_c t} = \begin{bmatrix} \cos\frac{\theta}{2} & -ie^{i\phi}\sin\frac{\theta}{2} \\ -ie^{-i\phi}\sin\frac{\theta}{2} & \cos\frac{\theta}{2} \end{bmatrix}, \qquad (8.49)$$

where $\theta \equiv \Omega t$ and the matrix is written in the $\{|e,n\rangle, |g,n\rangle\}$ basis. In particular, the transition $|g,n\rangle \leftrightarrow |e,n\rangle$ is obtained when $\theta = \pi$, for any ϕ (up to a phase factor determined by ϕ). More generally, Eq. (8.49) describes Rabi oscillations between the states $|g,n\rangle$ and $|e,n\rangle$.

2. *First red sideband:* in this case $\omega = \omega_a - \omega_t$ (red detuned laser) and the trapped ion-laser resonant interaction Hamiltonian is

$$H_- = \tfrac{1}{2}\hbar\Omega\eta\big(a\sigma_+ e^{-i\phi} + a^\dagger\sigma_- e^{i\phi}\big), \qquad (8.50)$$

where a and $a^\dagger$ are lowering and raising operators for the harmonic trapping potential and $\eta = 2\pi z_0/\lambda$ is the *Lamb–Dicke parameter*, with $z_0 = \langle 0|z^2|0\rangle^{1/2}$ spatial extension of the motional ground state and λ laser wavelength. Note that $z_0 = \sqrt{\hbar/(2NM\omega_t)}$, where M is the ion mass and N the number of ions in the string. The width of the ground state oscillations scales $\propto 1/\sqrt{NM}$ since the effective mass of the collective centre-of-mass motion is NM. Hamiltonian (8.50) generates the unitary evolution

$$R_-(\theta, \phi) = e^{-\frac{i}{\hbar}H_- t} = \begin{bmatrix} \cos\frac{\theta}{2} & -ie^{i\phi}\sin\frac{\theta}{2} \\ -ie^{-i\phi}\sin\frac{\theta}{2} & \cos\frac{\theta}{2} \end{bmatrix}, \qquad (8.51)$$

where $\theta = \eta\Omega\sqrt{n}t$ and the matrix is written in the $\{|g, n\rangle, |e, n-1\rangle\}$ basis. Therefore, Hamiltonian (8.50) gives rise to $|g, n\rangle \leftrightarrow |e, n-1\rangle$ transitions with Rabi frequency $\eta\Omega\sqrt{n}$. Note that (8.50) is formally equivalent to the resonant Jaynes–Cummings Hamiltonian. There is, however, a different physical interpretation: a phonon and not a photon is absorbed while the ion goes to the excited state. Moreover, the electromagnetic field is not quantized as in the Jaynes–Cummings model.

3. *First blue sideband:* we have $\omega = \omega_a + \omega_t$ (blue detuned laser) and resonant interaction Hamiltonian

$$H_+ = \tfrac{1}{2}\hbar\Omega\eta\left(a^\dagger\sigma_+ e^{-i\phi} + a\sigma_- e^{i\phi}\right). \qquad (8.52)$$

The unitary evolution $R_+(\theta, \phi) = e^{-\frac{i}{\hbar}H_+ t}$, with $\theta = \eta\Omega\sqrt{n+1}t$ has the same matrix representation as R_c and R_- but with respect to the $\{|g, n\rangle, |e, n+1\rangle\}$ basis. Therefore, Hamiltonian (8.52) induces $|g, n\rangle \leftrightarrow |e, n+1\rangle$ oscillations with frequency $\eta\Omega\sqrt{n+1}$. Such oscillations have no direct analogue in the CQED realm since a process in which the atom transits to an excited state while at the same time a photon is emitted would violate energy conservation. Hamiltonian (8.52) is known as the resonant *anti*-Jaynes–Cummings Hamiltonian.

Exercise 8.14 Give a quantum protocol to build a generic motional superposition state $\sum_{n=0}^{N} c_n |g, n\rangle$ starting from the ground state $|g, 0\rangle$.

The derivation of Hamiltonians (8.48), (8.50) and (8.52) can be found, for instance, in Leibfried *et al.* (2003a), see also exercise 8.15. Three important conditions must be fulfilled: (i) the Lamb–Dicke parameter $\eta \ll 1$ (values of $\eta \sim 0.2$ are typical in experiments); (ii) the laser must be on resonance to avoid undesired excitations of phonons; more precisely, we need $|\omega - \omega_a| \ll \omega_t$ for the carrier transition, $|\omega - (\omega_a - \omega_t)| \ll \omega_t$ for the first red sideband transition and $|\omega - (\omega_a + \omega_t)| \ll \omega_t$ for the first blue sideband transition; (iii) the pulses must be longer than $1/\omega_t$, so that their Fourier spectrum does not extend over the sidebands.

Exercise 8.15 The Hamiltonian describing the interaction of a trapped two-level ion with a laser field is

$$H_I = \hbar\Omega \left(\sigma_+ e^{i(kz-\omega t+\phi)} + \sigma_- e^{-i(kz-\omega t+\phi)} \right), \qquad (8.53)$$

where Ω is the Rabi frequency, ϕ the phase of the laser, $\sigma_+ = |e\rangle\langle g|$, $\sigma_- = \sigma_+^\dagger$, and the motion is restricted to one dimension, the position of the ion in the harmonic trap being $z = z_0(a^\dagger + a)$, with $z_0 = \sqrt{\hbar/(2M\omega_t)}$ (M is the mass of the ion and ω_t the angular frequency of the trap, typically of the order of $10\,\mathrm{MHz}$). The harmonic motion is quantized and described by the Hamiltonian $H_{\mathrm{osc}} = \hbar\omega_t \left(a^\dagger a + \frac{1}{2} \right)$. Study the effect of the interaction (8.53). In particular:
(i) find the Rabi frequency for the resonant transitions $|g, n\rangle \leftrightarrow |e, n'\rangle$, with $|n\rangle, |n'\rangle$ eigenstates of the Hamiltonian H_{osc};
(ii) derive Hamiltonians (8.48), (8.50) and (8.52) from (8.53) in the limit in which the Lamb–Dicke parameter $\eta = kz_0 \ll 1$.

Laser cooling. Laser cooling relies on the mechanical effect of light in a photon–ion scattering process, that is , on the fact that photons carry not only energy, but also momentum $p = h/\lambda$, where h is the Planck constant and λ the wavelength of the light. If an ion is moving along the light beam, it sees a Doppler-shifted light frequency, the frequency being higher if the ion moves towards the laser beam and lower if the atom moves away from the beam. The physical principle of *Doppler cooling* is to compensate the Doppler shift for ions approaching the laser beam by means of a red detuned laser. Then these ions are slowed down owing to the photons kicking them. Typically, Doppler cooling allows cooling down to an average motional quantum state $\langle n \rangle \sim 10$ for trap frequencies in the MHz range. The ultimate limit for Doppler cooling is due to the fact that the ions are excited to a strong (usually dipole) transition with natural linewidth (spontaneous emission rate) $\Gamma > \omega_t$.

The motional ground state $|n = 0\rangle$ can then be prepared by *sideband cooling*; that is, by exciting the ions to a narrow transition ($\Gamma \ll \omega_t$) (a forbidden optical line or a Raman transition). The laser is tuned into the $|g, n\rangle$ to $|e, n - 1\rangle$ first red sideband transition. Subsequent spontaneous emission occurs predominantly at the carrier frequency if the recoil energy of the atom is negligible compared with the vibrational quantum energy (this is the case if the Lamb–Dicke parameter $\eta \ll 1$). In this case, spontaneous emission induces the transition $|e, n - 1\rangle \to |g, n - 1\rangle$. The red detuned laser then leads to $|g, n - 1\rangle \to |e, n - 2\rangle$, and so on. At the end of the process, the state $|g, 0\rangle$ is reached with a high probability (preparation of the motional ground state for the centre-of-mass mode has been achieved with ground state occupation $> 99.9\%$).

Quantum gates. Single-qubit gates are obtained by tuning the laser to the carrier resonance. Indeed, it is clear from Eq. (8.49) that, starting from the ground state $|g\rangle$, a generic single-qubit state is obtained by means of a laser pulse of appropriate duration and phase.

The CNOT gate can be obtained following the proposal of Cirac and Zoller (1995). The basic idea is to employ the motional state of the string of ions as a "bus" to transfer quantum information between two qubits (ions). Therefore, the qubit–qubit interaction, which is necessary to implement controlled two-qubit operations, is mediated by the collective vibrational motion of the trapped ions.

Let us describe the Cirac–Zoller CNOT quantum gate between ions l (control qubit) and m (target qubit). We start from the initial state $|i_1, \ldots, i_l, \ldots, i_m, \ldots, i_N; n = 0\rangle$, which we simply write as $|i_l, i_m; 0\rangle$ since the other qubits are not affected by the quantum protocol described in what follows. The use of an auxiliary level ($|a\rangle$) helps in performing the CNOT gate.[12] The following sequence of laser pulses is applied:

1. A red detuned laser acts on ion l. The unitary evolution $R_-(\theta = \pi, \phi = 0)$ changes the states $\{|g_l, g_m\rangle, |g_l, e_m\rangle, |e_l, g_m\rangle, |e_l, e_m\rangle\}$ of the two-qubit computational basis as follows:

$$\begin{cases} |g_l, g_m; 0\rangle \rightarrow |g_l, g_m; 0\rangle, \\ |g_l, e_m; 0\rangle \rightarrow |g_l, e_m; 0\rangle, \\ |e_l, g_m; 0\rangle \rightarrow -i|g_l, g_m; 1\rangle, \\ |e_l, e_m; 0\rangle \rightarrow -i|g_l, e_m; 1\rangle. \end{cases} \qquad (8.54)$$

As a result of this laser pulse, the quantum information of the control ion is mapped onto the vibrational mode.

2. A red detuned laser is applied to ion m. The corresponding unitary evolution, written in the $\{|g_m, n = 1\rangle, |a_m, n = 0\rangle\}$ basis, is $R_-(\theta = 2\pi, \phi = 0)$, note that the auxiliary level $|a_m\rangle$ is involved. Since $\theta = 2\pi$, the state $|g_m, 1\rangle$ is mapped into $-|g_m, 1\rangle$. Therefore, the states obtained

---

[12]The auxiliary level could be a third level (in addition to $|g\rangle$ and $|e\rangle$) in the hyperfine structure of the ground state.

at the end of (8.54) are modified as follows:

$$
\left\{
\begin{array}{l}
|g_l, g_m; 0\rangle \rightarrow |g_l, g_m; 0\rangle, \\
|g_l, e_m; 0\rangle \rightarrow |g_l, e_m; 0\rangle, \\
-i|e_l, g_m; 1\rangle \rightarrow i|e_l, g_m; 1\rangle, \\
-i|e_l, e_m; 1\rangle \rightarrow -i|e_l, e_m; 1\rangle.
\end{array}
\right.
\tag{8.55}
$$

3. A red detuned laser is applied to ion l, inducing again the unitary evolution $R_-(\theta = \pi, \phi = 0)$. This leads to

$$
\left\{
\begin{array}{l}
|g_l, g_m; 0\rangle \rightarrow |g_l, g_m; 0\rangle, \\
|g_l, e_m; 0\rangle \rightarrow |g_l, e_m; 0\rangle, \\
i|e_l, g_m; 1\rangle \rightarrow |g_l, g_m; 0\rangle, \\
-i|e_l, e_m; 1\rangle \rightarrow -|g_l, e_m; 0\rangle.
\end{array}
\right.
\tag{8.56}
$$

The effect of this pulse is to map the state of the vibrational mode back onto the control qubit.

The global effect of the three laser pulses is to induce a controlled phase-shift gate CMINUS = CPHASE(π). The CNOT gate is then obtained from CMINUS after application of single-qubit (Hadamard) gates (see exercise 3.10).

It should be remarked that, although the vibrational mode could be regarded as an additional qubit (spanned by the phonon states $|0\rangle$ and $|1\rangle$), in practice it is only used as a bus to transfer quantum information between ions. Indeed, the "vibrational qubit" cannot be measured independently, as is the case for the internal electronic states of the ions.

Quantum-jump detection. After a quantum computation, the state of each ion can be measured using *quantum-jump* detection: each ion is illuminated with laser light of polarization and frequency such that it absorbs and then re-emits photons only if it is in one particular qubit level (say, the state $|e\rangle$). In contrast, if it is in the other ($|g\rangle$) state, the laser frequency is out of resonance and does not induce any transition. Thus, the detection of scattered fluorescence photons indicates that the ion was in the state $|e\rangle$. This is a projective measurement and a state discrimination efficiency above 99% can be reached. Moreover, it is possible to measure several ions in a trap individually. In order to uncover the average populations of the states $|g\rangle$ and $|e\rangle$ for each ion, one has to repeat the quantum computation and the final measurement a sufficient number of times.

Optical and hyperfine qubits. In closing this section, we briefly discuss the choice of the states $|g\rangle$ and $|e\rangle$ in the experiments. One needs two "stable" levels; that is, two levels whose decay rates are much smaller than the Rabi frequencies associated with the laser-induced transition $|g\rangle \leftrightarrow |e\rangle$. Two different strategies have been followed: $|g\rangle$ and $|e\rangle$ are either the ground state and a metastable excited state connected by a forbidden optical transition (*optical qubit*, as in Schmidt-Kaler *et al.*, 2003 for ^{40}Ca$^+$) or two hyperfine sub-levels of the ground state (*hyperfine qubit*, as in Turchette *et al.*, 1998 for ^{9}Be$^+$). An optical transition is driven by a single laser, while for a hyperfine transition two lasers are used, far detuned from an intermediate level $|c\rangle$. Such a Raman configuration (see Fig. 8.16 and exercise 8.16) is used because the frequency ω_a for a hyperfine transition is $O(\text{GHz})$ and can be driven resonantly by a single electromagnetic wave with wavelength of order 10^{-1} m. It is clear that such a wavelength, much larger than the distance between two nearby ions in a trap (approximately $10\,\mu$m), would not allow single-ion addressing.

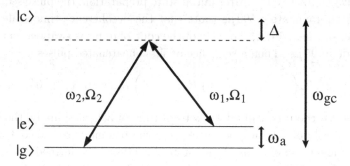

Fig. 8.16   A schematic diagram of a Raman transition. The two lasers have frequencies ω_1 and ω_2, while Ω_1 and Ω_2 are the Rabi frequencies for the transitions $|e\rangle \leftrightarrow |c\rangle$ and $|g\rangle \leftrightarrow |c\rangle$, respectively.

Exercise 8.16   Show that in the Raman configuration drawn in Fig. 8.16, the transition $|g\rangle \leftrightarrow |e\rangle$ takes place with Rabi frequency

$$\Omega_R \approx \frac{2\Omega_1\Omega_2}{\Delta}, \tag{8.57}$$

where Ω_1 and Ω_2 are the Rabi frequencies of the two applied laser fields and $\Delta \gg \Omega_1, \Omega_2$ is the detuning with respect to the transitions $|g\rangle \leftrightarrow |c\rangle$ and $|e\rangle \leftrightarrow |c\rangle$ (note that we have set $\hbar = 1$). We assume that the initial wave function $|\psi(0)\rangle = |g\rangle$. Besides Eq. (8.57), the Raman approximation also predicts that level $|c\rangle$ remains essentially unpopulated. Check the validity of the Raman approximation by direct numerical integration of the equations of motion for the overall three-level ($|g\rangle$, $|e\rangle$ and $|c\rangle$) system.

8.3.3   *Realization of the Cirac–Zoller CNOT gate*

In this section, we describe the implementation of the Cirac–Zoller CNOT quantum gate, reported in Schmidt-Kaler *et al.* (2003).[13] Two ^{40}Ca ions are held in a linear Paul trap. The state of each ion encodes a qubit, $|g\rangle$ and $|e\rangle$ corresponding to the $S_{1/2}$ ground state and to the metastable $D_{5/2}$ state (with lifetime approximately 1 s). The qubits are manipulated on the $S_{1/2}$ to $D_{5/2}$ quadrupole transition near 729 nm, by means of a laser tightly focused onto individual qubits. The inter-ion distance is 5.3 μm and the laser beam has a width around 2.5 μm.[14] The addressing beam can be switched from one ion to the other within 15 μs. Doppler cooling (for 2 ms) and sideband cooling (for 8 ms) prepare the vibrational mode in the $|n = 0\rangle$ state with 99.9% fidelity. After initial state preparation, the pulse sequence requires approximately 500 μs (note that the decoherence time scale is of the order of 1 ms). First, the state of the control qubit is swapped onto the bus mode in 95 μs. Then a sequence of six concatenated pulses,

$$R_+\!\left(\frac{\pi}{2},\pi\right) R_+\!\left(\frac{\pi}{\sqrt{2}},\frac{\pi}{2}\right) R_+(\pi,0)\, R_+\!\left(\frac{\pi}{\sqrt{2}},\frac{\pi}{2}\right) R_+(\pi,0)\, R_+\!\left(\frac{\pi}{2},0\right), \qquad (8.58)$$

is applied to the target qubit for a total time of 380 μs. Finally, the state of the control qubit is swapped back from the bus mode to the control ion with a single pulse of 95 μs. Final state detection is performed taking advantage of the quantum-jump technique: the $S_{1/2}$ to $P_{1/2}$ dipole transition near 397 nm is excited and the resulting fluorescence monitored by a CCD camera which resolves the individual ions. Fluorescence is collected for approximately 20 ms and state detection of each qubit is performed with approximately 98% efficiency (errors result from spurious fluorescence from the adjacent ion or from spontaneous decay of the ion within the detection time).

The experimentally observed truth table for the CNOT gate is shown in Fig. 8.17. The fidelity of the gate in this experiment is 71%. Sources of er-

---

[13] A different implementation of a two-ion gate has been realized in Leibfried *et al.* (2003b).

[14] The finite beam width introduces small addressing errors; that is, the neighbouring ion can also be excited. Such systematic errors can be compensated by adjustments to pulse lengths and phases.

rors in ion-trap quantum computation are the heating, due to stochastically fluctuating electric fields, and laser frequency noise.

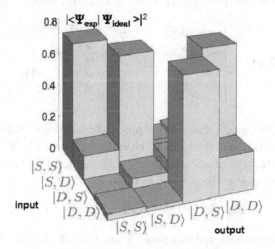

Fig. 8.17 The experimentally observed truth table of the Cirac–Zoller CNOT gate operation, using two ^{40}Ca$^+$ ions held in a linear trap and individually addressed using laser beams. The $S_{1/2}$ and $D_{5/2}$ states are denoted by $|S\rangle$ and $|D\rangle$ in the figure, with $|S\rangle \equiv |g\rangle \equiv |0\rangle$ and $|D\rangle \equiv |e\rangle \equiv |1\rangle$. We can see that the CNOT truth table is implemented: for instance, the input state $|DD\rangle = |11\rangle$ is, to a good approximation, mapped onto the output state $|DS\rangle = |10\rangle$. The figure is reprinted with permission from Schmidt-Kaler *et al.* (2003). Copyright (2003) by Macmillan Publishers Ltd.

The main difference between the original Cirac–Zoller proposal and the actual experimental implementation in Schmidt-Kaler *et al.* (2003) is that the technique of composite pulses, borrowed from NMR, is used instead of working with a third auxiliary level. Let us briefly illustrate the composite pulse method. The purpose of this technique is to avoid sideband pulses coupling to states outside the computational subspace $\{|g,0\rangle, |e,0\rangle, |g,1\rangle, |e,1\rangle\}$. This constitutes a problem due to the harmonic oscillator level structure: since the spacing between consecutive levels is always the same, resonant sideband pulses work simultaneously on all levels. Therefore, a blue sideband pulse induces a transition $|g,1\rangle \to |e,2\rangle$, so that population leaks from the computational subspace. Similarly, a red sideband pulse induces a $|e,1\rangle \to |g,2\rangle$ leakage transition. A key point in composite pulses is that the frequency of Rabi oscillations $|g,n\rangle \leftrightarrow |e,n-1\rangle$ and $|g,n-1\rangle \leftrightarrow |e,n\rangle$ is proportional to $\sqrt{n}$. Therefore, if we define $\theta = \eta\Omega t$ in Eq.(8.51); that is, $R_-(\theta,\phi)$ in this equation is written in the $\{|g,1\rangle, |e,0\rangle\}$

basis, and we choose $\theta = j\pi\sqrt{2}$, with j integer, then there is no probability transfer between the states $|e, 1\rangle$ and $|g, 2\rangle$. Indeed, the matrix representation of R_- in the subspace $\{|g, 1\rangle, |e, 0\rangle, |g, 2\rangle, |e, 1\rangle\}$ reads

$$
R_-(\theta = j\pi\sqrt{2}, \phi) = \begin{bmatrix} \cos\frac{j\pi}{\sqrt{2}} & -ie^{i\phi}\sin\frac{j\pi}{\sqrt{2}} & 0 & 0 \\ -ie^{-i\phi}\sin\frac{j\pi}{\sqrt{2}} & \cos\frac{j\pi}{\sqrt{2}} & 0 & 0 \\ 0 & 0 & (-1)^j & 0 \\ 0 & 0 & 0 & (-1)^j \end{bmatrix}.
$$

(8.59)

Hence, when $\theta = j\pi\sqrt{2}$ there is no probability leakage, induced by R_-, from the computational subspace. The same holds for $R_+(\theta, \phi)$ when the angle θ of the Rabi oscillation $|g, 0\rangle \leftrightarrow |e, 1\rangle$ is equal to $j\pi\sqrt{2}$. Next, it is easy to check that also the composite pulse

$$
R_\pm(-\theta', \phi') \, R_\pm(j\pi\sqrt{2}, \phi) \, R_\pm(\theta', \phi')
$$

(8.60)

preserves the computational subspace. Thus, the gate $R_\pm(j\pi\sqrt{2}, \phi)$ forbids probability leakage, while θ', ϕ and ϕ' can be tuned in order to implement the desired transformation.

As shown in exercise 8.17, it is possible to implement a CMINUS gate between the bus qubit and a ion qubit by means of the following composite pulse:

$$
R_+\left(\frac{\pi}{\sqrt{2}}, \frac{\pi}{2}\right) R_+(\pi, 0) \, R_+\left(\frac{\pi}{\sqrt{2}}, \frac{\pi}{2}\right) R_+(\pi, 0).
$$

(8.61)

This sequence of pulses substitutes the 2π-rotation of (8.55), with the advantage that there is no need of a third auxiliary level. Note that the six-pulses sequence (8.58) differs from (8.61) by the addition of two Ramsey pulses (essentially, Hadamard gates), required to map a CMINUS gate into a CNOT gate.

Exercise 8.17 Show that the composite pulse (8.61) implements a CMINUS gate between the bus qubits and the ion qubit, without probability leakage.

8.3.4   *Entanglement generation*

Trapped ions have been used to create and characterize multi-ion entangled states (so far, with up to eight qubits). A very important feature of this experiments is that entangled states are *engineered deterministically*. That is to say, the entanglement generation does not rely on random processes as in experiments with photons (for instance, the creation of entangled photons in parametric down-conversion, see Sec. 8.5.2). This is important because the production of entangled state *on demand* is crucial for the realization of large-scale quantum computers.

Two-ion entanglement. We first briefly describe the entanglement of two trapped ions reported in Turchette *et al.* (1998). The purpose of this experiment was to generate the state

$$|\psi_e(\phi)\rangle = \tfrac{3}{5}|ge\rangle - e^{i\phi}\tfrac{4}{5}|eg\rangle, \qquad (8.62)$$

where ϕ is a controllable phase factor. The state $|\psi_e(\phi)\rangle$ is a good approximation to the Bell state $|\psi^-\rangle = \frac{1}{\sqrt{2}}(|ge\rangle - |eg\rangle)$ for $\phi = 0$ and to the Bell state $|\psi^+\rangle = \frac{1}{\sqrt{2}}(|ge\rangle + |eg\rangle)$ for $\phi = \pi$. More precisely, the fidelity $|\langle\psi^-|\psi_e(0)\rangle|^2 = |\langle\psi^+|\psi_e(\pi)\rangle|^2 = 0.98$ and the entanglement (measured according to Eq. (5.231)) $E(|\psi_e(\phi)\rangle) \approx 0.94$. In the experiment, the fidelity obtained was $\langle\psi_e(0)|\rho^-|\psi_e(0)\rangle \approx \langle\psi_e(\pi)|\rho^+|\psi_e(\pi)\rangle \approx \langle\psi^\pm|\rho^\pm|\psi^\pm\rangle \approx 0.70$, with $\rho^\pm$ the density matrix describing the generated state.

The two-qubit states are two levels of the hyperfine structure of $^9\text{Be}^+$: $^2S_{1/2}|F=2, m_F=2\rangle \equiv |g\rangle$ and $^2S_{1/2}|F=1, m_F=1\rangle \equiv |e\rangle$. State readout is performed by observing the fluorescence of the $|g\rangle \rightarrow {}^2P_{3/2}|F=3, m_F=3\rangle$ transition driven by a resonant laser. Two $^9\text{Be}^+$ ions are confined in a Paul trap, with ion spacing $\approx 2\mu\text{m}$. Two types of transitions are driven: carrier and red sideband. As we know, sideband transitions involve the motional state of the ions. For $N = 2$ ions, the only two modes of the motion along z are the centre-of-mass mode, at frequency ω_z and the stretch mode, at frequency $\sqrt{3}\,\omega_z$ (see exercise 8.13). The stretch mode was used in this experiment since it was possible to cool it down to the ground state with higher probability (99%) than for the centre-of-mass mode. An interesting aspect of this experiment is that the distance between ions is too small to address them individually. Therefore, the following technique was pursued: different Rabi frequencies for the carrier transitions were obtained by applying a static electric field to push the ions along z. In this manner, the two ions couple differently to the laser beam and the two corresponding Rabi frequencies, Ω_1 and Ω_2, can be significantly different. In Turchette *et al.* (1998), $\Omega_1 = 2\Omega_2$ was chosen, so that, starting from the state $|g, g; 0\rangle$ ($|0\rangle$ is the ground state for the stretch mode) and driving the carrier transition for a time $t = \frac{\Omega_1}{\pi}$, the state $|g, e; 0\rangle$ is obtained (indeed, the first qubit is not flipped because $\Omega_1 t = \pi$, while the second is flipped because $\Omega_2 t = \frac{\pi}{2}$). The red sideband transition is then applied for an appropriate time, leading to the final state $|\psi_e(\phi); 0\rangle$, where the phase ϕ is due to the fact that the two ions see different laser phases.

Quantum-jump detection allows one to distinguish the state $|gg\rangle$ from the couple $\{|ge\rangle, |eg\rangle\}$ and the state $|ee\rangle$, just by looking at the intensity I_f of the fluorescence signal, which is proportional to the number of ions in the $|g\rangle$ state. Given a two-ion state $|\psi\rangle$, $I_f \propto 2p_{gg} + p_{ge} + p_{eg}$, where $p_{ij} \equiv \langle ij|\psi\rangle$ with $i, j = g, e$. In order to distinguish between the states $|\psi_e(0)\rangle$ and $|\psi_e(\pi)\rangle$, we exploit the fact that the singlet state $|\psi^-\rangle$ (which is very similar to $|\psi_e(0)\rangle$) is invariant under rotation (see Sec. 2.5), while this is not the case for the "triplet" state $|\psi^+\rangle$ (similarly to $|\psi_e(\pi)\rangle$). If both qubits are rotated, by means of a laser pulse, through the same angle θ about the x-axis of the Bloch sphere (see Eq. 8.24), it can be seen that we map the state $|\psi^+\rangle$ onto

$$\frac{1}{\sqrt{2}} \left[\cos\theta \big(|ge\rangle + |eg\rangle\big) - i\sin\theta \big(|gg\rangle + |ee\rangle\big) \right], \qquad (8.63)$$

while the state $|\psi^-\rangle$ is unchanged. Therefore, it is possible to distinguish between $|\psi_e(0)\rangle$ and $|\psi_e(\pi)\rangle$ (which are very similar to the Bell state $|\psi^-\rangle$ and $|\psi^+\rangle$) by looking at the evolution of the probabilities $p_{gg} + p_{ee}$ and $p_{ge} + p_{eg}$ as a function of $t = \theta/\Omega_r$, where Ω_r is the period of the Rabi pulse that implements the rotation of angle θ. The experimental results are shown in Fig. 8.18.

Multiparticle entanglement of trapped ions. Two different classes of many-ion entangled states have been prepared and characterized in ion-trap experiments: the Schrödinger cat states (also known as GHZ states)

$$|\psi_{\mathrm{cat}}(N)\rangle = \frac{1}{\sqrt{2}} \big(|g, g, \ldots, g\rangle + e^{i\theta}|e, e, \ldots, e\rangle\big) \qquad (8.64)$$

with up to $N = 6$ qubits (Leibfried *et al.*, 2005) and the so-called W states (see Dür *et al.*, 2000)

$$|\psi_W(N)\rangle = \frac{1}{\sqrt{N}} \big(|e, \ldots, e, g\rangle + |e, \ldots, e, g, e\rangle + \cdots + |g, e, \ldots, e\rangle\big) \qquad (8.65)$$

with up to $N = 8$ qubits (the first "quantum byte", see Häffner *et al.*, 2005). Cat states have been prepared with fidelities ranging from 0.76 (for $N = 4$) to 0.51 (for $N = 6$) and the presence of N-particle entanglement proved experimentally. On the other hand, W states have been generated with fidelities from 0.85 (for $N = 4$) to 0.72 (for $N = 8$) and fully characterized by means of quantum state tomography. Indeed, as shown in Fig. 8.19, the $N \times N$ density matrix is reconstructed. This result is achieved by

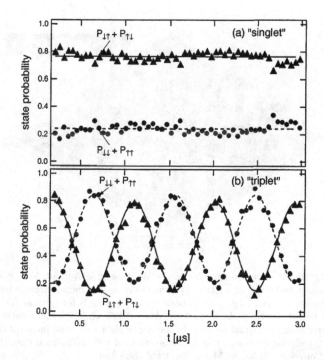

Fig. 8.18 Probabilities $p_{gg} + p_{ee}$ and $p_{ge} + p_{eg}$ as functions of time t for the state $|\psi_e(0)\rangle$ (above) and $|\psi_e(\pi)\rangle$ (below). The rotation angle $\theta = \Omega_r t$, with $\Omega_r/2\pi \approx 200\,\text{kHz}$. Note that in the figure $\downarrow$ stands for g and $\uparrow$ for e. The figure is reprinted with permission from Turchette *et al.* (1998). Copyright (1998) by the American Physical Society.

repeating the experiment (preparation of the W state) several times, each experimental run finishing with the measurement of σ_z for each qubit. The measurement basis is rotated prior to measurement by appropriate laser pulses: 3^N different bases are used and the experiment is repeated at least 100 times for each basis (for $N = 8$, this amounts to approximately 6.5×10^5 experimental runs, leading to the results shown in Fig. 8.19). Note that this procedure is a generalization to many qubits of the state reconstruction technique discussed in Sec. 5.5 for a single qubit.

Deterministic teleportation. Two experimental implementations of the quantum teleportation protocol using three trapped ions have been reported in Riebe *et al.* (2004) and Barrett *et al.* (2004). In both cases, the initial quantum state, $\alpha|g\rangle + \beta|e\rangle$, is prepared and then teleported from one ion to another, following the teleportation protocol described in Sec. 4.5. A

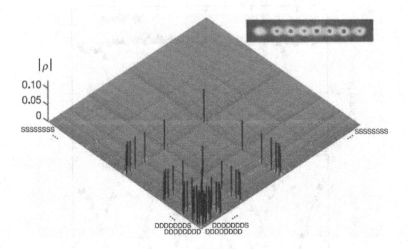

Fig. 8.19   Quantum state tomography of the W state $|\psi_W(N)\rangle$ for $N = 8$ ions in a trap. The absolute values $|\rho|$ of the experimentally reconstructed density matrix are shown in the $\{|g,\ldots,g\rangle,\ldots,|e,\ldots,e\rangle\}$ basis (here $|S\rangle$ stands for $|g\rangle$ and $|D\rangle$ for $|e\rangle$). Ideally, the height of the peaks should be $\frac{1}{N} = 0.125$ while all other entries of the density matrix should be equal to zero. In the upper right corner an image of the string of $N = 8$ trapped ions is shown. The figure is reprinted with permission from Häffner *et al.* (2005). Copyright (2005) by Macmillan Publishers Ltd.

fidelity of the teleported state around 0.75 is achieved, thus exceeding the maximum value of $2/3$ that could be reached without taking advantage of quantum entanglement. We stress that, as we shall see later in this chapter, in contrast to the teleportation experiments with entangled photons there is no post-selection of data after completion of the experiment.

It is interesting to note that quite different experimental techniques have been used in Riebe *et al.* (2004) and Barrett *et al.* (2004) to obtain similar experimental results. In the first case, optical qubits (calcium ions) are used, two-qubit gates closely follow the proposal of Cirac and Zoller and ions are addressed individually by tightly focused laser beams. In the latter case, hyperfine qubits (beryllium ions) are used, the two-qubit gates are performed following a geometric method and individual qubit addressing is possible thanks to a *segmented trap*. That is to say, the control electrodes are segmented, providing a total of six trapping zones, and the potentials applied to these electrodes can be varied in time, so shuttling ions between different zones. Thus, the ion on which we wish to shine a laser beam can be isolated, while still maintaining entanglement with the other ions. This result can be considered as a first step towards a scalable architecture of interconnected ion traps.

8.4 Solid state qubits

Qubits made out of solid-state devices may offer great advantages since fabrication by established lithographic methods allows for *scalability* (at least in principle). Moreover, another important feature of solid-state devices is their *flexibility* in design and manipulation schemes. Indeed, in contrast to "natural" atoms, "artificial" solid-state atoms can be lithographically designed to have specific characteristics such as a particular transition frequency. This tunability is an important advantage over natural atoms. Finally, solid-state qubits are easily embedded in electronic circuits and can take advantage of the rapid technological progress in solid-state devices as well as of continuous progress in the field of nanostructures. On the other hand, it should be remarked that there is a great variety of decoherence mechanisms, still not well understood, in solid-state devices.

Two main strategies have been followed for making solid-state qubits. In the first strategy, the qubits are single particles, such as nuclear spins in semiconductors or single electron spins in semiconductor quantum dots. In the second strategy, qubits are constructed from superconducting nanocircuits based on the Josephson effect.

8.4.1 *Spins in semiconductors*

A proposal by Kane (1998) is sketched in Fig. 8.20. The qubits are the $S = \frac{1}{2}$ nuclear spins of ^{31}P impurities in silicon. Gate operations are performed by means of magnetic fields (NMR techniques) and static electric fields. Each qubit is controlled through the hyperfine interaction between the nucleus of ^{31}P and the bound electron around it. Such an interaction is due to the coupling between the nuclear spin S_n and the electronic spin S_e and its strength is proportional to $|\psi(0)|^2$; that is, to the probability density of the electron wave function at the nucleus position. The hyperfine coupling can be controlled by an applied electric field (A-gates in Fig. 8.20) that shifts the electron wave function from the phosphorus nucleus, thus reducing the hyperfine interaction. The transition frequency of each qubit (^{31}P nucleus) is therefore determined by both the static magnetic field B applied to it and the hyperfine interaction. Thus, the A-gates can control the transition frequency of each single qubit and bring them into resonance with the oscillating magnetic field B_{AC}. In this manner, arbitrary single-qubit quantum gates can be realized with resonant pulses, such as in NMR implementations discussed in Sec. 8.1. Two-qubit quantum gates would be implemented using the J-gates of Fig. 8.20, which control the exchange interaction between two neighbouring bound electrons. Indeed, the exchange interaction depends on the overlap of the electron wave functions and can be controlled by the J-gates bringing the two electrons closer. Since the hyperfine interaction couple each qubit with its bound electron, the qubit–qubit interaction is mediated by the exchange interaction between the electrons. This proposal requires nanofabrication on the atomic scale, to place phosphorus impurities (and gates) in a silicon crystal in an ordered array with separation of order 10 nm. This is beyond the reach of current technology. Nevertheless, one should consider the fact that silicon technology is a very rapidly developing field.

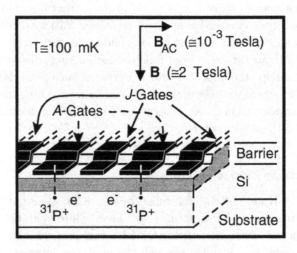

Fig. 8.20   A schematic drawing of Kane's proposal. The figure is reprinted with permission from Kane (1998). Copyright (1998) by Macmillan Publishers Ltd.

8.4.2   *Quantum dots*

Quantum dots are structures fabricated from semiconductor materials, in which electrostatic potentials confine electrons inside small "boxes". When the size of the box is comparable to the wavelength of the electrons by which it is occupied, then the system exhibits a sequence of discrete energy levels, quite as in atoms. For this reason quantum dots are also known as artificial atoms. Typical binding energies and size of the orbits are 1 meV and 50 nm, to be compared with 10 eV and 0.05 nm (Bohr radius) for natural atoms. Quantum dots are fabricated starting with a semiconductor heterostructure, a sandwich of different layers of semiconducting materials (such as GaAs and AlGaAs), which are grown on top of each other using molecular-beam epitaxy. By doping the AlGaAs layer with Si, free electrons are introduced, which accumulate at the interface between GaAs and AlGaAs, thus forming a two-dimensional gas of electrons that move along the interface. Metal gate electrodes applied on top of the heterostructure create an electric field that locally depletes the two-dimensional electron gas, creating one or more small islands (quantum dots) of confined electrons in an otherwise depleted region (see Fig. 8.21). Note that two sufficiently close quantum dots can be coupled through the overlapping of their electron wave functions, thus creating artificial molecules. At present, single and double quantum dots can be made, with a number of electrons controllable (by means of an applied voltage) down to just one electron.

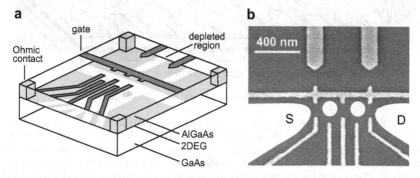

Fig. 8.21 A schematic drawing (a) and scanning electron micrograph (b) of a semiconductor heterostructure with two coupled quantum dots. In the left-hand figure, negative voltages applied to metal gate electrodes (dark gray) lead to depleted regions (white) in the two-dimensional electron gas (light gray). Electric contacts to reservoirs are obtained through ohmic contacts. In the right-hand figures, the gate electrodes (light gray) are shown on top of the surface of the heterostructure (dark gray). The source (S) and drain (D) reservoirs are connected to the two quantum dots (white circles) via tunnel barriers. The two upper electrodes can be used to measure changes in the number of electrons in the dots. The figure is taken from Elzerman *et al.* (2006). Copyright (2006) by the Italian Physical Society.

The qubit can be realized as the electronic spin of a single-electron quantum dot, the use of electron spins as qubits being attractive due to their long decoherence time. In the proposal of Loss and DiVincenzo (1998) (see Fig. 8.22) the dots that hold the electron spins (qubits) are placed in an array on top of a semiconductor heterostructure. A static magnetic field B induces an energy gap (Zeeman splitting) between the states $|0\rangle$ (spin up) and $|1\rangle$ (spin down) of each qubit. The Zeeman splitting is $\Delta E = g\mu_B B$, where $g \approx -0.44$ is the Landé g-factor of GaAs and $\mu_B \approx 9.27\times^{-24}$ Joule/Tesla is the Bohr magneton. The spin state of single qubits can then be controlled by applying an oscillating magnetic field B_{ac} in resonance with the Zeeman splitting (that is, with angular frequency $\Delta E/\hbar$). This technique is known as electron-spin resonance. A local difference in the Zeeman splittings could be obtained by means of gate potentials applied between the top and the bottom of the heterostructure. Each electron could then be shifted individually towards a layer of the heterostructure with a different g-factor. This would allow resonant addressing of individual qubits.

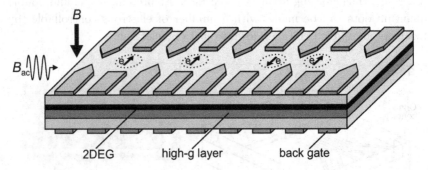

Fig. 8.22   A schematic picture of an array of quantum-dot spin qubits as proposed by Loss and DiVincenzo (1998). The quantum dots (circles) are created by metal electrodes on top of a semiconductor heterostructure containing a two-dimensional electron gas (2DEG). Each dot holds a single electron, whose spin state is pictorially represented by an arrow. The back gates can modify the Zeeman splitting by pulling the electron wave function into a layer with a large g-factor. The figure is taken from Elzerman *et al.* (2006). Copyright (2006) by the Italian Physical Society.

The interaction between two spins, S_i and S_j, can be modelled by the Heisenberg exchange interaction $J_{ij} S_i \cdot S_j$, where J_{ij} depends on the overlap of the electronic wavefunctions. The coupling J_{ij} is relevant only between nearest neighbour qubits, provided each electron is well localized in a single quantum dot. Applying a gate voltage at the surface, the potential barrier between adjacent dots can be increased, thus reducing drastically the Heisenberg coupling. It is therefore possible to switch on and off the coupling between qubits and this provides a clear mechanism for the implementation of two-qubit gates. In particular, it can be checked that, if the interaction between two neighbouring qubits is switched on for a specific duration t_s, then the SWAP gate is realized. If on the other hand the duration is $t_s/2$, then by definition the $\sqrt{\text{SWAP}}$ gate is implemented. The important point is that the $\sqrt{\text{SWAP}}$ and single-qubit gates constitute a universal set of quantum gates (indeed, as shown in exercise 8.20, the CMINUS gate can be obtained from $\sqrt{\text{SWAP}}$ and single-qubit gates).[15]

Readout is possible if the information contained in the spin is converted to information contained in the charge by a spin-dependent tunnelling process. First, the gate voltage is modified so that the electron stays in the dot if it has spin up, while it leaves the dot (tunnelling to a reservoir) if it has spin down. Detection of the charge of the dot is then possible using devices such as quantum point contacts (see Elzerman *et al.*, 2006). In this manner, the difficult problem of measuring the polarization of a single spin has been replaced by a much easier charge measurement.

Coherent control of two coupled electron spins in a double quantum dot was demonstrated by Petta *et al.* (2005). In this experiment, the qubit was encoded in the spin of two electron states with one electron charge in each dot. The two states of the qubit are therefore the single state ($S = 0$) and the triplet state $S = 1$ with $S_z = 0$. Coherent qubit manipulation (up to times larger than $1\,\mu$s, using echo techniques borrowed from NMR) was achieved by controlling the exchange interaction between the two dots.

Scalability is in principle possible since arrays of quantum dots can be produced with present technology. However, it should be taken into account that there are a great variety of possible decoherence processes in quantum dots and our knowledge of them is still very limited.

Exercise 8.18 The simplest example to study the bound states of a particle is the infinitely deep one-dimensional square-well potential

$$V(x) = \begin{cases} 0, & 0 < x < a, \\ +\infty, & x \leq 0, \ a \leq x. \end{cases} \tag{8.66}$$

Find the stationary states and the energy levels for this model.

Exercise 8.19 A more realistic case useful for the study of bound states

---

[15]Note that, by properly encoding each logical qubit into three spins instead of one, it is possible to perform universal quantum computation using only the Heisenberg exchange interaction (DiVincenzo *et al.*, 2000). This possibility may be useful as it avoids the implementation of single-spin rotations, which is difficult in quantum-dot arrays.

is the well of finite depth:

$$V(x) = \begin{cases} -V_0, & -a < x < a, \\ 0, & x \le -a, \ a \le x, \end{cases} \tag{8.67}$$

with $V_0 > 0$. Find the bound stationary states and energy levels for this model.

Exercise 8.20 Show that the CMINUS gate can be obtained from the $\sqrt{\text{SWAP}}$ and single-qubit gates as follows:

$$\text{CMINUS} = \left(I \otimes R_z(\pi)\right)\left(\sqrt{\text{SWAP}}\right)^{-1}\left(I \otimes R_z(\pi/2)\right)$$
$$\times \text{SWAP}\left(I \otimes R_z(-\pi/2)\right)\sqrt{\text{SWAP}}. \tag{8.68}$$

8.4.3 *Superconducting qubit circuits*

Superconductors have the ability to conduct electricity without loss of energy. In superconductors, pairs of electrons are bound together to form objects of twice the electron charge, known as Cooper pairs. A Josephson junction consists of two superconductors separated by a thin insulating barrier (see, *e.g.*, Tinkham, 1996). Cooper pairs can tunnel through the barrier, this being a dissipationless process. Note that quantum tunnelling allows transport through regions that are classically forbidden owing to potential barriers (see exercise 8.21).

Exercise 8.21 Study the transmission properties of a square barrier, described by the potential

$$V(x) = \begin{cases} V_0, & 0 < x < a, \\ 0, & x \le 0, \ a \le x, \end{cases} \tag{8.69}$$

with $V_0 > 0$. Consider the case where the energy $E < V_0$ (the *tunnel effect*).

Two energy scales determine the behaviour of a Josephson-junction circuit: the *Josephson energy* E_J and the electrostatic *charging energy* E_C for a single Cooper pair. The Josephson energy is related to the critical current I_J (the maximum current that can flow through the junction without dissipation) by the relation $E_J = I_J \hbar/2e$. Depending on the ratio E_J/E_C, one can distinguish between charge qubits ($E_J \ll E_J$, typically $E_J/E_C \sim 0.1$), charge-flux qubits ($E_J/E_C \sim 1$), flux qubits ($E_J/E_C \sim 10$) and phase qubits ($E_J \gg E_C$, typically $E_J/E_C \sim 10^6$). Single-qubit coherent control has been demonstrated in all these regimes. In the remaining part of this section, we shall limit ourselves to the discussion of two relevant examples, trying to give a flavour of the flexibility in the design and manipulation of superconducting qubits.

Charge qubits. Electrostatic potentials can confine Cooper pairs in a "box" of micron size. In a Josephson junction a *Cooper-pair box*, known as the island, is connected by a thin insulator (tunnel junction) to a superconducting reservoir (see Fig. 8.23). Cooper pairs can move from the island to the reservoir and *vice versa* by quantum tunnelling effects. They enter the island one-by-one when a control-gate electrode (voltage U), capacitively coupled to the island (capacitance C_g), is varied. The island has discrete quantum states and, as we shall see below, under appropriate experimental conditions the two lowest energy states form a two-level system appropriate for a qubit. The charging energy is $E_C = (2e)^2/2C$, where $C = C_J + C_g$ is the total capacitance of the island, C_J being the tunnel junction capacitance. If the capacitance C is in the range of a femtofarad or smaller, then $E_C/k_B \geq 1\,\mathrm{K}$. Typical values of E_J/k_B in the circuits considered here are instead $0.1\,\mathrm{K}$, so that $E_J/E_C \ll 1$.

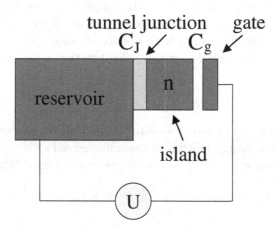

Fig. 8.23   A schematic drawing of a Josephson-junction qubit in its simplest design: a small superconducting island with n excess Cooper pairs (relative to some reference state) is connected by a tunnel junction with capacitance C_J and Josephson coupling energy E_J to a superconducting reservoir. The junction is biased by a gate voltage U with gate capacitance C_g.

The Cooper-pair box is described by the Hamiltonian

$$H = E_C(n - n_g)^2 - E_J \cos\phi, \qquad (8.70)$$

where n is the number of extra Cooper pairs in the island and ϕ the phase drop of the superconducting order parameter across the junction. The variables ϕ and n are conjugate; that is, $[\phi, n] = i$. The dimensionless gate charge $n_g = C_g U/2e$ can be controlled by tuning the gate voltage U. In the regime $E_J/E_C \ll 1$ a convenient basis is the basis of the eigenstates $|n\rangle$ of the number operator n. The Josephson term $E_J \cos\phi$ is not diagonal in this basis since

$$\cos\phi|n\rangle = \tfrac{1}{2}\left(e^{i\phi} + e^{-i\phi}\right)|n\rangle = \tfrac{1}{2}\big(|n+1\rangle + |n-1\rangle\big). \qquad (8.71)$$

Therefore, in the n-basis Hamiltonian (8.70) reads

$$H = E_C \sum_n (n - n_g)^2 |n\rangle\langle n| - \tfrac{1}{2}E_J \sum_n \big(|n+1\rangle\langle n| + |n\rangle\langle n+1|\big). \qquad (8.72)$$

For $E_J \ll E_C$ the charge states $|n\rangle$ are weakly mixed by the Josephson term, except near the "optimal" operating points with n_g half-integer, where the electrostatic charging energy of the states $\left|n_g - \tfrac{1}{2}\right\rangle$ and $\left|n_g + \tfrac{1}{2}\right\rangle$ is the same and the Josephson coupling mixes them strongly (see Fig. 8.24). Therefore, the dynamics at low temperatures ($k_B T \ll E_C$) is essentially limited to these two charge states. To simplify writing, we assume n_g around $\tfrac{1}{2}$, so that the two relevant charge states are $|0\rangle$ and $|1\rangle$. Projection of the Hamiltonian (8.72) onto the subspace spanned by these two states leads (neglecting an irrelevant energy offset) to the single-qubit Hamiltonian

$$H_Q = \tfrac{1}{2}\epsilon\sigma_z - \tfrac{1}{2}\Delta\sigma_x, \qquad (8.73)$$

where $\epsilon = E_C(2n_g - 1)$ and $\Delta = E_J$.[16] This Hamiltonian can be easily diagonalized. The energy splitting between its eigenvalues is $\Omega = \sqrt{\epsilon^2 + \Delta^2}$. The eigenvalues are $\lambda_\pm = \pm\tfrac{\Omega}{2}$ and the corresponding eigenstates read

$$\begin{aligned}
|+\rangle &= \cos\tfrac{\theta}{2}|0\rangle - \sin\tfrac{\theta}{2}|1\rangle, \\
|-\rangle &= \sin\tfrac{\theta}{2}|0\rangle + \cos\tfrac{\theta}{2}|1\rangle,
\end{aligned} \qquad (8.74)$$

where we have introduced the mixing angle θ, defined by $\tan\theta = \Delta/\epsilon$. At the degeneracy point, $\theta = \tfrac{\pi}{2}$, the eigenstates are equal superpositions of

---

[16]Note that Hamiltonian (8.73) is generic for a real Hamiltonian in the vicinity of an avoided crossing, see also the discussion on avoided crossings in Sec. 6.5.6.

the states $|0\rangle$ and $|1\rangle$ and the energy splitting $\Omega = \Delta = E_J$. Far from the degeneracy point the eigenstates $|\pm\rangle$ reduce to $|0\rangle$ and $|1\rangle$, as the charging energy is the dominant term in the Hamiltonian H_Q. Typical frequencies are $\Omega/2\pi \sim 10\,\mathrm{GHz}$.

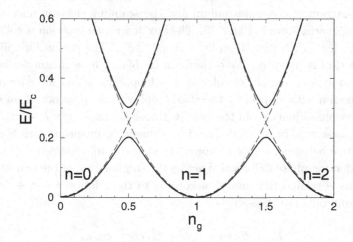

Fig. 8.24   The lowest energy levels of a Cooper-pair box, for $E_J/E_C = 0.1$ (solid curves). The dashed curves show the energy levels for $E_J = 0$.

It is now clear that generic single-qubit operations can be implemented by properly switching the gate voltage (thus tuning the Hamiltonian H_Q) for a given time (see exercise 8.22). For instance, one can start far from the degeneracy point, move the system quickly to the degeneracy point (by means of a change in the gate voltage) for a time T and then back to the initial value of the gate voltage. This pulse implements a rotation about the x-axis of the Bloch sphere and was realized by Nakamura *et al.* (1999). Protocols alternating pulses (rotations about the x-axis) and evolutions far from the operating point (that is, rotations about the z-axis) were also implemented. This is sufficient to obtain any unitary single-qubit transformation.[17]   Finally, we point out that conditional gate operation (CNOT gate) has been achieved (Yamamoto *et al.*, 2003) using a pair of superconducting charge qubits connected by a capacitor.

Exercise 8.22   For a two-level system, study the effect of a pulse of

---

[17]According to the Landau–Zehner effect, the level crossings shown in Fig. 8.24 produce tunable rotations of the single-qubit wave function. This could be used for the implementation of general single-qubit gates, see Benza and Strini (2003).

duration T, described by the Hamiltonian (8.73).

The "quantronium" circuit. The superconducting qubit demonstrated by the Saclay group (see Vion *et al.*, 2002 and Esteve and Vion, 2005) is an improved Cooper-pair box circuit, dubbed a "quantronium". Both the idealized diagram of the quantronium and the scanning electron micrograph of a sample are shown in Fig. 8.25. The box Josephson junction is split into two junctions, each with Josephson energy $\frac{1}{2}E_J$. The reason for splitting the junction is to form a loop that can be biased by a magnetic flux Φ, produced by a current I_Φ circulating in a loop. The main difference of the quantronium with respect to the charge qubit is the presence of an extra large Josephson junction in the loop (with Josephson energy $E_{J0} \approx 15E_J$), whose phase γ is in principle an extra dynamical quantum variable, but in practice behaves as an additional classical control parameter. The superconducting phase difference δ across the combination of the two smaller junctions is related to γ and Φ according to the relation $\delta = \gamma + 2e\Phi/\hbar$. The Hamiltonian of the split box reads

$$H = E_C(n - n_g)^2 - E_J \cos\frac{\delta}{2} \cos\phi, \qquad (8.75)$$

where ϕ is the superconducting phase of the island. Note that, unlike the Cooper-pair box Hamiltonian (8.70), the effective Josephson energy

$$E_J^\star = E_J \cos\frac{\delta}{2} \qquad (8.76)$$

can be tuned by changing the magnetic flux Φ via the current I_Φ. The circuit is operated at $E_J \sim E_C$, with $E_J/k_B \sim E_C/k_B \sim 1\,\mathrm{K}$, so that the charge states, unlike the case of charge qubits, are not good approximations to the eigenstates of Hamiltonian (8.75). As experiments are performed at low temperatures ($20\ \mathrm{mK} \ll E_J/k_B, E_C/k_B$) and the spectrum is sufficiently anharmonic, the dynamics is restricted to at most the two lowest energies eigenstates.[18] These are the states $|0\rangle$ and $|1\rangle$ of the superconducting qubit. The corresponding eigenvalues, E_0 and $E_1 = E_0 + \hbar\Omega$, depend on the two control parameters, n_g and δ. At the optimal working point ($n_g = \frac{1}{2}$, $\delta = 0$), both $\partial\Omega/\partial n_g$ and $\partial\Omega/\partial\delta$ vanish, so that to first order the quantronium is insensitive to noise in the control parameters n_g and δ. At the optimal working point, $\Omega/2\pi \approx 16\,\mathrm{GHz}$.

---

[18]The anharmonicity is an important requirement to avoid the population, under resonant pulses, of states outside the two-dimensional subspace used for the qubit; see a discussion of this problem in Sec. 8.3.3.

The two lowest energy states; that is, the states $|0\rangle$ and $|1\rangle$ of the qubit, are discriminated by applying a trapezoidal current pulse $I_b(t)$ to the large (readout) junction and by monitoring the voltage $V(t)$ across it. The peak value of $I_b(t)$ is slightly below the critical current of the readout junction. As this bias current adds to the loop current in the readout junction, then the switching of the readout junction to a finite voltage state occurs, for state $|1\rangle$, with probability p_1 larger than the switching probability p_0 for state $|0\rangle$. Note that this is in principle a standard projective measurement, even though the fidelity $p_1 - p_0$ of the measurement achieves a maximum value 0.4, much smaller than the unit fidelity of an ideal projective measurement.

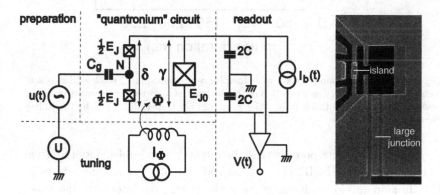

Fig. 8.25   Left: A scheme of the quantronium circuit. The black node denotes the superconducting island, delimited by two small Josephson junctions (crossed boxes) in a superconducting loop including also a third, larger, Josephson junction. Right: A scanning electron micrograph of a sample. The figure is taken from Vion *et al.* (2003).

The manipulation of the qubit state is achieved by applying time-dependent control parameters $I_b(t)$ and $n_g(t) = C_g(U + u(t))/2e$, where $u(t)$ is a microwave pulse. The controlled manipulation of the single-qubit state is shown in Fig. 8.26: a microwave resonant pulse of duration τ induces controlled Rabi oscillations between the states $|0\rangle$ and $|1\rangle$. If τ is appropriate, the NOT gate ($|0\rangle \rightarrow |1\rangle$, $|1\rangle \rightarrow |0\rangle$) is implemented. A Ramsey fringe experiment also allowed measurement of the decoherence time scale $t_d \approx 500\,\text{ns}$ for this circuit, see Vion *et al.* (2002). This time is much longer than the time required to implement a single-qubit gate, so that an arbitrary evolution of the two-level system can be implemented with a series of microwave pulses. Note that the time for a single qubit operation can be made as short as 2 ns.

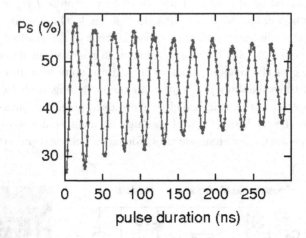

Fig. 8.26   Coherent manipulation of a Josephson-junction qubit: Rabi oscillations of the switching probability as a function of the duration of a microwave resonant pulse are observed. The figure is taken from Ithier *et al.* (2006). Copyright (2006) by the Italian Physical Society.

A very interesting point is that the quantronium implements the usual Hamiltonian of an NMR system. Indeed, for $n_g = \frac{1}{2}$ the first term in Hamiltonian (8.75) vanishes, so that the charge operator and the time-dependent control part of the Hamiltonian (the microwave pulse) are purely off-diagonal in the basis of the two lowest energy eigenstates of the Hamiltonian. Therefore, the quantronium implements the usual Hamiltonian of an NMR system, where the time-dependent RF field is orthogonal to the static magnetic field (see Sec. 8.1). This has allowed the successful implementation of manipulation methods inspired from NMR, such as spin echoes and composite pulse techniques (see Collin *et al.*, 2004).

Finally, we point out the analogy with atomic physics: the pulsed microwave generator plays the role of a laser resonant (or near-resonant) with the transition frequency of the two level-systems, the transition frequency is tuned by varying the voltage U or the magnetic flux Φ, similarly to the Stark and Zeeman effects in atomic physics, and the readout circuit is the analogue of a Stern–Gerlach apparatus.

8.5 Quantum communication with photons

At present, the only appropriate physical system for long-distance communication of quantum states is the photon. Photons can travel long distances with low loss in optical fibres or even in free space. Furthermore, the state of a single photon can be manipulated using basic linear optical components; that is, phase shifters and beam splitters, which we shall discuss in Sec. 8.5.1. The purpose of this section is to present the basic principles of the experimental implementations of teleportation and quantum cryptography with photons. Before doing so, a short introduction to linear optics is required.

8.5.1 *Linear optics*

An optical component is said to be linear if its output modes (with creation and annihilation operators $b_j^\dagger$ and b_j) are a linear combination of its input modes (with creation and annihilation operators $a_j^\dagger$ and a_j):

$$b_j^\dagger = \sum_k M_{jk} a_k^\dagger. \tag{8.77}$$

Phase shifter. This is defined by the transformation

$$U_P(\phi) = e^{i\phi m} = e^{i\phi a^\dagger a}. \tag{8.78}$$

Therefore, the Fock state $|m\rangle$ is mapped into $e^{i\phi m}|m\rangle$. In practice, a phase shifter is a slab of transparent medium with refractive index n different from the free space refractive index n_0. Hence the wave vectors in the medium and in free space are $k = n\omega/c$ and $k_0 = n_0\omega/c$, where $\omega/2\pi$ is the photon frequency and c the speed of light in the vacuum. If the photon travels a distance L through the medium, its phase changes by e^{ikL}, which is different from the phase change e^{ik_0L} for a photon travelling the same distance in free space. The phase shift ϕ in (8.78) is then kL for the photon travelling in the medium and k_0L for the photon travelling in free space.

Beam splitter. By definition, a beam splitter acts on two modes through the unitary transformation

$$U_B(\theta, \phi) = \begin{bmatrix} \cos\theta & -e^{i\phi}\sin\theta \\ e^{-i\phi}\sin\theta & \cos\theta \end{bmatrix}, \tag{8.79}$$

where the input and output modes are related through the linear mapping

$$a_l^\dagger |0\rangle \rightarrow \sum_m (U_B)_{ml} \, b_m^\dagger |0\rangle. \tag{8.80}$$

In particular, given the input state

$$|mn\rangle = \frac{(a_1^\dagger)^m}{\sqrt{m!}} \frac{(a_2^\dagger)^n}{\sqrt{n!}} |00\rangle, \tag{8.81}$$

we obtain the output state

$$U_B |mn\rangle = \frac{1}{\sqrt{m!n!}} \left[\sum_{i=1}^2 (U_B)_{i1} b_i^\dagger \right]^m \left[\sum_{j=1}^2 (U_B)_{j2} b_j^\dagger \right]^m |00\rangle$$

$$= \frac{1}{\sqrt{m!n!}} \left(\cos\theta b_1^\dagger + e^{-i\phi} \sin\theta b_2^\dagger \right)^m \left(-e^{i\phi} \sin\theta b_1^\dagger + \cos\theta b_2^\dagger \right)^n |00\rangle. \tag{8.82}$$

For instance,

$$U_B |00\rangle = |00\rangle,$$

$$U_B |10\rangle = \cos\theta \, |10\rangle + e^{-i\phi} \sin\theta \, |01\rangle,$$

$$U_B |01\rangle = -e^{i\phi} \sin\theta \, |10\rangle + \cos\theta \, |01\rangle,$$

$$U_B |11\rangle = -\sqrt{2} e^{i\phi} \sin\theta \cos\theta \, |20\rangle + \cos 2\theta \, |11\rangle + \sqrt{2} e^{-i\phi} \sin\theta \cos\theta \, |02\rangle,$$

$$U_B |20\rangle = \cos^2\theta \, |20\rangle + \sqrt{2} e^{-i\phi} \sin\theta \cos\theta \, |11\rangle + e^{-2i\phi} \sin^2\theta \, |02\rangle,$$

$$U_B |02\rangle = e^{2i\phi} \sin^2\theta \, |20\rangle - \sqrt{2} e^{i\phi} \sin\theta \cos\theta \, |11\rangle + \cos^2\theta \, |02\rangle. \tag{8.83}$$

Exercise 8.23 In the *dual-rail representation* a single photon can follow two different paths and the two states of the qubit ($|0\rangle$ and $|1\rangle$) correspond to the photon following one path or the other (see Fig. 8.27). The two logical states can be written as $|0\rangle = a_0^\dagger |0\rangle_0 |0\rangle_1 = |1\rangle_0 |0\rangle_1$ and $|1\rangle = a_1^\dagger |0\rangle_0 |0\rangle_1 = |0\rangle_0 |1\rangle_1$, where the operators $a_0^\dagger$ and $a_1^\dagger$ create a photon in the input modes 0 and 1 and $|0\rangle_0$, $|0\rangle_1$ are the vacuum states corresponding to these modes. A beam splitter (see Eq. (8.79)) with $\theta = \frac{\pi}{4}$ and $\phi = -\frac{\pi}{2}$) implements the transformation $|0\rangle \rightarrow \frac{1}{\sqrt{2}} \left(|0'\rangle + i |1'\rangle \right)$ and $|1\rangle \rightarrow \frac{1}{\sqrt{2}} \left(i |0'\rangle + |1'\rangle \right)$, where $|0'\rangle = b_{0'}^\dagger |0\rangle_{0'} |0\rangle_{1'} = |1\rangle_{0'} |0\rangle_{1'}$ and $|1'\rangle = b_{1'}^\dagger |0\rangle_{0'} |0\rangle_{1'} = |0\rangle_{0'} |1\rangle_{1'}$. Here $b_{0'}^\dagger$ and $b_{1'}^\dagger$ create a photon in the output modes $0'$ and $1'$. Show that this beam splitter, together with two $-\frac{\pi}{2}$ phase shifters, implements a Hadamard gate (see Fig. 8.27, left).

We can also introduce the *polarization qubit*: the two polarization states $|h\rangle$ and $|v\rangle$ stand for the states $|0\rangle$ and $|1\rangle$. Show that the CNOT gate is implemented (up to a sign factor) by the circuit in Fig. 8.27 (right), provided the dual-rail qubit is the control and the polarization qubit the target and that a polarization rotator ($|h\rangle \rightarrow |v\rangle$ and $|v\rangle \rightarrow -|h\rangle$) is placed in the upper ($1'$) path (see Cerf *et al.*, 1998).

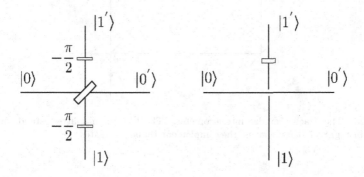

Fig. 8.27   Optical simulation of Hadamard (left) and CNOT (right) gates.

Exercise 8.24   The two beams emerging from the beam splitter in Fig. 8.27 can be recombined using perfectly reflecting mirrors and another beamsplitter. This is the principle of the *Mach–Zehnder interferometer* drawn in Fig. 8.28, an optical tool used to measure small phase shifts between the two paths connecting the two beam splitters. Show that, if a phase shifter is put into one arm of the interferometer, then the entire circuit is equivalent to a single beam splitter of arbitrary transmittivity (the transmittivity T and the reflectivity R in (8.79) are defined as $T = \cos^2 \theta$, $R = 1 - T = \sin^2 \theta$).

As discussed in Sec. 3.5, we can decompose any unitary operator U acting on a $N = 2^n$-dimensional Hilbert space into the product of $O(N^2)$ operations, each only acting non-trivially on two-dimensional subspaces. More precisely, we can write (see Reck *et al.*, 1994)

$$U = D V_{2,1} V_{3,1} V_{3,2} V_{4,1} \cdots V_{4,3} \cdots V_{N,1} V_{N,2} \cdots V_{N,N-2} V_{N,N-1}, \qquad (8.84)$$

where $V_{p,q}$ differs from the N-dimensional identity matrix only in the matrix elements qq, qp, pq, pp, here given by the beam splitter matrix (8.79), and D is a $N \times N$ diagonal matrix, with diagonal matrix elements of unit modulus. As shown in Fig. 8.29, transformation (8.84) can be implemented by means of a triangular array of $\frac{N(N-1)}{2}$ beam splitters plus N phase shifters. The top left beam splitter in this figure realizes $V_{N,N-1}$ and so on

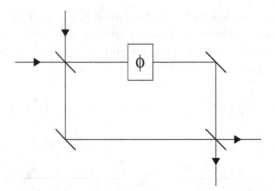

Fig. 8.28   The Mach–Zehnder interferometer. The two beam splitters stand for the circuit in Fig. 8.27 (left); that is, they implement Hadamard gates.

up to the top right beam splitter, realizing $V_{2,1}$. Finally, the phase shifters implement the diagonal matrix D. Therefore, any n-qubit quantum circuit can be simulated by a single-photon optical setup with $N = 2^n$ optical paths. Note that the number of optical devices (beam splitters and phase shifters) grows exponentially with n. This is the price to pay because qubit–qubit interactions are not included in this model; that is, entanglement is not generated (see also Sec. 3.2 for a discussion of the importance of entanglement in quantifying the resources required for computation with waves).

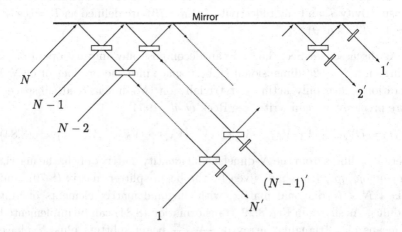

Fig. 8.29   A linear optics network implementing any $N \times N$ unitary matrix.

In *non-linear optics* the two-qubit CMINUS gate could, in principle, be implemented by taking advantage of the indirect interaction between photons, mediated by atoms in a *Kerr medium*. As a result, the refractive index n is a linear function of the total intensity I of light crossing the medium $(n(I) = n_0 + n_2 I)$, so that an extra phase shift $\phi \propto n_2 L$ is acquired if two photons propagate simultaneously through a Kerr medium of length L. If the medium is long enough to obtain $\phi = \pi$ and the case in which both photons cross the medium corresponds, in the dual-rail representation, to the two-qubit state $|11\rangle$, then the CMINUS = CPHASE(π) gate is realized. The drawback is that in Kerr media it is difficult to obtain $\phi = \pi$ before photon loss due to absorption becomes important.

As shown in Knill *et al.* (2001), see also Raussendorf and Briegel (2001), linear optics could be used in principle to implement an efficient quantum computation, provided we can detect photons and feed the results of measurements back to control future linear gates. This leads to *probabilistic gates*, see for instance exercises 8.25 and 8.26. Even though these gates are not unitary, it is possible, using quantum teleportation and quantum error correction as basic ingredients, to approximate unitary operations efficiently.

Exercise 8.25 *Non-linear sign shift.* Let us consider the quantum circuit drawn in Fig. 8.30, where the initial state is

$$|\psi\rangle|1\rangle|0\rangle = \left(\alpha|0\rangle + \beta|1\rangle + \gamma|2\rangle\right)|1\rangle|0\rangle = \left(\alpha + \beta a_1^\dagger + \gamma \frac{(a_1^\dagger)^2}{\sqrt{2}}\right) a_2^\dagger |000\rangle \quad (8.85)$$

and the unitary transformation

$$U = \begin{bmatrix} 1-\sqrt{2} & \frac{1}{\sqrt{\sqrt{2}}} & \sqrt{\frac{3}{\sqrt{2}}-2} \\ \frac{1}{\sqrt{\sqrt{2}}} & \frac{1}{2} & \frac{1}{2}-\frac{1}{\sqrt{2}} \\ \sqrt{\frac{3}{\sqrt{2}}-2} & \frac{1}{2}-\frac{1}{\sqrt{2}} & \sqrt{2}-\frac{1}{2} \end{bmatrix}. \quad (8.86)$$

Note that U can be realized using beam splitters and phase shifters as in Fig. 8.29. The circuit is *probabilistic*; that is, we accept the output $|\psi'\rangle$ if and only if we measure a single photon in mode 2 (second line in the circuit) and vacuum in mode 3 (lower line). Show that this measurement outcome is obtained with probability $\frac{1}{4}$ and that

$$|\psi'\rangle = \alpha|0\rangle + \beta|1\rangle - \gamma|2\rangle. \quad (8.87)$$

The state $|\psi'\rangle$ only differs from $|\psi\rangle$ in the sign of the coefficient in front of the two-photon state $|2\rangle$. The transformation $\alpha|0\rangle + \beta|1\rangle + \gamma|2\rangle \rightarrow \alpha|0\rangle + \beta|1\rangle - \gamma|2\rangle$ is known as a non-linear sign-shift gate.

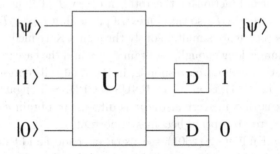

Fig. 8.30   A quantum circuit implementing the non-linear sign-shift gate.

Exercise 8.26   Show that the circuit in Fig. 8.31 implements a probabilistic CMINUS gate, the probability of success being $\frac{1}{16}$.

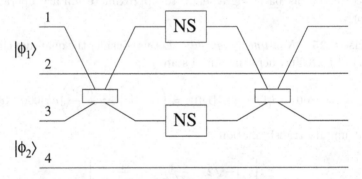

Fig. 8.31   A quantum circuit implementing a probabilistic CMINUS gate using two non-linear sign shift gates (NS) and two beam splitters with $\phi = 0$ and $\theta = \frac{\pi}{4}$ (left) or $\theta = -\frac{\pi}{4}$ (right). The initial states of the two qubits are encoded in the dual-rail representation: $|\phi_1\rangle = \alpha|0\rangle + \beta|1\rangle = \alpha|0\rangle_1|1\rangle_2 + \beta|1\rangle_1|0\rangle_2$ and $|\phi_2\rangle = \gamma|0\rangle + \delta|1\rangle = \gamma|0\rangle_3|1\rangle_4 + \delta|1\rangle_3|0\rangle_4$.

8.5.2   *Experimental quantum teleportation*

In this section we describe two experimental implementations of the teleportation protocol based on linear optics. In both cases the EPR states required by the protocol are photon pairs generated by *parametric down-conversion*. This phenomenon takes place when a laser beam passes through a non-linear crystal such as β-barium borate (BBO). Inside the crystal, an incoming pump photon can be converted into two photons of lower energy, one polarized vertically and the other polarized horizontally, conserving total energy and momentum.[19] In so-called type II down-conversion the photons are emitted along two cones (one photon per cone, see Fig. 8.32), corresponding to horizontally and vertically polarized photons. If the two photons travel along the cone intersections, neither photon has definite polarization. This corresponds to the entangled state

$$\frac{1}{\sqrt{2}}\left(|v\rangle_1|h\rangle_2 + e^{i\alpha}|h\rangle_1|v\rangle_2\right), \qquad (8.88)$$

where $|h\rangle_i$ and $|v\rangle_i$ denote the horizontal and vertical polarization states of photon i ($i = 1, 2$) and the relative phase α arises from the crystal birefringence.

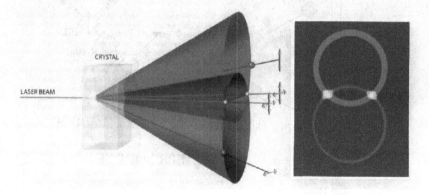

Fig. 8.32 Generation of entangled states by parametric down-conversion. The figure is taken from Zeilinger (2000).

The Rome experiment. In the Rome teleportation experiment (Boschi *et al.*, 1998) a total of two particles (photons) instead of three are used. The EPR state is realized by spatial entanglement, while the state to be teleported is encoded in the polarization degree of freedom of one of the two photons.[20]

---

[19]If $(\omega_p, \boldsymbol{k}_p)$, $(\omega_1, \boldsymbol{k}_1)$, $(\omega_2, \boldsymbol{k}_2)$ denote angular frequencies and wave vectors of the pump photon and of the two down-converted photons, then the relations $\omega_p = \omega_1 + \omega_2$ and $\boldsymbol{k}_p = \boldsymbol{k}_1 + \boldsymbol{k}_2$ hold.

[20]Note that two-photon *hyperentangled* states can also be generated; that is, states exhibiting entanglement in different degrees of freedom (for instance, both in polarization and spatial degrees of freedom), see Cinelli *et al.* (2005) and Barreiro *et al.* (2005).

The experimental setup is sketched in Fig. 8.33. Polarization entangled photons are created by parametric down-conversion, using a β-barium borate (BBO) crystal pumped by an ultraviolet (UV) laser with wavelength 351.1 nm. The down-converted photons have a wavelength of 702.2 nm and their EPR state is

$$\frac{1}{\sqrt{2}}\left(|v\rangle_1|h\rangle_2 + |h\rangle_1|v\rangle_2\right). \tag{8.89}$$

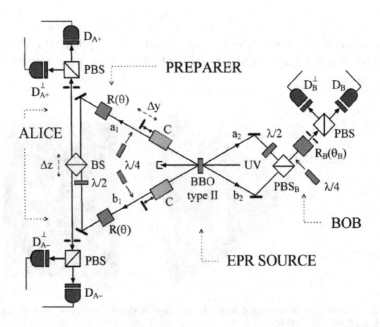

Fig. 8.33   A scheme showing the Rome teleportation experiment. The figure is reprinted with permission from Boschi *et al.* (1998). Copyright (1998) by the American Physical Society.

The two photons follow paths a_1 and b_1 and pass through calcite crystals (C), whose purpose is to transform the entangled polarization state (8.89) into an entangled spatial (path) state. Indeed, crystals C split the two polarization components of the photons: the vertical components are transmitted while the horizontal components are reflected, pass through the BBO crystal and follow paths b_2 and a_2. This corresponds to the following

---

Hyperentangled photons exhibit quantum stronger non-locality effects than EPR photon pairs, see Barbieri *et al.* (2006).

substitutions in (8.89):

$$|v\rangle_1 \rightarrow |a_1\rangle |v\rangle_1, \qquad |h\rangle_1 \rightarrow |b_2\rangle |h\rangle_2,$$
$$|v\rangle_2 \rightarrow |b_1\rangle |v\rangle_1, \qquad |h\rangle_2 \rightarrow |a_2\rangle |h\rangle_2, \tag{8.90}$$

where $|a_1\rangle |v\rangle_1$, for instance, represents the state of photon 1 in path a_1 and having vertical polarization. Note that from now on index 1 refers to the photon directed to Alice's laboratory, while index 2 denotes the photon travelling to Bob's laboratory. Using (8.90), state (8.89) becomes

$$\frac{1}{\sqrt{2}} \big(|a_1\rangle |a_2\rangle + |b_1\rangle |b_2\rangle \big) |v\rangle_1 |h\rangle_2, \tag{8.91}$$

so that the entanglement has been transferred from polarization to spatial degrees of freedom.

In order to prepare a generic state to be teleported, $\alpha |v\rangle_1 + \beta |h\rangle_1$, polarization rotators ($\lambda/4$ plates and $R(\theta)$ Fresnel rhombuses in Fig. 8.33) act in the same manner on the two paths a_1 and b_1 that can be followed by photon 1. Therefore, the state of the entire system reads

$$\frac{1}{\sqrt{2}} \big(|a_1\rangle |a_2\rangle + |b_1\rangle |b_2\rangle \big) \big(\alpha |v\rangle_1 + \beta |h\rangle_1 \big) |h\rangle_2, \tag{8.92}$$

As described in Sec. 4.5, a Bell measurement performed by Alice is required at this stage of the teleportation protocol. The analogous of the Bell states $|\phi^{\pm}\rangle$, $|\psi^{\pm}\rangle$ are

$$|c_{\pm}\rangle = \frac{1}{\sqrt{2}} \big(|a_1\rangle |v\rangle_1 \pm |b_1\rangle |h\rangle_1 \big), \quad |d_{\pm}\rangle = \frac{1}{\sqrt{2}} \big(|a_1\rangle |h\rangle_1 \pm |b_1\rangle |v\rangle_1 \big). \tag{8.93}$$

Alice must perform a (Bell) measurement on the basis $\{|c_{\pm}\rangle, |d_{\pm}\rangle\}$. For this purpose, the polarization of path b_1 is further rotated by 90° (by means of the $\lambda/2$ plate in Fig. 8.33), so that

$$|b_1\rangle |v\rangle_1 \rightarrow -|b_1\rangle |h\rangle_1, \quad |b_1\rangle |h\rangle_1 \rightarrow |b_1\rangle |v\rangle_1. \tag{8.94}$$

Therefore, the states of the Bell basis are transformed as follows:

$$|c_{\pm}\rangle \rightarrow \frac{1}{\sqrt{2}} \big(|a_1\rangle \pm |b_1\rangle \big) |v\rangle_1, \qquad |d_{\pm}\rangle \rightarrow \frac{1}{\sqrt{2}} \big(|a_1\rangle \mp |b_1\rangle \big) |h\rangle_1. \tag{8.95}$$

Paths a_1 and b_1 impinge on the beam splitter BS. This is a 50:50 beam splitter, namely $\theta = \frac{\pi}{4}$ in (8.79). The position Δz of BS is set so that the two output states are $\frac{1}{\sqrt{2}} \big(|a_1\rangle + |b_1\rangle \big)$ and $\frac{1}{\sqrt{2}} \big(|a_1\rangle - |b_1\rangle \big)$. Note that at the beam splitter BS the two polarizations h and v interfere independently. They are separated by the two polarizing beam splitters PBS. In this manner all four Bell states $|c_\pm\rangle$ and $|d_\pm\rangle$ are measured by detectors $D^\perp_{A_\pm}$ and $D_{A_\mp}$, respectively.

Bob's photon is reconstructed on a single path by means of a plate $(\lambda/2)$ and a polarizing beam splitter (PBS_B) oriented to transmit horizontal and reflect vertical polarizations. It can be seen that the state of the entire system (before Alice's measurement) becomes

$$
\begin{aligned}
&\tfrac{1}{2}|c_+\rangle\big(\alpha|v\rangle_2 + \beta|h\rangle_2\big) + \tfrac{1}{2}|c_-\rangle\big(\alpha|v\rangle_2 - \beta|h\rangle_2\big) \\
&+ \tfrac{1}{2}|d_+\rangle\big(\beta|v\rangle_2 + \alpha|h\rangle_2\big) + \tfrac{1}{2}|d_-\rangle\big(\beta|v\rangle_2 - \alpha|h\rangle_2\big).
\end{aligned}
\tag{8.96}
$$

The original state $\alpha|v\rangle + \beta|h\rangle$ is teleported without need of an additional unitary transformation when Alice detects $|c_+\rangle$; that is, when detector $D^\perp_{A_+}$ clicks. This is in agreement with the experimental results of Fig. 8.34. Bob's measuring axis is changed by means of a plate $(\lambda/4)$ and a polarization rotator $(R_B(\theta_B))$. The four coincidence experiments between Alice's detectors $D_{A_\pm}$ and $D^\dagger_{A_\pm}$ and Bob's detector $D_B(\theta)$ are shown in Fig. 8.34. The initial state is linearly polarized with $\theta = 22.5°$ ($\alpha = \sin\theta$ and $\beta = \cos\theta$). All maxima in the coincidence rates are compatible with Eq. (8.96).

We stress that in the Rome experiment it is impossible to teleport a part of an entangled state; that is, the scheme of Boschi *et al.* (1998) cannot be used as a primitive for quantum computation in a larger quantum network, as proposed by Gottesman and Chuang (1999). This would become possible if one had the ability to swap any unknown state onto the polarization degree of freedom of Alice's member of the EPR pair. However, this requires a two-qubit gate; that is, qubit–qubit interactions. This is beyond linear optics, which deals with non-interacting photons.[21]

The Innsbruck experiment. We now discuss the Innsbruck teleportation experiment (Bouwmeester *et al.*, 1997). A schematic drawing of the experimental setup is shown in Fig. 8.35. A pulse of ultraviolet radiation passes through a non-linear crystal and by parametric down-conversion creates two entangled photons in the state

$$
|\psi^-\rangle_{23} = \frac{1}{\sqrt{2}}\big(|h\rangle_2|v\rangle_3 - |v\rangle_2|h\rangle_3\big),
\tag{8.97}
$$

where the indices 2 and 3 refer to the paths followed by the two photons (see Fig. 8.35). After retroflection the beam passes through the crystal a

---

[21] As mentioned in Sec. 8.5.1, two-qubit gates can be implemented with linear optics, provided that the results of photon measurements are used to control future quantum gates.

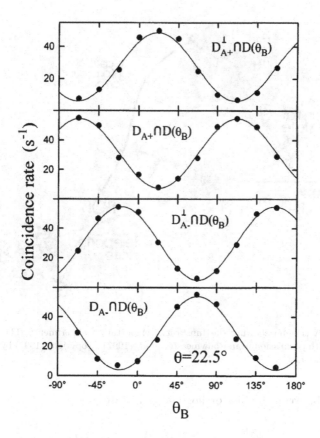

Fig. 8.34 Coincidences between Alice's four detectors $D_{A_\pm}$, $D_{A_\pm}^\perp$ and Bob's detector $D_B(\theta_B)$ as a function of Bob's measurement angle θ_B. The figure is reprinted with permission from Boschi *et al.* (1998). Copyright (1998) by the American Physical Society.

second time and, again by parametric down-conversion, generates a second entangled pair of photons, this time propagating along paths 1 and 4. The detection of photon 4 allows us to know when the second pair is emitted and projects photon 1 into a single-qubit state. A polarization rotator prepares the initial state $|\psi\rangle_1 = \alpha|h\rangle_1 + \beta|v\rangle_1$ of photon 1. The purpose of the experiment is to teleport this state to Bob's laboratory.

A complete Bell measurement is required in the teleportation protocol By definition, such measurement should be able to distinguish with 100%

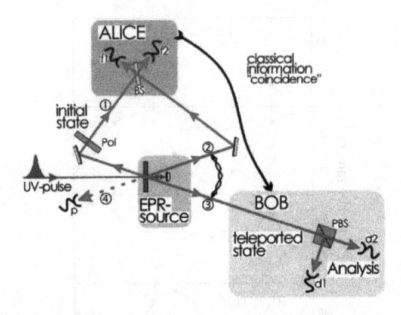

Fig. 8.35   A scheme showing the Innsbruck teleportation experiment. The figure is reprinted with permission from Bouwmeester *et al.* (1997). Copyright (1997) by Macmillan Publishers Ltd.

efficiency between the four orthogonal Bell states

$$
\begin{aligned}
|\phi^+\rangle_{12} &= \tfrac{1}{\sqrt{2}}\big(|h\rangle_1|h\rangle_2 + |v\rangle_1|v\rangle_2\big)\,, \\
|\phi^-\rangle_{12} &= \tfrac{1}{\sqrt{2}}\big(|h\rangle_1|h\rangle_2 - |v\rangle_1|v\rangle_2\big)\,, \\
|\psi^+\rangle_{12} &= \tfrac{1}{\sqrt{2}}\big(|h\rangle_1|v\rangle_2 + |v\rangle_1|h\rangle_2\big)\,, \\
|\psi^-\rangle_{12} &= \tfrac{1}{\sqrt{2}}\big(|h\rangle_1|v\rangle_2 - |v\rangle_1|h\rangle_2\big)\,.
\end{aligned}
\tag{8.98}
$$

A complete Bell measurement is impossible with only linear optical elements (see exercise 8.27). In particular, in Bouwmeester *et al.* (1997) only the Bell state $|\psi^-\rangle_{12}$ is identified. Since the four Bell states are found with equal probability (see Eq. (4.37)), teleportation is achieved in only a quarter of the cases.

Let us explain why it is possible to single out the state $|\psi^-\rangle_{12}$. We first consider the effect of Alice's beamsplitter on $|\psi^-\rangle_{12}$. We call $a_{i,h}^\dagger$ and $a_{i,v}^\dagger$ the operators creating a horizontally or vertically polarized photon in the spatial mode i ($i = 1, 2$). Therefore,

$$|\psi^-\rangle_{12} = \frac{1}{\sqrt{2}}\left(a^\dagger_{1,h}a^\dagger_{2,v} - a^\dagger_{1,v}a^\dagger_{2,h}\right)|0\rangle, \qquad (8.99)$$

where $|0\rangle$ is the vacuum states. A beam splitter (see Eq. (8.79)) with $\theta = \frac{\pi}{4}$ and $\phi = -\frac{\pi}{2}$ transforms this state into

$$\frac{1}{\sqrt{2}}\left(b^\dagger_{1,h}b^\dagger_{2,v} - b^\dagger_{1,v}b^\dagger_{2,h}\right)|0\rangle, \qquad (8.100)$$

where the operators $b^\dagger_{i,h}$ and $b^\dagger_{i,v}$ create photons in the output modes. (operators a and b are related via the mapping (8.80)). Therefore, Alice's detectors f1 and f2 both register a photon. For the other Bell states ($|\psi^+\rangle$, $|\phi^+\rangle$, $|\phi^-\rangle$) the two photons leave the same output port (see exercise 8.27). This is a manifestation of the fact that the singlet state $|\psi^-\rangle$ is the only Bell state that is antisymmetric under exchange of the polarization states of the two photons. Since photons are bosons, the wave function must be globally symmetric and therefore it must also be antisymmetric under exchange of the spatial variables: due to the antisymmetric nature of the spatial part of the wave function the photons emerge one on each side of the beam splitter. In contrast, the triplet states ($|\psi^+\rangle$, $|\phi^+\rangle$, $|\phi^-\rangle$) are symmetric under exchange of polarization or spatial states of the two photons. Owing to the symmetric nature of the spatial part of the wave function, in this case both photons leave the same output port.

Exercise 8.27 Show that $|\phi^+\rangle$ and $|\phi^-\rangle$ cannot be distinguished after a beam splitter. Can they be distinguished from the other Bell states?

In summary, if the Bell state $|\psi^-\rangle_{23}$ is prepared and Alice measures $|\psi^-\rangle_{12}$, then teleportation succeeds and Bob ends up with the state $|\psi\rangle_3 = \alpha|h\rangle_3 + \beta|v\rangle_3$. The polarization of Bob's photon is analyzed by passing it through a polarizing beam splitter. The experimental results in the cases in which the photon state to be teleported is polarized at $\pm 45°$ are shown in Fig. 8.36. Bob's polarizing beam splitter selects $-45°$ (detector d1) and $+45°$ (detector d2) polarization. Teleportation succeeds, for an initial state polarized at $+45°$, if only detector d2 clicks when both f1 and f2 click. On the other hand, only three-fold coincidences f1f2d1 should be registered when the initial polarization of photon 1 is $-45°$. This argument is valid if photons 1 and 2 cannot be distinguished at Alice's beam splitter by their arrival times; that is, if they are generated within a time smaller than the coherence time of the source (approximately 500 fs). This condition can be met or violated by changing the delay between the first and the second parametric down-conversion by moving the retroflection mirror (see Fig. 8.35). Outside the region of teleportation, photons 1 and 2 go either to f1 or to f2 independently of each other, so that the probability of detecting an f1f2 coincidence is 50%. In this case, photon 3 is completely unpolarized because it is a part of a Bell pair and therefore d1 and d2 have the same chance of receiving a photon. This analysis predict a 25% probability both for the three-fold coincidence f1f2d1 and f1f2d2, independently of the polarization state of photon 1. In conclusion, a dip in one of the two possible three-fold coincidences is expected in the teleportation region, while outside this region both coincidences have the same probability. This expectation is confirmed by the experimental data of Fig. 8.36.

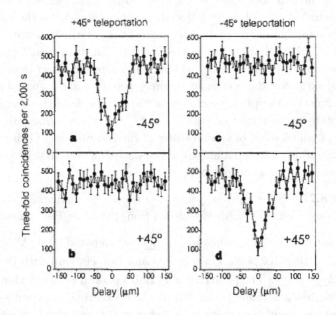

Fig. 8.36 Measured three-fold coincidence rates between Alice's detectors f1 and f2 and Bob's detectors d1 (+45°, top) and d2 (−45°, bottom), in the cases in which the photon to be teleported is polarized at +45° (left) or −45° (right). The coincidence rates are plotted as a function of the delay between the two-photon pairs, changed by translating the retroflection mirror in the setup of Fig. 8.35. The figure is reprinted with permission from Bouwmeester *et al.* (1997). Copyright (1997) by Macmillan Publishers Ltd.

In closing this section, we remark that both the Rome and Innsbruck experiments were performed in a single laboratory, so that teleportation was limited to a distance of the order of 1 m. Long-distance teleportation has been more recently reported in Marcikic *et al.* (2003) and in Ursin *et al.* (2004). In this latter experiment a member of the required EPR pair was transmitted through an 800-metre-long optical fibre installed underneath the Danube river in Vienna.

Exercise 8.28 *Universal NOT gate.* By definition, the universal NOT gate maps any one-qubit input state $|\psi\rangle$ into its perpendicular state $|\psi^\perp\rangle$. In the Bloch sphere representation, the states $|\psi\rangle$ and $|\psi^\perp\rangle$ are antipodes; that is, the universal NOT maps the Bloch coordinates (x, y, z) of the state $|\psi\rangle$ into $(-x, -y, -z)$. Note that a perfect universal NOT operation is prohibited by the laws of quantum mechanics. However, there exists an optimal approximation with fidelity $F = \frac{2}{3}$ (see Bužek *et al.*, 1999).

Let us consider the following modification of the teleportation protocol described in Sec. 4.5: Alice sends Bob a single bit of classical information, saying if she obtained $|\psi^+\rangle$ or one of the states $|\psi^-\rangle$, $|\phi^+\rangle$, $|\phi^-\rangle$ from her Bell measurement. Show that in the second case Bob ends up with an approximate universal NOT transformation of the state $|\psi\rangle$, with fidelity $F = \frac{2}{3}$. Note that here, in contrast to the original teleportation protocol, Bob does not perform any operation on his qubit.

Exercise 8.29 Consider the teleportation protocol with initial condition $|\psi\rangle|\psi^+\rangle$ (see Sec. 4.5) and assume that Alice performs a measurement capable of distinguishing between the Bell state $|\psi^+\rangle$ (first outcome) and the subspace spanned by the Bells states $|\psi^-\rangle$, $|\phi^+\rangle$, $|\phi^-\rangle$ (second outcome). Show that, if the second outcome occurs, then the post-measurement state of each of Alice's two qubits is a clone of the initial single-qubit state $|\psi\rangle$, with fidelity $F = \frac{5}{6}$ (see Ricci *et al.*, 2004).

8.5.3 *Experimental quantum-key distribution*

Quantum cryptography (or more precisely, quantum-key distribution) promises to become the first quantum-information protocol to find commercial applications, thanks to the enormous progress in the technology of optical-fibres and free-space optical communication.

Optical quantum cryptography is based on single-photon Fock states, emitted on demand. Unfortunately, these states are difficult to realize experimentally. However, single-photon Fock states can be approximated by means of *faint laser pulses*: a laser produces a coherent state, given by Eq. (8.36), and this state is attenuated to a very low mean photon number $\bar{n} \ll 1$. For instance, if $\bar{n} = 0.1$ (a value chosen by most experimentalists in quantum cryptography implementations), the coherent state (8.36) reads $|\alpha\rangle \approx 0.95|0\rangle + 0.30|1\rangle + 0.07|2\rangle + \ldots$, where $\alpha = \sqrt{\bar{n}} = \sqrt{0.1}$. This means that most pulses are empty: the probability that the attenuated coherent state contains no photons is $p_0 \approx (0.95)^2 \approx 0.90 \approx 1 - \bar{n}$. A single photon is found with probability $p_1 \approx (0.30)^2 = 0.09$. Therefore, the probability of having a non-empty pulse is $p(n>0) = 1 - p_0$ and the probability that a non-empty pulse contains more than one photon is $p(n>1|n>0) = p(n>1)/p(n>0) = (1 - p_0 - p_1)/(1 - p_0) \approx \bar{n}/2 = 0.05$ (this means that 5% of the non-empty pulses contain more than one photon). Single photons are typically detected by means of semiconductor avalanche photodiodes (APD's). Note that the detectors must be active for all pulses, including the empty ones. Therefore, the problem of dark counts (that is, when there is a click in a detector without an arriving photon) becomes more important when $\bar{n}$ is small.

Fibre-based and free-space systems. Photons can be transmitted from the sender (Bob) to the receiver (Alice) using *optical fibres* or *free space* as quantum channels (of course, such channels are only described as quantum because they are intended to transmit the quantum information encoded in single photons). Let us briefly discuss the advantages and drawbacks of both approaches. Long-distance optical-fibre transmissions exploit the low loss of silica fibres in the 1.3 and 1.55 μm wavelength bands. A further advantage is the possibility to employ standard fibres installed for classical communications. On the other hand, free-space applications are also possible. In this case, the emitter and the receiver are connected by telescopes pointing at each other (spectral filtering is used by the receiver to cut light outside the transmission bandwidth). A significant advantage of free-space quantum cryptography is that transmission over long distances is possible in a transmission window around 800 nm. In this window commercial photodetectors (silicon avalanche photodiodes) have high detection efficiency, up to approximately 70%, and low noise. In contrast, at the wavelengths used in optical fibres silicon APD's are not efficient and one can take advantage of APD's made from germanium or indium gallium arsenide (with detection efficiency < 15%). A disadvantage of free-space quantum cryptography is that its performance strongly depends on weather conditions and air pollution. On the other hand, a major advantage of this approach is that it could offer the possibility to overcome the distance limitations of fibre-based quantum cryptography. Actually, the main drawback with quantum communication via optical fibres is that the probability for photon-absorption losses grows exponentially with the length of the fibre. On the basis of present technology, it appears difficult to employ optical fibres for quantum communication over distances of more than 100 kilometres.[22] In contrast, the free-space approach could be extended to much longer distances, provided earth-to-satellite links were established. A significant advantage of satellite links is that the photon attenuation of a link from the earth directly upwards to a satellite is comparable to approximately 1.5 km horizontal transmission on ground. These latter transmissions are possible and this suggests the possibility to employ free-space photon transmission to distribute secret keys between parties located very far apart (say, in two different continents), using satellite-based links.

Polarization and phase coding. A natural method to code the four states of the BB84 protocol is to employ photons polarized at $-45°$, $0°$, $+45°$ and $90°$. For each pulse, Alice can rotate the polarization of one of these four states by means of electro-optic crystals (Pockels cells). Bob analyzes each photon in the vertical–horizontal basis or in the diagonal basis. If, for instance, a photon polarized at $+45°$ is sent and the measurement takes place in the diagonal basis, then the outcome is deterministic. On the other hand, if Bob chooses the horizontal-vertical basis, he randomly obtains one of the two possible outcomes. The main difficulty of this scheme is to maintain the photon polarization through the quantum channel connecting Alice and Bob. It is difficult to compensate for the polarization transformation induced by a long optical fibre since it is unstable over time, due, for instance, to temperature variations. Note that polarization coding is instead successful when used for free-space transmission over long distances.

Phase coding has proved to be more convenient for fibre-based implementations. The basic setup, drawn in Fig. 8.37, is an optical-fibre version of the Mach–Zehnder interferometer. It consists of two symmetric couplers (the equivalent of 50:50 beamsplitters) connected by two arms, each with a phase modulator (that is, a phase shifter). Alice and Bob can tune the phase shifts ϕ_A and ϕ_B, respectively. The "letters" used in the BB84 protocol (see Sec. 4.3.1) correspond to $\phi = 0, \pi$ (first "alphabet") and $\phi = \frac{\pi}{2}, \frac{3\pi}{2}$ (second alphabet). Alice randomly applies one of the above four phase shifts to encode a bit value (she associates 0 and $\frac{\pi}{2}$ with bit value 0 and π and $\frac{3}{2}\pi$

---

[22]Quantum repeaters; that is, quantum purification schemes aimed at improving the fidelity of the transmitted photons (see Briegel *et al.*, 1998), would overcome this limitation. In principle, quantum repeaters could extend quantum communication to arbitrarily long distances. However, they have not yet been demonstrated experimentally.

with 1). On the other hand, Bob randomly chooses the measurement basis by applying a phase shift of either 0 or $\frac{\pi}{2}$. When $|\phi_A - \phi_B| = 0, \pi$, then Bob obtains with unit probability a deterministic output (see exercise 8.24). On the other hand, when the phase difference is equal to $\frac{\pi}{2}$ or $\frac{3}{2}\pi$, then the photon is found with equal probability in one of Bob's two detectors.

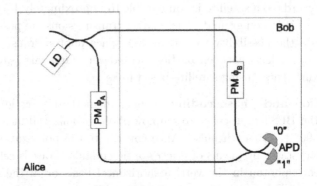

Fig. 8.37   A schematic drawing of an optical-fibre Mach–Zehnder interferometric setup for quantum cryptography. Photon pulses are emitted by a laser diode (LD) and then attenuated and sent from Alice to Bob by means of optical fibres; phase modulators (PM) of phase ϕ_A and ϕ_B are used in Alice's and Bob's laboratories; an avalanche photodiode (APD) is used to detect the photon in port 0 or 1. The figure is reprinted with permission from Gisin *et al.* (2002). Copyright (2002) by the American Physical Society.

Note that the phase-coding scheme works inasmuch as the path mismatch $k\Delta L$ (k is the wave number and ΔL the difference between the lengths of the two arms) is much smaller than the photon wavelength (of order $1\,\mu$m). This condition cannot be fulfilled when Alice and Bob are separated by long distances. For this reason the configurations in Fig. 8.38 with two unbalanced Mach–Zehnder interferometers is used. In this case the two interferometers, one in Alice's laboratory and the other in Bob's, are connected by *a single* optical fibre. When monitoring counts as a function of time from photon emission, Bob observes three peaks (see the inset in Fig. 8.38). The left/right peak corresponds to photons that travel along the short/long path in both Alice's and Bob's interferometers. The central peak is instead associated with photons that choose the long path in Alice's interferometer and the short path in Bob's or *vice versa*. As these two processes are indistinguishable, they produce the interference required in the phase-coding scheme. The advantage of this system is that it is sufficient to keep stable within a small fraction of the photon wavelength the imbalances of Alice's and Bob's interferometers and not the path difference over a long distance as in the previous scheme of Fig. 8.37. Using this approach, quantum-key distribution over 67 km (between Geneva and Lausanne) with a net key rate of approximately 50 Hz was reported in Stucki *et al.* (2002).

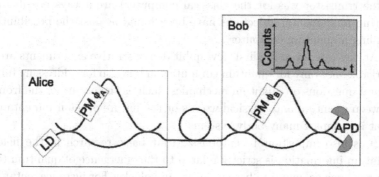

Fig. 8.38 A schematic drawing of a double Mach–Zehnder interferometer for quantum cryptography. The inset shows the temporal count distribution recorded as a function of the time passed since emission of a pulse by Alice (interference is observed in the central peak). The figure is reprinted with permission from Gisin *et al.* (2002). Copyright (2002) by the American Physical Society.

In closing this section, we note that quantum-key distribution experiments using entangled photon pairs have also been performed. For instance, Jennewein *et al.* (2000) reported key generation over a distance of 360 m with a rate of 400 − 800 Hz and a bit error rate of 3%.

8.6 Problems and prospects

It appears probable that in the near future quantum cryptography will be the first quantum-information protocol to find commercial applications. Here the question is how extensive the market will be and this will largely depend on the transmission rates, at present limited to the kHz range. The development of fast single-photon sources and high-efficiency detectors is required to improve significantly the transmission rates, thus broadening the prospects of quantum cryptography.

With regard to quantum computation, the situation is much more difficult. It is not clear if and when we shall be able to build a useful quantum computer; that is, a quantum computer capable of outperforming existing classical computers in important computational tasks. When the problem of decoherence is taken into account for a complex many-qubit system, which we require to perform coherent controlled evolution, then large-scale quantum computers appear unrealistic with present technology. On the other hand, we should bear in mind that technical breakthroughs (such as the transistor was for the classical computer) are always possible and that no fundamental objections have been found against the possibility of building a quantum computer.

At any rate, even the first, few-qubit demonstrative experiments are remarkable, not only for quantum computation but also for addressing fundamental questions on quantum mechanics, such as the nature of the frontier between quantum and classical worlds or the nature of quantum entanglement in complex many-body systems.

It is also important to emphasize that basic research in the field of quantum information is strictly related to the emergence of quantum technologies such as quantum based sensors and clocks. For instance, entangled states could be used to improve the resolution of optical lithography and interferometric measurements.

The time when a quantum computer will be on the desk in our office is uncertain. What is certain is that we are witnessing the emergence of a new and very promising field of investigation in physics, mathematics and computer science.

8.7   A guide to the bibliography

The experimental effort in the field of quantum-information processing is huge and has produced beautiful experimental results. In the following, we shall limit our references to review papers that might be used by the reader as an entry point.

NMR quantum-information processing is reviewed in Jones (2001), Laflamme *et al.* (2002) and Ramanathan *et al.* (2004). NMR quantum-control techniques (many of them also useful in other implementations of quantum computation) are discussed in Vandersypen and Chuang (2004).

Cavity quantum electrodynamics experiments manipulating the entanglement of Rydberg atoms and photons are reviewed in Raimond *et al.* (2001); for discussions of CQED in the optical domain see Mabuchi and Doherty (2002), Miller *et al.* (2005) and Raimond and Rempe (2005). Quantum computation with trapped ions is discussed in Wineland *et al.* (2003), Blatt *et al.* (2004) and Eschner (2006). Quantum computing and quantum communication with quantum optical methods are discussed in Cirac *et al.* (2002). An introduction to the experimental aspects of quantum information with neutral atoms in optical lattices is provided in Bloch (2006); see also Arimondo *et al.* (2005). The prospect of cold atoms in atom chips is discussed in Schmiedmayer and Hinds (2005). Tutorial reviews on quantum information processing with atoms, ions and photons are Monroe (2002) and Cirac and Zoller (2004).

Linear optic quantum computation is discussed in Myers and Laflamme (2006) and Kok *et al.* (2005). Quantum optics implementations with continuous variables are reviewed in Braunstein and van Loock (2005).

Spin qubits in semiconductor quantum dots are described in Elzerman *et al.* (2006). A very readable introduction is Burkard and Loss (2002). Superconducting quantum bits are reviewed in Averin (2000), Makhlin *et al.* (2001), Devoret *et al.* (2004), Esteve and Vion (2005), Wendin and Shumeiko (2005) and Falci and Fazio (2006); for a simple introduction see You and Nori (2005); for a discussion of the state of the art and prospects of this implementation see Mooij (2005).

Quantum cryptography is reviewed in Gisin *et al.* (2002).

Finally, we point out that the prospects of quantum computation and quantum cryptography are discussed in the roadmaps available at http://qist.lanl.gov/. A report on current status and prospects of quantum information processing and communication is available at http://qist.ect.it/.

Solutions to the exercises

Chapter 5

Exercise 5.1   The most general 2×2 Hermitian matrix A can be written as

$$A = \begin{bmatrix} a & b + ic \\ b - ic & d \end{bmatrix}, \tag{B.1}$$

where a, b, c and d are real parameters. We have $A = \alpha I + \beta \sigma_x + \gamma \sigma_y + \delta \sigma_z$, provided $\alpha = (a + d)/2$, $\delta = (a - d)/2$, $\beta = b$ and $\gamma = -c$. Therefore, α, β, γ and δ are all real.

Exercise 5.2   We have $\sigma_x^2 = \sigma_y^2 = \sigma_z^2 = I$ and $\sigma_x \sigma_y = i\sigma_z$, $\sigma_y \sigma_z = i\sigma_x$ and $\sigma_z \sigma_x = i\sigma_y$. A compact method to express these relations is

$$\sigma_j \sigma_k = \delta_{jk} I + i \sum_{l=1}^{3} \epsilon_{jkl} \sigma_l, \tag{B.2}$$

where $\sigma_1 \equiv \sigma_x$, $\sigma_2 \equiv \sigma_y$, $\sigma_3 \equiv \sigma_z$ and ϵ_{jkl} is the Ricci antisymmetric tensor, with $\epsilon_{jkl} = 0$ if the three indices are not all different, $\epsilon_{123} = \epsilon_{231} = \epsilon_{312} = 1$ and $\epsilon_{213} = \epsilon_{321} = \epsilon_{132} = -1$. As $\text{Tr} \, I = 2$ and $\text{Tr} \, \sigma_l = 0$, we obtain from Eq. (B.2) $\text{Tr}(\sigma_j \sigma_k) = 2\delta_{jk}$.

Exercise 5.3   Let $\mathbf{r}_1 = (r_{11}, r_{12}, r_{13}) = (x_1, y_1, z_1)$ and $\mathbf{r}_2 = (r_{21}, r_{22}, r_{23}) = (x_2, y_2, z_2)$ denote the Bloch vectors associated with the density matrices ρ_1 and

ρ_2, respectively. We obtain by direct computation

$$
\begin{aligned}
[\rho_1, \rho_2] &= \tfrac{1}{2}\left(I + \boldsymbol{r}_1 \cdot \boldsymbol{\sigma}\right)\tfrac{1}{2}\left(I + \boldsymbol{r}_2 \cdot \boldsymbol{\sigma}\right) - \tfrac{1}{2}\left(I + \boldsymbol{r}_2 \cdot \boldsymbol{\sigma}\right)\tfrac{1}{2}\left(I + \boldsymbol{r}_1 \cdot \boldsymbol{\sigma}\right) \\
&= \tfrac{1}{4}\left[(\boldsymbol{r}_1 \cdot \boldsymbol{\sigma})(\boldsymbol{r}_2 \cdot \boldsymbol{\sigma}) - (\boldsymbol{r}_2 \cdot \boldsymbol{\sigma})(\boldsymbol{r}_1 \cdot \boldsymbol{\sigma})\right] = \tfrac{1}{4}\sum_{jk} r_{1j}r_{2k}(\sigma_j\sigma_k - \sigma_k\sigma_j) \\
&= \tfrac{1}{4}\sum_{jk} r_{1j}r_{2k}[\sigma_j, \sigma_k] = \tfrac{1}{4}\sum_{jk} 2i\epsilon_{jkl}r_{1j}r_{2k}\sigma_l \\
&= \tfrac{1}{2}\left(x_1y_2 - y_1x_2\right)\sigma_z + \tfrac{1}{2}\left(y_1z_2 - z_1y_2\right)\sigma_x + \tfrac{1}{2}\left(z_1x_2 - x_1z_2\right)\sigma_y. \quad \text{(B.3)}
\end{aligned}
$$

Thus, $[\rho_1, \rho_2] = 0$ when the conditions $x_1y_2 - y_1x_2 = 0$, $y_1z_2 - z_1y_2 = 0$ and $z_1x_2 - x_1z_2 = 0$ are simultaneously satisfied; that is, when $\boldsymbol{r}_1$ and $\boldsymbol{r}_2$ are parallel.

Exercise 5.4   The reduced density matrix ρ_1, defined by Eq (5.51), is Hermitian since

$$
\left(\rho_1\right)_{ji}^{*} = \sum_{\alpha} \rho_{j\alpha;i\alpha}^{*} = \sum_{\alpha} \rho_{i\alpha;j\alpha} = \left(\rho_1\right)_{ij}. \quad \text{(B.4)}
$$

To show that ρ_1 is non-negative, it is convenient to decompose the total density matrix ρ in the basis of its eigenvectors $\{|u_k\rangle_1|v_\gamma\rangle_2\}$:

$$
\rho = \sum_{k,\gamma} \rho_{k\gamma;k\gamma}|u_k\rangle_1|v_\gamma\rangle_2 \,_1\langle u_k|_2\langle v_\gamma|, \quad \text{(B.5)}
$$

with eigenvalues $\rho_{k\gamma;k\gamma} \geq 0$, as the density matrix ρ is non-negative. Then, for any $|\phi\rangle_1 \in \mathcal{H}_1$,

$$
{}_1\langle\phi|\rho_1|\phi\rangle_1 = \sum_{\gamma}\sum_{k} \rho_{k\gamma;k\gamma}\left|{}_1\langle\phi|k\rangle_1\right|^2 \geq 0. \quad \text{(B.6)}
$$

Finally, ρ_1 has unit trace since

$$
\sum_{i}\left(\rho_1\right)_{ii} = \sum_{i,\alpha} \rho_{i\alpha;i\alpha} = 1. \quad \text{(B.7)}
$$

Exercise 5.5   In the teleportation protocol, as a result of Alice's measurement, the state of Bob's qubit is left, with equal probability $p = \tfrac{1}{4}$, in one out of the four following possibilities: $\alpha|0\rangle + \beta|1\rangle$, $\alpha|0\rangle - \beta|1\rangle$, $\alpha|1\rangle + \beta|0\rangle$ and $\alpha|1\rangle - \beta|0\rangle$, where $\alpha|0\rangle + \beta|1\rangle$ is the state to be teleported. Therefore, Bob's density matrix is given by

$$
\begin{aligned}
\rho_B &= \tfrac{1}{4}\begin{bmatrix} |\alpha|^2 & \alpha\beta^{\star} \\ \beta\alpha^{\star} & |\beta|^2 \end{bmatrix} + \tfrac{1}{4}\begin{bmatrix} |\alpha|^2 & -\alpha\beta^{\star} \\ -\beta\alpha^{\star} & |\beta|^2 \end{bmatrix} \\
&\quad + \tfrac{1}{4}\begin{bmatrix} |\beta|^2 & \beta\alpha^{\star} \\ \beta^{\star}\alpha & |\alpha|^2 \end{bmatrix} + \tfrac{1}{4}\begin{bmatrix} |\beta|^2 & -\beta\alpha^{\star} \\ -\beta^{\star}\alpha & |\alpha|^2 \end{bmatrix} = \tfrac{1}{2}I. \quad \text{(B.8)}
\end{aligned}
$$

As $\rho_B = \frac{1}{2}I$, Bob cannot obtain any information on the state to be teleported from any measurement performed on the qubit in his possession.

Exercise 5.6   For a pure bipartite separable state, we can write

$$|\psi\rangle = |\alpha\rangle_1 \otimes |\beta\rangle_2. \qquad (B.9)$$

Hence, the corresponding density matrix is given by

$$\rho = |\psi\rangle\langle\psi| = |\alpha\rangle_1 |\beta\rangle_2 \, {}_1\langle\alpha|_2\langle\beta| = |\alpha\rangle_1 \, {}_1\langle\alpha| \otimes |\beta\rangle_2 \, {}_2\langle\beta|. \qquad (B.10)$$

After partial tracing, we obtain

$$\begin{aligned}
\rho_1 &= \mathrm{Tr}_2\left(|\alpha\rangle_1 \, {}_1\langle\alpha| \otimes |\beta\rangle_2 \, {}_2\langle\beta|\right) = |\alpha\rangle_1 \, {}_1\langle\alpha|, \\
\rho_2 &= \mathrm{Tr}_1\left(|\alpha\rangle_1 \, {}_1\langle\alpha| \otimes |\beta\rangle_2 \, {}_2\langle\beta|\right) = |\beta\rangle_2 \, {}_2\langle\beta|,
\end{aligned} \qquad (B.11)$$

and therefore

$$\rho = \rho_1 \otimes \rho_2. \qquad (B.12)$$

Exercise 5.7   Let us first consider the preparation of the two bottom qubits in Fig. 5.1, initially prepared in the state $|00\rangle$. To simplify writing, we adopt $C_i \equiv \cos\theta_i$ and $S_i \equiv \sin\theta_i$. After application of the first rotation matrix we obtain the state

$$C_1|00\rangle + S_1|10\rangle. \qquad (B.13)$$

The first CNOT gate leads to

$$C_1|00\rangle + S_1|11\rangle, \qquad (B.14)$$

the second rotation to

$$C_1C_2|00\rangle + C_1S_2|01\rangle - S_1S_2|10\rangle + S_1C_2|11\rangle, \qquad (B.15)$$

the second CNOT to

$$C_1C_2|00\rangle + C_1S_2|11\rangle - S_1S_2|10\rangle + S_1C_2|01\rangle, \qquad (B.16)$$

and the third rotation to

$$\begin{aligned}
& C_1C_2C_3|00\rangle + C_1C_2S_3|10\rangle - C_1S_2S_3|01\rangle + C_1S_2C_3|11\rangle \\
& + S_1S_2S_3|00\rangle - S_1S_2C_3|10\rangle + S_1C_2S_3|01\rangle + S_1C_2S_3|11\rangle.
\end{aligned} \qquad (B.17)$$

From (5.63) we have $C_1 = \sqrt{\frac{\sqrt{5}+1}{2\sqrt{5}}}$, $C_2 = \sqrt{\frac{3+\sqrt{5}}{6}}$, $C_3 = \sqrt{\frac{\sqrt{5}+2}{2\sqrt{5}}}$, $S_1 = \sqrt{\frac{\sqrt{5}-1}{2\sqrt{5}}}$, $S_2 = \sqrt{\frac{3-\sqrt{5}}{6}}$, $S_3 = \sqrt{\frac{\sqrt{5}-2}{2\sqrt{5}}}$. Inserting these numerical values in (B.17) we obtain the state vector (5.64).

We now discuss the copying part of the circuit in Fig. 5.1. The four CNOT gates map the state

$$(\alpha|0\rangle + \beta|1\rangle)\tfrac{1}{\sqrt{6}}(2|00\rangle + |01\rangle + |11\rangle) \tag{B.18}$$

into (5.65). Tracing over the bottom qubit in Fig. 5.1 we obtain the two-qubit density matrix

$$\rho_{12} = \left[\alpha\sqrt{\tfrac{2}{3}}|00\rangle + \beta\sqrt{\tfrac{1}{6}}(|10\rangle + |01\rangle)\right]\left[\alpha^\star\sqrt{\tfrac{2}{3}}\langle00| + \beta^\star\sqrt{\tfrac{1}{6}}(\langle10| + \langle01|)\right]$$

$$+ \left[\beta\sqrt{\tfrac{2}{3}}|11\rangle + \alpha\sqrt{\tfrac{1}{6}}(|10\rangle + |01\rangle)\right]\left[\beta^\star\sqrt{\tfrac{2}{3}}\langle11| + \alpha^\star\sqrt{\tfrac{1}{6}}(\langle10| + \langle01|)\right]. \tag{B.19}$$

Hence

$$\rho_1 = \mathrm{Tr}_2\,\rho_{12}$$

$$= \tfrac{2}{3}|\alpha|^2|0\rangle\langle0| + \tfrac{1}{3}\alpha\beta^\star|0\rangle\langle1| + \tfrac{1}{6}|\beta|^2|0\rangle\langle0| + \tfrac{1}{3}\beta\alpha^\star|1\rangle\langle0| + \tfrac{1}{6}|\beta|^2|1\rangle\langle1|$$

$$+ \tfrac{2}{3}|\beta|^2|1\rangle\langle1| + \tfrac{1}{3}\beta\alpha^\star|1\rangle\langle0| + \tfrac{1}{6}|\alpha|^2|0\rangle\langle0| + \tfrac{1}{3}\alpha\beta^\star|0\rangle\langle1| + \tfrac{1}{6}|\alpha|^2|1\rangle\langle1|. \tag{B.20}$$

It is easy to see that this expression is equal to (5.68). Since the two-qubit density matrix ρ_{12} is symmetric under exchange of the two qubits, we have $\rho_2 = \rho_1$.

Exercise 5.8   If the state $|\psi\rangle$ has Schmidt decomposition (5.82), then

$$\rho_1 = \sum_i p_i\,|i\rangle_1\,_1\langle i|, \quad \rho_2 = \sum_i p_i\,|i'\rangle_2\,_2\langle i'|, \quad \rho_3 = \sum_i p_i\,|i''\rangle_3\,_3\langle i''|. \tag{B.21}$$

This means that the reduced density matrices ρ_1, ρ_2 and ρ_3 should have the same spectrum. Therefore, to solve this exercise, it will be sufficient to provide a counterexample. For instance, the state

$$|\psi\rangle = \tfrac{1}{\sqrt{2}}\left(|01\rangle_{12} + |10\rangle_{12}\right) \otimes |0\rangle_3 \tag{B.22}$$

does not satisfy the requirement of equal spectra for ρ_1, ρ_2 and ρ_3. Indeed, this state is the tensor product of a Bell state for the first two qubits and a separable state for the third qubit, and therefore $\rho_1 = \tfrac{1}{2}I_1$, $\rho_2 = \tfrac{1}{2}I_2$, $\rho_3 = |0\rangle_3\,_3\langle0|$. This means that ρ_1 and ρ_2 have eigenvalues $p_1 = p_2 = \tfrac{1}{2}$, while ρ_3 has a single eigenvalue equal to 1.

Exercise 5.9   Let us consider the system of equations (5.87), for the case in which there is a two-qubit system and we add two ancillary qubits to purify it. The two-qubit density matrix ρ_1 is a 4×4 matrix. Taking into account that $(\rho_1)_{ji} = (\rho_1)_{ij}^\star$ $(i,j = 1,\ldots,4)$ and that $\mathrm{Tr}\,\rho_1 = 1$, ρ_1 is determined by 15 independent real parameters. The pure state of the extended 4-qubit system depends on 16 complex coefficients $c_{i\alpha}$ $(i, \alpha = 1, \ldots, 4)$, namely on 32 real parameters, 30 of which are independent, taking into account the normalization condition and the existence of an arbitrary global phase. Therefore, we have the freedom to set $30 - 15 = 15$ real parameters. We take

$$c_{12} = c_{13} = c_{14} = c_{23} = c_{24} = c_{34} = 0 \tag{B.23}$$

and the coefficients c_{22}, c_{33} and c_{44} real and positive. We now solve Eq. (5.87); that is, we determine the coefficients $c_{i\alpha}$ as a function of the matrix elements $(\rho_1)_{ij}$. We obtain

$$(\rho_1)_{11} = \sum_\alpha c_{1\alpha} c_{1\alpha}^\star = c_{11} c_{11}^\star. \tag{B.24}$$

We can exploit the existence of an arbitrary global phase to choose also c_{11} real and positive, so that

$$c_{11} = \sqrt{(\rho_1)_{11}}. \tag{B.25}$$

Then we obtain

$$(\rho_1)_{12} = \sum_\alpha c_{1\alpha} c_{2\alpha}^\star = c_{11} c_{21}^\star, \tag{B.26}$$

and therefore

$$c_{21} = \frac{(\rho_1)_{12}^\star}{c_{11}} = \frac{(\rho_1)_{12}^\star}{\sqrt{(\rho_1)_{11}}}. \tag{B.27}$$

Similarly, we obtain

$$c_{31} = \frac{(\rho_1)_{13}^\star}{\sqrt{(\rho_1)_{11}}}, \qquad c_{41} = \frac{(\rho_1)_{14}^\star}{\sqrt{(\rho_1)_{11}}}. \tag{B.28}$$

Then we use

$$(\rho_1)_{22} = \sum_\alpha c_{2\alpha} c_{2\alpha}^\star = \frac{|(\rho_1)_{12}|^2}{(\rho_1)_{11}} + |c_{22}|^2 \tag{B.29}$$

to extract

$$c_{22} = \sqrt{(\rho_1)_{22} - \frac{|(\rho_1)_{12}|^2}{(\rho_1)_{11}}}. \tag{B.30}$$

We can now derive c_{32} and c_{42} from the equations

$$(\rho_1)_{23} = \frac{(\rho_1)_{12}^\star (\rho_1)_{13}}{(\rho_1)_{11}} + \sqrt{(\rho_1)_{22} - \frac{|(\rho_1)_{12}|^2}{(\rho_1)_{11}}} \, c_{32}^\star$$

$$(\rho_1)_{24} = \frac{(\rho_1)_{12}^\star (\rho_1)_{14}}{(\rho_1)_{11}} + \sqrt{(\rho_1)_{22} - \frac{|(\rho_1)_{12}|^2}{(\rho_1)_{11}}} \, c_{42}^\star. \tag{B.31}$$

Finally, we obtain c_{33}, c_{43} and c_{44} from the equations

$$(\rho_1)_{33} = |c_{31}|^2 + |c_{32}|^2 + |c_{33}|^2,$$

$$(\rho_1)_{34} = c_{31}c_{41}^* + c_{32}c_{42}^* + c_{33}c_{43}^*,$$

$$(\rho_1)_{44} = \sum_{\alpha=1}^{4} |c_{4\alpha}|^2. \tag{B.32}$$

Exercise 5.10 T is positive since $\rho_1' = T(\rho_1) = \rho_1^T$ has the same eigenvalues as ρ_1, and is therefore non-negative as is ρ_1. Let us now show that T is not completely positive. The density operator ρ corresponding to the state (5.106) is given by

$$\rho_{1E} = \tfrac{1}{2}\Big(|0\rangle_{1}\,_{1}\langle 0| \otimes |1\rangle_{EE}\langle 1| + |1\rangle_{1}\,_{1}\langle 1| \otimes |0\rangle_{EE}\langle 0|$$
$$+ |0\rangle_{1}\,_{1}\langle 1| \otimes |1\rangle_{EE}\langle 0| + |1\rangle_{1}\,_{1}\langle 0| \otimes |0\rangle_{EE}\langle 1|\Big). \tag{B.33}$$

Therefore,

$$\rho_{1E}' \equiv (T \otimes \mathcal{I}_E)(\rho_{1E}) = \tfrac{1}{2}\Big(|0\rangle_{1}\,_{1}\langle 0| \otimes |1\rangle_{E}\,_{E}\langle 1| + |1\rangle_{1}\,_{1}\langle 1| \otimes |0\rangle_{E}\,_{E}\langle 0|$$
$$+ |1\rangle_{1}\,_{1}\langle 0| \otimes |1\rangle_{E}\,_{E}\langle 0| + |0\rangle_{1}\,_{1}\langle 1| \otimes |0\rangle_{E}\,_{E}\langle 1|\Big).$$
$$\tag{B.34}$$

The matrix representation of ρ_{1E}' in the computational basis is given by

$$\rho_{1E}' = \frac{1}{2}\begin{bmatrix} 0 & 0 & 0 & 1 \\ 0 & 1 & 0 & 0 \\ 0 & 0 & 1 & 0 \\ 1 & 0 & 0 & 0 \end{bmatrix}, \tag{B.35}$$

whose eigenvalues are $\lambda_0 = -\tfrac{1}{2}$ and $\lambda_1 = \lambda_2 = \lambda_3 = \tfrac{1}{2}$. Therefore, $T \otimes I_E$ is not positive, implying that T is not completely positive.

Exercise 5.11 The state of the system after having obtained n times the outcome 0 from the generalized measurement (5.134) is given by

$$|\psi^{(n)}\rangle = \alpha^{(n)}|0\rangle + \beta^{(n)}|1\rangle$$
$$\approx \alpha \left(1 + \tfrac{1}{2}|\beta|^2\theta^2\right)^n |0\rangle + \beta \left(1 - \tfrac{1}{2}|\alpha|^2\theta^2\right)^n |1\rangle. \tag{B.36}$$

We therefore obtain

$$\frac{\beta^{(n)}}{\alpha^{(n)}} = \frac{\left(1 - \tfrac{1}{2}|\alpha|^2\,\theta^2\right)^n}{\left(1 + \tfrac{1}{2}|\beta|^2\,\theta^2\right)^n}\,\frac{\beta}{\alpha} \approx \exp\!\left(-\tfrac{1}{2}\,n\theta^2\right)\frac{\beta}{\alpha} \tag{B.37}$$

and, since $|\alpha^{(n)}|^2 + |\beta^{(n)}|^2 = 1$,

$$|\alpha^{(n)}|^2 \approx \frac{|\alpha|^2}{|\alpha|^2 + |\beta|^2 \exp(-n\theta^2)}, \quad |\beta^{(n)}|^2 \approx \frac{|\beta|^2 \exp(-n\theta^2)}{|\alpha|^2 + |\beta|^2 \exp(-n\theta^2)}. \quad (B.38)$$

The probability of obtaining outcome 0 at the n-th weak measurement, provided the same outcome 0 was obtained in all previous measurements, is

$$p_0^{(n)} = |\alpha^{(n)}|^2 + |\beta^{(n)}|^2 \cos^2 \theta \approx 1 - |\beta^{(n)}|^2 \theta^2. \quad (B.39)$$

The probability that the outcome 0 is always obtained is

$$\begin{aligned}
p_0 &= \prod_{n=0}^{+\infty} p_0^{(n)} \approx \prod_{n=0}^{+\infty} \left(1 - |\beta^{(n)}|^2 \theta^2\right) \\
&\approx \prod_{n=0}^{+\infty} \left(1 - \frac{|\beta|^2 \exp(-n\theta^2)}{|\alpha|^2 + |\beta|^2 \exp(-n\theta^2)} \theta^2\right) \\
&= \prod_{n=0}^{+\infty} \left(\frac{|\alpha|^2 + |\beta|^2 \exp(-n\theta^2)(1 - \theta^2)}{|\alpha|^2 + |\beta|^2 \exp(-n\theta^2)}\right) \\
&\approx \prod_{n=0}^{+\infty} \frac{|\alpha|^2 + |\beta|^2 \exp[-(n+1)\theta^2]}{|\alpha|^2 + |\beta|^2 \exp(-n\theta^2)} \\
&= \left(\frac{|\alpha|^2 + |\beta|^2 \exp(-\theta^2)}{|\alpha|^2 + |\beta|^2}\right) \left(\frac{|\alpha|^2 + |\beta|^2 \exp(-2\theta^2)}{|\alpha|^2 + |\beta|^2 \exp(-\theta^2)}\right) \cdots \\
&= \lim_{n \to +\infty} \frac{|\alpha|^2 + |\beta|^2 \exp[-(n+1)\theta^2]}{|\alpha|^2 + |\beta|^2} = |\alpha|^2. \quad (B.40)
\end{aligned}$$

Exercise 5.12 To find the maximum Shannon entropy, defined by Eq. (5.150), we must solve the system

$$\begin{cases}
\dfrac{\partial}{\partial p_j} \left[-\displaystyle\sum_{i=1}^{k} p_i \log p_i - \lambda \left(\sum_{i=1}^{k} p_i - 1\right)\right] = 0, \quad (j = 1, \dots, k), \\
\displaystyle\sum_{i=1}^{k} p_i = 1,
\end{cases} \quad (B.41)$$

where the Lagrange multiplier λ is introduced to take into account the constraint $\sum_i p_i = 1$. The system of equations (B.41) is solved for $p_1 = \cdots = p_k = 1/k$.

Exercise 5.13

$$\log\left(\frac{n!}{\prod_{i=1}^{k}(np_i)!}\right) = \log n! - \sum_{i=1}^{k}\log(np_i)!$$

$$= n\log n - \frac{n}{\ln 2} - \sum_{i=1}^{k}\left[np_i\log(np_i) - \frac{np_i}{\ln 2}\right]$$

$$= -n\sum_{i=1}^{k}p_i\log p_i = nH(p_1,\ldots,p_k). \tag{B.42}$$

Note that, in order to apply Stirling's formula, we have assumed that $np_i \gg 1$, for all i.

Exercise 5.14 This quantum channel does not change the states sent by Alice, and therefore we recover the special case $\theta = 0$ of the example discussed in Sec. 5.11.2.

Exercise 5.15 The Bloch vectors $\boldsymbol{r}_0 = (0,0,1)$ and $\boldsymbol{r}_1 = (0,0,-1)$, corresponding to the states $|0\rangle$ and $|1\rangle$ sent by Alice, are modified by the quantum channel as follows:

$$\boldsymbol{r}_0 \to \tilde{\boldsymbol{r}}_0 = \boldsymbol{r}_0, \qquad \boldsymbol{r}_1 \to \tilde{\boldsymbol{r}}_1 = (0,0,\sin^2\theta - \cos^2\theta). \tag{B.43}$$

Let $\tilde{\rho}_0$ and $\tilde{\rho}_1$ denote the density matrices corresponding to the Bloch vectors $\tilde{\boldsymbol{r}}_0$ and $\tilde{\boldsymbol{r}}_1$, respectively. Notice that for $\theta = 0$ this quantum channel reduces to the identity.

In order to compute the mutual information of Alice and Bob, we first need the conditional probabilities

$$p(y|x) = \text{Tr}(\tilde{\rho}_x F_y) \qquad (x,y = 0,1,2), \tag{B.44}$$

where the von Neumann projectors F_0 and F_1 are given by Eq. (5.207). If we choose the measurement axis $\hat{n}$ in the (x,z) plane of the Bloch sphere (see Fig. 5.11); that is, $\hat{n} = (\sin\bar{\theta}, 0, \cos\bar{\theta})$, we obtain

$$\begin{aligned}
p(0|0) &= \tfrac{1}{2}(1 + \cos\bar{\theta}), & p(1|0) &= \tfrac{1}{2}(1 - \cos\bar{\theta}), \\
p(0|1) &= \tfrac{1}{2}[1 - \cos 2\theta\cos\bar{\theta}], & p(1|1) &= \tfrac{1}{2}[1 + \cos 2\theta\cos\bar{\theta}].
\end{aligned} \tag{B.45}$$

We now compute $p(x,y) = p(x)p(y|x)$. We know that the states $|0\rangle$ and $|1\rangle$ are sent by Alice with probabilities $p(X = 0) = p$ and $p(X = 1) = 1 - p$, respectively. Therefore, we have

$$\begin{aligned}
p(0,0) &= \tfrac{1}{2}p(1 + \cos\bar{\theta}), & p(0,1) &= \tfrac{1}{2}p(1 - \cos\bar{\theta}), \tag{B.46} \\
p(1,0) &= \tfrac{1}{2}(1 - p)[1 - \cos 2\theta\cos\bar{\theta}], & p(1,1) &= \tfrac{1}{2}(1 - p)[1 + \cos 2\theta\cos\bar{\theta}].
\end{aligned}$$

Then we compute $p(y) = \sum_x p(x,y)$ and obtain

$$p(Y = 0) = \tfrac{1}{2}\left\{1 + \cos\bar{\theta}\left[p - (1-p)\cos 2\theta\right]\right\},$$
$$p(Y = 1) = \tfrac{1}{2}\left\{1 - \cos\bar{\theta}\left[p - (1-p)\cos 2\theta\right]\right\}. \tag{B.47}$$

Finally, we insert the expressions derived for $p(x)$, $p(y)$, and $p(x,y)$ into (5.201), obtaining the mutual information $I(X{:}Y)$.

A few examples of $I(X{:}Y)$ as a function of the parameters p, θ, and $\bar{\theta}$ are shown in Figs. B.1-B.3. As a general result, the mutual information is maximized for measurements along the z-axis; that is, when $\bar{\theta} = 0, \pi$. This result is demonstrated in Fig. B.1 for fixed p and θ. In Fig. B.2, we show that $I = 0$ when $\theta = \pi/2$. This is quite natural since for $\theta = \pi/2$ the quantum channel maps the entire Bloch sphere onto a single point, namely the north pole of the sphere. Therefore, whatever state Alice sends, Bob always receive the state $|0\rangle$. Hence, there is no transmission of information. Finally, in Fig. B.3 we show the mutual information as a function of p, for given θ and $\bar{\theta}$.

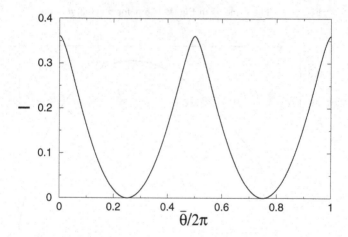

Fig. B.1 The mutual information $I(X{:}Y)$ for a message transmitted as described in exercise 5.15, for $p = 0.9$, $\theta = \pi/8$.

Exercise 5.16 For a separable state decomposition (5.232) holds. Therefore,

$$\langle(\Sigma_i)^2\rangle = \text{Tr}\left[\sum_k p_k \rho_{Ak} \otimes \rho_{Bk}(\Sigma_i)^2\right]$$
$$= \sum_k p_k \,\text{Tr}\left[\rho_{Ak} \otimes \rho_{Bk}(\Sigma_i)^2\right] = \sum_k p_k \langle(\Sigma_i)^2\rangle_k, \tag{B.48}$$

where $\langle\dots\rangle_k$ denotes the average over the density operator $\rho_{kA} \otimes \rho_{kB}$. We then

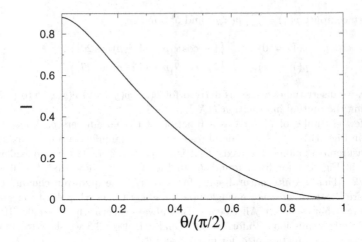

Fig. B.2   The same as in Fig. B.1, but for $p = 0.3$, $\bar{\theta} = 0$.

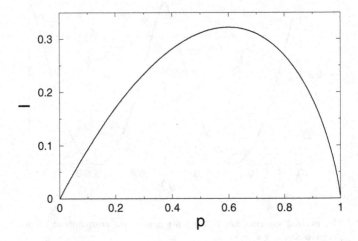

Fig. B.3   The same as in Fig. B.1, but for $\theta = \pi/4$, $\bar{\theta} = 0$.

obtain

$$
\begin{aligned}
\langle(\Delta\Sigma_i)^2\rangle &= \sum_k p_k \langle(\Sigma_i)^2\rangle_k - \langle\Sigma_i\rangle^2 \\
&= \sum_k p_k \left[\langle(\sigma_i^{(A)})^2\rangle_k + 2\langle\sigma_i^{(A)}\rangle_k\langle\sigma_i^{(B)}\rangle_k + \langle(\sigma_i^{(B)})^2\rangle_k\right] - \langle\Sigma_i\rangle^2 \\
&= \sum_k p_k \left[\langle(\Delta\sigma_i^{(A)})^2\rangle_k + \langle(\Delta\sigma_i^{(B)})^2\rangle_k + \langle\Sigma_i\rangle_k^2\right] - \left(\sum_k p_k\langle\Sigma_i\rangle_k\right)^2.
\end{aligned}
$$

$$(B.49)$$

Finally, we apply the Cauchy–Schwarz inequality (see Sec. 2.3)

$$\sum_l p_l \sum_k p_k \langle \Sigma_i \rangle_k^2 \geq \left(\sum_k p_k \langle \Sigma_i \rangle \right)^2 \tag{B.50}$$

to see that the last two terms in (B.49) are bounded from below by zero. Hence,

$$\langle (\Delta \Sigma_i)^2 \rangle \geq \sum_k p_k \left[\langle (\Delta \sigma_i^{(A)})^2 \rangle_k + \langle (\Delta \sigma_i^{(B)})^2 \rangle_k \right]. \tag{B.51}$$

As this latter inequality is valid for $i = x, y, z$, we obtain

$$\langle (\Delta \Sigma_x)^2 \rangle + \langle (\Delta \Sigma_y)^2 \rangle + \langle (\Delta \Sigma_z)^2 \rangle \geq$$

$$\sum_k p_k \left[\langle (\Delta \sigma_x^{(A)})^2 \rangle_k + \langle (\Delta \sigma_y^{(A)})^2 \rangle_k + \langle (\Delta \sigma_z^{(A)})^2 \rangle_k \right.$$

$$\left. + \langle (\Delta \sigma_x^{(B)})^2 \rangle_k + \langle (\Delta \sigma_y^{(B)})^2 \rangle_k + \langle (\Delta \sigma_z^{(B)})^2 \rangle_k \right]. \tag{B.52}$$

Note that we have bounded from below a sum of variances of a two-qubit state by a sum of variances of single-qubit states, which are quantities much easier to compute experimentally. We now compute, for a single qubit, $\langle (\Delta \sigma_x)^2 \rangle_k + \langle (\Delta \sigma_y)^2 \rangle_k + \langle (\Delta \sigma_z)^2 \rangle_k$. After writing the single-qubit density matrix as in (5.32), it is easy to check that

$$\langle (\Delta \sigma_x)^2 \rangle_k + \langle (\Delta \sigma_y)^2 \rangle_k + \langle (\Delta \sigma_z)^2 \rangle_k = 3 - (x^2 + y^2 + z^2) \geq 2. \tag{B.53}$$

Finally, we substitute this inequality into (B.52) and obtain (5.238). If a given state ρ_{AB} does not satisfy the inequality (5.238), we can therefore conclude that ρ_{AB} is entangled.

In the case of the Werner state (5.235) we obtain by direct computation

$$\langle \Sigma_i \rangle = \text{Tr}\left[(\rho_W)_{AB} \Sigma_i \right] = 0, \quad \langle (\Sigma_i)^2 \rangle = \text{Tr}\left[(\rho_W)_{AB} (\Sigma_i)^2 \right] = 2 - 2p, \tag{B.54}$$

with $i = x, y, z$. Therefore,

$$\langle (\Delta \Sigma_x)^2 \rangle + \langle (\Delta \Sigma_y)^2 \rangle + \langle (\Delta \Sigma_z)^2 \rangle = 6 - 6p, \tag{B.55}$$

so that the inequality (5.238) is satisfied when $p \leq \frac{1}{3}$. We can therefore conclude that the Werner state is entangled for $\frac{1}{3} < p \leq 1$. Note that the separability criterion (5.238) is much easier to use in experiments than the Peres criterion since it only requires that the single particle polarizations are measured. The Peres criterion can instead be applied only after that the entire two-qubit density matrix is reconstructed. Finally, we point out that a separability criterion for continuous variable systems can be derived following the same procedure used in this exercise (see Duan *et al.*, 2000).

Chapter 6

Exercise 6.1 The condition $\sum_k E_k^\dagger E_k = I$ implies that

$$\sum_k (\gamma_k^* I + a_k^* \cdot \sigma)(\gamma_k I + a_k \cdot \sigma) = I. \tag{B.56}$$

Using the relation (see exercise 3.6)

$$(a \cdot \sigma)(b \cdot \sigma) = (a \cdot b)I + i\sigma \cdot (a \times b), \tag{B.57}$$

we obtain

$$\sum_k |\gamma_k|^2 I + \sum_k [\gamma_k^* a_k + \gamma_k a_k^*] \cdot \sigma + \sum_k (a_k^* \cdot \sigma)(a_k \cdot \sigma)$$

$$= \sum_k |\gamma_k|^2 I + \sum_k [\gamma_k^* a_k + \gamma_k a_k^*] \cdot \sigma + \sum_k (a_k^* \cdot a_k)I + i \sum_k (a_k^* \times a_k) \cdot \sigma$$

$$= I. \tag{B.58}$$

This implies that

$$\sum_k |\gamma_k|^2 + \sum_k (a_k^* \cdot a_k) = 1 \tag{B.59}$$

$$\sum_k (\gamma_k^* a_k + \gamma_k a_k^*) = i \sum_k a_k \times a_k^*. \tag{B.60}$$

It is convenient to employ the Bloch sphere representation (6.6) for ρ and $\rho' = \sum_k E_k \rho E_k^\dagger$. We have

$$I + r' \cdot \sigma = \sum_k (\gamma_k I + a_k \cdot \sigma)(I + r \cdot \sigma)(\gamma_k^* I + a_k^* \cdot \sigma)$$

$$= \sum_k \Big[|\gamma_k|^2 I + \gamma_k a_k^* \cdot \sigma + |\gamma_k|^2 (r \cdot \sigma) + \gamma_k (r \cdot \sigma)(a_k^* \cdot \sigma) + \gamma_k^* a_k \cdot \sigma$$

$$+ (a_k \cdot \sigma)(a_k^* \cdot \sigma) + \gamma_k^* (a_k \cdot \sigma)(r \cdot \sigma) + (a_k \cdot \sigma)(r \cdot \sigma)(a_k^* \cdot \sigma) \Big]. \tag{B.61}$$

Let us examine the eight terms of the right-hand side of this equations separately. The sum of the fourth and the seventh term simplifies as follows:

$$\sum_k \left[\gamma_k (r \cdot \sigma)(a_k^* \cdot \sigma) + \gamma_k^* (a_k \cdot \sigma)(r \cdot \sigma) \right]$$

$$= \sum_k \left[\gamma_k (r \cdot a_k^*)I + i\gamma_k \sigma \cdot r \times a_k^* + \gamma_k^* (r \cdot a_k)I + i\gamma_k^* \sigma \cdot a_k \times r \right]$$

$$= \sum_k \left[(\gamma_k a_k^* + \gamma_k^* a_k) \cdot r I + i(\gamma_k^* a_k - \gamma_k a_k^*) \times r \cdot \sigma \right]. \tag{B.62}$$

The sixth term can be written as

$$\sum_k (a_k \cdot \sigma)(a_k^* \cdot \sigma) = \sum_k \left[(a_k \cdot a_k^*) I + i\sigma \cdot a_k \times a_k^* \right]. \tag{B.63}$$

Finally, for the eighth term we have

$$\sum_k (a_k \cdot \sigma)(r \cdot \sigma)(a_k^* \cdot \sigma) = \sum_k \left\{ (a_k \cdot \sigma)\left[(r \cdot a_k^*) I + i\sigma \cdot r \times a_k^* \right] \right\}$$

$$= \sum_k \left[(r \cdot a_k^*)(a_k \cdot \sigma) + i(a_k \cdot \sigma)(r \times a_k^* \cdot \sigma) \right]$$

$$= \sum_k \left\{ (r \cdot a_k^*)(a_k \cdot \sigma) + i\left[(a_k \cdot r \times a_k^*) I + i\sigma \cdot a_k \times (r \times a_k^*) \right] \right\}$$

$$= \sum_k \left\{ (r \cdot a_k^*)(a_k \cdot \sigma) + i(a_k \cdot r \times a_k^*) I - \sigma \cdot \left[(a_k \cdot a_k^*) r - (a_k \cdot r) a_k^* \right] \right\}. \tag{B.64}$$

Note that we have used the relation

$$a \times (b \times c) = (a \cdot c) b - (a \cdot b) c, \tag{B.65}$$

with $a \to a_k$, $b \to r$, and $c \to a_k^*$. After insertion of Eqs. (B.62–B.64) into Eq. (B.61), we obtain

$$I + r' \cdot \sigma = \sum_k \left\{ |\gamma_k|^2 I + (\gamma_k a_k^* + \gamma_k^* a_k) \cdot \sigma + |\gamma_k|^2 (r \cdot \sigma) + (\gamma_k a_k^* + \gamma_k^* a_k) \cdot r I \right.$$

$$+ i(\gamma_k^* a_k - \gamma_k a_k^*) \times r \cdot \sigma + (a_k \cdot a_k^*) I + i\sigma \cdot a_k \times a_k^* + (r \cdot a_k^*)(a_k \cdot \sigma)$$

$$\left. + i(r \cdot a_k^* \times a_k) I - \sigma \cdot \left[(a_k \cdot a_k^*) r - (a_k \cdot r) a_k^* \right] \right\}. \tag{B.66}$$

We can simplify Eq. (B.66) taking advantage of Eqs. (B.59–B.60). We obtain

$$r' \cdot \sigma = \sum_k \left\{ 2i(a_k \times a_k^*) \cdot \sigma + \left[|\gamma_k|^2 - (a_k \cdot a_k^*) \right] (r \cdot \sigma) + i(\gamma_k^* a_k - \gamma_k a_k^*) \times r \cdot \sigma \right.$$

$$\left. + \left[(a_k \cdot \sigma)(r \cdot a_k^*) + (a_k^* \cdot \sigma)(r \cdot a_k) \right] \right\}. \tag{B.67}$$

Therefore, $r' = Mr + c$, with

$$c = 2i \sum_k (a_k \times a_k^*), \tag{B.68}$$

that is

$$c_j = 2i \sum_{klm} \epsilon_{jlm} a_{kl} a_{km}^*. \tag{B.69}$$

Similarly, one can see that M is given by Eq. (6.9).

Exercise 6.2   Let us define $S \equiv \sqrt{M^T M}$. The matrix S is by definition symmetric. Therefore, it is diagonalizable and we can write its spectral decomposition $S = \sum_i s_i |i\rangle\langle i|$. As S is a non-negative operator, we have $s_i \geq 0$. We now define $|\psi_i\rangle \equiv M|i\rangle$. We see that $\langle \psi_i|\psi_i\rangle = \langle i|M^T M|i\rangle = s_i^2$. For $s_i \neq 0$, we define $|\alpha_i\rangle \equiv |\psi_i\rangle/s_i$. The vectors $|\alpha_i\rangle$ are normalized and orthogonal. We employ the Gram–Schmidt decomposition to complete the orthonormal basis $\{|\alpha_i\rangle\}$. Finally, let us define the operator $O \equiv \sum_i |\alpha_i\rangle\langle i|$. This operator is orthogonal since $O^T O = \sum_{i,j} |j\rangle\langle\alpha_j|\alpha_i\rangle\langle i| = \sum_i |i\rangle\langle i| = I$. When $s_i \neq 0$ we have $OS|i\rangle = s_i O|i\rangle = s_i|\alpha_i\rangle = |\psi_i\rangle = M|i\rangle$; when $s_i = 0$, $OS|i\rangle = 0 = |\psi_i\rangle = M|i\rangle$. Since the actions of the linear operators M and OS on the basis $\{|i\rangle\}$ coincide, then $M = OS$.

Exercise 6.3   Let A, B, C and B denote the quantum gates of Fig. 6.6, from left to right. We have

$$
A = \begin{bmatrix} C & 0 & -S & 0 \\ 0 & C & 0 & -S \\ S & 0 & C & 0 \\ 0 & S & 0 & C \end{bmatrix}, \quad
B = \begin{bmatrix} 1 & 0 & 0 & 0 \\ 0 & 0 & 0 & 1 \\ 0 & 0 & 1 & 0 \\ 0 & 1 & 0 & 0 \end{bmatrix},
$$

$$
C = \begin{bmatrix} C & 0 & S & 0 \\ 0 & C & 0 & S \\ -S & 0 & C & 0 \\ 0 & -S & 0 & C \end{bmatrix}, \tag{B.70}
$$

where $C \equiv \cos\frac{\theta}{2}$ and $S \equiv \sin\frac{\theta}{2}$. The action of the circuit in Fig. 6.6 is described by the unitary operator

$$
U = BCBA = \begin{bmatrix} 1 & 0 & 0 & 0 \\ 0 & C^2 - S^2 & 0 & -2CS \\ 0 & 0 & 1 & 0 \\ 0 & 2CS & 0 & C^2 - S^2 \end{bmatrix}. \tag{B.71}
$$

We have $\rho_{\text{fin}}^{(\text{tot})} = U \rho_{\text{in}}^{(\text{tot})} U^\dagger$, with

$$
\rho_{\text{in}}^{(\text{tot})} = \begin{bmatrix} \rho & 0 \\ 0 & 0 \end{bmatrix}. \tag{B.72}
$$

The Kraus operator F_k is defined as $F_k = {}_e\langle k|U|0\rangle_e$, where the subscript e refers to the environmental qubit. Since $(F_k)_{ij} = \langle ki|U|0j\rangle$, it is easy to see that U is represented as the block matrix

$$
U = \begin{bmatrix} F_0 & .. \\ F_1 & .. \end{bmatrix}, \tag{B.73}
$$

where

$$F_0 = \begin{bmatrix} 1 & 0 \\ 0 & C^2-S^2 \end{bmatrix} = \begin{bmatrix} 1 & 0 \\ 0 & \cos\theta \end{bmatrix}, \quad F_1 = \begin{bmatrix} 0 & 0 \\ 0 & 2CS \end{bmatrix} = \begin{bmatrix} 0 & 0 \\ 0 & \sin\theta \end{bmatrix}. \quad \text{(B.74)}$$

Exercise 6.4 Following the same procedure of Sec. 6.1.3, we obtain

$$\rho' = |\alpha|^2 U\rho U^\dagger + |\beta|^2 \rho. \quad \text{(B.75)}$$

Therefore, the Kraus operators are given by $E_0 = |\beta|I$ and $E_1 = |\alpha|U$.

Exercise 6.5 Let A, B, C, and D denote the quantum gates of Fig. 6.9, from left to right. We have

$$A = \begin{bmatrix} C & 0 & -S & 0 \\ 0 & C & 0 & -S \\ S & 0 & C & 0 \\ 0 & S & 0 & C \end{bmatrix}, \quad B = \begin{bmatrix} 1 & 0 & 0 & 0 \\ 0 & 0 & 0 & 1 \\ 0 & 0 & 1 & 0 \\ 0 & 1 & 0 & 0 \end{bmatrix},$$

$$C = \begin{bmatrix} C & 0 & S & 0 \\ 0 & C & 0 & S \\ -S & 0 & C & 0 \\ 0 & -S & 0 & C \end{bmatrix}, \quad D = \begin{bmatrix} 1 & 0 & 0 & 0 \\ 0 & 1 & 0 & 0 \\ 0 & 0 & 0 & 1 \\ 0 & 0 & 1 & 0 \end{bmatrix}, \quad \text{(B.76)}$$

where $C \equiv \cos\frac{\theta}{2}$ and $S \equiv \sin\frac{\theta}{2}$. The action of the circuit in Fig. 6.9 is described by the unitary operator

$$U = DCBA = \begin{bmatrix} 1 & 0 & 0 & 0 \\ 0 & 2CS & 0 & C^2-S^2 \\ 0 & C^2-S^2 & 0 & -2CS \\ 0 & 0 & 1 & 0 \end{bmatrix}. \quad \text{(B.77)}$$

As discussed in the solution of exercise 6.3, the Kraus operators are read from the first two columns of the matrix U; that is ,

$$E_0 = \begin{bmatrix} 1 & 0 \\ 0 & 2CS \end{bmatrix} = \begin{bmatrix} 1 & 0 \\ 0 & \sin\theta \end{bmatrix}, \quad E_1 = \begin{bmatrix} 0 & C^2-S^2 \\ 0 & 0 \end{bmatrix} = \begin{bmatrix} 0 & \cos\theta \\ 0 & 0 \end{bmatrix}. \quad \text{(B.78)}$$

It is instructive to derive the transformation of the Bloch sphere. We have

$$
\rho' = \frac{1}{2}\begin{bmatrix} 1+z' & x'-iy' \\ x'+iy' & 1-z' \end{bmatrix} = E_0\,\rho\,E_0^\dagger + E_1\,\rho\,E_1^\dagger
$$

$$
= \begin{bmatrix} 1 & 0 \\ 0 & \sin\theta \end{bmatrix}\frac{1}{2}\begin{bmatrix} 1+z & x-iy \\ x+iy & 1-z \end{bmatrix}\begin{bmatrix} 1 & 0 \\ 0 & \sin\theta \end{bmatrix}
$$

$$
+ \begin{bmatrix} 0 & \cos\theta \\ 0 & 0 \end{bmatrix}\frac{1}{2}\begin{bmatrix} 1+z & x-iy \\ x+iy & 1-z \end{bmatrix}\begin{bmatrix} 0 & 0 \\ \cos\theta & 0 \end{bmatrix}
$$

$$
= \frac{1}{2}\begin{bmatrix} 1+\cos^2\theta+z\sin^2\theta & (x-iy)\sin\theta \\ (x+iy)\sin\theta & 1-(\cos^2\theta+z\sin^2\theta) \end{bmatrix}. \tag{B.79}
$$

Therefore,

$$
x' = x\sin\theta, \qquad y' = y\sin\theta, \qquad z' = \cos^2\theta + z\sin^2\theta. \tag{B.80}
$$

These equations show that the Bloch sphere $x^2 + y^2 + z^2$ is deformed into the ellipsoid

$$
\frac{x'^2 + y'^2}{\sin^2\theta} + \frac{(z' - \cos^2\theta)^2}{\sin^4\theta}. \tag{B.81}
$$

This ellipsoid has z as symmetry axis and centre $(0, 0, \cos^2\theta)$. A displacement of the centre of the Bloch sphere necessarily demands a deformation of the sphere, if we wish ρ' to still represent a density matrix. Note that Eq. (B.81) corresponds to the minimum deformation required to the Bloch sphere in order to displace its centre along the z-axis by $\cos^2\theta$. Indeed, the ellipsoid (B.81) and the Bloch sphere have a higher order tangency (see Fig. B.4).

Exercise 6.6   By direct computation we obtain

$$
R_z\rho R_z^\dagger = \begin{bmatrix} e^{-i\frac{\theta}{2}} & 0 \\ 0 & e^{i\frac{\theta}{2}} \end{bmatrix}\begin{bmatrix} p & \alpha \\ \alpha^\star & 1-p \end{bmatrix}\begin{bmatrix} e^{i\frac{\theta}{2}} & 0 \\ 0 & e^{-i\frac{\theta}{2}} \end{bmatrix} = \begin{bmatrix} p & \alpha e^{-i\theta} \\ \alpha^\star e^{i\theta} & 1-p \end{bmatrix}. \tag{B.82}
$$

It is now sufficient to use

$$
\frac{1}{\sigma\sqrt{2\pi}}\int_{-\infty}^{\infty} e^{-i\theta}e^{-\frac{\theta^2}{2\sigma^2}}\,d\theta = e^{-\frac{\sigma^2}{2}}, \tag{B.83}
$$

with $\sigma = \sqrt{2\lambda}$, to obtain

$$
\rho' = \begin{bmatrix} p & \alpha e^{-\lambda} \\ \alpha^\star e^{-\lambda} & 1-p \end{bmatrix}. \tag{B.84}
$$

As $p = \frac{1}{2}(1 + z') = \frac{1}{2}(1 + z)$ and $\alpha e^{-\lambda} = \frac{1}{2}(x' - iy') = e^{-\lambda}\frac{1}{2}(x - iy)$, Eq. (6.54) immediately follows.

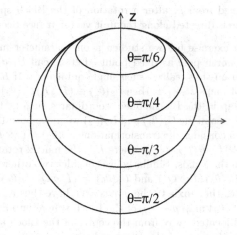

Fig. B.4 A visualization of the minimum deformation required to displace the centre of the Bloch sphere along the z-axis. The horizontal axis may be any axis in the (x, y) plane.

Exercise 6.7 Let ρ_1 denote the density matrix describing the less significant qubit in Fig. 6.10 after the action of the two-qubit unitary transformation D. We have

$$\rho_1 = E_0 \rho E_0^\dagger + E_1 \rho E_1\dagger, \tag{B.85}$$

where the Kraus operators E_0 and E_1 are read directly from the first two columns of D:

$$E_0 = \begin{bmatrix} C_0 & 0 \\ 0 & C_1 \end{bmatrix}, \quad E_1 = \begin{bmatrix} S_0 & 0 \\ 0 & S_1 \end{bmatrix}. \tag{B.86}$$

Therefore, we obtain from (B.85) that

$$
\begin{aligned}
\rho_1 &= \frac{1}{2} \begin{bmatrix} 1+z_1 & x_1-iy_1 \\ x_1+iy_1 & 1-z_1 \end{bmatrix} \\
&= \begin{bmatrix} C_0 & 0 \\ 0 & C_1 \end{bmatrix} \frac{1}{2} \begin{bmatrix} 1+z & x-iy \\ x+iy & 1-z \end{bmatrix} \begin{bmatrix} C_0 & 0 \\ 0 & C_1 \end{bmatrix} \\
&\quad + \begin{bmatrix} S_0 & 0 \\ 0 & S_1 \end{bmatrix} \frac{1}{2} \begin{bmatrix} 1+z & x-iy \\ x+iy & 1-z \end{bmatrix} \begin{bmatrix} S_0 & 0 \\ 0 & S_1 \end{bmatrix} \\
&= \frac{1}{2} \begin{bmatrix} 1+z & (x-iy)(C_0C_1+S_0S_1) \\ (x+iy)(C_0C_1+S_0S_1) & 1-z \end{bmatrix}.
\end{aligned}
\tag{B.87}
$$

Thus,

$$x_1 = (C_0C_1 + S_0S_1)x = \cos(\theta_0 - \theta_1)x, \quad y_1 = \cos(\theta_0 - \theta_1)y, \quad z_1 = z. \tag{B.88}$$

Finally, ρ' is obtained from ρ_1 after a rotation of the Bloch sphere through an angle ξ about the axis directed along the unit vector n (see Sec. 3.3.1).

Exercise 6.8 In exercise 6.5 we studied a one-parameter map of the Bloch sphere having the north pole as fixed point. Let us call $U_1(\theta_1)$ the two-qubit unitary transformation that realizes such single-qubit map. If R denotes a Bloch sphere rotation that maps $z \rightarrow -z$, then $U_2(\theta_2) \equiv (I \otimes R) U_1(\theta_2) (I \otimes R^\dagger)$ induces a one-parameter map in the Bloch-sphere coordinates such as the south pole is the fixed point. If we combine $U_1(\theta_1)$ and $U_2(\theta_2)$ we move both the north and the south pole. We then consider the transformations $U_3(\theta_3) = (I \otimes P) U_1(\theta_3)(I \otimes P^\dagger)$ and $U_4(\theta_4) = (I \otimes P) U_2(\theta_4) (I \otimes P^\dagger)$, where P is the matrix rotating the z-axis of the Bloch sphere to the x-axis. Similarly, we consider two other transformations, $U_5(\theta_5) = (I \otimes Q) U_1(\theta_5) (I \otimes Q^\dagger)$ and $U_6(\theta_6) = (I \otimes Q) U_2(\theta_6) (I \otimes Q^\dagger)$, where the matrix Q rotates the z-axis to the y-axis. We have thus generated a generic 6-parameter $(\theta_1, \ldots, \theta_6)$ map $\rho \rightarrow \rho_E$ of the Bloch sphere into an ellipsoid whose centre is in general located away from the centre of the Bloch sphere but whose axes are parallel to the axes of the Bloch sphere. If we add two generic three-parameter rotations $W_1(\theta_7, \theta_8, \theta_9)$ and $W_2(\theta_{10}, \theta_{11}, \theta_{12})$, we obtain a generic 12-parameter affine shift of the Bloch sphere: $\rho \rightarrow \rho' = W_2 \rho_E W_1$.

Exercise 6.9 As the unitary operator V acts only on the environmental qubit, it does not modify the system density matrix. Indeed, we have

$$
\begin{aligned}
\rho' &= \mathrm{Tr}_{\mathrm{env}} \left[(V \otimes I) U(|0\rangle\langle 0| \otimes \rho) U^\dagger (V^\dagger \otimes I) \right] \\
&= \mathrm{Tr}_{\mathrm{env}} \left[U(|0\rangle\langle 0| \otimes \rho) U^\dagger (V^\dagger \otimes I)(V \otimes I) \right] \\
&= \mathrm{Tr}_{\mathrm{env}} \left[U(|0\rangle\langle 0| \otimes \rho) U^\dagger \right].
\end{aligned} \tag{B.89}
$$

Exercise 6.10 To solve this exercise, it is useful to remember that any 2×2 unitary matrix U can be seen (up to an overall phase factor) as a rotation through an angle δ about some axis of the Bloch sphere (see Sec. 3.3.1). Hence, we have

$$
U = \cos \tfrac{\delta}{2} I - i \sin \tfrac{\delta}{2} (n \cdot \sigma), \tag{B.90}
$$

where n is the unit vector directed along the rotation axis.
(i) For $U = \sigma_x$, $\delta = -\pi$ and $n = (1, 0, 0)$. Thus, σ_x maps a vector $r = (x, y, z)$ of the Bloch sphere into $r_1 = (x_1, y_1, z_1) = (x, -y, -z)$. As discussed in exercise 6.5, the next four quantum gates map r_1 into $r_2 = (x_2, y_2, z_2) = (x_1 \sin \theta, y_1 \sin \theta, \cos^2 \theta + z_1 \sin^2 \theta)$. Finally, $U^\dagger$ maps r_2 into $r' = (x', y', z') = (x_2, -y_2, -z_2)$. The composition of the above three unitary transformations leads to $r' = (x \sin \theta, y \sin \theta, -\cos^2 \theta + z \sin^2 \theta)$. The fixed point of the transformation $r \rightarrow r'$ is the south pole of the Bloch sphere; that is, $r = (0, 0, -1)$. Note that, if the transformations U and $U^\dagger$ had not been applied, the fixed point would have been the north pole of the Bloch sphere, $r = (0, 0, 1)$.
(ii) We can see from (B.90) that $U = \frac{1}{\sqrt{2}}(I \pm i\sigma_j)$ induces a rotation through an angle $\mp\pi/2$ about the j-axis of the Bloch sphere. Let us consider the three cases separately.

a) For $U = \frac{1}{\sqrt{2}}(I \pm i\sigma_x)$ we have $\boldsymbol{r}_1 = (x, \pm z, \mp y)$, $\boldsymbol{r}_2 = (x_1 \sin\theta, y_1 \sin\theta, \cos^2\theta + z_1 \sin^2\theta)$, $\boldsymbol{r}' = (x_2, \mp z_2, \pm y_2)$. Therefore, $\boldsymbol{r}' = (x \sin\theta, \mp \cos^2\theta + y \sin^2\theta, z \sin\theta)$.

b) For $U = \frac{1}{\sqrt{2}}(I \pm i\sigma_y)$ we have $\boldsymbol{r}_1 = (\mp z, y, \pm x)$, $\boldsymbol{r}_2 = (x_1 \sin\theta, y_1 \sin\theta, \cos^2\theta + z_1 \sin^2\theta)$, $\boldsymbol{r}' = (\pm z_2, y_2, \mp x_2)$. Therefore, $\boldsymbol{r}' = (\pm \cos^2\theta + x \sin^2\theta, y \sin\theta, z \sin\theta)$.

c) For $U = \frac{1}{\sqrt{2}}(I \pm i\sigma_z)$ we have $\boldsymbol{r}_1 = (\pm y, \mp x, z)$, $\boldsymbol{r}_2 = (x_1 \sin\theta, y_1 \sin\theta, \cos^2\theta + z_1 \sin^2\theta)$, $\boldsymbol{r}' = (\mp y_2, \pm x_2, z_2)$. Therefore, $\boldsymbol{r}' = (x \sin\theta, y \sin\theta, \cos^2\theta + z \sin^2\theta)$.

Exercise 6.11 Let us first discuss the teleportation protocol (see Fig. B.5). The state $|\psi\rangle$ to be teleported and the imperfect Bell states $\rho_{\text{Bell}}^{(\text{imp})}$ are given by $|\psi\rangle = \alpha|0\rangle + \beta|1\rangle$ and

$$
\rho_{\text{Bell}}^{(\text{imp})} = \frac{1}{2} \begin{bmatrix} 0 & 0 & 0 & 0 \\ 0 & 1 & C & 0 \\ 0 & C & 1 & 0 \\ 0 & 0 & 0 & 0 \end{bmatrix},
\tag{B.91}
$$

where $C = \cos\theta$. The matrix representation of the Bell measurement B is given by Eq. (4.30). We have

$$
\rho_B = (B \otimes I)\rho_A (B^\dagger \otimes I),
\tag{B.92}
$$

where

$$
\rho_A = |\psi\rangle\langle\psi| \otimes \rho_{\text{Bell}}^{(\text{imp})}
\tag{B.93}
$$

and

$$
B \otimes I = \frac{1}{\sqrt{2}} \begin{bmatrix} I & 0 & 0 & I \\ 0 & I & I & 0 \\ I & 0 & 0 & -I \\ 0 & I & -I & 0 \end{bmatrix}.
\tag{B.94}
$$

We obtain

$$
\rho_B = \frac{1}{4} \begin{bmatrix} D & .. & .. & .. \\ .. & E & .. & .. \\ .. & .. & F & .. \\ .. & .. & .. & G \end{bmatrix},
\tag{B.95}
$$

where

$$
D = \begin{bmatrix} |\beta|^2 & \alpha^\star\beta C \\ \alpha\beta^\star C & |\alpha|^2 \end{bmatrix}, \qquad
E = \begin{bmatrix} |\alpha|^2 & \alpha\beta^\star C \\ \alpha^\star\beta C & |\beta|^2 \end{bmatrix},
$$
$$
F = \begin{bmatrix} |\beta|^2 & -\alpha^\star\beta C \\ -\alpha\beta^\star C & |\alpha|^2 \end{bmatrix}, \qquad
G = \begin{bmatrix} |\alpha|^2 & -\alpha\beta^\star C \\ -\alpha^\star\beta C & |\beta|^2 \end{bmatrix},
\tag{B.96}
$$

and we have denoted by .. the 2×2 matrix blocks whose expressions are not needed in our subsequent calculations. The outcome of the measurement performed on the first two qubits (by means of the detectors D_0 and D_1) determines the state of the third qubit (D if the outcome is 00, E if the outcome is 01, F if the outcome is 10, and G if the outcome is 11). As discussed in Sec. 4.5, in all cases the unitary operator U recovers the state $\rho_f = E$. The teleportation fidelity is

$$F = \langle\psi|\rho_f|\psi\rangle = |\alpha|^4 + |\beta|^4 + 2C|\alpha|^2|\beta|^2 = 1 - 2(1-C)(|\alpha|^2 - |\alpha|^4). \quad (B.97)$$

Note that teleportation is perfect only for $C = 1$, otherwise $F < 1$.

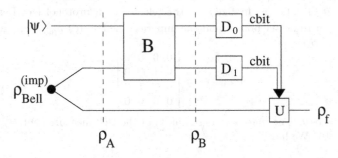

Fig. B.5   A schematic drawing of the teleportation protocol with an imperfect Bell state.

Let us now discuss the dense-coding protocol (see Fig. B.6). As we saw in Sec. 4.4, Alice applies the unitary transformation $U \in \{I, \sigma_x, \sigma_y, \sigma_z\}$ to her half of the (imperfect) Bell state. We have

$$\rho_A = (I \otimes U)\rho_{\mathrm{Bell}}^{(\mathrm{imp})}(I \otimes U^\dagger) = \begin{bmatrix} U & 0 \\ 0 & U \end{bmatrix} \frac{1}{2} \begin{bmatrix} 0 & 0 & 0 & 0 \\ 0 & 1 & C & 0 \\ 0 & C & 1 & 0 \\ 0 & 0 & 0 & 0 \end{bmatrix} \begin{bmatrix} U^\dagger & 0 \\ 0 & U^\dagger \end{bmatrix}. \quad (B.98)$$

The final density matrix is then given by $\rho_{\mathrm{fin}} = B\rho_A B^\dagger$, where B is defined in Eq. (4.30). If Alice applies $U = I$ we obtain

$$\frac{1}{2} \begin{bmatrix} 0 & 0 & 0 & 0 \\ 0 & 1{+}C & 0 & 0 \\ 0 & 0 & 0 & 0 \\ 0 & 0 & 0 & 1{-}C \end{bmatrix}. \quad (B.99)$$

If instead $U = \sigma_x$, then

$$\frac{1}{2} \begin{bmatrix} 1{+}C & 0 & 0 & 0 \\ 0 & 0 & 0 & 0 \\ 0 & 0 & 1{-}C & 0 \\ 0 & 0 & 0 & 0 \end{bmatrix}. \quad (B.100)$$

If $U = \sigma_y$, then

$$\frac{1}{2} \begin{bmatrix} 1-C & 0 & 0 & 0 \\ 0 & 0 & 0 & 0 \\ 0 & 0 & 1+C & 0 \\ 0 & 0 & 0 & 0 \end{bmatrix}. \qquad \text{(B.101)}$$

Finally, if Alice applies $U = \sigma_z$, we obtain

$$\frac{1}{2} \begin{bmatrix} 0 & 0 & 0 & 0 \\ 0 & 1-C & 0 & 0 \\ 0 & 0 & 0 & 0 \\ 0 & 0 & 0 & 1+C \end{bmatrix}. \qquad \text{(B.102)}$$

In all cases, Bob correctly recovers the two classical bits transmitted by Alice with probability $p = \frac{1}{2}(1 + C)$. Only when $C = 1$ the error probability is zero; that is, $p = 1$.

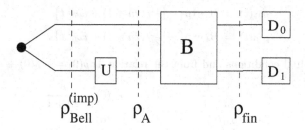

Fig. B.6   A schematic drawing of the dense-coding protocol with an imperfect Bell state.

Exercise 6.12   It is convenient to write the master equation (6.100) in the $\{|0\rangle, |1\rangle\}$ basis. In this basis we have

$$\sigma_- = \begin{bmatrix} 0 & 0 \\ 1 & 0 \end{bmatrix}, \quad \sigma_+ = \begin{bmatrix} 0 & 1 \\ 0 & 0 \end{bmatrix}. \qquad \text{(B.103)}$$

If we write the density matrix ρ as

$$\rho = \begin{bmatrix} \rho_{00} & \rho_{01} \\ \rho_{10} & \rho_{11} \end{bmatrix}, \qquad \text{(B.104)}$$

we obtain

$$-\frac{i}{\hbar} [H, \rho] = -i\omega_0 \begin{bmatrix} 0 & \rho_{01} \\ -\rho_{10} & 0 \end{bmatrix}, \qquad \text{(B.105)}$$

$$\sigma_+ \rho \sigma_- - \tfrac{1}{2}\sigma_-\sigma_+\rho - \tfrac{1}{2}\rho\sigma_-\sigma_+ = \tfrac{1}{2}\begin{bmatrix} 2\rho_{11} & -\rho_{01} \\ -\rho_{10} & -2\rho_{11} \end{bmatrix},\tag{B.106}$$

$$\sigma_- \rho \sigma_+ - \tfrac{1}{2}\sigma_+\sigma_-\rho - \tfrac{1}{2}\rho\sigma_+\sigma_- = \tfrac{1}{2}\begin{bmatrix} -2\rho_{00} & -\rho_{01} \\ -\rho_{10} & 2\rho_{00} \end{bmatrix}.\tag{B.107}$$

After insertion of (B.105–B.106) into (6.100) we obtain the following equations:

$$\begin{aligned}
\dot{\rho}_{00} &= \gamma\,(\bar{n}+1)\,\rho_{11} - \gamma\,\bar{n}\,\rho_{00},\\
\dot{\rho}_{11} &= -\gamma\,(\bar{n}+1)\,\rho_{11} + \gamma\,\bar{n}\,\rho_{00},\\
\dot{\rho}_{01} &= -\left[\tfrac{1}{2}\gamma\,(2\bar{n}+1) - i\omega_0\right]\rho_{01},\\
\dot{\rho}_{10} &= -\left[\tfrac{1}{2}\gamma\,(2\bar{n}+1) + i\omega_0\right]\rho_{10}.
\end{aligned}\tag{B.108}$$

Their solution is

$$\begin{aligned}
\rho_{00}(t) &= B_1 - B_2 \exp\left[-\gamma\,(2\bar{n}+1)\,t\right],\\
\rho_{11}(t) &= \frac{\bar{n}}{\bar{n}+1}\,B_1 + B_2 \exp\left[-\gamma\,(2\bar{n}+1)\,t\right],\\
\rho_{01}(t) &= B_3 \exp\left\{-\left[\tfrac{1}{2}\gamma\,(2\bar{n}+1) - i\omega_0\right]t\right\},\\
\rho_{10}(t) &= B_4 \exp\left\{-\left[\tfrac{1}{2}\gamma\,(2\bar{n}+1) + i\omega_0\right]t\right\}.
\end{aligned}\tag{B.109}$$

From the initial conditions and from the relation $\operatorname{Tr}\rho(0) = \rho_{00}(0) + \rho_{11}(0) = 1$ we obtain

$$\begin{aligned}
B_1 &= \frac{\bar{n}+1}{2\bar{n}+1}, & B_2 &= \rho_{11}(0) - \frac{\bar{n}}{2\bar{n}+1},\\
B_3 &= \rho_{01}(0), & B_4 &= \rho_{10}(0).
\end{aligned}\tag{B.110}$$

It is clear from Eq. (B.109) that the asymptotic density matrix is given by

$$\rho_s \equiv \lim_{t\to\infty}\rho(t) = \begin{bmatrix} \frac{\bar{n}+1}{2\bar{n}+1} & 0 \\ 0 & \frac{\bar{n}}{2\bar{n}+1} \end{bmatrix}.\tag{B.111}$$

We note that the diagonal terms approach equilibrium on a time scale $\tau_d = [\gamma(2\bar{n}+1)]^{-1}$ while the off-diagonal terms require a time scale $\tau_{nd} = 2\tau_d$. The stationary density matrix ρ_s corresponds to the Bloch vector $r_s = (0, 0, \frac{1}{2\bar{n}+1})$.

Exercise 6.13 We expand the commutators in Eq. (6.102) obtaining

$$\dot{\rho} = -\frac{i}{\hbar}\,[H,\rho] + \frac{1}{2}\sum_{i,j=1}^{N^2-1} A_{ij}\big\{2\sigma_i\rho\sigma_j^{\dagger} - \rho\sigma_j^{\dagger}\sigma_i - \sigma_j^{\dagger}\sigma_i\rho\big\}.\tag{B.112}$$

After substitution of Eq. (6.105), namely of the expression

$$A_{ij} = \sum_{k=1}^{N^2-1} S_{ki}^{\star}\tilde{A}_{kk}S_{kj},\tag{B.113}$$

into (B.112), we have

$$\dot{\rho} = -\frac{i}{\hbar}[H, \rho] + \sum_{i,j,k=1}^{N^2-1} S_{ki}^{\star} \tilde{A}_{kk} S_{kj} \left\{ \sigma_i \rho \sigma_j^{\dagger} - \frac{1}{2} \rho \sigma_j^{\dagger} \sigma_i - \frac{1}{2} \sigma_j^{\dagger} \sigma_i \rho \right\}. \qquad (B.114)$$

Finally, we define

$$L_k = \sqrt{\tilde{A}_{kk}} \sum_i S_{ki}^{\star} \sigma_i, \qquad (B.115)$$

which implies that

$$L_k^{\dagger} = \sqrt{\tilde{A}_{kk}} \sum_j S_{kj} \sigma_j^{\dagger}. \qquad (B.116)$$

Substitution of (B.115) and (B.116) into (B.114) finally leads to the GKSL master equation (6.99).

Exercise 6.14 Equation (6.117) can be written as

$$\begin{bmatrix} \dot{x} \\ \dot{y} \\ \dot{z} \end{bmatrix} = \begin{bmatrix} 0 & -\omega_0 & \Delta' \\ \omega_0 & 0 & -\Delta \\ -\Delta' & \Delta & 0 \end{bmatrix} \begin{bmatrix} x \\ y \\ z \end{bmatrix}. \qquad (B.117)$$

This matrix is antisymmetric and therefore corresponds to a rotation. Indeed, we can write

$$\dot{r} = \Omega \times r, \qquad (B.118)$$

with

$$\Omega = (\Delta, \Delta', \omega_0). \qquad (B.119)$$

Therefore, we immediately obtain the rotation axis (6.118) and the rotation frequency $\Omega = \sqrt{\Delta^2 + \Delta'^2 + \omega_0^2}$.

Exercise 6.15 The wave function (6.134) is a solution of the Schrödinger equation for a one-dimensional free particle since

$$i\hbar \frac{\partial}{\partial t} \psi(x, t) = -\frac{\hbar^2}{2m} \frac{\partial^2}{\partial x^2} \psi(x, t)$$

$$= -\frac{\hbar^2}{2m} \left\{ \left[-\frac{\left(x - x_0 - p_0 \frac{t}{m}\right)}{\delta^2 \left(1 + i\frac{\hbar t}{m\delta^2}\right)} + i\frac{p_0}{\hbar} \right]^2 - \frac{1}{\delta^2 \left(1 + i\frac{\hbar t}{m\delta^2}\right)} \right\}. \qquad (B.120)$$

The average value of a generic observable is given by the formula in the footnote on page 367. In particular, we have

$$\langle x(t) \rangle = \int_{-\infty}^{+\infty} dx\, x \big| \psi(x,t) \big|^2. \tag{B.121}$$

If we define $x' = x - x_0 - \frac{p}{m}t$, we obtain

$$\langle x(t) \rangle = \int_{-\infty}^{+\infty} dx' \left(x' + x_0 + \frac{p}{m}t \right) f(x'), \tag{B.122}$$

where

$$f(x) = \frac{1}{\sigma\sqrt{2\pi}} \exp\left(-\frac{x^2}{2\sigma^2} \right), \tag{B.123}$$

with $\sigma = \frac{\delta}{\sqrt{2}} \sqrt{1 + \frac{\hbar^2 t^2}{m^2 \delta^4}}$. Using

$$\int_{-\infty}^{+\infty} dx\, f(x) = 1, \qquad \int_{-\infty}^{+\infty} dx\, x f(x) = 0, \tag{B.124}$$

we have

$$\langle x(t) \rangle = x_0 + \frac{p_0}{m}t. \tag{B.125}$$

Similarly, we obtain

$$\langle p(t) \rangle = \int_{-\infty}^{+\infty} dx\, \psi^*(x,t) \left(-i\hbar\frac{\partial}{\partial x} \right) \psi(x,t) = p_0 \tag{B.126}$$

The variances $(\Delta x)^2 = \langle (x - \langle x \rangle)^2 \rangle$ and $(\Delta p)^2 = \langle (p - \langle p \rangle)^2 \rangle$ are computed in the same manner: the result of Eq. (6.137) is obtained using (B.124) and

$$\int_{-\infty}^{+\infty} dx\, x^2 f(x) = \sigma^2. \tag{B.127}$$

Exercise 6.16   The canonical commutation relation is modified by (6.193):

$$[z,p] \rightarrow [n_0^{-2}z, n_0 p] = i n_0^{-1}. \tag{B.128}$$

Therefore, in quantum mechanics there is an additional parameter that is absent in classical mechanics, the effective Planck constant $\hbar_{\text{eff}} = \frac{1}{n_0}$. The classical limit is obtained when $\hbar_{\text{eff}} \rightarrow 0$ ($n_0 \rightarrow \infty$), keeping ϵ_0 and ω_0 constant.

Exercise 6.17   The conditional probability of finding a level in the interval $[e + s, e + s + ds]$ given a level at e is $\frac{1}{\langle s \rangle}ds$, where $\langle s \rangle$ is the average spacing. We also note that the probability that there are no levels in the interval $[e, e + s]$ is $\int_s^\infty ds' P(s')$. Therefore,

$$P(s)ds = \left(\int_s^\infty ds' P(s') \right) \frac{1}{\langle s \rangle}ds. \tag{B.129}$$

After differentiating Eq. (B.129), we obtain $\frac{dP}{ds} = -\frac{1}{\langle s \rangle} P$ and therefore

$$P(s) = P(0) \exp\left(-\frac{s}{\langle s \rangle}\right). \tag{B.130}$$

Note that $P(0) = \frac{1}{\langle s \rangle}$ in order to assure that $\int_0^\infty ds P(s) = 1$. Thus, (B.130) coincides with the Poisson distribution (6.204), provided the spacing s is measured in units of the mean level spacing $\langle s \rangle$.

Exercise 6.18   The matrix

$$H = \begin{bmatrix} H_{11} & H_{12} \\ H_{12} & H_{22} \end{bmatrix} \tag{B.131}$$

has eigenvalues

$$E_\pm = \frac{H_{11} + H_{22} \pm \sqrt{(H_{11} - H_{22})^2 + 4H_{12}^2}}{2}. \tag{B.132}$$

Thus, the spacing $s = E_+ - E_-$ between these eigenvalues is

$$s = \sqrt{(H_{11} - H_{22})^2 + 4H_{12}^2}. \tag{B.133}$$

If we introduce the (random) variables $x = H_{11} - H_{22}$ and $y = 2H_{12}$, then $s = \sqrt{x^2 + y^2}$ represents the distance from the origin of a point in the (x, y) plane. Thus, for small ds, the probability of finding the eigenvalue spacing in the interval $[s, s + ds]$ is proportional to the area of the region in the (x, y) plane from s to $s + ds$; that is, $P(s)ds = Cs ds$, where C is a constant. If $F(s)$ denote the probability of finding no spacings from 0 to s, then

$$F(s + ds) = F(s) - P(s)ds = F(s)(1 - Cs ds). \tag{B.134}$$

After expanding $F(s + ds)$ in a Taylor series for small ds, we obtain

$$\begin{aligned} F(s) &= \exp\left(-\tfrac{1}{2} Cs^2\right), \\ P(s) &= -\frac{dF(s)}{ds} = Cs \exp\left(-\tfrac{1}{2} Cs^2\right). \end{aligned} \tag{B.135}$$

It is then sufficient to express the constant C in terms of the average spacing

$$\langle s \rangle = \int_0^\infty ds \, s P(s) = \sqrt{\frac{\pi}{2C}} \tag{B.136}$$

and to measure s in units of $\langle s \rangle$ to obtain the Wigner surmise (6.212).

Exercise 6.19   For an all-to-all interaction we have $O(n^2)$ states directly coupled inside the central band, so that the chaos border $J_c \sim \Delta_c \sim \delta/n^2$. Therefore, the scaling of J_c with n is polynomial.

Exercise 6.20   Let us write

$$\rho = \tfrac{1}{2}\,(I + \boldsymbol{r}\cdot\boldsymbol{\sigma}), \qquad \sigma = \tfrac{1}{2}\,(I + \boldsymbol{s}\cdot\boldsymbol{\sigma}). \qquad (B.137)$$

We obtain

$$D(\rho,\sigma) = \tfrac{1}{4}\,\mathrm{Tr}\,|(\boldsymbol{r}-\boldsymbol{s})\cdot\boldsymbol{\sigma}| = \tfrac{1}{2}\,|\boldsymbol{r}-\boldsymbol{s}|, \qquad (B.138)$$

where we have used the fact that $(\boldsymbol{r}-\boldsymbol{s})\cdot\boldsymbol{\sigma}$ has eigenvalues $\lambda_\pm = \pm|\boldsymbol{r}-\boldsymbol{s}|$, so that $\mathrm{Tr}\,|(\boldsymbol{r}-\boldsymbol{s})\cdot\boldsymbol{\sigma}| = |\lambda_+| + |\lambda_-| = 2|\boldsymbol{r}-\boldsymbol{s}|$.

Chapter 7

Exercise 7.1   The quantum circuit extracting the error syndrome without auxiliary qubits is shown in Fig. B.7. It can be readily checked that the output $x_0 = 0$, $x_1 = 0$ corresponds to no error, $x_0 = 1$, $x_1 = 1$ to error on the first qubit, $x_0 = 1$, $x_1 = 0$ to error on the second qubit and $x_0 = 0$, $x_1 = 1$ to error on the third qubit. In the case $x_0 = x_1 = 1$ the logical state $\alpha|0\rangle + \beta|1\rangle$ is recovered after a bit flip of the first qubit, in all the other cases the state of the first qubit is correct and no further action is required. At the end the three qubits are left in the state $(\alpha|0\rangle + \beta|1\rangle)|x_0\rangle|x_1\rangle$.

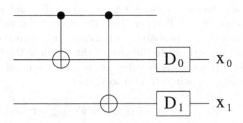

Fig. B.7   A quantum circuit extracting the error syndrome in the case of the three-qubit bit-flip errors, without using any ancillary qubits.

Exercise 7.2   If a single qubit, prepared in the state $|\psi\rangle = \alpha|0\rangle + \beta|1\rangle$, is sent through a bit-flip noisy channel, the final state of the qubit is described by

$$\rho = (1-\epsilon)\big(\alpha|0\rangle + \beta|1\rangle\big)\big(\alpha^*\langle 0| + \beta^*\langle 1|\big) + \epsilon\big(\alpha|1\rangle + \beta|0\rangle\big)\big(\alpha^*\langle 1| + \beta^*\langle 0|\big), \quad (B.139)$$

where ϵ is the bit-flip probability. Hence,

$$\rho = \begin{bmatrix} (1-\epsilon)|\alpha|^2 + \epsilon|\beta|^2 & (1-\epsilon)\alpha\beta^* + \epsilon\alpha^*\beta \\ (1-\epsilon)\alpha^*\beta + \epsilon\alpha\beta^* & (1-\epsilon)|\beta|^2 + \epsilon|\alpha|^2 \end{bmatrix} = \frac{1}{2}\begin{bmatrix} 1+z' & x'-iy' \\ x'+iy' & 1-z' \end{bmatrix}, \quad (B.140)$$

where $x' = x$, $y' = (1 - 2\epsilon)y$, $z' = (1 - 2\epsilon)z$ while (x, y, z) and (x', y', z') are the Bloch-sphere coordinates corresponding to the initial density matrix $\rho_0 = |\psi\rangle\langle\psi|$ and to ρ, respectively. Therefore, Eq. (6.29) for a bit-flip channel is recovered. The fidelity of the transmitted state is given by Eq. (5.60), namely,

$$
F = \langle\psi|\rho|\psi\rangle = \mathrm{Tr}(\rho_0\rho) = \mathrm{Tr}\left(\frac{1}{2}\begin{bmatrix} 1+z & x-iy \\ x+iy & 1-z \end{bmatrix}\frac{1}{2}\begin{bmatrix} 1+z' & x'-iy' \\ x'+iy' & 1-z' \end{bmatrix}\right)
$$
$$
= \tfrac{1}{2}\left[1 + x^2 + (1 - 2\epsilon)(y^2 + z^2)\right] = 1 - \epsilon\,(y^2 + z^2). \tag{B.141}
$$

Note that unit fidelity is recovered when $\epsilon = 0$ or when the initial state is an eigenstate of σ_x ($x = \pm1$, $y = z = 0$). For a generic initial state $F = 1 - O(\epsilon)$.

When the three qubit bit-flip code is applied the initial logical qubit is encoded into three physical qubits, which are sent through the bit-flip channel. After error correction and decoding the initial state is recovered, unless two or more qubits were flipped. Therefore, following Eq. (7.8) and the subsequent discussion, we can see that the final state of the logical qubit is given by

$$
\rho = \left[(1 - \epsilon)^3 + 3\epsilon\,(1 - \epsilon)^2\right]\left(\alpha|0\rangle + \beta|1\rangle\right)\left(\alpha^*\langle0| + \beta^*\langle1|\right)
$$
$$
+ \left[3\epsilon^2(1 - \epsilon) + \epsilon^3\right]\left(\alpha|1\rangle + \beta|0\rangle\right)\left(\alpha^*\langle1| + \beta^*\langle0|\right). \tag{B.142}
$$

Therefore, the fidelity of the transmitted state can be computed as in Eq. (B.141), after the substitution $\epsilon \to \epsilon_c \equiv 3\epsilon^2 - 2\epsilon^3$. We have $F = 1 - \epsilon_c(y^2 + z^2) = 1 - (3\epsilon^2 - 2\epsilon^3)(y^2 + z^2)$. For a generic state, $F = 1 - O(\epsilon^2)$. This has to be compared with $F = 1 - O(\epsilon)$, obtained by sending a single qubit without error correction. Therefore, the error correction procedure greatly improves the fidelity of the transmitted quantum information for $\epsilon \ll 1$.

Exercise 7.3   Let us consider the state $|0_L\rangle$ and an error affecting the first qubit (the state $|1_L\rangle$ and errors affecting other qubits are treated in the same manner). We have

$$
U|0_L\rangle|0\rangle_E
$$
$$
= \tfrac{1}{\sqrt{2}}\left(|0\rangle|e_0\rangle_E + |1\rangle|e_1\rangle_E\right)|00\rangle(\ldots)(\ldots)
$$
$$
+ \tfrac{1}{\sqrt{2}}\left(|0\rangle|e_2\rangle_E + |1\rangle|e_3\rangle_E\right)|11\rangle(\ldots)(\ldots)
$$
$$
= \tfrac{1}{\sqrt{2}}\left(|000\rangle|e_0\rangle_E + |100\rangle|e_1\rangle_E + |011\rangle|e_2\rangle_E + |111\rangle|e_3\rangle_E\right)(\ldots)(\ldots), \tag{B.143}
$$

where the state of the last six qubits has been simply denoted as $(\ldots)(\ldots)$ since they remain de-entangled from the environment. The density matrix describing the first three qubits plus the environment reads

$$
\rho = \tfrac{1}{2}\left(|000\rangle|e_0\rangle_E + |100\rangle|e_1\rangle_E + |011\rangle|e_2\rangle_E + |111\rangle|e_3\rangle_E\right)
$$
$$
\times \left(\langle000|_E\langle e_0| + \langle100|_E\langle e_1| + \langle011|_E\langle e_2| + \langle111|_E\langle e_3|\right). \tag{B.144}
$$

After tracing over the three qubits, we obtain the density matrix of the environment:

$$\rho_E = \tfrac{1}{2}\big(|e_0\rangle_{EE}\langle e_0| + |e_1\rangle_{EE}\langle e_1| + |e_2\rangle_{EE}\langle e_2| + |e_3\rangle_{EE}\langle e_3|\big). \tag{B.145}$$

Exercise 7.4   The correctable errors are

$$\begin{aligned}
E_1 &= \sigma_x^{(1)} \otimes I^{(2)} \otimes I^{(3)} = E_1^\dagger, \\
E_2 &= I^{(1)} \otimes \sigma_x^{(2)} \otimes I^{(3)} = E_2^\dagger, \\
E_3 &= I^{(1)} \otimes I^{(2)} \otimes \sigma_x^{(3)} = E_3^\dagger.
\end{aligned} \tag{B.146}$$

Since $\sigma_x^2 = I$, we have $E_1^\dagger E_1 = E_2^\dagger E_2 = E_3^\dagger E_3 = I$. Thus, we obtain

$$\langle i_L|E_a^\dagger E_a|j_L\rangle = \delta_{ij}, \tag{B.147}$$

where $a = 1, 2, 3$, $i = 0, 1$ ($|0_L\rangle = |000\rangle$, $|1_L\rangle = |111\rangle$). Finally, it is easy to check that, for $a \neq b$, $\langle i_L|E_a^\dagger E_b|i_L\rangle = 0$. For instance, for $a = 1$, $b = 2$ we have

$$\langle 0_L|E_1^\dagger E_2|0_L\rangle = \langle 000|\sigma_x^{(1)} \otimes \sigma_x^{(2)} \otimes I^{(3)}|000\rangle = \langle 000|110\rangle = 0. \tag{B.148}$$

Similarly, we obtain $\langle 1_L|E_1^\dagger E_2|1_L\rangle = 0$. Thus, condition 7.29 is fulfilled with $C_{ab} = \delta_{ab}$.

Exercise 7.5   The non-degeneracy of the three-qubit code was verified in the previous exercise (we have seen that $C_{ab} = \delta_{ab}$). To show the degeneracy of the nine-qubit Shor code it is sufficient to find a case in which $C_{ab} \neq \delta_{ab}$. Let E_1 and E_2 denote the phase errors affecting the first and on the second qubit, respectively. We obtain

$$\begin{aligned}
|0'_L\rangle &\equiv E_1|0_L\rangle = \tfrac{1}{\sqrt{8}}(|000\rangle - |111\rangle)(|000\rangle + |111\rangle)(|000\rangle + |111\rangle), \\
|1'_L\rangle &\equiv E_1|1_L\rangle = \tfrac{1}{\sqrt{8}}(|000\rangle + |111\rangle)(|000\rangle - |111\rangle)(|000\rangle - |111\rangle).
\end{aligned} \tag{B.149}$$

Similarly, we obtain

$$\begin{aligned}
|0''_L\rangle &\equiv E_2|0_L\rangle = \tfrac{1}{\sqrt{8}}(|000\rangle - |111\rangle)(|000\rangle + |111\rangle)(|000\rangle + |111\rangle), \\
|1''_L\rangle &\equiv E_2|1_L\rangle = \tfrac{1}{\sqrt{8}}(|000\rangle + |111\rangle)(|000\rangle - |111\rangle)(|000\rangle - |111\rangle).
\end{aligned} \tag{B.150}$$

Since $|0''_L\rangle = |0'_L\rangle$ and $|1''_L\rangle = |1'_L\rangle$, we have $\langle i_L|E_1^\dagger E_2|i_L\rangle = 1$, with $i = 0, 1$. Thus, $C_{12} = 1 \neq \delta_{12}$.

Exercise 7.6   As an example, we consider the second line of Table 7.1. The effect of the error $\sigma_x^{(3)}\sigma_z^{(3)}$ on the encoded states (7.35) leads to

$$
\begin{aligned}
\sigma_x^{(3)}\sigma_z^{(3)}|0_L\rangle &= \tfrac{1}{\sqrt{8}}\big(|00100\rangle + |01011\rangle - |10111\rangle - |11000\rangle \\
&\quad - |00010\rangle + |01101\rangle - |10001\rangle + |11110\rangle\big), \\
\sigma_x^{(3)}\sigma_z^{(3)}|1_L\rangle &= \tfrac{1}{\sqrt{8}}\big(-|11011\rangle - |10100\rangle - |01000\rangle - |00111\rangle \\
&\quad + |11101\rangle - |10010\rangle - |01110\rangle + |00001\rangle\big).
\end{aligned}
\tag{B.151}
$$

It can be checked by direct computation that the quantum circuit in Fig. 7.8 maps the error-affected state $\sigma_x^{(3)}\sigma_z^{(3)}(\alpha|0_L\rangle + \beta|1_L\rangle)$ into $-\alpha|11101\rangle + \beta|11001\rangle$. Therefore, the detectors D_a, D_b, D_c, and D_d in Fig. 7.8 give outcome $a = 1$, $b = 1$, $c = 0$, and $d = 1$ and the third qubit is in the state $|\psi'\rangle = -\alpha|0\rangle + \beta|1\rangle$, in agreement with Table 7.1.

Exercise 7.7   We have

$$
G = \begin{bmatrix} 1 \\ 1 \\ 1 \end{bmatrix},
\tag{B.152}
$$

so that the codewords are

$$
y_0 = Gx_0 = \begin{bmatrix} 1 \\ 1 \\ 1 \end{bmatrix}[0] = \begin{bmatrix} 0 \\ 0 \\ 0 \end{bmatrix}, \quad y_1 = Gx_1 = \begin{bmatrix} 1 \\ 1 \\ 1 \end{bmatrix}[1] = \begin{bmatrix} 1 \\ 1 \\ 1 \end{bmatrix}.
\tag{B.153}
$$

The lines of the parity-check matrix H must be linearly independent and orthogonal to the columns of G. These conditions are fulfilled by taking

$$
H = \begin{bmatrix} 0 & 1 & 1 \\ 1 & 0 & 1 \end{bmatrix}.
\tag{B.154}
$$

The set of correctable errors consists of

$$
e_1 = \begin{bmatrix} 1 \\ 0 \\ 0 \end{bmatrix}, \quad e_2 = \begin{bmatrix} 0 \\ 1 \\ 0 \end{bmatrix}, \quad e_3 = \begin{bmatrix} 0 \\ 0 \\ 1 \end{bmatrix}.
\tag{B.155}
$$

The corresponding error syndromes are given by

$$He_1 = \begin{bmatrix} 0 & 1 & 1 \\ 1 & 0 & 1 \end{bmatrix} \begin{bmatrix} 1 \\ 0 \\ 0 \end{bmatrix} = \begin{bmatrix} 0 \\ 1 \end{bmatrix},$$

$$He_2 = \begin{bmatrix} 0 & 1 & 1 \\ 1 & 0 & 1 \end{bmatrix} \begin{bmatrix} 0 \\ 1 \\ 0 \end{bmatrix} = \begin{bmatrix} 1 \\ 0 \end{bmatrix}, \tag{B.156}$$

$$He_3 = \begin{bmatrix} 0 & 1 & 1 \\ 1 & 0 & 1 \end{bmatrix} \begin{bmatrix} 0 \\ 0 \\ 1 \end{bmatrix} = \begin{bmatrix} 1 \\ 1 \end{bmatrix}.$$

Exercise 7.8   For instance, we can consider the two-qubit error

$$e_{12} = \begin{bmatrix} 1 \\ 1 \\ 0 \\ 0 \\ 0 \\ 0 \\ 0 \end{bmatrix}. \tag{B.157}$$

The error syndrome is

$$H(C_1)\, e_{12} = \begin{bmatrix} 0 & 0 & 0 & 1 & 1 & 1 & 1 \\ 0 & 1 & 1 & 0 & 0 & 1 & 1 \\ 1 & 0 & 1 & 0 & 1 & 0 & 1 \end{bmatrix} \begin{bmatrix} 1 \\ 1 \\ 0 \\ 0 \\ 0 \\ 0 \\ 0 \end{bmatrix} = \begin{bmatrix} 0 \\ 1 \\ 1 \end{bmatrix}. \tag{B.158}$$

As shown in Eq. (7.48), such an error syndrome would be erroneously interpreted as a single error affecting the third qubit.

Exercise 7.9   Given the parity-check matrix (7.47), the quantum circuit mapping the state (7.56) into (7.57) is readily obtained and shown in Fig. B.8. The error syndrome is then obtained after measurement of the three ancillary qubits.

Exercise 7.10   First of all we note that, for any $w \in C$ and $z \in C^\perp$, $w \cdot z = 0$. In order to prove this relation, we write $w = Gx$ and $z = H^T t$, with x and t column vectors of dimension k and $n - k$, respectively. Hence, we have $w \cdot z = w^T z = x^T G^T H^T t = 0$ because $G^T H^T = 0$. This implies that, for any $w \in C$ and $z \in C^\perp$, $\sum_{w \in C} (-)^{w \cdot z} = \sum_{w \in C} 1 = 2^k$.

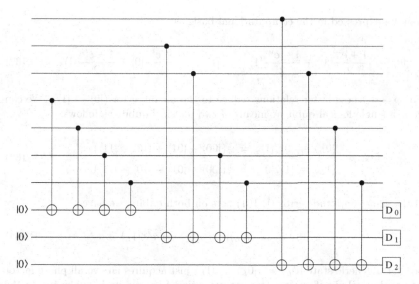

Fig. B.8 A quantum circuit extracting the error syndrome (for amplitude errors) in the CSS code, for the case in which C_1 is the $[7, 4, 1]$ Hamming code and $C_2 = C_1^\perp$. The three auxiliary qubits are initially prepared in the state $|0\rangle$.

To solve the second part of this exercise ($z \notin C^\perp$), we use the identity

$$\sum_a (-)^{a \cdot b} = 0, \tag{B.159}$$

where a and b are k-bit strings, with $b \neq 0$; that is, b is different from the string with all bits equal to 0. We then have

$$\sum_{w \in C} (-)^{w \cdot z} = \sum_{w \in C} (-)^{(Gx) \cdot z} = \sum_{w \in C} (-)^{(Gx)^T z}$$
$$= \sum_{w \in C} (-)^{x^T (G^T z)} = \sum_{w \in C} (-)^{x \cdot (G^T z)}. \tag{B.160}$$

Since G^T is the parity-check matrix for $C^\perp$ and we assume that z does not reside in $C^\perp$, then $G^T z \neq 0$. Finally, using (B.159), we obtain $\sum_{w \in C} (-)^{w \cdot z} = 0$ for $z \notin C^\perp$.

Exercise 7.11 It is sufficient to remember that an amplitude error in the computational basis $\{|0\rangle, |1\rangle\}$ becomes a phase error in the basis $\{|0\rangle_x, |1\rangle_x\}$. Indeed, the mapping

$$|0\rangle_x \rightarrow |0\rangle_x, \qquad |1\rangle_x \rightarrow e^{i\phi}|1\rangle_x \tag{B.161}$$

may be expressed in the computational basis as

$$|0\rangle \;\rightarrow\; \frac{1+e^{i\phi}}{2}|0\rangle + \frac{1-e^{i\phi}}{2}|1\rangle, \qquad |1\rangle \;\rightarrow\; \frac{1-e^{i\phi}}{2}|0\rangle + \frac{1+e^{i\phi}}{2}|1\rangle. \qquad \text{(B.162)}$$

For $\phi = \pi$, these latter relations reduce to the bit-flip error ($|0\rangle \leftrightarrow |1\rangle$). We can code a single logical qubit by means of two physical qubits as follows:

$$
\begin{aligned}
|0_L\rangle &\equiv |0\rangle_x |1\rangle_x = \tfrac{1}{2}\big(|00\rangle - |01\rangle + |10\rangle - |11\rangle\big) \\
|1_L\rangle &\equiv |1\rangle_x |0\rangle_x = \tfrac{1}{2}\big(|00\rangle + |01\rangle - |10\rangle - |11\rangle\big).
\end{aligned}
\qquad \text{(B.163)}
$$

If the same amplitude error (B.161) acts on both qubits, we have

$$|0_L\rangle \;\rightarrow\; e^{i\phi}|0_L\rangle, \qquad |1_L\rangle \;\rightarrow\; e^{i\phi}|1_L\rangle, \qquad \text{(B.164)}$$

so that a generic state $|\psi_L\rangle = \alpha|0_L\rangle + \beta|1_L\rangle$ just acquires an overall phase factor of no physical significance. The generalization to n qubits is analogous to the discussion in Sec. 7.8.

Exercise 7.12 In the basis of the eigenstates of σ_z the Hamiltonian reads as follows:

$$H = \tfrac{1}{2}\hbar \begin{bmatrix} \omega & \Omega \\ \Omega & -\omega \end{bmatrix}. \qquad \text{(B.165)}$$

The corresponding Schrödinger equation can be solved as discussed in exercise 3.22. Alternatively, we can write the density matrix as $\rho(t) = \tfrac{1}{2}(I + \boldsymbol{r}(t) \cdot \boldsymbol{\sigma})$ and obtain from the von Neumann equation (5.15) the following equations for components (x, y, z) of the Bloch vector $\boldsymbol{r}$:

$$
\begin{cases}
\dot{x} = -\omega y, \\
\dot{y} = \omega x - \Omega z, \\
\dot{z} = \Omega y.
\end{cases}
\qquad \text{(B.166)}
$$

The solution to this linear system reads

$$
\begin{bmatrix} x(t) \\ y(t) \\ z(t) \end{bmatrix}
= A \begin{bmatrix} 1 \\ 0 \\ \frac{\omega}{\Omega} \end{bmatrix}
+ B \begin{bmatrix} 1 \\ -i\frac{\sqrt{\omega^2+\Omega^2}}{\omega} \\ -\frac{\Omega}{\omega} \end{bmatrix} e^{i\sqrt{\omega^2+\Omega^2}\,t}
+ C \begin{bmatrix} 1 \\ i\frac{\sqrt{\omega^2+\Omega^2}}{\omega} \\ -\frac{\Omega}{\omega} \end{bmatrix} e^{-i\sqrt{\omega^2+\Omega^2}\,t},
$$

$$\text{(B.167)}$$

with the constants A, B and C determined by the initial condition. Starting from the state $|0\rangle$ ($x(0) = y(0) = 0$, $z(0) = 1$) we obtain $A = \frac{\omega\Omega}{\omega^2+\Omega^2}$ and $B = C = -\frac{A}{2}$, so that

$$
\begin{cases}
x(t) = \dfrac{\omega\Omega}{\omega^2+\Omega^2}\left[1-\cos\!\left(\sqrt{\omega^2+\Omega^2}\,t\right)\right], \\[2mm]
y(t) = -\dfrac{\Omega}{\sqrt{\omega^2+\Omega^2}}\sin\!\left(\sqrt{\omega^2+\Omega^2}\,t\right); \\[2mm]
z(t) = \dfrac{\omega^2}{\omega^2+\Omega^2}+\dfrac{\Omega^2}{\omega^2+\Omega^2}\cos\!\left(\sqrt{\omega^2+\Omega^2}\,t\right).
\end{cases}
\tag{B.168}
$$

The survival probability is given by

$$
p(t) = \mathrm{Tr}\big(\rho(t)|0\rangle\langle 0|\big) = \tfrac{1}{2}\big[1+z(t)\big] = 1-\dfrac{\Omega^2}{\omega^2+\Omega^2}\sin^2\!\left(\sqrt{\omega^2+\Omega^2}\,\dfrac{t}{2}\right). \tag{B.169}
$$

The short-time expansion of this equation leads to $p(t) = 1 - \dfrac{t^2}{t_Z^2}$, with the Zeno time

$$
t_Z = \frac{2}{\Omega} = \frac{\hbar}{\sqrt{\langle 0|H_{\mathrm{int}}^2|0\rangle}}. \tag{B.170}
$$

A graphical visualization of the Zeno effect is shown in Fig. B.9: the survival probability is enhanced by frequent measurements of σ_z, performed at time intervals $\tau \ll t_Z$. In such a case, the survival probability is bounded below (interpolated at $t = N\tau$) by the curve

$$
p(t) = \exp\!\left(-\frac{\tau}{t_Z^2}t\right). \tag{B.171}
$$

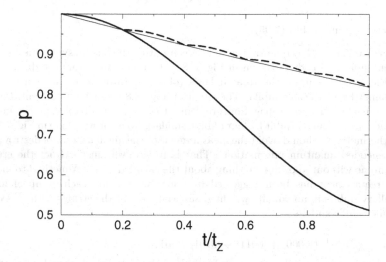

Fig. B.9 Evolution of the survival probability with (above) and without (below) measurements, with $\omega = \Omega$ and $\tau = \tfrac{1}{5}t_Z$. The thin solid line shows the exponential decay (B.171).

Exercise 7.13   The eigenvalues E_j and the corresponding eigenvectors $|\varphi_j\rangle$ of Hamiltonian (7.94) read

$$E_0 = 0, \quad E_1 = \sqrt{\Omega_1^2 + \Omega_2^2}, \quad E_2 = -\sqrt{\Omega_1^2 + \Omega_2^2}, \tag{B.172}$$

$$|\varphi_0\rangle = \begin{bmatrix} \frac{\Omega_2}{\sqrt{\Omega_1^2+\Omega_2^2}} \\ 0 \\ -\frac{\Omega_1}{\sqrt{\Omega_1^2+\Omega_2^2}} \end{bmatrix}, \quad |\varphi_1\rangle = \begin{bmatrix} \frac{\Omega_1}{\sqrt{2(\Omega_1^2+\Omega_2^2)}} \\ \frac{1}{\sqrt{2}} \\ \frac{\Omega_2}{\sqrt{2(\Omega_1^2+\Omega_2^2)}} \end{bmatrix}, \quad |\varphi_2\rangle = \begin{bmatrix} \frac{\Omega_1}{\sqrt{2(\Omega_1^2+\Omega_2^2)}} \\ -\frac{1}{\sqrt{2}} \\ \frac{\Omega_2}{\sqrt{2(\Omega_1^2+\Omega_2^2)}} \end{bmatrix}.$$

$$\tag{B.173}$$

The wave function $|\psi(0)\rangle = |0\rangle$ may be expanded over the (orthonormal) basis $\{|\varphi_0\rangle, |\varphi_1\rangle, |\varphi_2\rangle\}$:

$$|\psi(0)\rangle = \sum_{j=0}^{2} c_j |\varphi_j\rangle, \quad c_j = \langle \varphi_j |\psi(0)\rangle = \langle \varphi_j |0\rangle. \tag{B.174}$$

We then obtain

$$|\psi(t)\rangle = \sum_{j=0}^{2} c_j \exp\left(-\frac{i}{\hbar} E_j t\right) |\varphi_j\rangle. \tag{B.175}$$

Therefore, the survival probability at time t reads

$$p(t) = \left|\langle 0|\psi(t)\rangle\right|^2 = \left|\sum_{j=0}^{2} \exp\left(-\frac{i}{\hbar} E_j t\right) |\langle 0|\varphi_j\rangle|^2\right|^2, \tag{B.176}$$

directly leading to Eq. (7.95).

Exercise 7.14   The quantum circuit for computing the bit-flip syndrome for the seven-qubit CSS code is shown in Fig. B.8. The phase-flip errors are diagnosed by the same circuit in the rotated basis (obtained through application of the Hadamard gate to each qubit). The circuit in Fig. B.8 is not fault-tolerant owing to the backward sign propagation problem. To make it fault-tolerant, we must replace each ancillary qubit by four qubits, similarly to what was done in Fig. 7.10. Furthermore, we should avoid the measurement of the ancillary qubits destroying the encoded quantum information. That is to say, we must extract the error syndrome without knowing anything about the encoded state. A method to meet this requirement has been suggested by Shor: we prepare each group of four ancillary qubits in an equally weighted superposition of the strings with an even number of 0's and 1's:

$$|\Psi\rangle_{\text{Shor}} = \tfrac{1}{\sqrt{8}} \big(|0000\rangle + |0011\rangle + |0101\rangle + |0110\rangle$$
$$+ |1001\rangle + |1010\rangle + |1100\rangle + |1111\rangle\big). \tag{B.177}$$

It can be checked that, if no error has corrupted the data qubits, then the circuit in Fig. B.8 will leave the state $|\Psi\rangle_{\text{Shor}}$ unchanged. On the other hand, if a correctable single-qubit error has occurred, then the state $|\Psi\rangle_{\text{Shor}}$ will be transformed into an equally weighted superposition of the states with an odd number of 0's and 1's. Thus, the parity of the four ancillary qubits tells us whether or not there is a single-qubit error but does not reveal any information on the encoded state $\alpha|0_L\rangle + \beta|1_L\rangle$.

Exercise 7.15   Let us first consider the case in which the initial state of Bob's qubit is pure, $|\psi\rangle = \mu|0\rangle + \nu|1\rangle$, where μ and ν are complex numbers, with $|\mu|^2 + |\nu|^2 = 1$. The unitary transformation W in Fig. 7.13 maps the state $|\psi\rangle|\Phi\rangle$ (where $|\Phi\rangle$ is given by Eq. (7.115)) onto the state

$$|\Psi\rangle = \mu\big(\alpha|000\rangle + \beta|101\rangle + \gamma|110\rangle + \delta|011\rangle\big)$$
$$+ \nu\big(\alpha|111\rangle + \beta|010\rangle + \gamma|001\rangle + \delta|100\rangle\big). \quad \text{(B.178)}$$

We then obtain the density matrix ρ_B after tracing the density matrix $|\Psi\rangle\langle\Psi|$ over Eve's qubit and the ancillary qubit. We have

$$\rho_B = \begin{bmatrix} |\mu|^2(\alpha^2 + \delta^2) & 2\mu\nu^*\alpha\delta \\ + |\nu|^2(\beta^2 + \gamma^2) & + 2\mu^*\nu\beta\gamma \\ & \\ 2\mu^*\nu\alpha\delta & |\mu|^2(\beta^2 + \gamma^2) \\ + 2\mu\nu^*\beta\gamma & + |\nu|^2(\alpha^2 + \delta^2) \end{bmatrix}. \quad \text{(B.179)}$$

In the same manner we obtain the density matrix ρ_E after tracing over Bob's qubit and the ancillary qubit:

$$\rho_E = \begin{bmatrix} |\mu|^2(\alpha^2 + \beta^2) & 2\mu\nu^*\alpha\beta \\ + |\nu|^2(\gamma^2 + \delta^2) & + 2\mu^*\nu\gamma\delta \\ & \\ 2\mu^*\nu\alpha\beta & |\mu|^2(\gamma^2 + \delta^2) \\ + 2\mu\nu^*\gamma\delta & + |\nu|^2(\alpha^2 + \beta^2) \end{bmatrix}. \quad \text{(B.180)}$$

Let us call (x, y, z), (x_B, y_B, z_B) and (x_E, y_E, z_E) the Bloch-sphere coordinates corresponding to $|\psi\rangle\langle\psi|$, ρ_B and ρ_E. We have

$$\mu\nu^* = \tfrac{1}{2}(x - iy), \quad |\mu|^2 = \tfrac{1}{2}(1 + z), \quad |\nu|^2 = \tfrac{1}{2}(1 - z). \quad \text{(B.181)}$$

After setting $\gamma = 0$, we obtain

$$\begin{cases} \tfrac{1}{2}(x_B - iy_B) = (\rho_B)_{01} = (x - iy)\alpha\delta, \\ \\ \tfrac{1}{2}(1 + z_B) = (\rho_B)_{00} = \tfrac{1}{2}(1 + z)(\alpha^2 + \delta^2) + \tfrac{1}{2}(1 - z)\beta^2, \end{cases} \quad \text{(B.182)}$$

which imply

$$\begin{cases} x_B = 2\alpha\delta x, \\ y_B = 2\alpha\delta y, \\ z_B = (\alpha^2 + \delta^2 - \beta^2)z. \end{cases} \quad \text{(B.183)}$$

The state ρ_B is an isotropic cloning of $|\psi\rangle\langle\psi|$ when $R_B = x_B/x = y_B/y = z_B/z$. Therefore, we obtain

$$\begin{cases} 2\alpha\delta = \alpha^2 + \delta^2 - \beta^2, \\ \alpha^2 + \beta^2 + \delta^2 = 1, \end{cases} \tag{B.184}$$

so that

$$\delta = \tfrac{1}{2}\alpha \pm \sqrt{\tfrac{1}{2} - \tfrac{3}{4}\alpha^2}. \tag{B.185}$$

In the same manner we obtain

$$\begin{cases} x_E = 2\alpha\beta x, \\ y_E = 2\alpha\beta y, \\ z_E = (\alpha^2 + \beta^2 - \delta^2)z. \end{cases} \tag{B.186}$$

Isotropic cloning ($R_E = x_E/x = y_E/y = z_E/z$) is obtained when

$$\begin{cases} 2\alpha\beta = \alpha^2 + \beta^2 - \delta^2, \\ \alpha^2 + \beta^2 + \delta^2 = 1, \end{cases} \tag{B.187}$$

so that

$$\beta = \tfrac{1}{2}\alpha \pm \sqrt{\tfrac{1}{2} - \tfrac{3}{4}\alpha^2}. \tag{B.188}$$

Note that, if we choose the plus sign in (B.185), then the minus sign has to be taken in (B.188) in order that the normalization condition $\alpha^2 + \beta^2 + \delta^2$ be satisfied. This choice corresponds to Eq. (7.116).

Note that the cloning is also isotropic in the case in which the initial state ρ of Bob's qubits is mixed. In this case we may write $\rho = \sum_i p_i \rho_i$, with $\rho_i = |\psi_i\rangle\langle\psi_i|$ a pure state. The Bloch vector $\mathbf{r}$ associated with ρ is the weighted sum of the Bloch vectors $\mathbf{r}_i$ associated with the density matrices ρ_i: $\mathbf{r} = \sum_i p_i \mathbf{r}_i$. Since we have seen that for pure initial states $(\mathbf{r}_i)_B = R_B \mathbf{r}_i$ and $(\mathbf{r}_i)_E = R_E \mathbf{r}_i$, then $\mathbf{r}_B = \sum_i p_i (\mathbf{r}_i)_B = R_B \mathbf{r}$ and $\mathbf{r}_E = \sum_i p_i (\mathbf{r}_i)_E = R_E \mathbf{r}$.

Chapter 8

Exercise 8.1   We substitute $|\psi\rangle = \exp\left(-\tfrac{i}{\hbar}\omega t S_z\right)|\psi\rangle_r$ into the Schrödinger equation $i\hbar\tfrac{d}{dt}|\psi\rangle = H|\psi\rangle$, thus obtaining

$$i\hbar\frac{d}{dt}\left[\exp\left(-\frac{i}{\hbar}\omega t S_z\right)|\psi\rangle_r\right] = H\left[\exp\left(-\frac{i}{\hbar}\omega t S_z\right)|\psi\rangle_r\right]. \tag{B.189}$$

A straightforward calculation then gives

$$i\hbar\frac{d}{dt}|\psi\rangle_r = \left[\exp\left(\frac{i}{\hbar}\omega t S_z\right)\left(-\omega S_z + H\right)\exp\left(-\frac{i}{\hbar}\omega t S_z\right)\right]|\psi\rangle_r = H_r|\psi\rangle_r, \tag{B.190}$$

where the Hamiltonian H_r reads

$$H_r = (\omega_0 - \omega)S_z + \omega_1 \left[\cos(\phi)S_x + \sin(\phi)S_y\right]. \tag{B.191}$$

Note that the Hamiltonian H_r is time-independent; that is, the rotating field lies along a fixed axis in the rotating frame. Moreover, at resonance ($\omega = \omega_0$) the first term in (B.191) disappears, so that, in the rotating frame, the spin simply precesses with frequency ω_1 about the axis directed along the unit vector $(\cos\phi, \sin\phi, 0)$.

If the oscillating field is off-resonance by an amount $\Delta\omega = \omega_0 - \omega$, then in the rotating frame the spin precesses with frequency $\omega_1' = \sqrt{(\Delta\omega)^2 + \omega_1^2}$ about the axis $\frac{1}{\omega_1'}(\omega_1\cos\phi, \omega_1\sin\phi, \Delta\omega)$. Note that, if $|\Delta\omega| \gg \omega_1$; that is, we are far from resonance, then the oscillating field is unable to flip the state of a spin ($|0\rangle \leftrightarrow |1\rangle$). Indeed, in this case the rotation axis in practice coincides with the z-axis.

Exercise 8.2 We obtain

$$\begin{aligned}
R_{x_1}(\pi) U_I\left(\frac{t}{2}\right) R_{x_1}(\pi) &= -\sigma_x^{(1)} \left[\cos\left(\frac{\alpha t}{2\hbar}\right)I - i\sin\left(\frac{\alpha t}{2\hbar}\right)\sigma_z^{(1)} \otimes \sigma_z^{(2)}\right]\sigma_x^{(1)} \\
&= -\cos\left(\frac{\alpha t}{2\hbar}\right)I + i\sin\left(\frac{\alpha t}{2\hbar}\right)\sigma_x^{(1)}\left(\sigma_z^{(1)} \otimes \sigma_z^{(2)}\right)\sigma_x^{(1)} \\
&= -\cos\left(\frac{\alpha t}{2\hbar}\right)I - i\sin\left(\frac{\alpha t}{2\hbar}\right)\sigma_z^{(1)} \otimes \sigma_z^{(2)} = U_I^\dagger\left(\frac{t}{2}\right),
\end{aligned} \tag{B.192}$$

where we have used $\sigma_x\sigma_z\sigma_x = -\sigma_z$. Therefore,

$$U_I\left(\tfrac{t}{2}\right) R_{x_1}(\pi) U_I\left(\tfrac{t}{2}\right) R_{x_1}(\pi) = U_I\left(\tfrac{t}{2}\right) U_I^\dagger\left(\tfrac{t}{2}\right) = I. \tag{B.193}$$

Exercise 8.3 The phase accumulated due to the Hamiltonian evolution of the spin up to time t can be eliminated if two π pulse are applied at times 0 and t and then the evolution continues for another time t. Indeed, we have

$$\begin{aligned}
&\exp\left(-\frac{i}{\hbar}Ht\right) \exp\left(-i\frac{\pi}{2}\sigma_x\right) \exp\left(-\frac{i}{\hbar}Ht\right) \exp\left(-i\frac{\pi}{2}\sigma_x\right) \\
&= -\exp\left(-\frac{i}{2}\omega_0 t\sigma_z\right) \sigma_x \exp\left(-\frac{i}{2}\omega_0 t\sigma_z\right) \sigma_x = I. \tag{B.194}
\end{aligned}$$

Exercise 8.4 By means of refocusing techniques the contributions to the temporal evolution due to one- and two-qubit terms in the Hamiltonian (8.21) can be switched on and off at will. We can then implement the CMINUS gate as

$$\text{CMINUS} = \sqrt{i}\,\exp\left(i\frac{\pi}{4}\sigma_z^{(1)} \otimes \sigma_z^{(2)}\right) \exp\left(-i\frac{\pi}{4}\sigma_z^{(1)}\right) \exp\left(-i\frac{\pi}{4}\sigma_z^{(2)}\right). \tag{B.195}$$

Indeed, we have

$$\exp\left(i\tfrac{\pi}{4}\sigma_z^{(1)}\otimes\sigma_z^{(2)}\right) = \tfrac{1}{\sqrt{2}}\begin{bmatrix} 1+i & 0 & 0 & 0 \\ 0 & 1-i & 0 & 0 \\ 0 & 0 & 1-i & 0 \\ 0 & 0 & 0 & 1+i \end{bmatrix}, \quad (B.196)$$

$$\exp\left(-i\tfrac{\pi}{4}\sigma_z^{(1)}\right) = \tfrac{1}{\sqrt{2}}\begin{bmatrix} 1-i & 0 \\ 0 & 1+i \end{bmatrix}\otimes I^{(2)} = \tfrac{1}{\sqrt{2}}\begin{bmatrix} 1-i & 0 & 0 & 0 \\ 0 & 1-i & 0 & 0 \\ 0 & 0 & 1+i & 0 \\ 0 & 0 & 0 & 1+i \end{bmatrix}, \quad (B.197)$$

$$\exp\left(-i\tfrac{\pi}{4}\sigma_z^{(2)}\right) = I^{(1)}\otimes\tfrac{1}{\sqrt{2}}\begin{bmatrix} 1-i & 0 \\ 0 & 1+i \end{bmatrix} = \tfrac{1}{\sqrt{2}}\begin{bmatrix} 1-i & 0 & 0 & 0 \\ 0 & 1+i & 0 & 0 \\ 0 & 0 & 1-i & 0 \\ 0 & 0 & 0 & 1+i \end{bmatrix}. \quad (B.198)$$

The decomposition (B.195) may be verified by direct multiplication of the matrices (B.196–B.198). The CNOT gate can then be obtained from CMINUS and Hadamard gates as shown in Fig. 3.6. Note that the waiting time for the implementation of the unitary transformation $\exp\left(i\tfrac{\pi}{4}\sigma_z^{(1)}\otimes\sigma_z^{(2)}\right)$ is proportional to $\frac{1}{J_{12}}$. Therefore, if $J_{12}\sim 1\,\text{kHz}$, the realization of a CNOT gates requires a time of the order of milliseconds.

Exercise 8.5 It is convenient to project the solution $\psi(\boldsymbol{r},t) = \langle\boldsymbol{r}|\psi(t)\rangle$ of the Schrödinger equation (8.25) onto the basis of eigenfunctions of H_0. We obtain

$$\psi(\boldsymbol{r},t) = \sum_i c_i(t)\phi_i(\boldsymbol{r})\exp\left(-i\frac{E_i}{\hbar}t\right), \quad (B.199)$$

where $\phi_i(\boldsymbol{r})$ is the eigenfunction of H_0 corresponding to the eigenvalue E_i (i.e., $H_0\phi_i(\boldsymbol{r}) = E_i\phi_i(\boldsymbol{r})$). We are interested in the case in which only two states of the atom ($\phi_g(\boldsymbol{r})$ and $\phi_e(\boldsymbol{r})$) are relevant for its dynamical evolution, the corresponding energies being $E_g = \hbar\omega_g$ and $E_e = \hbar\omega_e$. After substitution of (B.199) into (8.25) we obtain

$$i\hbar\frac{\partial}{\partial t}\psi(\boldsymbol{r},t) = i\hbar\sum_{i=g,e}\left[\dot{c}_i(t) - i\,\omega_i c_i(t)\right]\phi_i(\boldsymbol{r})\,e^{-i\omega_i t}$$

$$= \sum_{k=g,e}(H_0 + H_I)\,c_k(t)\,\phi_k(\boldsymbol{r})\,e^{-i\omega_k t} = \sum_{k=g,e}(\hbar\omega_k + H_I)\,c_k(t)\,\phi_k(\boldsymbol{r})\,e^{-i\omega_k t}.$$

$$(B.200)$$

We now write

$$E(t) = E_0\cos(\omega t + \phi) = \tfrac{1}{2}E_0\left(e^{i(\omega t+\phi)} + e^{-i(\omega t+\phi)}\right) = \alpha e^{i\omega t} + \alpha^\star e^{-i\omega t}, \quad (B.201)$$

where $\alpha = \frac{1}{2}E_0\,e^{i\phi}$. Thus, Eq. (B.200) reads

$$i\hbar \sum_{i=g,e} \phi_i(\boldsymbol{r})\,e^{-i\omega_i t}\dot{c}_i(t) = \sum_{k=g,e} c_k(t)\,e^{-i\omega_k t}\left[-ez\left(\alpha\,e^{i\omega t} + \alpha^\star\,e^{-i\omega t}\right)\right]\phi_k(\boldsymbol{r}),$$

(B.202)

corresponding to two first-order ordinary differential equations in the variables c_0 and c_1. After multiplication on the left of both members of (B.202) by $\phi_j^\star(\boldsymbol{r})$ and integration over $\boldsymbol{r}$ we obtain, for $j = g,e$,

$$i\hbar e^{-i\omega_j t}\dot{c}_j(t) = \sum_{k=g,e} c_k(t)\,e^{-i\omega_k t}\int d\boldsymbol{r}\,\phi_j^\star(\boldsymbol{r})\left[-ez\left(\alpha\,e^{i\omega t} + \alpha^\star\,e^{-i\omega t}\right)\right]\phi_k(\boldsymbol{r}),$$

(B.203)

where we have taken advantage of the standard orthonormality relation $\int d\boldsymbol{r}\,\phi_j^\star(\boldsymbol{r})\phi_i(\boldsymbol{r}) = \delta_{ij}$. We must evaluate the four integrals

$$D_{jk} = -e\int d\boldsymbol{r}\,\phi_j^\star(\boldsymbol{r})\,z\,\phi_k(\boldsymbol{r}), \qquad (j,k = g,e).$$

(B.204)

Since $V(r)$ is spherically symmetric, the eigenfunctions $\phi_i(\boldsymbol{r})$ are symmetric or antisymmetric under the inversion $\boldsymbol{r} \to -\boldsymbol{r}$. Thus, $D_{gg} = D_{ee} = 0$ and Eq. (B.203) reads

$$\begin{cases} i\hbar\dot{c}_g(t) = c_e(t)\left[D\,\alpha\,e^{i(\omega-\omega_a)t} + D\,\alpha^\star e^{-i(\omega+\omega_a)t}\right], \\ i\hbar\dot{c}_e(t) = c_g(t)\left[D^\star\alpha\,e^{i(\omega+\omega_a)t} + D^\star\alpha^\star e^{-i(\omega-\omega_a)t}\right], \end{cases}$$

(B.205)

where we have defined $D = D_{ge} = D_{eg}^\star$ and $\omega_a = \omega_e - \omega_g$. The terms depending on $\omega + \omega_a$ oscillate very rapidly and can therefore be neglected (*rotating wave approximation*). Setting

$$\Delta = \omega - \omega_a, \qquad \Omega = \frac{D\alpha}{\hbar},$$

(B.206)

we obtain

$$\begin{cases} i\dot{c}_g(t) = \Omega\,e^{i\Delta t}c_e, \\ i\dot{c}_e(t) = \Omega^\star e^{-i\Delta t}c_g. \end{cases}$$

(B.207)

To solve this system, it is convenient to differentiate the first equation and substitute the result into the second. We have

$$i\ddot{c}_g = i\Delta\Omega\,e^{i\Delta t}c_e + \Omega\,e^{i\Delta t}\dot{c}_e$$

$$= i\Delta i\dot{c}_g + \Omega\,e^{i\Delta t}\frac{1}{i}\Omega^\star e^{-i\Delta t}c_g,$$

(B.208)

that is,

$$\ddot{c}_g - i\Delta\dot{c}_g + |\Omega|^2 c_g = 0.$$

(B.209)

This last equation is easily solved by setting $c_0 = Ae^{i\xi t}$, which leads to the algebraic equation

$$\xi^2 - \Delta\xi - |\Omega|^2 = 0, \tag{B.210}$$

whose solutions are

$$\xi_\pm = \frac{\Delta}{2} \pm \sqrt{\frac{\Delta^2}{4} + |\Omega|^2}. \tag{B.211}$$

Therefore, the general solution of (B.205) is

$$\begin{cases} c_g(t) = A_+ e^{i\xi_+ t} + A_- e^{i\xi_- t}, \\ c_e(t) = -\dfrac{1}{\Omega} \left(\xi_+ A_+ e^{i(\xi_+ - \Delta)t} + \xi_- A_- e^{i(\xi_- - \Delta)t} \right). \end{cases} \tag{B.212}$$

The constants A_+ and A_- are determined from the initial conditions $c_g^{(0)} = c_g(t = 0)$ and $c_e^{(0)} = c_e(t = 0)$. We obtain

$$A_+ = \frac{\xi_- c_g^{(0)} + \Omega c_e^{(0)}}{\xi_- - \xi_+}, \qquad A_- = \frac{\xi_+ c_g^{(0)} + \Omega c_e^{(0)}}{\xi_+ - \xi_-}. \tag{B.213}$$

We finally substitute these relations into (B.212) and obtain

$$\begin{bmatrix} c_g(t) \\ c_e(t) \end{bmatrix} = U \begin{bmatrix} c_g^{(0)} \\ c_e^{(0)} \end{bmatrix} = \begin{bmatrix} U_{gg} & U_{ge} \\ U_{eg} & U_{ee} \end{bmatrix} \begin{bmatrix} c_g^{(0)} \\ c_e^{(0)} \end{bmatrix}. \tag{B.214}$$

Here U is a unitary matrix with matrix elements

$$U_{gg} = e^{i\frac{\Delta}{2}t} \left[\cos(at) - \frac{i\Delta \sin(at)}{2a} \right] = U_{ee}^\star,$$

$$U_{ge} = e^{i\frac{\Delta}{2}t} \left[\frac{-i\Omega \sin(at)}{a} \right] = -U_{eg}^\star, \tag{B.215}$$

where

$$a = \sqrt{\frac{\Delta^2}{4} + |\Omega|^2}, \qquad \Delta = \omega - \omega_a. \tag{B.216}$$

We are particularly interested in the resonant case in which the detuning parameter $\Delta = 0$. It is easy to see that in this case the unitary matrix U reduces to (8.24), with $|\Omega|t = \frac{\theta}{2}$.

Exercise 8.6   From the properties of the Hermite polynomials, given in the formulæ in the footnote on page 534, we derive that

$$H_{n+1}(\xi) = \left(2\xi - \frac{d}{d\xi} \right) H_n(\xi). \tag{B.217}$$

We have

$$a = \frac{1}{\sqrt{2}} \left(\xi + \frac{d}{d\xi} \right), \qquad a^\dagger = \frac{1}{\sqrt{2}} \left(\xi - \frac{d}{d\xi} \right), \tag{B.218}$$

where $\xi = \sqrt{\frac{m\omega}{\hbar}}\, x$. It follows immediately that

$$a^\dagger \phi_n(\xi) = \sqrt{n+1}\,\phi_{n+1}(\xi), \qquad a\phi_n(\xi) = \sqrt{n}\,\phi_{n-1}(\xi). \tag{B.219}$$

The iterative application of the first of these relations leads to

$$\phi_n(\xi) = \frac{a^\dagger}{\sqrt{n!}}\,\phi_0(\xi). \tag{B.220}$$

Exercise 8.7 We can expand the solution to the Schrödinger equation for the Jaynes–Cummings model as

$$|\psi(t)\rangle = \sum_{i=g,e} \sum_{n=0}^{\infty} c_{i,n}(t)|i,n\rangle, \tag{B.221}$$

where the indices i and n label the atomic state and the number of photons. Therefore, the Schrödinger equation reads

$$i\hbar \sum_{i=g,e} \sum_{n=0}^{\infty} \dot{c}_{i,n}(t)|i,n\rangle = \sum_{i'=g,e} \sum_{n'=0}^{\infty} H c_{i',n'}(t)|i',n'\rangle, \tag{B.222}$$

which implies

$$i\hbar \dot{c}_{i,n}(t)|i,n\rangle = \sum_{i'=g,e} \sum_{n'=0}^{\infty} \langle i,n|H|i',n'\rangle c_{i',n'}(t). \tag{B.223}$$

The matrix elements of the Jaynes–Cummings Hamiltonian can be easily computed and we obtain

$$\begin{aligned}
\langle g,n|H|g,n'\rangle &= \hbar\left[-\frac{\omega_a}{2} + \omega\left(n+\frac{1}{2}\right)\right]\delta_{n,n'}, \\
\langle g,n|H|e,n'\rangle &= \lambda^*\sqrt{n}\,\delta_{n,n'+1}, \\
\langle e,n|H|g,n'\rangle &= \lambda\sqrt{n+1}\,\delta_{n,n'-1}, \\
\langle e,n|H|e,n'\rangle &= \hbar\left[\frac{\omega_a}{2} + \omega\left(n+\frac{1}{2}\right)\right]\delta_{n,n'}.
\end{aligned} \tag{B.224}$$

After substitution of these matrix elements into (B.223) we obtain

$$\begin{cases}
i\hbar\dot{c}_{g,n} = \hbar\left[-\dfrac{\omega_a}{2} + \omega\left(n+\dfrac{1}{2}\right)\right]c_{g,n}(t) + \lambda^*\sqrt{n}\,c_{e,n-1}(t), \\[4mm]
i\hbar\dot{c}_{e,n} = \hbar\left[\dfrac{\omega_a}{2} + \omega\left(n+\dfrac{1}{2}\right)\right]c_{e,n}(t) + \lambda\sqrt{n+1}\,c_{g,n+1}(t).
\end{cases} \tag{B.225}$$

It is convenient to write the second equation for $n \to n - 1$. Then system (B.225) reads

$$\begin{cases} i\dot{c}_{g,n} = \left[-\frac{\omega_a}{2} + \omega \left(n + \frac{1}{2} \right) \right] c_{g,n}(t) + \frac{\lambda^*}{\hbar} \sqrt{n}\, c_{e,n-1}(t), \\[2mm] i\dot{c}_{e,n-1} = \left[\frac{\omega_a}{2} + \omega \left(n - \frac{1}{2} \right) \right] c_{e,n-1}(t) + \frac{\lambda}{\hbar} \sqrt{n}\, c_{g,n}(t). \end{cases} \tag{B.226}$$

It is now clear that the level $|g, n\rangle$ is only coupled to $|e, n - 1\rangle$. In order to solve these equations, we first separate the time dependence due to H_0, setting

$$\begin{cases} c_{g,n}(t) = \exp\left\{ -i \left[-\frac{\omega_a}{2} + \omega \left(n + \frac{1}{2} \right) \right] t \right\} \tilde{c}_{g,n}(t), \\[2mm] c_{e,n-1}(t) = \exp\left\{ -i \left[\frac{\omega_a}{2} + \omega \left(n - \frac{1}{2} \right) \right] t \right\} \tilde{c}_{e,n-1}(t). \end{cases} \tag{B.227}$$

We insert these relations into (B.226) and, after defining

$$\Delta = \omega - \omega_a, \qquad \Omega_n = \frac{\lambda^*}{\hbar} \sqrt{n}, \tag{B.228}$$

we obtain

$$\begin{cases} i\dot{\tilde{c}}_{g,n} = \Omega_n \exp(i\Delta t)\, \tilde{c}_{e,n-1}(t), \\[2mm] i\dot{\tilde{c}}_{e,n-1} = \Omega_n^* \exp(-i\Delta t)\, \tilde{c}_{g,n}(t). \end{cases} \tag{B.229}$$

These equations are analogous to (B.207), obtained for a two-level atom interacting with a classical field and can be solved in the same manner. Thus, in the resonant case $\Delta = 0$ there are Rabi oscillations between the states $|g, n\rangle$ and $|e, n-1\rangle$. The frequency $|\Omega_n|$ ($\Omega_n = |\Omega_n| e^{i\phi_n}$) of these oscillations is proportional to the atom–field interaction strength $|\lambda|$ and to the square root of the number n of photons in the resonant cavity.

Exercise 8.8   It is clear from exercise 8.7 that the overall state of the atom–field system at time t is given by

$$|\psi(t)\rangle = c_{g,n}(t)|g, n\rangle + c_{e,n-1}(t)|e, n - 1\rangle. \tag{B.230}$$

The coefficients $c_{g,n}(t)$ and $c_{e,n-1}(t)$ are determined by the initial conditions $c_{g,n}^{(0)} = c_{g,n}(t = 0)$, $c_{e,n-1}^{(0)} = c_{e,n-1}(t = 0)$. At resonance ($\omega = \omega_a$) we have

$$\begin{bmatrix} c_{g,n}(t) \\ c_{e,n-1}(t) \end{bmatrix} = \begin{bmatrix} \cos(|\Omega_n|t) & -ie^{i\phi_n} \sin(|\Omega_n|t) \\ -ie^{-i\phi_n} \sin(|\Omega_n|t) & \cos(|\Omega_n|t) \end{bmatrix} \begin{bmatrix} c_{g,n}^{(0)} \\ c_{e,n-1}^{(0)} \end{bmatrix}. \tag{B.231}$$

Since $|\psi_0\rangle = |g, n\rangle$, then

$$|\psi(t)\rangle = \cos(|\Omega_n|t)|g, n\rangle - ie^{-i\phi_n} \sin(|\Omega_n|t)|e, n - 1\rangle. \tag{B.232}$$

The overall atom–field system at time t is in a pure state, $\rho(t) = |\psi(t)\rangle\langle\psi(t)|$. Thus, the entanglement between the atom and the field is quantified by the reduced von Neumann entropy

$$S_a(t) = -\text{Tr}[\rho_a(t) \log \rho_a(t)], \tag{B.233}$$

where

$$\rho_a(t) = \text{Tr}_f [\rho(t)] = \cos^2(|\Omega_n|t)|g\rangle\langle g| + \sin^2(|\Omega_n|t)|e\rangle\langle e| \tag{B.234}$$

is the reduced density matrix describing the state of the two-level atom (the trace is taken over the field degree of freedom). Therefore, $S_a(t) = -\sum_{i=1}^{2} \lambda_i(t) \log \lambda_i(t)$, where $\lambda_1(t) = \cos^2(|\Omega_n|t)$ and $\lambda_2 = \sin^2(|\Omega_n|t)$. are the eigenvalues of ρ_a. The temporal evolution of $\rho_a(t)$ is shown in Fig. B.10: it oscillates between $S_a = 0$ (for separable states) and $S_a = 1$ (for maximally entangled Bell states).

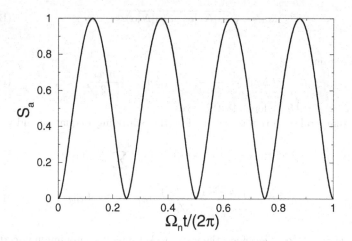

Fig. B.10 The evolution of the entropy S_a for a two-level atom coupled to a single mode of the electromagnetic field at resonance ($\omega = \omega_a$).

Exercise 8.9 (i) We have

$$\langle n|a|\alpha\rangle = \alpha\langle n|\alpha\rangle = \sqrt{n+1}\,\langle n+1|\alpha\rangle. \tag{B.235}$$

This relation can be iterated, thus obtaining

$$\langle n|\alpha\rangle = \frac{\alpha^n}{\sqrt{n!}}\langle 0|\alpha\rangle, \tag{B.236}$$

which implies

$$|\alpha\rangle = \sum_{n=0}^{\infty} |n\rangle\langle n|\alpha\rangle = \langle 0|\alpha\rangle \sum_{n=0}^{\infty} \frac{\alpha^n}{\sqrt{n!}}|n\rangle. \tag{B.237}$$

Normalization of this state ($\langle\alpha|\alpha\rangle = 1$) leads to $|\langle 0|\alpha\rangle|^2 = e^{-|\alpha|^2}$, so that state (8.36) is obtained.

(ii) We have

$$p(n) \equiv |\langle n|\alpha\rangle|^2 = \frac{|\alpha|^{2n}}{n!}e^{-|\alpha|^2}. \tag{B.238}$$

Therefore, the probability $p(n)$ of finding n photons in $|\alpha\rangle$ is given by a Poisson distribution and the mean photon number

$$\bar{n} = \langle\alpha|N|\alpha\rangle = \sum_{n=0}^{\infty} np(n) = e^{-|\alpha|^2}\sum_{n=0}^{\infty} n\frac{\left(|\alpha|^2\right)^n}{n!} = e^{-|\alpha|^2}|\alpha|^2 e^{|\alpha|^2} = |\alpha|^2, \tag{B.239}$$

where $N = a^\dagger a$ is the number operator. We also obtain

$$\Delta n = \sqrt{\langle\alpha|N^2|\alpha\rangle - \langle\alpha|N|\alpha\rangle} = \sqrt{\bar{n}}. \tag{B.240}$$

(iii) We obtain

$$(\Delta x)^2 = \frac{\hbar}{2m\omega}, \qquad (\Delta p)^2 = \frac{m\hbar\omega}{2}, \tag{B.241}$$

so that $\Delta x \Delta p = \frac{1}{2}\hbar$, independently of α.

The temporal evolution is governed by the Schrödinger equation and we have

$$\begin{aligned}|\psi(t)\rangle &= e^{-\frac{i}{\hbar}Ht}|\alpha\rangle = e^{-i\omega\left(N+\frac{1}{2}\right)t}e^{-\frac{1}{2}|\alpha|^2}\sum_{n=0}^{\infty}\frac{\alpha^n}{\sqrt{n!}}|n\rangle \\ &= e^{-\frac{i}{2}\omega t}e^{-\frac{1}{2}|\alpha|^2}\sum_{n=0}^{\infty}\frac{\left(\alpha e^{-i\omega t}\right)^n}{\sqrt{n!}}|n\rangle = e^{-\frac{i}{2}\omega t}|\alpha(t)\rangle,\end{aligned} \tag{B.242}$$

where $\alpha(t) = e^{-i\omega t}\alpha$. Therefore, the wave packet remains a minimum uncertainty coherent state at all times, and its centre in phase space follows the classical path in time. This is made more explicitly by computing the expectation values of x and p:

$$\begin{aligned}\langle x(t)\rangle &= \sqrt{\frac{2\hbar}{m\omega}}\,\mathrm{Re}(\alpha(t)) = \langle x(0)\rangle\cos(\omega t) + \frac{\langle p(0)\rangle}{m\omega}\sin(\omega t), \\ \langle p(t)\rangle &= \sqrt{2m\hbar\omega}\,\mathrm{Im}(\alpha(t)) = \langle p(0)\rangle\cos(\omega t) - m\omega\langle x(0)\rangle\sin(\omega t).\end{aligned} \tag{B.243}$$

Exercise 8.10   The initial atom–field state is

$$|\psi_0\rangle = |g\rangle \otimes \sum_{n=0}^{\infty} c_n|n\rangle, \tag{B.244}$$

where

$$c_n = e^{-|\alpha|^2/2} \frac{\alpha^n}{\sqrt{n!}}, \qquad \alpha \in \mathbb{C}. \tag{B.245}$$

The temporal evolution of the atom–field state in the Jaynes–Cummings model was discussed in exercise 8.7. Given the initial condition (B.244), we obtain

$$|\psi(t)\rangle = \sum_{n=0}^{\infty} \left[c_n \cos(\lambda\sqrt{n}t)|g\rangle - ic_{n+1}\sin(\lambda\sqrt{n+1}t)|e\rangle \right] |n\rangle, \tag{B.246}$$

where we have assumed λ real, the detuning parameter $\Delta = 0$ and set $\hbar = 1$. The density matrix describing the atomic state $\rho^{(a)}$ at time t is obtained after tracing over the field degree of freedom the overall density matrix $|\psi(t)\rangle\langle\psi(t)|$. The matrix elements of $\rho^{(a)}(t)$ in the $\{|g\rangle, |e\rangle\}$ basis read as follows:

$$\rho_{gg}^{(a)} = \sum_{n=0}^{\infty} |c_n|^2 \cos^2(\lambda\sqrt{n}t),$$

$$\rho_{ee}^{(a)} = 1 - \rho_{gg}^{(a)}, \tag{B.247}$$

$$\rho_{ge}^{(a)} = \left(\rho_{eg}^{(a)}\right)^* = i \sum_{n=0}^{\infty} c_n^* c_{n+1} \cos(\lambda\sqrt{n}t) \sin(\lambda\sqrt{n+1}t).$$

After writing explicitly the coefficients c_n, we obtain

$$\rho_{gg}^{(a)} = \sum_{n=0}^{\infty} e^{-|\alpha|^2} \frac{|\alpha|^{2n}}{n!} \cos^2(\lambda\sqrt{n}t)$$

$$\rho_{ee}^{(a)} = \sum_{n=0}^{\infty} e^{-|\alpha|^2} \frac{|\alpha|^{2n}}{n!} \sin^2(\lambda\sqrt{n}t) \tag{B.248}$$

$$\rho_{ge}^{(a)} = \left(\rho_{eg}^{(a)}\right)^* = i \sum_{n=0}^{\infty} e^{-|\alpha|^2} \frac{|\alpha|^{2n}}{n!} \frac{\alpha^*}{\sqrt{n+1}} \cos(\lambda\sqrt{n}t) \sin(\lambda\sqrt{n+1}t).$$

Given the density matrix $\rho^{(a)}(t)$, it is easy to determine the Bloch-sphere coordinates $(x(t), y(t), z(t))$ of the two-level atom and the entropy $S(t) = -\sum_{i=1}^{2} \lambda_i(t) \log \lambda_i(t)$, where $\lambda_1(t), \lambda_2(t)$ are the eigenvalues of $\rho^{(a)}(t)$. The von Neumann entropy $S(t)$ is a measure of the atom–field entanglement.

Examples of the temporal evolution of a two-level atom interacting with a single-mode field, initially prepared in a coherent state, are shown in Figs. B.11–B.12. The first figure refers to the short-time motion. The representative point describing the atomic state exhibits a motion similar to a spiral and *collapses* to the centre of the Bloch sphere. Thus, the state of the atom is no longer pure and its entropy is non-zero. The usual Rabi oscillations between the atomic states $|g\rangle$ and $|e\rangle$ are damped. Note that the number of oscillations before damping dominates increases when the average number of photons $\bar{n}$ in the field is larger. This is quite natural as the transition to the classical electromagnetic field takes place when the number of photons $\bar{n} \to \infty$. The behaviour of the atomic state at longer times is shown in Fig. B.12. The main feature of this figure is the existence of *revivals*; that is, at times much longer than the damping time, the amplitude of the Rabi oscillations increases.

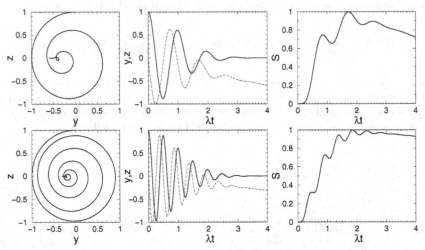

Fig. B.11 The temporal evolution of a two-level atom interacting with an initially coherent single-mode field with $\alpha = \sqrt{\bar{n}} = \sqrt{10}$ (top) and $\sqrt{40}$ (bottom): Bloch sphere trajectory (left), Bloch-sphere coordinates y (dashed curve) and z (full curve) versus time (middle) and entropy S versus time (right). Note that for the chosen initial conditions $x = 0$ at all times.

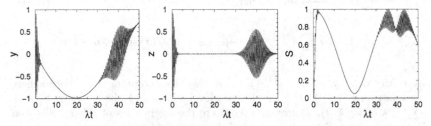

Fig. B.12 The same as in Fig. B.11 but at longer times, for $\bar{n} = 40$.

The phenomenon of collapse and revival is repeated with increasing time and can be qualitatively understood as follows. Rabi oscillations of $z(t)$ are determined by the sum of oscillating terms $\propto \cos^2(\lambda\sqrt{n}t)$ appearing in $\rho_{gg}(t) = \frac{1}{2}(1 + z(t))$. Each term in the summation represents Rabi oscillations with frequency $\propto \sqrt{n}$. At time $t = 0$ all these terms are correlated. As time goes on, the Rabi oscillations associated with different values of n have different frequencies and therefore become uncorrelated leading to the phenomenon of collapse. Correlation between these contributions is restored, at least partially, at longer times and revivals occur. Note that revivals are purely quantum phenomena and are due to the discrete structure of the photon distribution (only integer n values are allowed).

The relevant time scales can be estimated as follows: the period t_R of Rabi oscillations is given by the inverse of the Rabi frequency at $n = \bar{n}$; that is, $t_R \sim 1/\Omega_{\bar{n}} \sim 1/\lambda\sqrt{\bar{n}}$. These oscillations continue until the collapse time t_c when the terms associated with different n values become dephased. Since in the initial coherent state approximately $\Delta n = \sqrt{\bar{n}}$ Fock states are relevant, t_c can be estimated from the condition $(\Omega_{\bar{n}+\Delta n} - \Omega_{\bar{n}-\Delta n})t_c \sim 1$, leading to $t_c \sim 1/\lambda$. Finally, the revival times $t_r^{(m)}$ can be estimated by requiring that the phases of the oscillations corresponding to neighbouring photon numbers differ by an integral multiple of 2π; that is, when $(\Omega_{\bar{n}} - \Omega_{\bar{n}-1})t_r^{(m)} = 2\pi m$. This implies $t_r^{(m)} \sim m\sqrt{\bar{n}}/\lambda$. In particular, the time of the first revival $(m = 1)$ is $\sim \sqrt{\bar{n}}/\lambda$. Therefore, revivals, which are a purely quantum phenomenon, require longer and longer times when $\bar{n} \to \infty$.

It is instructive to evaluate the entropy at the first Rabi oscillation; that is, for $\lambda\sqrt{\bar{n}}\,\tilde{t} = \frac{\pi}{2}$, in the limit of large mean photon number $\bar{n}$ (note that $\tilde{t}$ corresponds to half of the period of Rabi oscillation; that is, $\tilde{t} = \frac{1}{2}t_R$). We first evaluate the matrix elements of $\rho^{(a)}(t)$ in (B.248), assuming that only the terms with $n-\bar{n} \ll \bar{n}$ contribute significantly. This is the case as for a coherent state the root mean square deviation Δn in the photon number is equal to $\sqrt{\bar{n}} \ll \bar{n}$. We then obtain $\rho_{gg}^{(a)}(\tilde{t}) \approx \frac{\pi^2}{16\bar{n}}$, $|\rho_{ge}^{(a)}(\tilde{t})| \approx \frac{\pi}{8\bar{n}}$, from which we can compute $\lambda_1(\tilde{t}) = \frac{\pi^2}{16\bar{n}^2}$ and $\lambda_2(\tilde{t}) = 1 - \frac{\pi^2}{16\bar{n}^2}$. Note that $S(\tilde{t}) \propto (1/\bar{n}^2)\log(1/\bar{n}^2) \to 0$ when $\bar{n} \to \infty$. This is expected since there is no decoherence induced by a classical electromagnetic field.

Exercise 8.11 We apply the unitary transformation (8.24) to both atoms, with $\theta = \frac{\pi}{2}$ and phases ϕ_1 and ϕ_2. The global unitary transformation is given, in the $\{|g_1, g_2\rangle, |g_1, e_2\rangle, |e_1, g_2\rangle, |e_1, e_2\rangle\}$ basis, by

$$U = \frac{1}{\sqrt{2}} \begin{bmatrix} 1 & -ie^{i\phi_1} \\ -ie^{-i\phi_1} & 1 \end{bmatrix} \otimes \frac{1}{\sqrt{2}} \begin{bmatrix} 1 & -ie^{i\phi_2} \\ -ie^{-i\phi_2} & 1 \end{bmatrix}. \tag{B.249}$$

After application of U to the Bell state $\frac{1}{\sqrt{2}}(|e_1, g_2\rangle - |g_1, e_2\rangle)$ we obtain the state

$$\frac{1}{2\sqrt{2}} \Big[-i\big(e^{i\phi_2} - e^{i\phi_1}\big)|g_1, g_2\rangle - \big(1 + e^{i(\phi_1 - \phi_2)}\big)|g_1, e_2\rangle$$
$$+ \big(1 + e^{i(\phi_2 - \phi_1)}\big)|e_1, g_2\rangle + i\big(e^{-i\phi_1} - e^{-i\phi_2}\big)|e_1, e_2\rangle \Big], \tag{B.250}$$

from which the probabilities (8.42) directly follow.

Exercise 8.12 (i) The electric field is given by $\boldsymbol{E} = -\nabla\Phi$ and the equations of motion for the ion are $M\ddot{\boldsymbol{r}} = q\boldsymbol{E}$. Thus, we obtain

$$\ddot{x} = -\frac{q}{M}\left[\frac{U_0}{R^2}\cos(\omega_{RF}t) - \epsilon\,\frac{V_0}{R^2}\right]x, \qquad \text{(B.251a)}$$

$$\ddot{y} = \frac{q}{M}\left[\frac{U_0}{R^2}\cos(\omega_{RF}t) + (1-\epsilon)\,\frac{V_0}{R^2}\right]y, \qquad \text{(B.251b)}$$

$$\ddot{z} = -\frac{q}{M}\frac{V_0}{R^2}z. \qquad \text{(B.251c)}$$

Therefore, the motion along z is harmonic, with frequency

$$\omega_z = \sqrt{\frac{qV_0}{MR^2}}, \qquad \text{(B.252)}$$

while the motion along y and z is governed by the Mathieu equation (8.45), where

$$a_x = -\frac{4q\epsilon V_0}{M\omega_{RF}^2 R^2}, \quad q_x = \frac{2qU_0}{M\omega_{RF}^2 R^2} \qquad \text{(B.253)}$$

for $\xi = x$ and

$$a_y = -\frac{4q(1-\epsilon)V_0}{M\omega_{RF}^2 R^2}, q_y = -\frac{2qU_0}{M\omega_{RF}^2 R^2} \qquad \text{(B.254)}$$

for $\xi = y$. The analytic treatment of the Mathieu equation can be found, for instance, in Leibfried *et al.* (2003a). Here in Fig. B.13 we simply draw, for small q, a, the stability region in the q–a plane, as obtained from the numerical integration of the Mathieu equation.

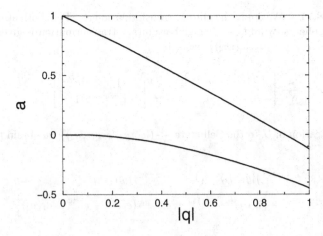

Fig. B.13 The stability diagram for the Mathieu equation. The two solid curves bound the stability region.

(ii) As an example, we compare in Fig. B.14 the approximate solution (8.46) to the Mathieu equation (8.45) for $q = 0.3$ and various a values. It is clear that (8.46) is a good approximation for small a (and q), while, as expected, it differs more and more from the exact solution when the stability border is approached.

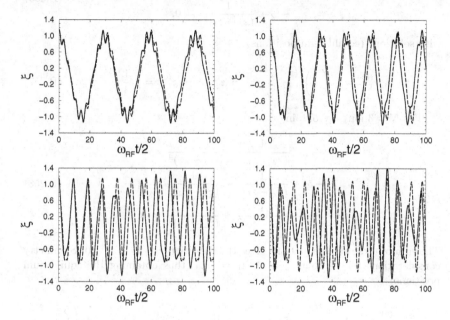

Fig. B.14 A comparison between the numerical integration of Eq. (8.45) (solid line) and the approximate solution (8.46) (dashed line), for $q = 0.3$, $a = 0$ (top left), 0.1 (top right), 0.4 (bottom left) and 0.65 (bottom right).

It is interesting that the approximate solution (8.46) gives harmonic oscillations of size ξ_0, at a frequency $\omega_\xi = \beta_\xi \omega_{RF}/2 \ll \omega_{RF}$ (the *secular motion*), superposed with smaller driven excursions of size $\xi_0 q_\xi/2 \ll \xi_0$, at a frequency ω_{RF} (the *micromotion*). If the fast and small oscillations of the micromotion are neglected, the motion can be approximated by that of a harmonic oscillator at a frequency ω_ξ. Such a harmonic approximation is used in the rest of this section, starting from the ion trap Hamiltonian (8.47).

Exercise 8.13 The balance between the harmonic and the Coulomb forces gives the equilibrium positions. This leads, for $N = 2$, to the following equations:

$$-M\omega_z^2 z_1 - \frac{q^2}{4\pi\epsilon_0(z_2 - z_1)^2} = 0,$$
$$-M\omega_z^2 z_2 + \frac{q^2}{4\pi\epsilon_0(z_2 - z_1)^2} = 0.$$
(B.255)

It is convenient to employ units in which $q^2/(4\pi\epsilon_0 M\omega_z^2) = 1$, so that the above equations become

$$-z_1 - \frac{1}{(z_2 - z_1)^2} = 0,$$
$$-z_2 + \frac{1}{(z_2 - z_1)^2} = 0,$$

(B.256)

with solutions (equilibrium positions)

$$z_1 = z_1^{(0)} = -\left(\frac{1}{4}\right)^{1/3}, \qquad z_2 = z_2^{(0)} = -z_1^{(0)} = \left(\frac{1}{4}\right)^{1/3}.$$

(B.257)

For $N = 3$ ions, we obtain

$$-z_1 - \frac{1}{(z_2 - z_1)^2} - \frac{1}{(z_3 - z_1)^2} = 0,$$
$$-z_2 + \frac{1}{(z_2 - z_1)^2} - \frac{1}{(z_3 - z_2)^2} = 0,$$
$$-z_3 + \frac{1}{(z_3 - z_1)^2} + \frac{1}{(z_3 - z_2)^2} = 0.$$

(B.258)

The sum of these three equations gives the condition $z_1 + z_2 + z_3 = 0$. Symmetry considerations tell us that $z_2 = 0$, $z_1 = -z_3$, thus obtaining the equilibrium positions

$$z_1 = z_1^{(0)} = -\left(\frac{5}{4}\right)^{1/3}, \quad z_2 = z_2^{(0)} = 0, \quad z_3 = z_3^{(0)} = -z_1^{(0)} = \left(\frac{5}{4}\right)^{1/3}.$$

(B.259)

Let us now compute the normal modes of vibration of the string about the equilibrium positions. Starting from Hamiltonian (8.47) we obtain the equations of motion

$$\ddot{z}_i = -z_i + \sum_{k=1}^{i-1} \frac{1}{(z_i - z_k)^2} - \sum_{k=i+1}^{N} \frac{1}{(z_k - z_i)^2}, \quad (i = 1, \ldots, N),$$

(B.260)

where we have set the unit of time in such a manner that $\omega_z = 1$. By linearizing (B.260) around equilibrium positions and setting

$$z_i = z_i^{(0)} + \xi_i, \quad (i = 1, \ldots, N),$$

(B.261)

we obtain

$$\ddot{\xi}_i = -\xi_i - 2 \sum_{k=1}^{i-1} \frac{\xi_i - \xi_k}{\left(z_i^{(0)} - z_k^{(0)}\right)^3} + 2 \sum_{k=i+1}^{N} \frac{\xi_k - \xi_i}{\left(z_k^{(0)} - z_i^{(0)}\right)^3}, \quad (i = 1, \ldots, N).$$

(B.262)

For $N = 2$ Eqs. (B.262) read as follows:

$$\ddot{\xi}_1 = -\xi_1 + 2\frac{\xi_2 - \xi_1}{\left(z_2^{(0)} - z_1^{(0)}\right)^3},$$

$$\ddot{\xi}_2 = -\xi_2 - 2\frac{\xi_2 - \xi_1}{\left(z_2^{(0)} - z_1^{(0)}\right)^3}. \tag{B.263}$$

Looking for normal modes, $\xi_i(t) = a_i e^{i\omega t}$, we obtain

$$-\omega^2 \xi_1 + \xi_1 - 2\frac{\xi_2 - \xi_1}{\left(z_2^{(0)} - z_1^{(0)}\right)^3} = 0,$$

$$-\omega^2 \xi_2 + \xi_2 + 2\frac{\xi_2 - \xi_1}{\left(z_2^{(0)} - z_1^{(0)}\right)^3} = 0. \tag{B.264}$$

After substitution of the equilibrium positions (B.257) into (B.264) we arrive to the eigenvalue equation $(2 - \omega^2)^2 - 1 = 0$, whose solutions are

$$\omega_1 = 1, \quad \omega_2 = \sqrt{3}. \tag{B.265}$$

The corresponding eigenstates are ($\xi_1 = 1$, $\xi_2 = 1$) (centre-of-mass mode) and $(-1, 1)$ (stretch mode).

For $N = 3$, we similarly derive the eigenvalue equation

$$\det \begin{bmatrix} -\omega^2 + \frac{14}{5} & -\frac{8}{5} & -\frac{1}{5} \\ -\frac{8}{5} & -\omega^2 + \frac{21}{5} & -\frac{8}{5} \\ -\frac{1}{5} & -\frac{8}{5} & -\omega^2 + \frac{14}{5} \end{bmatrix} = 0, \tag{B.266}$$

whose solutions are

$$\omega_1 = 1, \quad \omega_2 = \sqrt{3}, \quad \omega_3 = \sqrt{\frac{29}{5}}. \tag{B.267}$$

The corresponding eigenstates are ($\xi_1 = 1$, $\xi_2 = 1$, $\xi_3 = 1$) (centre-of-mass mode), $(-1, 0, 1)$ (stretch mode) and $(1, -2, 1)$.

Exercise 8.14 We look for a unitary operator U such that $U|g, 0\rangle = \sum_{n=0}^{N} c_n |g, n\rangle$. We first construct U^{-1} step-by-step; that is, starting from a generic superposition $\sum_{n=0}^{N} c_n |g, n\rangle$, the operator U^{-1} maps this state into the ground state $|g, 0\rangle$. As a first step, a red detuned laser transfers the entire population of the state $|g, N\rangle$ into $|e, N - 1\rangle$. Then a tuned laser moves the entire population of $|e, N - 1\rangle$ into $|g, N - 1\rangle$. We then transfer $|g, N - 1\rangle$ into $|e, N - 2\rangle$ and so on. Note that at each transformation we must keep track of what is happening to all populated levels of the trapped ion. Inverting the procedure illustrated above, we obtain the operator U. Note that $2N$ single-qubit (-ion) gates are required to construct a generic N-level state. Further details can be found in Gardiner *et al.* (1997).

Exercise 8.15 (i) The term e^{ikz} in (8.53) couples the radiation to the motion of the trapped ion. Therefore, at resonance it is possible to induce transitions $|g, n\rangle \leftrightarrow |e, n'\rangle$, with renormalized Rabi frequency

$$\Omega_{n,n'} = \Omega |\langle n' | e^{ikz} | n \rangle|. \tag{B.268}$$

In order to evaluate the matrix element $\langle n' | e^{ikz} | n \rangle$, we employ the formula

$$e^{A+B} = e^A e^B e^{-\frac{[A,B]}{2}}, \tag{B.269}$$

which is valid when the operators A and B are such that

$$[A, [A, B]] = 0 = [B, [A, B]]. \tag{B.270}$$

Therefore, taking $A = i\eta a^\dagger$ and $B = i\eta a$, we obtain

$$e^{ikz} = e^{ikz_0(a^\dagger + a)} = e^{-\frac{1}{2}\eta^2} e^{i\eta a^\dagger} e^{i\eta a}. \tag{B.271}$$

Since

$$e^{i\eta a} |n\rangle = \sum_{m=0}^{n} \frac{1}{m!} (i\eta)^m \sqrt{\frac{n!}{(n-m)!}} |n-m\rangle, \tag{B.272}$$

we obtain

$$\langle n' | e^{ikz} | n \rangle = e^{-\frac{1}{2}\eta^2} \sum_{m'=0}^{n'} \frac{(-i\eta)^{m'}}{m'!} \sqrt{\frac{n'!}{(n'-m')!}}$$

$$\times \langle n' - m' | \sum_{m=0}^{n} \frac{(i\eta)^m}{m!} \sqrt{\frac{n!}{(n-m)!}} |n-m\rangle. \tag{B.273}$$

Due to the orthogonality of the Fock states, $\langle n' - m' | n - m \rangle = \delta_{n'-m', n-m}$. Thus, Eq. (B.273) becomes

$$\langle n' | e^{ikz} | n \rangle = e^{-\frac{1}{2}\eta^2} \sum_{m=\max(0, n-n')}^{n} \frac{(-i\eta)^{(m+n'-n)}(i\eta)^m}{m!(m+n'-n)!} \frac{\sqrt{n'! n!}}{(n-m)!}. \tag{B.274}$$

(ii) When $\eta \ll 1$; that is, the wavelength of the laser is much larger than the extension of the ion wave function, we can expand the exponential in (8.53) to the first order in η:

$$e^{ikz} = e^{i\eta(a^\dagger + a)} \approx 1 + i\eta a^\dagger + i\eta a. \tag{B.275}$$

The three terms on the right-hand side of this equation are associated with the carrier resonance and the first blue and red sidebands.

Exercise 8.16   First of all, we write down the equations of motion for the three-level system:

$$\dot{c}_g = i c_c \Omega_2 \exp[-i(\omega_{gc} - \omega_2)t + i\phi_2],$$
$$\dot{c}_e = i c_c \Omega_1 \exp[-i(\omega_{gc} - \omega_a - \omega_1)t + i\phi_1], \tag{B.276}$$
$$\dot{c}_c = i c_g \Omega_2 \exp[i(\omega_{gc} - \omega_2)t - i\phi_2] + i c_e \Omega_1 \exp[i(\omega_{gc} - \omega_a - \omega_1)t - i\phi_1].$$

Note that, when $\Omega_1 = 0$ (or $\Omega_2 = 0$), we recover the well-known equations of motion for a two-level system in a classical electromagnetic field.

We now make the following assumptions:

$$\Delta \equiv \omega_{gc} - \omega_2 \gg \Omega_1, \Omega_2, \qquad \omega_{gc} - \omega_2 \approx \omega_{gc} - \omega_a - \omega_1. \tag{B.277}$$

Therefore, $c_g(t)$ and $c_e(t)$ change in time much more slowly than $c_c(t)$ and we can integrate the last equation in (B.276) neglecting the variation in time of c_g and c_e. We also assume that $c_c(t = 0) = 0$. This leads to

$$c_c = \frac{c_g \Omega_2}{\Delta} e^{i(\Delta t - \phi_2)} + \frac{c_e \Omega_1}{\Delta} e^{i(\Delta t - \phi_1)}. \tag{B.278}$$

After substitution of this expression for c_c into the first two equations of (B.276), we obtain

$$\dot{c}_g = i \frac{\Omega_2^2}{\Delta} c_g + i \frac{\Omega_1 \Omega_2}{\Delta} e^{i(\phi_2 - \phi_1)} c_e,$$
$$\dot{c}_e = i \frac{\Omega_1 \Omega_2}{\Delta} e^{-i(\phi_2 - \phi_1)} c_g + i \frac{\Omega_1^2}{\Delta} c_e. \tag{B.279}$$

It is clear that the Rabi frequency Ω_R for this two-level system is given by Eq. (8.57).

Note that, since the exact equation for $\dot{c}_c$ (B.276) contains terms oscillating with frequency Δ, the Raman approximation is valid when

$$\Omega_R = \frac{2\Omega_1 \Omega_2}{\Delta} \ll \Delta. \tag{B.280}$$

Such a condition is fulfilled when $\Omega_1, \Omega_2 \ll \Delta$. Moreover, we can estimate from (B.278) that

$$|c_c| \approx \frac{\Omega_2}{\Delta} |c_g|, \qquad |c_c| \approx \frac{\Omega_1}{\Delta} |c_e|. \tag{B.281}$$

Since $|c_g|, |c_e| \leq 1$, we have that the population of level $|c\rangle$ does not exceed

$$|c_c| \approx \frac{\Omega}{\Delta} \ll 1, \tag{B.282}$$

where we have considered the special case $\Omega \equiv \Omega_1 = \Omega_2$. Therefore, level c is very weakly populated.

A comparison between the Raman approximation and the direct numerical integration of the equations of motion (B.276) is shown in Fig. B.15. It is clear that the Raman approximation is quite good in the case $\Omega/\Delta = 0.1 \ll 1$.

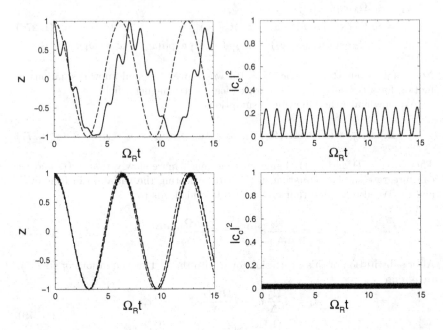

Fig. B.15   A comparison between the numerical integration of Eqs.(B.276) (solid line) and the Raman approximation (dashed line), for $\Omega_1 = \Omega_2 = 0.1$, $\phi_1 = \phi_2 = 0$, initial conditions $c_g = 1$, $c_e = c_c = 0$, $\Delta = 0.3$ (above) and $\Delta = 1$ (below): temporal evolution of the z coordinate of the Bloch sphere for the qubit spanned by the levels $|g\rangle$ and $|e\rangle$ (left figures) and population of level $|c\rangle$ (right figures).

Exercise 8.17   The matrix representation of $R_+(\theta, \phi)$ in the subspace $\{|g, 0\rangle, |e, 1\rangle, |g, 1\rangle, |e, 2\rangle\}$ reads

$$
R_+(\theta, \phi) = \begin{bmatrix} \cos\frac{\theta}{2} & -ie^{i\phi}\sin\frac{\theta}{2} & 0 & 0 \\ -ie^{-i\phi}\sin\frac{\theta}{2} & \cos\frac{\theta}{2} & 0 & 0 \\ 0 & 0 & \cos\frac{\theta}{\sqrt{2}} & -ie^{i\phi}\sin\frac{\theta}{\sqrt{2}} \\ 0 & 0 & -ie^{-i\phi}\sin\frac{\theta}{\sqrt{2}} & \cos\frac{\theta}{\sqrt{2}} \end{bmatrix}. \quad \text{(B.283)}
$$

It is then easy to check by direct matrix multiplications that the global effect of the sequence of pulses (8.61) is

$$
R_+\left(\frac{\pi}{\sqrt{2}}, \frac{\pi}{2}\right) R_+(\pi, 0)\, R_+\left(\frac{\pi}{\sqrt{2}}, \frac{\pi}{2}\right) R_+(\pi, 0) = -I. \quad \text{(B.284)}
$$

Therefore, there is no coupling between the states $|g, 1\rangle$ and $|e, 2\rangle$ and three states $(|g, 0\rangle, |e, 1\rangle, |g, 1\rangle)$ of the computational basis acquire a phase factor of -1. On the other hand, the forth state, $|e, 0\rangle$, of the computational basis is not affected by the blue side band pulses and its phase does not change. Therefore, the composite pulse (8.61) acts as a CMINUS gate (up to an overall phase factor of no physical significance).

Exercise 8.18 The motion of the particle is bounded within the interval $[0, a]$; that is, its wave function $\phi(x)$ must be zero outside this interval. Continuity of the wave function at $x = 0$ and $x = a$ implies

$$\lim_{x \to 0^+} \phi(x) = \lim_{x \to a^-} \phi(x) = 0. \tag{B.285}$$

The solutions to the stationary Schrödinger equation $H\phi(x) = E\phi(x)$ inside the interval $[0, a]$ can be written as

$$\phi(x) = Ae^{ikx} + A'e^{-ikx}, \tag{B.286}$$

where A and A' are complex constants. Since $\phi(0) = 0$, it follows that $A' = -A$, and therefore

$$\phi(x) = 2iA \sin(kx). \tag{B.287}$$

Moreover, $\phi(a) = 0$, which leads to

$$k = \frac{n\pi}{a}, \tag{B.288}$$

where n is an arbitrary positive integer. If we normalize (B.287); that is, we require $\int_{-\infty}^{+\infty} dx |\phi(x)|^2 = \int_0^a dx |\phi(x)|^2 = 1$, and we take into account (B.288), we then obtain the stationary wave functions

$$\phi_n(x) = \sqrt{\frac{2}{a}} \sin\left(\frac{\pi n}{a} x\right), \tag{B.289}$$

with energies

$$E_n = \frac{\pi^2 \hbar^2}{2ma^2} n^2. \tag{B.290}$$

The general solution of the Schrödinger equation $i\hbar \frac{\partial}{\partial t} \psi(x, t) = H\psi(x, t)$ is

$$\psi(x, t) = \sum_{n=1}^{\infty} c_n e^{-\frac{i}{\hbar} E_n t} \phi_n(x), \tag{B.291}$$

where the coefficients c_n are determined by the initial condition $\psi(x, 0) = \psi_0(x)$:

$$c_n = \int_0^a dx \psi_0(x) \phi_n(x). \tag{B.292}$$

Exercise 8.19   Since we are looking for bound states, we limit ourselves to studying the case $-V_0 < E < 0$. Taking into account the boundary condition $\lim_{x \to \pm\infty} \phi(x) = 0$, we obtain

$$\phi(x) = \begin{cases} A \exp(kx), & x < -a, \\ B \cos(k'x) + C \sin(k'x), & -a \leq x \leq a, \\ D \exp(-kx), & x > a, \end{cases} \quad \text{(B.293)}$$

with

$$k = \frac{1}{\hbar}\sqrt{-2mE}, \qquad k' = \frac{1}{\hbar}\sqrt{2m(V_0 + E)}. \quad \text{(B.294)}$$

By requiring the continuity of ϕ and $d\phi/dx$ at $x = \pm a$ we derive four linear homogeneous equations in the variables A, B, C, D:

$$\begin{aligned} A \exp(-ka) &= B \cos(k'a) - C \sin(k'a), \\ D \exp(-ka) &= B \cos(k'a) + C \sin(k'a), \\ kA \exp(-ka) &= k'B \sin(k'a) + k'C \cos(k'a), \\ -kD \exp(-ka) &= -k'B \sin(k'a) + k'C \cos(k'a). \end{aligned} \quad \text{(B.295)}$$

After appropriately adding and subtracting these relations we have

$$\begin{aligned} (A + D) \exp(-ka) - 2B \cos(k'A) &= 0, \\ k(A + D) \exp(-ka) - 2k'B \sin(k'a) &= 0, \\ (A - D) \exp(-ka) + 2C \sin(k'A) &= 0, \\ k(A - D) \exp(-ka) - 2k'C \cos(k'a) &= 0. \end{aligned} \quad \text{(B.296)}$$

Non-trivial solutions are obtained when

$$\det \begin{bmatrix} \exp(-ka) & -2\cos(k'a) & 0 & 0 \\ k \exp(-ka) & -2k' \sin(k'a) & 0 & 0 \\ 0 & 0 & \exp(-ka) & 2\sin(k'a) \\ 0 & 0 & k \exp(-ka) & -2k' \cos(k'a) \end{bmatrix}, \quad \text{(B.297)}$$

that is, when one of the following two equations is satisfied:

$$k' \sin(k'a) - k \cos(k'a) = 0, \quad \text{(B.298a)}$$

$$k' \cos(k'a) + k \sin(k'a) = 0. \quad \text{(B.298b)}$$

If (B.298a) is satisfied, then

$$D = A, \qquad C = 0, \qquad B = \frac{\exp(-ka)}{\cos(k'a)} A, \quad \text{(B.299)}$$

with A determined from the normalization condition $\int_{-\infty}^{+\infty} dx |\phi(x)|^2 = 1$. Note that this solution is even, namely $\phi(-x) = \phi(x)$. On the other hand, if (B.298b) is satisfied we obtain

$$D = -A, \qquad B = 0, \qquad C = -\frac{\exp(-ka)}{\cos(k'a)} A \qquad \text{(B.300)}$$

and again A is determined from the normalization of the wave function. This solution is odd, $\phi(-x) = -\phi(x)$.

We now find the energy levels. First of all, we observe that (B.294) leads to

$$k^2 + k'^2 = \frac{2mV_0}{\hbar^2} \equiv k_0^2. \qquad \text{(B.301)}$$

In the case of even eigenfunctions, the energy levels are determined by the intersections of (B.301) with (B.298a), in the case of odd eigenfunctions by the intersections of (B.301) with (B.298b). The solutions can be found graphically, as shown in Fig. B.16. Note that the number of bound states depends on the parameter $\sqrt{k_0}a$, that is , on the depth V_0 of the square well. If $k_0 a \leq \frac{\pi}{2}$, namely $V_0 \leq \bar{V} \equiv \frac{\pi^2 \hbar^2}{8ma^2}$, there exists only one bound state of the particle, corresponding to an even wave function. If $\frac{\pi}{2} < k_0 a \leq \pi$, that is, $\bar{V} < V_0 \leq 4\bar{V}$, we have two bound states since a level corresponding to an odd wave function appears, and so on. Note that energy levels corresponding to even and odd wave functions appear alternatively as V_0 increases.

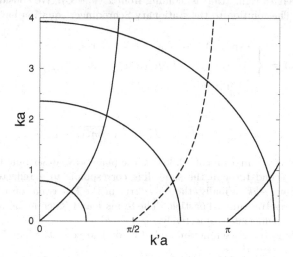

Fig. B.16 The energy levels of a particle in a square well potential determined graphically. The circular arcs have radius $k_0 a = \frac{\pi}{4}, \frac{3\pi}{4}, \frac{5\pi}{4}$. The intersections of these arcs with the solid and the dashed curves determine the energy levels corresponding to even and odd eigenfunctions, respectively.

Exercise 8.20   It is useful to define $\alpha = \frac{1+i}{\sqrt{2}}$ and $\beta = \frac{1+i}{2}$. Then we have

$$
\sqrt{\text{SWAP}} = \begin{bmatrix} 1 & 0 & 0 & 0 \\ 0 & \beta^{\star} & \beta & 0 \\ 0 & \beta & \beta^{\star} & 0 \\ 0 & 0 & 0 & 1 \end{bmatrix}, \qquad (\sqrt{\text{SWAP}})^{-1} = \begin{bmatrix} 1 & 0 & 0 & 0 \\ 0 & \beta & \beta^{\star} & 0 \\ 0 & \beta^{\star} & \beta & 0 \\ 0 & 0 & 0 & 1 \end{bmatrix},
$$

$$
\text{SWAP} = (\sqrt{\text{SWAP}})^2 = \begin{bmatrix} 1 & 0 & 0 & 0 \\ 0 & 0 & 1 & 0 \\ 0 & 1 & 0 & 0 \\ 0 & 0 & 0 & 1 \end{bmatrix}, \tag{B.302}
$$

$$
I \otimes R_z(\pi/2) = \text{diag}(\alpha^{\star}, \alpha, \alpha^{\star}, \alpha), \qquad I \otimes R_z(-\pi/2) = (I \otimes R_z(\pi/2))^{\star},
$$

$$
I \otimes R_z(\pi) = \text{diag}(-i, i, -i, i). \tag{B.303}
$$

If we multiply the above matrices, as in (8.68), up to an irrelevant global phase factor, we obtain the CMINUS quantum gate.

Exercise 8.21   We wish to compute the transmission and reflection probabilities for an incident particle, with momentum $\hbar k$ (and energy $E = \frac{\hbar^2 k^2}{2m}$), propagating from left to right (that is, coming from $x \to -\infty$). We consider the case $0 < E < V_0$. The solution to the stationary Schrödinger equation has the form

$$
\phi(x) = \begin{cases} \exp(ikx) + R\exp(-ikx), & x < 0, \\ A\cosh(k'x) + B\sinh(k'x), & 0 \le x \le a, \\ T\exp(ikx), & x > a, \end{cases} \tag{B.304}
$$

with

$$
k = \frac{1}{\hbar}\sqrt{2mE}, \qquad k' = \frac{1}{\hbar}\sqrt{2m(V_0 - E)}. \tag{B.305}
$$

The first term in the first line of (B.304) is the plane wave describing the incident particle, the second term in the same line corresponds to a reflected particle, with momentum $-\hbar k$. Finally, the only term in the last equation of (B.304) is associated with a transmitted particle. The terms T and R define the transmission and reflection coefficients, respectively.

Continuity of the wave function and of its derivative at $x = 0$ and $x = a$ leads to the following equations:

$$
\begin{aligned}
1 + R &= A, \\
A\cosh(k'a) + B\sinh(k'a) &= T\exp(ika), \\
ik(1 - R) &= k'B, \\
k'A\sinh(k'a) + k'B\cosh(k'a) &= ikT\exp(ika).
\end{aligned} \tag{B.306}
$$

We can solve these equations, thus obtaining

$$T = \frac{2ikk' \exp(-ika)}{2ikk' \cosh(k'a) + [k^2 - k'^2] \sinh(k'a)},$$

$$R = \frac{[k^2 + k'^2] \sinh(k'a)}{2ikk' \cosh(k'a) + [k^2 - k'^2] \sinh(k'a)}, \tag{B.307}$$

$$A = 1 + R,$$

$$B = i\frac{k}{k'}(1 - R).$$

It is easy to verify that $|R|^2 + |T|^2 = 1$; namely, the sum of the reflection and transmission probabilities is equal to unity. We emphasize that, in contrast to the classical predictions, the particle has a non-zero probability of crossing the potential barrier even though its energy E is smaller than the height V_0 of the barrier (tunnel effect).

Exercise 8.22   The time-evolution operator over a time interval T reads

$$U(T) = \exp\left[-\frac{iT}{\hbar}\left(\frac{\omega_0 \sigma_z}{2} + \Delta\sigma_x\right)\right] = \cos\left(\frac{\theta(T)}{2}\right)I - i\sin\left(\frac{\theta(T)}{2}\right)\mathbf{n} \cdot \boldsymbol{\sigma}, \tag{B.308}$$

where we have defined

$$\theta(T) = \frac{2T}{\hbar}\sqrt{\frac{1}{4}\omega_0^2 + \Delta^2}, \tag{B.309}$$

$$\mathbf{n} = \left(\frac{\Delta}{\sqrt{\frac{1}{4}\omega_0^2 + \Delta^2}}, 0, \frac{\omega_0}{2\sqrt{\frac{1}{4}\omega_0^2 + \Delta^2}}\right). \tag{B.310}$$

Therefore, the pulse induces a rotation through an angle $\theta(T)$ about the $\mathbf{n}$-axis of the Bloch sphere.

Exercise 8.23   Let us first consider the circuit in Fig. 8.27 (left). The sequence of gates $U_P\left(\phi = -\frac{\pi}{2}\right) U_B\left(\theta = \frac{\pi}{4}, \phi = -\frac{\pi}{2}\right) U_P\left(\phi = -\frac{\pi}{2}\right)$ transforms the input states $|0\rangle$ and $|1\rangle$ as follows:

$$|0\rangle = |1\rangle_0|0\rangle_1 \to |1\rangle_0|0\rangle_1 \to \frac{1}{\sqrt{2}}\left(|1\rangle_0|0\rangle_1 + i|0\rangle_0|1\rangle_1\right)$$

$$\to \frac{1}{\sqrt{2}}\left(|1\rangle_0|0\rangle_1 + |0\rangle_0|1\rangle_1\right) = \frac{1}{\sqrt{2}}\left(|0'\rangle + |1'\rangle\right),$$

$$|1\rangle = |0\rangle_0|1\rangle_1 \to -i|0\rangle_0|1\rangle_1 \to \frac{-i}{\sqrt{2}}\left(i|1\rangle_0|0\rangle_1 + |0\rangle_0|1\rangle_1\right) \tag{B.311}$$

$$\to \frac{1}{\sqrt{2}}\left(|1\rangle_0|0\rangle_1 - |0\rangle_0|1\rangle_1\right) = \frac{1}{\sqrt{2}}\left(|0'\rangle - |1'\rangle\right).$$

This is exactly a Hadamard transformation.

We now show that the circuit in Fig. 8.27 (right) implements the CNOT gate (up to a sign factor). Indeed, we have

$$
\begin{aligned}
|0\rangle|h\rangle &= |1\rangle_0|0\rangle_1|h\rangle \rightarrow |1\rangle_{0'}|0\rangle_{1'}|h\rangle = |0'\rangle|h\rangle, \\
|0\rangle|v\rangle &= |1\rangle_0|0\rangle_1|v\rangle \rightarrow |1\rangle_{0'}|0\rangle_{1'}|v\rangle = |0'\rangle|v\rangle, \\
|1\rangle|h\rangle &= |0\rangle_0|1\rangle_1|h\rangle \rightarrow |0\rangle_{0'}|1\rangle_{1'}|v\rangle = |1'\rangle|v\rangle, \\
|1\rangle|v\rangle &= |0\rangle_0|1\rangle_1|v\rangle \rightarrow -|0\rangle_{0'}|1\rangle_{1'}|h\rangle = -|1'\rangle|h\rangle.
\end{aligned}
\tag{B.312}
$$

Exercise 8.24   We obtain

$$
\begin{aligned}
HU_P(\phi)H &= \tfrac{1}{\sqrt{2}} \begin{bmatrix} 1 & 1 \\ 1 & -1 \end{bmatrix} \begin{bmatrix} 1 & 0 \\ 0 & e^{i\phi} \end{bmatrix} \tfrac{1}{\sqrt{2}} \begin{bmatrix} 1 & 1 \\ 1 & -1 \end{bmatrix} \\
&= e^{i\frac{\phi}{2}} \begin{bmatrix} \cos\frac{\phi}{2} & -i\sin\frac{\phi}{2} \\ -i\sin\frac{\phi}{2} & \cos\frac{\phi}{2} \end{bmatrix} = e^{i\frac{\phi}{2}} U_B\left(\tfrac{\phi}{2}, \tfrac{\pi}{2}\right),
\end{aligned}
\tag{B.313}
$$

with U_P and U_B defined in (8.78) and (8.79). Therefore, the entire Mach–Zehnder interferometer corresponds to a beam splitter of transmittance $T = \cos^2\left(\frac{\theta}{2}\right)$. Note that, if the phase shift $\phi = 0$, then the photon leaves the interferometer in the same direction as it entered, as expected from the fact that $H^2 = I$.

Exercise 8.25   It is convenient to write the matrix U in (8.86) as follows:

$$
U = \begin{bmatrix} A & B & C \\ D & E & F \\ G & H & K \end{bmatrix}.
\tag{B.314}
$$

The transformation of the creation operators is given by

$$
a_l^\dagger \rightarrow \sum_m U_{ml} a_m^\dagger,
\tag{B.315}
$$

namely

$$
\begin{aligned}
a_1^\dagger &\rightarrow A a_1^\dagger + D a_2^\dagger + G a_3^\dagger, \\
a_2^\dagger &\rightarrow B a_1^\dagger + E a_2^\dagger + H a_3^\dagger, \\
a_3^\dagger &\rightarrow C a_1^\dagger + F a_2^\dagger + K a_3^\dagger.
\end{aligned}
\tag{B.316}
$$

Therefore, the initial state $|\psi\rangle|1\rangle|0\rangle$ is mapped by U into

$$
\left[\alpha + \beta(A a_1^\dagger + D a_2^\dagger + G a_3^\dagger) + \tfrac{1}{\sqrt{2}}\gamma(A a_1^\dagger + D a_2^\dagger + G a_3^\dagger)^2\right](B a_1^\dagger + E a_2^\dagger + H a_3^\dagger)|000\rangle.
\tag{B.317}
$$

Since we accept only measurement outcomes with one photon in mode 2 and no photons in mode 3, in (B.317) we must keep only the terms with mode 2 in the state $|1\rangle$ and mode 3 in $|0\rangle$, thus obtaining

$$\left[\alpha E + \beta(AE + BD)a_1^\dagger + \tfrac{1}{\sqrt{2}}\gamma(A^2 E + 2ADB)(a_1^\dagger)^2\right]|010\rangle$$

$$= \tfrac{1}{2}\left[\alpha + \beta a_1^\dagger - \tfrac{1}{\sqrt{2}}\gamma(a_1^\dagger)^2\right]|010\rangle = \tfrac{1}{2}|\psi'\rangle|10\rangle. \quad (B.318)$$

Therefore, the final state $|\psi'\rangle$ is obtained with probability $\tfrac{1}{4}$.

Exercise 8.26   The initial state can be written as

$$\alpha\gamma|0101\rangle + \alpha\delta|0110\rangle + \beta\gamma|1001\rangle + \beta\delta|1010\rangle. \quad (B.319)$$

Note that, for the sake of simplicity, we omit the indices $1, \dots, 4$ specifying the modes. Using (8.83) we can see that the first beam splitter transforms (B.319) into

$$\alpha\gamma|0101\rangle - \tfrac{1}{\sqrt{2}}\alpha\delta|1100\rangle + \tfrac{1}{\sqrt{2}}\alpha\delta|0110\rangle + \tfrac{1}{\sqrt{2}}\beta\gamma|1001\rangle$$
$$+ \tfrac{1}{\sqrt{2}}\beta\gamma|0011\rangle - \tfrac{1}{\sqrt{2}}\beta\delta|2000\rangle + \tfrac{1}{\sqrt{2}}\beta\delta|0020\rangle. \quad (B.320)$$

After the two non-linear sign shift gates we have (with overall success probability $\left(\tfrac{1}{4}\right)^2 = \tfrac{1}{16}$)

$$\alpha\gamma|0101\rangle - \tfrac{1}{\sqrt{2}}\alpha\delta|1100\rangle + \tfrac{1}{\sqrt{2}}\alpha\delta|0110\rangle + \tfrac{1}{\sqrt{2}}\beta\gamma|1001\rangle$$
$$+ \tfrac{1}{\sqrt{2}}\beta\gamma|0011\rangle + \tfrac{1}{\sqrt{2}}\beta\delta|2000\rangle - \tfrac{1}{\sqrt{2}}\beta\delta|0020\rangle. \quad (B.321)$$

After the final beam splitter we obtain

$$\alpha\gamma|0101\rangle + \alpha\delta|0110\rangle + \beta\gamma|1001\rangle - \beta\delta|0101\rangle. \quad (B.322)$$

Exercise 8.27   A $\theta = \tfrac{\pi}{4}$, $\phi = -\tfrac{\pi}{2}$ beam splitter maps

$$|\psi^+\rangle_{12} = \tfrac{1}{\sqrt{2}}\left(a_{1,h}^\dagger a_{2,v}^\dagger + a_{1,v}^\dagger a_{2,h}^\dagger\right)|0\rangle \quad (B.323)$$

into

$$\tfrac{i}{\sqrt{2}}\left(b_{1,h}^\dagger b_{1,v}^\dagger + b_{2,h}^\dagger b_{2,v}^\dagger\right)|0\rangle \quad (B.324)$$

and

$$|\phi^\pm\rangle_{12} = \tfrac{1}{\sqrt{2}}\left(a_{1,h}^\dagger a_{2,h}^\dagger \pm a_{1,v}^\dagger a_{2,v}^\dagger\right)|0\rangle \quad (B.325)$$

into

$$\tfrac{i}{2\sqrt{2}}\left[\left(b_{1,h}^\dagger\right)^2 + \left(b_{2,h}^\dagger\right)^2 \pm \left(b_{1,v}^\dagger\right)^2 \pm \left(b_{2,v}^\dagger\right)^2\right]|0\rangle. \quad (B.326)$$

Therefore, in all the above three cases a single detector clicks. We can distinguish $|\psi^+\rangle$ from $|\phi^\pm\rangle$ since the photons have either orthogonal polarizations or the same polarization, respectively. Of course, we cannot distinguish, using polarization measurements, between $|\phi^+\rangle$ and $|\phi^-\rangle$. A proof that it is impossible to implement a complete Bell measurement using only linear optical elements can be found in Lütkenhaus *et al.* (1999).

Exercise 8.28   Let $|\psi\rangle = \alpha|0\rangle + \beta|1\rangle$ and $\rho = |\psi\rangle\langle\psi|$. If Alice measures $|\phi^+\rangle$, then the state of Bob's particle collapses onto $(\alpha|1\rangle+\beta|0\rangle)(\alpha^*\langle 1|+\beta^*\langle 0|) = \sigma_x\rho\sigma_x$. Analogously, if Alice measures $|\phi^-\rangle$ or $|\psi^-\rangle$, then the post-measurement state of Bob's qubit is $\sigma_y\rho\sigma_y$ or $\sigma_z\rho\sigma_z$, respectively. Taking into account all three (equiprobable) possibilities, Bob's state is described by the density matrix

$$\rho' = \tfrac{1}{3}(\sigma_x\rho\sigma_x + \sigma_y\rho\sigma_y + \sigma_z\rho\sigma_z) = \tfrac{1}{2}\begin{bmatrix} 1-\tfrac{1}{3}z & -\tfrac{1}{3}(x-iy) \\ -\tfrac{1}{3}(x+iy) & 1+\tfrac{1}{3}z \end{bmatrix}. \tag{B.327}$$

In the case of an ideal universal NOT transformation,

$$\rho'_{\text{ideal}} = \tfrac{1}{2}\begin{bmatrix} 1-z & -(x-iy) \\ -(x+iy) & 1+z \end{bmatrix}. \tag{B.328}$$

Given a pure initial state, ρ'_{ideal} is also pure and therefore the fidelity of the state ρ' is

$$F = \text{Tr}\big(\rho'_{\text{ideal}}\rho'\big) = \tfrac{2}{3}. \tag{B.329}$$

Exercise 8.29   The projector over the subspace spanned by $|\psi^-\rangle$, $|\phi^+\rangle$, $|\phi^-\rangle$ is

$$P = |\psi^-\rangle\langle\psi^-| + |\phi^+\rangle\langle\phi^+| + |\phi^-\rangle\langle\phi^-|. \tag{B.330}$$

A measurement projecting onto this subspace maps the initial state $\rho_{\text{in}}^{(\text{tot})} = |\psi\rangle\langle\psi| \otimes |\psi^+\rangle\langle\psi^+|$ into

$$\rho^{(\text{tot})} = \frac{(P \otimes I_{\text{Bob}})\,\rho_{\text{in}}^{(\text{tot})}\,(P \otimes I_{\text{Bob}})}{\text{Tr}\big[\rho_{\text{in}}^{(\text{tot})}(P \otimes I_{\text{Bob}})\big]}. \tag{B.331}$$

We can compute from $\rho^{(\text{tot})}$ the states of $\rho_1^{(A)}$ and $\rho_2^{(A)}$ of Alice's qubits:

$$\rho_1^{(A)} = \rho_2^{(A)} = \tfrac{1}{2}\begin{bmatrix} 1+\tfrac{2}{3}z & \tfrac{2}{3}(x-iy) \\ \tfrac{2}{3}(x+iy) & 1-\tfrac{2}{3}z \end{bmatrix}, \tag{B.332}$$

where x, y, z are the Bloch-sphere coordinates of the initial state $|\psi\rangle$. The fidelity of the clones is given by

$$F = \langle\psi|\rho_1^{(A)}|\psi\rangle = \langle\psi|\rho_2^{(A)}|\psi\rangle = \tfrac{5}{6}. \tag{B.333}$$

Bibliography

Alber, G., Beth, T., Horodecki, M., Horodecki, P., Horodecki, R., Rötteler, M., Weinfurter, H., Werner, R., and Zeilinger, A. (2001), Quantum information – An introduction to theoretical concepts and experiments, Springer–Verlag.

Alber, G., Beth, Th., Charnes, Ch., Delgado, A., Grassl, M., and Mussinger. M. (2001), Stabilizing distinguishable qubits against spontaneous decay by detected-jump correcting quantum codes, *Phys. Rev. Lett.* **86**, 4402.

Alekseev, V. M. and Jacobson, M. V. (1981), Symbolic dynamics and hyperbolic dynamic systems, *Phys. Rep.* **75**, 287.

Arimondo, E., Bloch, I., and Meschede, D. (2005), Atomic q-bits and optical lattices, in "Quantum information processing and communication in Europe", available at http://cordis.europa.eu/ist/fet/qipc.htm.

Arnold, V. I. (1997), Mathematical methods of classical mechanics (2nd Ed.), Springer–Verlag.

Averin, D. V. (2000), Quantum computing and quantum measurement with mesoscopic Josephson junctions, quant-ph/0008114, in *Fortschr. Phys.* **48**, 1055.

Bandyopadhyay, S. (2000), Qubit- and entanglement-assisted optimal entanglement concentration, *Phys. Rev. A* **62**, 032308.

Barbieri, M., De Martini, F., Mataloni, P., Vallone, G., and Cabello, A. (2006), Enhancing the violation of the Einstein–Podolsky–Rosen local realism by quantum hyperentanglement, *Phys. Rev. Lett.* **97**, 140407.

Barenco, A., Brun, T. A., Schack, R., and Spiller, T. P. (1997), Effects of noise on quantum error correction algorithms, *Phys. Rev. A* **56**, 1177.

Barnum, H., Fuchs, C. A., Jozsa, R., and Schumacher, B. (1996), General fidelity limit for quantum channels, *Phys. Rev. A* **54**, 4707.

Barreiro, J. T., Langford, N. K., Peters, N. A., and Kwiat, P. G. (2005), Generation of hyperentangled photon pairs, *Phys. Rev. Lett.* **95**, 260501.

Barrett, M. D., Chiaverini, J., Schaetz, T., Britton, J., Itano, W. M., Jost, J. D., Knill, E., Langer, C., Leibfried, D., Ozeri, R., and Wineland, D. J. (2004), Deterministic quantum teleportation of atomic qubits, *Nature* **429**, 737.

Bayfield, J. E. and Koch, P. M. (1974), Multiphoton ionization of highly excited hydrogen atoms, *Phys. Rev. Lett.* **33**, 258.

Bayfield, J. E., Casati, G., Guarneri, I., and Sokol, D. W. (1989), Localization of classically chaotic diffusion for hydrogen atoms in microwave fields, *Phys. Rev. Lett.* **63**, 364.

Beige, A., Braun, D., Tregenna, B., and Knight, P. L. (2000), Quantum computing using dissipation to remain in a decoherence-free subspace, *Phys. Rev. Lett.* **85**, 1762.

Benenti, G., Casati, and Shepelyansky, D. L. (2001), Emergence of Fermi–Dirac thermalization in the quantum computer core, *Eur. Phys. J. D* **17**, 265.

Benenti, G., Casati, G., Montangero, S., and Shepelyansky, D. L. (2001), Efficient quantum computing of complex dynamics, *Phys. Rev. Lett.* **87**, 227901.

Benenti, G., Casati, G., Montangero, S., and Shepelyansky, D. L. (2002), Eigenstates of an operating quantum computer: Hypersensitivity to static imperfections, *Eur. Phys. J. D* **20**, 293.

Benenti, G. and Casati, G. (2002), Quantum-classical correspondence in perturbed chaotic systems, *Phys. Rev. E* **65**, 066205.

Benenti, G., Casati, G., and Veble, G. (2003), Stability of classical chaotic motion under a system's perturbations, *Phys. Rev. E* **67**, 055202(R).

Benenti, G., Casati, G., and Montangero, S. (2004), Quantum computing and information extraction for dynamical quantum systems, *Quantum Information Processing* **3**.

Benenti, G., Felloni, S., and Strini, G. (2006), Effects of single-qubit quantum noise on entanglement purification, *Eur. Phys. J. D* **38**, 389.

Benenti, G. and Casati, G. (2006), Quantum chaos, decoherence and quantum computation, in Proceedings of the "E. Fermi" Varenna School on "Quantum computers, algorithms and chaos", Casati, G., Shepelyansky, D. L., Zoller, P., and Benenti G. (Eds.), IOS Press and SIF, Bologna.

Bennett, C. H., Bernstein, H. J., Popescu, S., and Schumacher, B. (1996), Concentrating partial entanglement by local operations, *Phys. Rev. A* **53**, 2046.

Bennett, C. H., Brassard, G., Popescu, S., Schumacher, B., Smolin, J. A., and Wootters, W. K. (1996), Purification of noisy entanglement and faithful teleportation via noisy channels, *Phys. Rev. Lett.* **76**, 722.

Bennett, C. H., DiVincenzo, D. P., Smolin, J. A., and Wootters, W. K. (1996), Mixed-state entanglement and quantum error correction, *Phys. Rev. A* **54**, 3824.

Benza, V. and Strini, G. (2003), A single-qubit Landau–Zener gate, *Fortschr. Phys.* **51**, 14.

Berman, G. P. and Zaslavsky, G. M. (1978), Condition of stochasticity in quantum nonlinear systems, *Physica A* **91**, 450.

Berry, M. V. and Tabor, M. (1977), Level clustering in the regular spectrum, *Proc. R. Soc. Lond. A* **356**, 375.

Berry, M. V. (1985), Semiclassical theory of spectral rigidity, *Proc. R. Soc. Lond. A* **400**, 229.

Blatt, R., Häffner, H., Roos, C. F., Becher, C., and Schmidt-Kaler, F. (2004), Ion trap quantum computing with Ca$^+$ ions, *Quantum Information Processing* **3**, 61.

Bloch, I. (2006), Engineering multi-particle entanglement with neutral atoms in optical lattices, in Proceedings of the "E. Fermi" Varenna School on "Quantum computers, algorithms and chaos", Casati, G., Shepelyansky, D. L., Zoller, P., and Benenti G. (Eds.), IOS Press and SIF, Bologna.

Bohigas, O., Haq, R. U., and Pandey, A. (1983), Fluctuation properties of nuclear energy levels and widths: Comparison of theory with experiment, in "Nuclear data for science and technology", Bockhoff K. H. (Ed.), Reidel, Dordrecht, p. 809.

Bohigas, O., Giannoni, M. J., and Schmit, C. (1984), Characterization of chaotic quantum spectra and universality of level fluctuation laws, *Phys. Rev. Lett.* **52**, 1.

Bohigas, O. (1991), Random matrix theories and chaotic dynamics, in "Chaos and quantum physics", Les Houches Summer Schools, Session LII, Giannoni, M.-J., Voros, A., and Zinn-Justin, J. (Eds.), North–Holland, Amsterdam.

Boschi, D., Branca, S., De Martini, F., Hardy, L., and Popescu, S. (1998), Experimental realization of teleporting an unknown pure quantum state via dual classical and Einstein–Podolsky–Rosen channels, *Phys. Rev. Lett.* **80**, 1121.

Bouwmeester, D., Pan, J.-W., Mattle, K., Eibl, M., Weinfurter, H., and Zeilinger, A. (1997), Experimental quantum teleportation, *Nature* **390**, 575.

Braginsky, V. B. and Khalili, F. Ya. (1992), Quantum measurement, Cambridge University Press, Cambridge.

Braunstein, S. L. and van Loock, P. (2005), Quantum information with continuous variables, *Rev. Mod. Phys.* **77**, 513.

Breuer, H.-P., Kappler, B., and Petruccione, F. (1999), Stochastic wavefunction method for non-Markovian quantum master equations, *Phys. Rev. A* **59**, 1633.

Briegel, H.-J., Dür, W., Cirac, J. I., and Zoller, P. (1998), Quantum repeaters: The role of imperfect local operations in quantum communication, *Phys. Rev. Lett.* **81**, 5932.

Brun, T. A. (2002), A simple model of quantum trajectories, *Am. J. Phys.* **70**, 719.

Bruß, D., DiVincenzo, D. P., Ekert, A., Fuchs, C. A., Macchiavello, C., and Smolin, J. A. (1998), Optimal universal and state-dependent quantum cloning, *Phys. Rev. A* **57**, 2368.

Bruß, D. (2002), Characterizing entanglement, *J. Math. Phys.* **43**, 4237.

Burkard, G. and Loss, D. (2002), Spin qubits in solid-state structures, *Europhysics News* **33/5**, 166.

Bužek, V. and Hillery, M. (1996), Quantum copying: Beyond the no-cloning theorem, *Phys. Rev. A* **54**, 1844; Universal optimal cloning of qubits and quantum registers, quant-ph/9801009.

Bužek, V., Hillery, M., and Werner, R. F. (1999), Optimal manipulation with qubits: Universal-NOT gate, *Phys. Rev. A* **60**, R2626.

Caldeira, A. O. and Leggett, A. J. (1983), Quantum tunnelling in a dissipative system, *Ann. Phys. (N.Y.)* **153**, 445.

Calderbank, A. R. and Shor, P. W. (1996), Good quantum error-correcting codes exist, *Phys. Rev. A* **54**, 1098.

Camarda, H. S. and Georgopulos, P. D. (1983), Statistical behavior of atomic energy levels: Agreement with random-matrix theory, *Phys. Rev. Lett.* **50**, 492.

Carlo, G. G., Benenti, G., and Casati, G. (2003), Teleportation in a noisy environment: A quantum trajectories approach, *Phys. Rev. Lett.* **91**, 257903.

Carlo, G. G., Benenti, G., Casati, G., and Mejía-Monasterio, C. (2004), Simulating noisy quantum protocols with quantum trajectories, *Phys. Rev. A* **69**, 062317.

Carmichael, H. J. (1993), An open systems approach to quantum optics, Lecture Notes in Physics, Springer, Berlin.

Carvalho, A. R. R., Mintert, F., and Buchleitner, A. (2004), Decoherence and multipartite entanglement, *Phys. Rev. Lett.* **93**, 230501.

Casati, G., Guarneri, I., and Shepelyansky, D. L. (1988), Hydrogen atom in monochromatic field: Chaos and dynamical photonic localization, *IEEE J. Quantum Electron.* **24**, 1420.

Casati, G., Guarneri, I., and Shepelyansky, D. L. (1990), Classical chaos, quantum localization and fluctuations: A unified view, *Physica A* **163**, 205.

Casati, G. and Chirikov, B. V. (Eds.) (1995), Quantum chaos: Between order and disorder, Cambridge University Press, Cambridge.

Casati, G. and Prosen, T. (2005), Quantum chaos and the double-slit experiment, *Phys. Rev. A* **72**, 032111.

Cerf, N. J., Adami, C., and Kwiat, P. G. (1998), Optical simulation of quantum logic, *Phys. Rev. A* **57**, R1477.

Chang, S.-J. and Shi, K.-J. (1986), Evolution and exact eigenstates of a resonant quantum system, *Phys. Rev. A* **34**, 7.

Chirikov, B. V., Izrailev, F. M., and Shepelyansky, D. L. (1981), Dynamical stochasticity in classical and quantum mechanics, *Sov. Sci. Rev.* **C2**, 209.

Cinelli, C., Barbieri, M., Perris, R., Mataloni, P., and De Martini, F. (2005), All-versus-nothing nonlocality test of quantum mechanics by two-photon hyperentanglement, *Phys. Rev. Lett.* **95**, 240405.

Cirac, J. I. and Zoller, P. (1995), Quantum computations with cold trapped ions, *Phys. Rev. Lett.* **74**, 4091.

Cirac, J. I., Duan, L. M., and Zoller, P. (2002), Quantum optical implementation of quantum information processing, quant-ph/0405030, in Proceedings of the "E. Fermi" Varenna School on "Experimental quantum computation and information", De Martini, F. and Monroe, C. (Eds.), IOS Press and SIF, Bologna.

Cirac, J. I. and Zoller, P. (2004), New frontiers in quantum information with atoms and ions, *Phys. Today*, March 2004, p. 38.

Coffman, V., Kundu, J., and Wootters, W. K. (2000), Distributed entanglement, *Phys. Rev. A* **61**, 052306.

Cohen-Tannoudji, C., Diu, B., and Laloë, F. (1977), Quantum mechanics, vols. I and II, Hermann, Paris.

Collin, E., Ithier, G., Aassime, A., Joyez, P., Vion, D., and Esteve, D. (2004), NMR-like control of a quantum bit superconducting circuit, *Phys. Rev. Lett.* **93**, 157005.

Cover, T. M. and Thomas, J. A. (1991), Elements of information theory, John Wiley & Sons, New York.

De Chiara, G., Fazio, R., Macchiavello, C., Montangero, S., and Palma, G.

M. (2004), Quantum cloning in spin networks, *Phys. Rev. A* **70**, 062308.

Delone, N. B., Krainov, V. P., and Shepelyansky, D. L. (1983), Highly excited atoms in the electromagnetic field, *Sov. Phys. Usp.* **26**, 551.

Deutsch, D., Ekert, A., Jozsa, R., Macchiavello, C., Popescu, S., and Sanpera, A. (1996), Quantum privacy amplification and the security of quantum cryptography over noisy channels, *Phys. Rev. Lett.* **77**, 2818.

Devoret, M. H., Wallraff, A., and Martinis, J. M. (2004), Superconducting qubits: A Short review, cond-mat/0411174.

Dittrich, T., Hänggi, P., Ingold, G.-L., Kramer, B., Schön, S., and Zwerger, W. (1998), Quantum transport and dissipation, Wiley–VCH, Weinheim.

DiVincenzo, D. P. (2000), The physical implementation of quantum computation, *Fortschr. Phys.* **48**, 771.

DiVincenzo, D. P., Bacon, D., Kempe, J., Burkard, G., and Whaley, K. B. (2000), Universal quantum computation with the exchange interaction, *Nature* **408**, 339.

Duan, L.-M., Giedke, G., Cirac, J. I., and Zoller, P. (2000), Inseparability criterion for continuous variable systems, *Phys. Rev. Lett.* **84**, 2722.

Dür, W., Vidal, G., and Cirac, J. I. (2000), Three qubits can be entangled in two inequivalent ways, *Phys. Rev. A* **62**, 062314.

Dyson, F. J. (1962), Statistical theory of the energy levels of complex systems I, *J. Math. Phys.* **3**, 140.

Elzerman, J. M., Kouwenhoven, L. P., and Vandersypen, L. M. K. (2006), Electron spin qubits in quantum dots, in Proceedings of the "E. Fermi" Varenna School on "Quantum computers, algorithms and chaos", Casati, G., Shepelyansky, D. L., Zoller, P., and Benenti G. (Eds.), IOS Press and SIF, Bologna.

Ekert, A., Hayden, P.M., and Inamori, H. (2001), Basic concepts in quantum computation, quant-ph/0011013, in "Coherent atomic matter waves", Les Houches Summer Schools, Session LXXII, Kaiser, R., Westbrook, C., and David, F. (Eds.), Springer–Verlag.

Eschner, J. (2006), Quantum computation with trapped ions, in Proceedings of the "E. Fermi" Varenna School on "Quantum computers, algorithms and chaos", Casati, G., Shepelyansky, D. L., Zoller, P., and Benenti G. (Eds.), IOS Press and SIF, Bologna.

Esteve, D. and Vion, D. (2005), Solid state quantum bits, cond-mat/0505676.

Facchi, P. and Pascazio, S. (2003), Three different manifestations of the quantum Zeno effect, quant-ph/0303161, in "Irreversible quantum dynamics", Benatti, F. and Floreanini, R. (Eds.), Lecture Notes in Physics,

Vol. 622, p. 141, Springer–Verlag.

Facchi, P., Lidar, D. A., and Pascazio, S. (2004), Unification of dynamical decoupling and the quantum Zeno effect, *Phys. Rev. A* **69**, 032314.

Facchi, P., Montangero, S., Fazio, R., and Pascazio, S. (2005), Dynamical imperfections in quantum computers, *Phys. Rev. A* **71**, 060306(R).

Falci, G. and Fazio, R. (2006), Quantum computation with Josephson qubits, in Proceedings of the "E. Fermi" Varenna School on "Quantum computers, algorithms and chaos", Casati, G., Shepelyansky, D. L., Zoller, P., and Benenti G. (Eds.), IOS Press and SIF, Bologna.

Frahm, K. M., Fleckinger, R., and Shepelyansky, D. L. (2004), Quantum chaos and random matrix theory for fidelity decay in quantum computations with static imperfections, *Eur. Phys. J. D* **29**, 139.

Friedrich, H. and Wintgen, F. (1989), The hydrogen atom in a uniform magnetic field – an example of chaos, *Phys Rep.* **183**, 37.

Fuchs, C. A. and Caves, C. M. (1994), Ensemble-dependent bounds for accessible information in quantum mechanics, *Phys. Rev. Lett.* **73**, 3047.

Galvez, E. J., Sauer, B. E., Moorman, L., Koch, P. M., and Richards, D. (1988), Microwave ionization of H atoms: Breakdown of classical dynamics for high frequencies, *Phys. Rev. Lett.* **61**, 2011.

Gardiner, C. W. and Zoller, P. (2000), Quantum noise (2nd. Ed.), Springer–Verlag.

Gardiner, S. A., Cirac, J. I., and Zoller, P. (1997), Nonclassical states and measurement of general motional observables of a trapped ion, *Phys. Rev. A* **55**, 1683.

Georgeot, B. and Shepelyansky, D. L. (2000), Quantum chaos border for quantum computing, *Phys. Rev. E* **62**, 3504; Emergence of quantum chaos in the quantum computer core and how to manage it, *Phys. Rev. E* **62**, 6366.

Georgeot, B. (2006), Quantum algorithms and quantum chaos, in Proceedings of the "E. Fermi" Varenna School on "Quantum computers, algorithms and chaos", Casati, G., Shepelyansky, D. L., Zoller, P., and Benenti G. (Eds.), IOS Press and SIF, Bologna.

Gisin, N. and Massar, S. (1997), Optimal quantum cloning machines, *Phys. Rev. Lett.* **79**, 2153.

Gisin, N., Ribordy, G., Tittel, W., and Zbinden, H. (2002), Quantum cryptography, *Rev. Mod. Phys.* **74**, 145.

Gorin, T., Prosen, T., Seligman, T. H., and Žnidarič, M. (2006), Dynamics of Loschmidt echoes and fidelity decay, *Phys. Rep.* **435**, 33.

Gorini, V., Kossakowski, A., and Sudarshan, E. C. G. (1976), Completely positive dynamical semigroups of N-level systems, *J. Math. Phys.* **17**, 821.

Gottesman, D. and Chuang, I. L. (1999), Demonstrating the viability of universal quantum computation using teleportation and single-qubit operations, *Nature* **402**, 390.

Gottesman, D. (2000), An introduction to quantum error correction, quant-ph/0004072.

Gray, R. M. (1990), Entropy and information theory, Springer–Verlag.

Guhr, T., Müller-Groeling, A., and Weidenmüller, H. A. (1998), Random-matrix theories in quantum physics: Common concepts, *Phys. Rep.* **299**, 189.

Haake, F. (2000), Quantum signatures of chaos (2nd. Ed.), Springer–Verlag.

Häffner, H., Hänsel, W., Roos, C. F., Benhelm, J., Chek-al-kar, D., Chwalla, M., Körber, T., Rapol, U. D., Riebe, M., Schmidt, P. O., Becher, C., Gühne, O., Dür, W., and Blatt, R. (2005), Scalable multiparticle entanglement of trapped ions, *Nature* **438**, 643.

Hagley, E., Maître, X., Nogues, G., Wunderlich, C., Brune, M., Raimond, J. M., and Haroche, S. (1997), Generation of Einstein–Podolsky–Rosen pairs of atoms, *Phys. Rev. Lett.* **79**, 1.

Heller, E. J. (1991), Wavepacket dynamics and quantum chaology, in "Chaos and quantum physics", Les Houches Summer Schools, Session LII, Giannoni, M.-J., Voros, A., and Zinn-Justin, J. (Eds.), North–Holland, Amsterdam.

Henry, M. K., Emerson, J., Martinez, R., and Cory, D. G. (2005), Study of localization in the quantum sawtooth map emulated on a quantum information processor, quant-ph/0512204.

Holevo, A. S. (1998), The capacity of the quantum channel with general signal states, *IEEE Trans. Inf. Theory*, **44**, 269.

Horodecki, M., Horodecki, P., and Horodecki, R. (1996), Separability of mixed states: Necessary and sufficient conditions, *Phys. Lett. A* **223**, 1.

Horodecki, M., Horodecki, P., and Horodecki, R. (1998), Mixed-state entanglement and distillation: Is there a "bound" entanglement in nature?, *Phys. Rev. Lett.* **80**, 5239 (1998).

Horodecki, P. (1997), Separability criterion and inseparable mixed states with positive partial transposition, *Phys. Lett. A* **232**, 333.

Huang, K. (1987), Statistical mechanics (2nd Ed.), John Wiley & Sons, New York.

Ithier, G., Nguyen, F., Collin, E., Boulant, N., Meeson, P. J., Joyez, P., Vion, D., and Esteve, D. (2006), Solid state quantum bit circuits, in

Proceedings of the "E. Fermi" Varenna School on "Quantum computers, algorithms and chaos", Casati, G.., Shepelyansky, D. L., Zoller, P., and Benenti G. (Eds.), IOS Press and SIF, Bologna.

Izrailev. F. M. (1990), Simple models of quantum chaos:spectrum and eigenfunctions, *Phys. Rep.* **196**, 299.

Jennewein, T., Simon, C., Weihs, G., Weinfurter, H., and Zeilinger, A. (2000), Quantum cryptography with entangled photons, *Phys. Rev. Lett.* **84**, 4729.

Jensen, R. V. (1982), Stochastic ionization of surface–state electrons, *Phys. Rev. Lett.* **49**, 1365.

Jones, J. A. (2001), NMR quantum computation, quant-ph/0009002, in *Prog. NMR Spectr.* **38**, 325.

Jozsa, R. and Linden, N. (2003), On the role of entanglement in quantum-computational speed-up, *Proc. R. Soc. Lond. A* **459**, 2011.

Kac, M. (1966), Can one hear the shape of a drum? *Amer. Math. Monthly* **73**, 1.

Kane, B. (1998), A silicon-based nuclear spin quantum computer, *Nature* **393**, 133.

Kern, O., Alber, G., and Shepelyansky D. L. (2005), Quantum error correction of coherent errors by randomization, *Eur. Phys. J. D* **32**, 153.

Kiefer, C. and Joos, E. (1998), Decoherence: Concepts and examples, quant-ph/9803052.

Knill, E. and Laflamme, R. (1997), Theory of quantum error-correcting codes, *Phys. Rev. A* **55**, 900.

Knill, E., Laflamme, R., and Viola, L. (2000), Theory of quantum error correction for general noise, *Phys. Rev. Lett.* **84**, 2525.

Knill, E., Laflamme, R., and Milburn, G. J. (2001), A scheme for efficient quantum computation with linear optics, *Nature* **409**, 46.

Knill, E., Laflamme, R., Ashikhmin, A., Barnum, H., Viola, L., and Zurek, W. H. (2002), Introduction to quantum error correction, quant-ph/0207170, *Los Alamos Science* **27**, 188.

Koch, P. M. and van Leeuwen, K. A. H. (1995), The importance of resonances in microwave "ionization" of excited hydrogen atoms, *Phys. Rep.* **255**, 289.

Kok, P., Munro, W. J., Nemoto, K., Ralph, T. C., Dowling, J. P., and Milburn, G. J. (2005), Linear optical quantum computing, quant-ph/0512071.

Kornfeld, I. P., Fomin, S. V., and Sinai, Ya. G. (1982), Ergodic theory, Springer–Verlag.

Kraus, K. (1983), States, effects, and operations: Fundamental notions of quantum theory, Lecture Notes in Physics, Vol. 190, Springer–Verlag.

Laflamme, R., Miquel, C., Paz, J. P., and Zurek, W. H. (1996), Perfect quantum error correcting code, *Phys. Rev. Lett.* **77**, 198.

Laflamme, R., Knill, E., Cory, D. G., Fortunato, E. M., Havel, T., Miquel, C., Martinez, R., Negrevergne, C. J., Ortiz, G., Pravia, M. A., Sharf, Y., Sinha, S., Somma, R., and Viola, L. (2002), NMR and quantum information processing, quant-ph/0207172, *Los Alamos Science* **27**, 226.

Latora, V. and Baranger, M. (1999), Kolmogorov–Sinai entropy rate versus physical entropy, *Phys. Rev. Lett.* **82**, 520.

Lee, P. A. and Ramakrishnan, T. V. (1985), Disordered electronic systems, *Rev. Mod. Phys.* **57**, 287.

Leibfried, D., Blatt, R., Monroe, C., and Wineland, D. (2003), Quantum dynamics of single trapped ions, *Rev. Mod. Phys.* **75**, 281.

Leibfried, D., DeMarco, B., Meyer, V., Lucas, D., Barrett, M., Britton, J., Itano, W. M., Jelenkovic, B., Langer, C., Rosenband, T., and Wineland, D. J. (2003), Experimental demonstration of a robust, high-fidelity geometric two ion-qubit phase gate, *Nature* **422**, 412.

Leibfried, D., Knill, E., Seidelin, S., Britton, J., Blakestad, R. B., Chiaverini, J., Hume, D. B., Itano, W. M., Jost, J. D., Langer, C., Ozeri, R., Reichle, R., and Wineland, D. J. (2005), Creation of a six-atom 'Schrödinger cat' state, *Nature* **438**, 638

Leopold, J. G. and Percival, I. C. (1978), Microwave ionization and excitation of Rydberg atoms, *Phys. Rev. Lett.* **41**, 944.

Li, B., Casati, G., Wang, J., and Prosen, T. (2004), Fourier law in the alternate-mass hard-core potential chain, *Phys. Rev. Lett.* **92**, 254301.

Lidar, D. A., Chuang, I. L., and Whaley, K. B. (1998), Decoherence-free subspaces for quantum computation, *Phys. Rev. Lett.* **81**, 2594.

Lidar, D. A., Bacon, D., Kempe, J., and Whaley, K. B. (2000), Protecting quantum information against exchange errors using decoherence free states, *Phys. Rev. A* **61**, 052307.

Lidar, D. A. and Whaley, B. (2003), Decoherence-free subspaces and subsystems, quant-ph/0301032, in "Irreversible quantum dynamics", Benatti, F. and Floreanini, R. (Eds.), Lecture Notes in Physics, Vol. 622, p. 83, Springer–Verlag.

Lindblad, G. (1976), On the generators of quantum dynamical semigroups, *Commun. Math. Phys.* **48**, 119.

Lichtenberg, A. and Lieberman, M. (1992), Regular and chaotic dynamics (2nd Ed.), Springer–Verlag.

Loss, D. and DiVincenzo, D. P. (1998), Quantum computation with quantum dots, *Phys. Rev. A* **57**, 120.

Lütkenhaus, N., Calsamiglia, J., and Suominen, K.-A. (1999), Bell measurements for teleportation, *Phys. Rev. A* **59**, 3295.

Mabuchi, H. and Doherty, A. C. (2002), Cavity quantum electrodynamics: Coherence in context, *Science* **298**, 1372.

Macchiavello, C. (1998), On the analytical convergence of the QPA procedure, *Phys. Lett. A* **246**, 385.

Macchiavello, C. and Palma, G. M. (2002), Entanglement-enhanced information transmission over a quantum channel with correlated noise, *Phys. Rev. A* **65**, 050301(R).

Makhlin, Y., Schön, G., and Shnirman, A. (2001), Quantum-state engineering with Josephson-junction devices, *Rev. Mod. Phys.* **73**, 357.

Marcikic, I., de Riedmatten, H., Tittel, W., Zbinden, H., and Gisin, N. (2003), Long-distance teleportation of qubits at telecommunication wavelengths, *Nature* **421**, 509.

McDonald, S. W. and Kaufman, A. N. (1979), Spectrum and eigenfunctions for a Hamiltonian with stochastic trajectories, *Phys. Rev. Lett.* **42**, 1189.

Mehta, M. L. (1991), Random matrices, Academic Press, San Diego.

Miller, R., Northup, T. E., Birnbaum, K. M., Boca, A., Boozer, A. D., and Kimble, H. J. (2005), Trapped atoms in cavity QED: Coupling quantized light and matter, *J. Phys. B* **38**, S551.

Miquel, C., Paz, J. P., and Perazzo, R. (1996), Factoring in a dissipative quantum computer, *Phys. Rev. A* **54**, 2605.

Monroe, C. (2002), Quantum information processing with atoms and photons, *Nature* **416**, 238.

Mooij, H. (2005), Superconducting qubits: Quantum mechanics by fabrication, in "Quantum information processing and communication in Europe", available at http://cordis.europa.eu/ist/fet/qipc.htm.

Moore, F. L., Robinson J. C., Barucha, C. F., Sundaram, B., and Raizen, M. G. (1995), Atom optics realization of the quantum delta-kicked rotor, *Phys. Rev. Lett.* **75**, 4598.

Müller, S., Heusler, S., Braun, P., Haake, F., and Altland, A. (2004), Semiclassical foundation of universality in quantum chaos, *Phys. Rev. Lett.* **93**, 014103.

Myers, C. R. and Laflamme, R. (2006), Linear optics quantum computation: An overview, quant-ph/0512104, in Proceedings of the "E. Fermi" Varenna School on "Quantum computers, algorithms and chaos", Casati, G., Shepelyansky, D. L., Zoller, P., and Benenti G. (Eds.), IOS Press and

Principles of Quantum Computation and Information. II

SIF, Bologna.

Nakamura, Y., Pashkin, Yu. A., and Tsai, J. S. (1999), Coherent control of macroscopic quantum states in a single-Cooper-pair box, *Nature* **398**, 786.

Namiki, M., Pascazio, S., and Nakazato, H. (1997), Decoherence and quantum measurements, World Scientific, Singapore.

Negrevergne, C., Mahesh, T. S., Ryan, C. A., Ditty, M., Cyr-Racine, F., Power, W., Boulant, N., Havel, T., Cory, D. G., and Laflamme, R. (2006), Benchmarking quantum control methods on a 12-qubit system, *Phys. Rev. Lett.* **96**, 170501.

Nielsen, M. A. and Chuang, I. L. (2000), Quantum computation and quantum information, Cambridge University Press, Cambridge.

Orús, R. and Latorre, J. I. (2004), Universality of entanglement and quantum-computation complexity, *Phys. Rev. A* **69**, 052308.

Osborne, T. J. and Nielsen, M. A. (2002), Entanglement in a simple quantum phase transition, *Phys. Rev. A* **66**, 032110.

Osterloh, A., Amico, L., Falci, G., and Fazio, R. (2002), Scaling of entanglement close to a quantum phase transition, *Nature* **416**, 608.

Ott, E. (2002), Chaos in dynamical systems (2nd. Ed.), Cambridge University Press, Cambridge.

Palma, G. M., Suominen, K.-A., and Ekert, A. K. (1996), Quantum computers and dissipation, *Proc. R. Soc. Lond. A* **452**, 567.

Peres, A. (1993), Quantum theory: Concepts and methods, Kluwer Academic, Dordrecht.

Peres, A. (1996), Separability criterion for density matrices, *Phys. Rev. Lett.* **77**, 1413.

Petta, J. R., Johnson, A. C., Taylor, J. M., Laird, E. A., Yacoby, A., Lukin, M. D., Marcus, C. M., Hanson, M. P., and Gossard, A. C. (2005), Coherent manipulation of coupled electron spins in semiconductor quantum dots, *Science* **309**, 2180.

Plenio, M. B. and Knight, P. L. (1998), The quantum-jump approach to dissipative dynamics in quantum optics, *Rev. Mod. Phys.* **70**, 101.

Plenio, M. B. and Virmani, S. (2007), An introduction to entanglement measures, *Quant. Inf. Comp.* **7**, 1.

Preskill, J. (1998), Lecture notes on quantum information and computation, available at http://theory.caltech.edu/people/preskill/.

Preskill, J. (1998), Fault-tolerant quantum computation, quant-ph/9712048, in "Introduction to quantum computation and information", Lo, H.-K., Popescu, S., and Spiller, T. (Eds.), World Scientific,

Singapore.

Preskill, J. (1999), Battling decoherence: The fault-tolerant quantum computer, *Phys. Today*, June 1999, p. 24.

Prokof'ev, N. V. and Stamp, P. C. E. (2000), Theory of the spin bath, *Rep. Prog. Phys.* **63**, 669.

Prosen, T. and Žnidarič, M. (2001), Can quantum chaos enhance the stability of quantum computation?, *J. Phys. A* **34**, L681.

Prosen, T. and Žnidarič, M. (2002), Stability of quantum motion and correlation decay, *J. Phys. A* **35**, 1455.

Raimond, J. M., Brune, M., and Haroche, S. (2001), Colloquium: Manipulating quantum entanglement with atoms and photons in a cavity, *Rev. Mod. Phys.* **73**, 565.

Raimond, J. M. and Rempe, R. (2005), Cavity quantum electrodynamics, in "Quantum information processing and communication in Europe", available at http://cordis.europa.eu/ist/fet/qipc.htm.

Raizen, M., Milner, V., Oskay, W. H., and Steck, D. A. (2000), Experimental study of quantum chaos with cold atoms, in Proceedings of the "E. Fermi" Varenna School on "New directions in quantum chaos", Casati, G., Guarneri, I., and Smilansky, U. (Eds.), IOS Press, Amsterdam.

Ramanathan, C., Boulant, N., Chen, Z., Cory, D. G., Chuang, I., and Steffen, M. (2004), NMR quantum information processing, *Quantum Information Processing* **3**, 15.

Rauschenbeutel, A., Nogues, G., Osnaghi, S., Bertet, P., Brune, M., Raimond, J. M., and Haroche, S. (1999), Coherent operation of a tunable quantum phase gate in cavity QED, *Phys. Rev. Lett.* **83**, 5166.

Raussendorf, R. and Briegel, H. J. (2001), A one-way quantum computer, *Phys. Rev. Lett.* **86**, 5188.

Reck, M., Zeilinger, A., Bernstein, H. J., and Bertani, P. (1994), Experimental realization of any discrete unitary operator, *Phys. Rev. Lett.* **73**, 58.

Ricci, M., Sciarrino, F., Sias, C., and De Martini, F. (2004), Teleportation scheme implementing the universal optimal quantum cloning machine and the universal NOT gate, *Phys. Rev. Lett.* **92**, 047901.

Riebe, M., Häffner, H., Roos, C. F., Hänsel, W., Benhelm, J., Lancaster, G. P. T., Körber, T. W., Becher, C., Schmidt-kaler, F., James, D. F. V., and Blatt, R. (2004), Deterministic quantum teleportation with atoms, *Nature* **429**, 734.

Roscilde, T., Verrucchi, P., Fubini, A., Haas, S., and Tognetti, V. (2004), Studying quantum spin systems through entanglement estimators, *Phys.*

Rev. Lett. **93**, 167203.

Rossini, D., Benenti, G., and Casati, G. (2006), Conservative chaotic map as a model of quantum many-body environment, *Phys. Rev. E* **74**, 036209.

Scarani, V., Iblisdir, S., Gisin, N., and Acín, A. (2005), Quantum cloning, *Rev. Mod. Phys.* **77**, 1225.

Schmidt-Kaler, F., Häffner, H., Riebe, M., Gulde, S., Lancaster, G. P. T., Deuschle, T., Becher, C., Roos, C. F., Eschner, J., and Blatt, R. (2003), Realization of the Cirac–Zoller controlled-NOT quantum gate, *Nature* **422**, 408.

Schmiedmayer, J. and Hinds, E. (2005), Atom chips, in "Quantum information processing and communication in Europe", available at http://cordis.europa.eu/ist/fet/qipc.htm.

Schumacher, B. (1995), Quantum coding, *Phys. Rev. A* **51**, 2738.

Schumacher, B. (1996), Sending entanglement through noisy quantum channels, *Phys. Rev. A* **54**, 2614.

Schumacher, B. and Westmoreland, M. D. (1997), Sending classical information via noisy quantum channels, *Phys. Rev. A* **56**, 131.

Schumacher, B. (1998), Lecture notes on quantum information theory, available at http://www2.kenyon.edu/people/schumab/.

Scully, M. O. and Zubairy, M. S. (1997), Quantum optics, Cambridge University Press, Cambridge.

Shannon, C. E. (1948), A mathematical theory of communication, *Bell System Tech. J.* **27**, 379; 623.

Shepelyansky, D. L. (1983), Some statistical properties of simple classically stochastic quantum systems, *Physica D* **8**, 208.

Shepelyansky, D. L. (1986), Localization of quasienergy eigenfunctions in action space, *Phys. Rev. Lett.* **56**, 677.

Shepelyansky, D. L. (1994), Kramers–map approach for stabilization of a hydrogen atom in a monochromatic field, *Phys. Rev. A* **50**, 575.

Shepelyansky, D. L. (2001), Quantum chaos and quantum computers, *Physica Scripta* **T90**, 112.

Shor, P. W. (1995), Scheme for reducing decoherence in quantum computer memory, *Phys. Rev. A* **52**, R2493.

Steane, A. M. (1996), Error correcting codes in quantum theory, *Phys. Rev. Lett.* **77**, 793.

Steane, A. M. (1996), Multiple particle interference and quantum error correction, *Proc. R. Soc. Lond. A* **452**, 2551.

Steane, A. M. (2006), A tutorial on quantum error correction, in Proceedings of the "E. Fermi" Varenna School on "Quantum computers, algorithms and chaos", Casati, G., Shepelyansky, D. L., Zoller, P., and Benenti G. (Eds.), IOS Press and SIF, Bologna.

Stöckmann, H.-J. (1999), Quantum chaos: An introduction, Cambridge University Press, Cambridge.

Strini, G. (2002), Error sensitivity of a quantum simulator I: A first example, *Fortschr. Phys.* **50**, 171.

Stucki, D., Gisin, N., Guinnard, O., Ribordy, G., and Zbinden, H. (2002), Quantum key distribution over 67 km with a plug&play system, *New J. Phys.* **4**, 41.

Tinkham, M. (1996), Introduction to superconductivity, McGraw–Hill, New York.

Toda, M., Kubo, R., and Saitô, N. (1983), Statistical physics I, Springer–Verlag.

Tombesi, P. (2006), Entanglement in quantum optics, in Proceedings of the "E. Fermi" Varenna School on "Quantum computers, algorithms and chaos", Casati, G., Shepelyansky, D. L., Zoller, P., and Benenti G. (Eds.), IOS Press and SIF, Bologna.

Turchette, Q. A., Wood, C. S., King, B. E., Myatt, C. J., Leibfried, D., Itano, W. M., Monroe, C., and Wineland, D. J. (1998), Deterministic entanglement of two trapped ions, *Phys. Rev. Lett.* **81**, 3631.

Ursin, R., Jennewein, T., Asplemeyer, M., Kaltenbaek, R., Lindenthal, M., Walther, P., and Zeilinger, A. (2004), Quantum teleportation across the Danube, *Nature* **430**, 849 (2004).

Vandersypen, L. M. K., Steffen, M., Breyta, G., Yannoni, C. S., Sherwood, M. H., and Chuang, I. L. (2001), Experimental realization of Shor's quantum factoring algorithm using nuclear magnetic resonance, *Nature* **414**, 883.

Varcoe, B. T. H., Brattke, S., Englert, B.-G., and Walther, H. (2000), Fock state Rabi oscillations; a building block for the observation of new phenomena in quantum optics, *Fortschr. Phys.* **48**, 679.

Verstraete, F., Porras, D., and Cirac, J. I. (2004), Density matrix renormalization group and periodic boundary conditions: A quantum information perspective, *Phys. Rev. Lett.* **93**, 227205.

Vidal, G., Latorre, J. I., Rico, E., and Kitaev, A. (2003), Entanglement in quantum critical phenomena, *Phys. Rev. Lett.* **90**, 227902.

Vidal, G. (2003), Efficient classical simulation of slightly entangled quantum computations, *Phys. Rev. Lett.* **91**, 147902.

Vidal, G. (2004), Efficient simulation of one-dimensional quantum many-body systems, *Phys. Rev. Lett.* **93**, 040502.

Viola, L. and Knill, E. (2005), Random decoupling schemes for quantum dynamical control and error suppression, *Phys. Rev. Lett.* **94**, 060502.

Vandersypen, L. M. K. and Chuang, I. L. (2004), NMR techniques for quantum control and computation, *Rev. Mod. Phys.* **76**, 1037.

Viola, L., Knill, E., and Laflamme, R. (2001), Constructing qubits in physical systems, *J. Phys. A* **34**, 7067.

Vion, D., Aassime, A., Cottet, A., Joyez, P., Pothier, H., Urbina, C., Esteve, D., and Devoret, M. H. (2002), Manipulating the quantum state of an electrical circuit, *Science*, **296**, 886.

Vion, D., Aassime, A., Cottet, A., Joyez, P., Pothier, H., Urbina, C., Esteve, D., and Devoret, M. H. (2003), Rabi oscillations, Ramsey fringes and spin echoes in an electrical circuit, *Fortschr. Phys.* **51**, 462.

Weinstein, Y. S., Pravia, M. A., Fortunato, E. M., Lloyd, S., and Cory, D. G. (2001), Implementation of the quantum Fourier transform, *Phys. Rev. Lett.* **86**, 1889.

Weinstein, Y. S., Havel, T. F., Emerson, J., Boulant, N., Saraceno, M., Lloyd, S., and Cory, D. G. (2004), Quantum process tomography of the quantum Fourier transform, *J. Chem. Phys.* **121**, 6117.

Weiss, U. (1999), Quantum dissipative systems (2nd Ed.), World Scientific, Singapore.

Wendin, G. and Shumeiko, V. S. (2005), Superconducting quantum circuits, qubits and computing, cond-mat/0508729.

Wineland, D. J., Barrett, M., Britton, J., Chiaverini, J., DeMarco, B.L., Itano, W. M., Jelenkovic, B. M., Langer, C., Leibfried, D., Meyer, V., Rosenband, T., and Schaetz, T. (2003), Quantum information processing with trapped ions, *Phil. Trans. R. Soc. Lond.* **A361**, 1349.

Wootters, W. K. (1998), Entanglement of formation of an arbitrary state of two qubits, *Phys. Rev. Lett.* **80**, 2245.

Yamamoto, T., Pashkin, Yu. A., Astafiev, O., Nakamura, Y., and Tsai, J. S., Demonstration of conditional gate operation using superconducting charge qubits, *Nature* **425**, 941.

You, J. Q. and Nori, F. (2005), Superconducting circuits and quantum information, *Phys. Today*, November 2005, p. 42.

Zanardi, P. and Rasetti. M. (1998), Noiseless quantum codes, *Phys. Rev. Lett.* **79**, 3306.

Zeilinger, A. (2000), Quantum teleportation, *Sci. Am.*, **282:4**, 32, April 2000.

Zimmermann, Th., Köppel, H., Cederbaum, L. S., Persch, G., and

Demtröder, W. (1988), Confirmation of random-matrix fluctuations in molecular spectra, *Phys. Rev. Lett.* **61**, 3.

Zurek, W. H. (1991), *Phys. Today*, October 1991, p. 36; see also quant-ph/0306072.

Zurek, W. H. (2003), Decoherence, einselection, and the quantum origins of the classical, *Rev. Mod. Phys.* **75**, 715.

Index